AN INTRODUCTION
TO COMPLEX ANALYSIS

AN INTRODUCTION TO COMPLEX ANALYSIS

O. Carruth McGehee

A Wiley-Interscience Publication

JOHN WILEY & SONS, INC.

New York • Chichester • Weinheim • Brisbane • Singapore • Toronto

Library of Congress Cataloging in Publication Data:

McGehee, O. Carruth.
    An introduction to complex analysis / O. Carruth McGehee.
    Includes bibliographical references and index.
    ISBN 0-471-33233-X (cloth : alk. paper)
    1. Mathematical analysis.   2. Functions of complex variables.   I. Title.

QA300 .M275 2000
515—dc21   00-042256

Printed in the United States of America
10 9 8 7 6 5 4 3 2 1

To Mary and Matt

Contents

Preface   *xiii*

Symbols and Terms   *xix*

1   *Preliminaries*   *1*

   1.1   *Preview*   *1*

     A   *It Takes Two Harmonic Functions*   *3*

     B   *Heat Flow*   *6*

     C   *A Geometric Rule*   *9*

     D   *Electrostatics*   *10*

     E   *Fluid Flow*   *13*

     F   *One Model, Many Applications*   *14*

       Exercises   *15*

   1.2   *Sets, Functions, and Visualization*   *18*

     A   *Terminology and Notation for Sets*   *18*

     B   *Terminology and Notation for Functions*   *20*

     C   *Functions from* $\mathbb{R}$ *to* $\mathbb{R}$ *25*

     D   *Functions from* $\mathbb{R}^2$ *to* $\mathbb{R}$ *27*

     E   *Functions from* $\mathbb{R}^2$ *to* $\mathbb{R}^2$ *29*

       Exercises   *30*

   1.3   *Structures on* $\mathbb{R}^2$*, and Linear Maps from* $\mathbb{R}^2$ *to* $\mathbb{R}^2$ *34*

	A	The Real Line and the Plane	34
	B	Polar Coordinates in the Plane	36
	C	When Is a Mapping $M : \mathbb{R}^2 \to \mathbb{R}^2$ Linear?	38
	D	Visualizing Nonsingular Linear Mappings	40
	E	The Determinant of a Two-by-Two Matrix	44
	F	Pure Magnifications, Rotations, and Conjugation	45
	G	Conformal Linear Mappings	46
		Exercises	48
1.4	Open Sets, Open Mappings, Connected Sets		51
	A	Distance, Interior, Boundary, Openness	51
	B	Continuity in Terms of Open Sets	55
	C	Open Mappings	56
	D	Connected Sets	57
		Exercises	58
1.5	A Review of Some Calculus		61
	A	Integration Theory for Real-Valued Functions	61
	B	Improper Integrals, Principal Values	63
	C	Partial Derivatives	66
	D	Divergence and Curl	68
		Exercises	70
1.6	Harmonic Functions		71
	A	The Geometry of Laplace's Equation	71
	B	The Geometry of the Cauchy-Riemann Equations	72
	C	The Mean Value Property	73
	D	Changing Variables in a Dirichlet or Neumann Problem	76
		Exercises	77
2	Basic Tools		83
2.1	The Complex Plane		83
	A	The Definition of a Field	83
	B	Complex Multiplication	84
	C	Powers and Roots	87
	D	Conjugation	89
	E	Quotients of Complex Numbers	90
	F	When Is a Mapping $L : \mathbb{C} \to \mathbb{C}$ Linear?	91
	G	Complex Equations for Lines and Circles	92

	H	*The Reciprocal Map, and Reflection in the Unit Circle*	*93*
	I	*Reflections in Lines and Circles*	*96*
		Exercises	*97*
2.2		*Visualizing Powers, Exponential, Logarithm, and Sine*	*102*
	A	*Powers of z*	*103*
	B	*Exponential and Logarithms*	*104*
	C	*Sin z*	*106*
	D	*The Cosine and Sine, and the Hyperbolic Cosine and Sine*	*110*
		Exercises	*111*
2.3		*Differentiability*	*115*
	A	*Differentiability at a Point*	*115*
	B	*Differentiability in the Complex Sense: Holomorphy*	*119*
	C	*Finding Derivatives*	*122*
	D	*Picturing the Local Behavior of Holomorphic Mappings*	*124*
		Exercises	*126*
2.4		*Sequences, Compactness, Convergence*	*128*
	A	*Sequences of Complex Numbers*	*128*
	B	*The Limit Superior of a Sequence of Reals*	*131*
	C	*Implications of Compactness*	*133*
	D	*Sequences of Functions*	*134*
		Exercises	*135*
2.5		*Integrals Over Curves, Paths, and Contours*	*138*
	A	*Integrals of Complex-Valued Functions*	*138*
	B	*Curves*	*138*
	C	*Paths*	*144*
	D	*Pathwise Connected Sets*	*147*
	E	*Independence of Path and Morera's Theorem*	*148*
	F	*Goursat's Lemma*	*150*
	G	*The Winding Number*	*153*
	H	*Green's Theorem*	*155*
	I	*Irrotational and Incompressible Fluid Flow*	*158*
	J	*Contours*	*161*
		Exercises	*162*
2.6		*Power Series*	*166*
	A	*Infinite Series*	*166*

	B	The Geometric Series	167
	C	An Improved Root Test	171
	D	Power Series and the Cauchy-Hadamard Theorem	172
	E	Uniqueness of the Power Series Representation	174
	F	Integrals That Give Rise to Power Series	178
		Exercises	180

3 The Cauchy Theory — 187

3.1		Fundamental Properties of Holomorphic Functions	188
	A	Integral and Series Representations	188
	B	Eight Ways to Say "Holomorphic"	193
	C	Determinism	193
	D	Liouville's Theorem	196
	E	The Fundamental Theorem of Algebra	196
	F	Subuniform Convergence Preserves Holomorphy	197
		Exercises	198
3.2		Cauchy's Theorem	204
	A	Černý's 1976 Proof	205
	B	Simply Connected Sets	208
	C	Subuniform Boundedness, Subuniform Convergence	209
3.3		Isolated Singularities	212
	A	The Laurent Series Representation on an Annulus	212
	B	Behavior Near an Isolated Singularity in the Plane	216
	C	Examples: Classifying Singularities, Finding Residues	219
	D	Behavior Near a Singularity at Infinity	225
	E	A Digression: Picard's Great Theorem	229
		Exercises	229
3.4		The Residue Theorem and the Argument Principle	236
	A	Meromorphic Functions and the Extended Plane	236
	B	The Residue Theorem	239
	C	Multiplicity and Valence	242
	D	Valence for a Rational Function	243

	E	*The Argument Principle: Integrals That Count*	*243*
		Exercises	*249*
3.5	*Mapping Properties*		*251*
		Exercises	*259*
3.6	*The Riemann Sphere*		*260*
		Exercises	*264*

4 *The Residue Calculus* *267*

4.1	*Integrals of Trigonometric Functions*		*268*
		Exercises	*270*
4.2	*Estimating Complex Integrals*		*273*
		Exercises	*276*
4.3	*Integrals of Rational Functions Over the Line*		*277*
		Exercises	*280*
4.4	*Integrals Involving the Exponential*		*282*
	A	*Integrals Giving Fourier Transforms*	*286*
		Exercises	*290*
4.5	*Integrals Involving a Logarithm*		*293*
		Exercises	*301*
4.6	*Integration on a Riemann Surface*		*302*
	A	*Mellin Transforms*	*306*
		Exercises	*307*
4.7	*The Inverse Laplace Transform*		*309*
		Exercises	*315*

5 *Boundary Value Problems* *317*

5.1	*Examples*		*318*
	A	*Easy Problems*	*318*
	B	*The Conformal Mapping Method*	*323*
		Exercises	*326*
5.2	*The Möbius Maps*		*327*
		Exercises	*338*
5.3	*Electric Fields*		*341*
	A	*A Point Charge in 3-Space*	*341*
	B	*Uniform Charge on One or More Long Wires*	*342*
	C	*Examples with Bounded Potentials*	*347*
		Exercises	*350*
5.4	*Steady Flow of a Perfect Fluid*		*350*

		Exercises	*354*
5.5		*Using the Poisson Integral to Obtain Solutions*	*355*
	A	*The Poisson Integral on a Disk*	*355*
	B	*Solutions on the Disk by the Poisson Integral*	*358*
	C	*Geometry of the Poisson Integral*	*361*
	D	*Harmonic Functions and the Mean Value Property*	*363*
	E	*The Neumann Problem on a Disk*	*364*
	F	*The Poisson Integral on a Half-Plane, and on Other Domains*	*365*
		Exercises	*366*
5.6		*When Is the Solution Unique?*	*368*
		Exercises	*370*
5.7		*The Schwarz Reflection Principle*	*370*
5.8		*Schwarz-Christoffel Formulas*	*374*
	A	*Triangles*	*375*
	B	*Rectangles and Other Polygons*	*385*
	C	*Generalized Polygons*	*389*
		Exercises	*390*
6		**Lagniappe**	*393*
6.1		*Dixon's 1971 Proof of Cauchy's Theorem*	*394*
6.2		*Runge's Theorem*	*398*
		Exercises	*403*
6.3		*The Riemann Mapping Theorem*	*404*
		Exercises	*405*
6.4		*The Osgood-Taylor-Carathéodory Theorem*	*406*
		References	*413*
		Index	*419*

Preface

To Students, About Prerequisites

This text is primarily for use in a one-quarter or one-semester undergraduate course. Such a course should cover a reasonable amount of material chosen from the first five chapters to fit the backgrounds and interests of the students. The prerequisites consist of the standard calculus sequence, including the differentiation and integration of functions of two or more real variables.

In teaching Mathematics 4036 here at Louisiana State University, I've found that I need to review and restate for my students some of the necessary ideas from the calculus while setting the stage for the problems and methods of complex analysis. Chapter 1 is devoted to such preparation. I offer the following remarks about specific background material that you need.

- From the very beginning, Chapter 1 assumes a certain familiarity with partial derivatives for real-valued functions of two real variables. Section 1.5 states and reviews the most advanced results concerning partial derivatives that you need to understand. In Section 1.6 the multidimensional Chain Rule is used twice in the text and in several of the problems.

- In Section 1.2, I present notation and terminology for sets and functions. The examples presented make use of certain real-valued functions of a real variable, which ought to be familiar from your previous studies: polynomials, exponential, logarithm, sine, cosine, hyperbolic cosine, and hyperbolic sine.

- In Section 1.3, I develop from scratch the modest amount of linear algebra that you need. It will be helpful if you have previously dealt with a system of two linear equations in two unknowns, and understand how a two-by-two matrix determines a linear transformation of the plane.

- I assume that you are familiar with the Riemann integral as taught in first-year calculus. Section 1.5 summarizes the integration theory that you need. It will be helpful if you have had some experience with integrals over curves and with integrals over two- and three-dimensional domains.

- Sections 2.4 and 2.6 present everything you need about complex-valued sequences and series. You may find this material easier if you take some time to review your previous studies of sequences, infinite series, and Taylor series.

From time to time, you may need to review or look up some topic you're not familiar with, even beyond the fairly complete list above. I have tried to be careful, when using prerequisite material, to make clear what I am assuming and how I am applying it.

This book uses some concepts which occur also in upper-division advanced calculus or real-variables courses, but I do not assume you have had those courses. Such topics as open sets, compact sets, and uniform convergence are important for making complex analysis simple and elegant. I develop those topics as needed.

All veterans of the calculus sequence have no doubt acquired a quite adequate mental concept of the real number system $\mathbb{R}$. It is good to be aware of the completeness property of $\mathbb{R}$, and to recognize it when it is used. I will state and explain this version of it in Section 2.4: Every nonempty set of real numbers which is bounded above has a least upper bound.

I have in mind an audience of students like the ones to whom I've taught the course here at LSU. Most have completed the calculus sequence at some time in the past. Most have taken other mathematics courses as well, such as differential equations. About half of them are majoring in some field of engineering or natural science; they take complex analysis as an elective, or perhaps as part of a minor or double major. They bring to the course an awareness of certain physical and computational problems, and they want this course to provide them some new and useful ideas and techniques. Some have a well-formed curiosity about, for example, the integration methods that come from the Residue Theorem, and the conformal mapping method for solving Dirichlet problems.

At the very beginning of the book, I try to respond to these students' curiosity and motivation by describing some of the results we'll obtain, and some of the physical problems we'll be able to solve. I present the physics briefly, just enough to explain how the mathematical models may be interpreted in physical terms.

The audience also includes mathematics majors whose priority is to master the theory and prepare for more advanced mathematics. But the physics is for their benefit too. The applications serve to illuminate the theory, provide sources of intuition, create historical awareness, and make a connection with other knowledge.

To All Readers, About the Book

I've thought about this project, the writing of this book, from time to time since the late 1960s when I first taught complex analysis at the University of California at Berkeley. Having taught the course often, I've encountered a number of good books, old and new, which are serviceable as undergraduate texts. Nevertheless, I have felt moved to write a new one. Perhaps teachers selecting a text will be glad to have one more choice. In recent decades, there have been developments in what we know, and in how we understand the subject, which need to be incorporated in a text. If I've done so successfully, then perhaps this book will have a good influence on the teaching of the subject. Beyond what I've already said, I offer the following remarks about the book I have aimed for.

- When teaching the one-semester course here at LSU, I have covered most of Chapters 1–3, presented just a sample of the integration techniques in Chapter 4, and devoted a few weeks to a careful treatment of selected topics from Chapter 5. When the students have especially good backgrounds, Chapter 1 can be covered quickly and Chapter 2 in less time than otherwise; and then more of Chapters 3–5 can be treated. For more suggestions for planning the course and using the book, visit the web site at http://www.math.lsu.edu/~mcgehee.

- The book is also usable for a first course in complex analysis for graduate students in mathematics, especially if one objective is to acquaint them with applications, such as they may wish to teach some day. Such a course might cover Chapter 3 thoroughly and also reach the advanced topics of Chapter 6.

- The book engages the students in a review, and in the active use, of their calculus background. Chapter 1 states typical boundary value problems, and previews the methods and solutions, in terms of the real-variable calculus. In Chapter 2, the complex-valued exponential, sine, and cosine are defined and developed in terms of the familiar real-valued versions of those functions. I frequently give specific page references to material in two currently popular calculus texts. I realize that those editions will be superseded in a few years, but I wish to make the point that this book is designed to fit and to follow what the students have studied before.

- Chapter 2 introduces complex numbers, and it soon becomes clear that a useful extra structure is being added to the already-understood real setting. By thus waiting to add the complex structure until the real preliminaries have been discussed, the book avoids the complex-from-the-first-day approach that often (I think) leaves a student feeling that real and complex methods are like oil and water, and wondering how they mix.

- I define conformality separately for linear maps, and emphasize the nature of differentiability as *approximability by linear maps*. I think this is the best way to remove the mystery about complex versus real differentiability for functions from the plane to the plane, a mystery which (I think) sometimes haunts a student throughout the course and for years afterward.

- The book follows an efficient path through the development of the theory. I've included what I consider important for understanding, and for use. I have tried to give short proofs, of the kind whose motivating idea is clear, and in which the techniques and the reasoning will be helpful later on. I think I've largely succeeded in doing so up to Section 5.8.

- It seems to me that the modern homology (or winding-number) version of Cauchy's Theorem deserves the spotlight in an introductory course, especially a course for engineering students. It is easy to state, easy to understand, and easy to prove. It is also stunningly powerful and convenient to use. This book does not speak of homotopies or deformations.

- I emphasize visualization. I often leave a drawing for the reader to do, saying "Make a sketch!" But there are many figures, chosen to provide helpful examples and to represent important ideas to the reader's mind. Some of them will require, and I hope reward, a bit of careful attention. I drew all the figures using the graphics commands of *Mathematica* 3.0. All show precisely rendered actual objects, with two exceptions: Figure 5.7-3, page 374, is frankly schematic; and in Figure 5.8-1, page 376, the shaded region is an "artist's conception."

Acknowledgments

Many people have influenced this work. Many have given me help, encouragement, and inspiration. I should like to mention just a few.

There are influences from long ago. In 1961–1962, I was taught an excellent one-year graduate course at Yale University by Gustav Hedlund, using the text by Einar Hille.

Although I never met Maurice Heins, I've benefited from having a set of notes taken from the lectures in an undergraduate course that he taught in the Fall of 1963.

On a personal level, I wish to thank most warmly three persons who have been good friends to me and to this work: Kevin Scott Brown, currently a graduate student in physics at Cornell University; the mathematician Robert B. Burckel, Professor at Kansas State University; and the mathematician David W. Kammler, Professor at Southern Illinois University.

I wish also to thank the several other people, some of whom I do not know by name, who reviewed early drafts of this book and gave encouragement and helpful advice.

I am grateful for the many students in my classes whom I've had the pleasure of watching as they learned and enjoyed complex analysis.

I thank my colleagues in the Mathematics Department at LSU for their consideration and support for this undertaking. Particular thanks to Professor Neal Stoltzfus for his advice and help with the production of this book on our computer system;

and also to Loc Thi Stewart, the Department's Computer Analyst, for her reliable technical assistance.

I'm of course thankful for the modern tools that make the life of a writer so much easier these days, LaTeX and *Mathematica*.

Finally, I thank the editors at John Wiley & Sons, who have been superb.

O. CARRUTH MCGEHEE

Baton Rouge, Louisiana

July 11, 2000

Symbols and Terms

$\emptyset$	the empty set; the set that has no elements.
$\implies$	If P and Q are statements, then $P \implies Q$ means that P implies Q. In other words, P only if Q; or, If P, then Q.
$\iff$	If P and Q are statements, then $P \iff Q$ means that P and Q are equivalent. In other words, P if and only if Q.
$:=$ or $=:$	An equality sign with a colon on one side of it indicates that the expression on that side is equal by *definition* to the one on the other side.
$\equiv$	is equivalent to; is identically equal to. For example, we will write "$z \equiv x + iy$" as a reminder that z and $x + iy$ are the same thing. If u is a function and α is a number, then "$u \equiv \alpha$" or "$u(z) \equiv \alpha$" means that $u(z) = \alpha$ for every z.
$\mapsto$	This symbol is called the maps-to arrow, and allows one to identify a function without giving it a name. For example, the function $t \mapsto t^2$ is the function whose value at t is t^2.
$A \times B$	the Cartesian product of the sets A and B; the set of ordered pairs (a, b) such that $a \in A$ and $b \in B$.
$A \cap B$	the intersection of the sets A and B; the set of points z such that both $z \in A$ and $z \in B$.
A^o	the interior of the set A. The point a belongs to A^o if, for some $r > 0$, A contains the disk $D(a, r)$.
$\mathrm{Ann}_a(r_1, r_2)$	the open annulus of center a, inner radius r_1, and outer radius r_2. See subsection 3.3A.
$\mathrm{Ann}(r_1, r_2)$	$:= \mathrm{Ann}_0(r_1, r_2)$.

$a \in S$	the element a belongs to the set S. The phrase "all $a \in S$" should be read, "all a in S."
$\text{Arc}_a(r, t_0, t_1)$	the curve given by $z(t) := a + re^{it}$ $(t_0 \le t \le t_1)$; the arc of the circle with center a and radius r, subtending the angle from t_0 to t_1. See 2.5B.
$\text{Arc}(r, t_0, t_1)$	$:= \text{Arc}_0(r, t_0, t_1)$.
arg	An argument is a continuous function $(x, y) \to \theta$ definable, under certain conditions, on a set O, such that if we let $r := \sqrt{x^2 + y^2}$, then at each point, $x = r\cos\theta$ and $y = r\sin\theta$. See 1.3B.
$\text{arg}_{p \mapsto q}$	The unique argument definable, under certain conditions, on a connected set O containing the point p such that $\arg p = q$. See 1.3B.
Arg	The principal argument, defined on V_π, is the argument taking on values in the interval $(-\pi, \pi)$. It is also denoted by $\text{arg}_{1 \mapsto 0}$. See 1.3B.
bA	the set of boundary points of the set A. The point a is a boundary point of A if, for every $r > 0$, the disk $D(a, r)$ contains at least one point of A and at least one point of the complement of A.
∂A	the oriented boundary of the set A in the plane, which means the set of boundary points of A together with a parametrization which makes it a path or contour that winds around every point of A once counterclockwise.
boundary	See bA or ∂A.
bounded	The set S is bounded if there is a constant M such that $|z| \le M$ for all $z \in S$. The function f is bounded if its range is a bounded set.
branch point	See 3.3.7.
Cauchy-Riemann	Two harmonic functions $u(x, y)$ and $v(x, y)$ satisfy the Cauchy-Riemann equations if $u_x = v_y$ and $u_y = -v_x$.
compact	The set S is compact if it is both closed and bounded. Compact sets may be characterized in terms of sequences; see Section 2.4.
composition	The composition of two functions f and g is the function $f \circ g$ given by $f \circ g(z) = f(g(z))$.
conformal	A function is conformal if it is holomorphic and has nonzero derivative. A one-to-one holomorphic function from O_1 onto O_2 is a conformal equivalence of the two open sets; its derivative will necessarily be nonzero everywhere.
contour	See subsection 2.5J.
$\mathbb{C}$	the complex number system; or complex field; or complex plane.
$\hat{\mathbb{C}}$	the extended plane, $\mathbb{C} \cup \{\infty\}$.
$\mathbb{C}'$	the punctured plane, $\mathbb{C} \setminus \{0\}$.

C^n	A C^n function is one whose partial derivatives of orders $\leq n$ all exist and are continuous.
C^∞	A C^∞ function is one whose partial derivatives of all orders exist and are continuous.
Circle	(with a capital C) a circle or a line. See p. 328.
D	the unit disk; the set $\{z \in \mathbb{C} \mid |z| < 1\}$; the open disk of center 0 and radius 1.
$D(a, r)$	the set $\{z \in \mathbb{C} \mid |z - a| < r\}$; the open disk of center $a \in \mathbb{C}$ and radius r. It is understood that $r > 0$. Note that $D(0, 1)$ is also denoted by D.
$D(\infty, r)$	the set $\{z \in \hat{\mathbb{C}} \mid |z| > r\}$. It is understood that $r \geq 0$.
$D'(a, r)$	the set $\{z \in \mathbb{C} \mid 0 < |z - a| < r\}$; the punctured disk of center $a \in \mathbb{C}$ and radius r. It is understood that $r > 0$.
$\bar{D}(a, r)$	the closed disk of center $a \in \mathbb{C}$ and radius $r > 0$; the set $\{z \in \mathbb{C} \mid |z - a| \leq r\}$.
disjoint	Two sets are disjoint if their intersection is empty, that is, if there is no point that is contained in both of them.
$\text{dist}(z, A)$	the distance from the point z to the set A, which is the infimum of the set of numbers $\{|z - a| \mid a \in A\}$.
$f(x)$	When f is a function and x is an element of its domain, $f(x)$ is the value of f at x.
$f[X]$	the image of the set X under the mapping f; the set of all values $f(x)$ with $x \in X$.
$f^{-1}[Y]$	the pre-image of the set Y under the mapping f; the set of all elements x of the domain of f such that $f(x)$ belongs to Y. The symbol f^{-1} is read "f-inverse" and should not be confused with the reciprocal $1/f$, the function whose value at x is $1/f(x)$. See subsection 1.2B.
$f^{-1}(y)$	When f is a one-to-one function and y is an element of its range, $f^{-1}(y)$ is the unique x such that $f(x) = y$. Then f^{-1} is a function whose domain is the range of f. The symbol f^{-1} is read "f-inverse" and should not be confused with the reciprocal $1/f$, the function whose value at x is $1/f(x)$.
H_α	the half-line $\{re^{i\alpha} \mid r \geq 0\}$. See V_α.
H'_α	the ray $\{re^{i\alpha} \mid r > 0\}$.
$\int_\gamma |dz|$	the length of a C^1 curve or path γ. See Section 2.5, subsections A, C, and J.
inside	The inside of a contour Γ is the set of points z for which the winding number $W_\Gamma(z)$ is nonzero.
locally	A property holds locally on the set O if, for every point $a \in O$, there is an open set O_1 such that $a \in O_1 \subseteq O$ and the property holds on O_1.
log	A logarithm $z \mapsto \log z$ is a continuous function definable, under certain conditions, on a set O, such that $e^{\log z} \equiv z$ for $z \in O$. See 2.2B.

$\log_{p \mapsto q}$	The unique logarithm definable, under certain conditions, on a connected set O containing the point p such that $\log p = q$. See 2.2B.
Log	The principal logarithm, defined on V_π, is the one whose imaginary part takes on values in the interval $(-\pi, \pi)$. It is also denoted by $\log_{1 \mapsto 0}$. See 2.2B.
m-to-one	When we say that a function f maps a set A m-to-one onto a set B, we mean that for each point b in B, there are exactly m distinct points a in A such that $f(a) = b$.
map, mapping	These words are synonyms for "function."
$\mathbb{N}$	the set of positive integers.
near a	To say that a condition holds near a is to say that for some $r > 0$, it holds on $D(a, r)$.
neighborhood	A set is a neighborhood of the point a if it contains an open set that contains a; equivalently, if, for some $r > 0$, it contains the disk $D(a, r)$.
potential	The function u is a potential for the vector field $\vec{\mathbf{F}}$ if $\vec{\mathbf{F}} = \mathbf{grad}\, u$. An alternative definition requires instead that $\vec{\mathbf{F}} = -\mathbf{grad}\, u$. The factor -1 is a reasonable convention in certain applications; for example, when u is temperature, the heat flow vector points from the warmer place toward the colder.
$[p, q]$	the line segment that runs from p to q, usually parametrized by the mapping $t \mapsto p(1 - t) + tq \;\; (0 \leq t \leq 1)$.
$[p_1, p_2, \cdots, p_n]$	the path comprising the line segments $[p_1, p_2]$, $[p_2, p_3]$, $\cdots$, $[p_{n-1}, p_n]$.
Q_I	the open first quadrant; the set $\{(x, y) \mid x > 0, y > 0\}$.
$\mathbb{R}$	the set of real numbers.
$\mathbb{R}^+$	the set of nonnegative real numbers.
$\mathbb{R}^2$	the set of ordered pairs (a_1, a_2) of real numbers; the Cartesian product $\mathbb{R} \times \mathbb{R}$; the Cartesian plane; the Euclidean plane.
$\mathbb{R}^n$	the set of ordered n-tuples $(a_1, a_2, \cdots, a_n)$ of real numbers; the Cartesian product $\mathbb{R} \times \mathbb{R} \cdots \times \mathbb{R}$; Euclidean n-space.
$((r, \theta))$	the point in the plane with polar coordinates r and θ; that is, the point (x, y) such that $x = r \cos \theta$ and $y = r \sin \theta$; in other words, the point $re^{i\theta}$.
rational function	the ratio of two polynomials. See 2.3.9.
restriction	Let $f : A \to B$ and $g : X \to B$, where X is a subset of A. If $g(x) = f(x)$ for $x \in X$, then g is the restriction of f to X.
subuniform	uniform on each compact subset.
$T(a, b, c)$	the triangle whose vertices are the three points in the plane $a, b,$ and c. It is the union of the three line segments $[a, b]$, $[b, c]$, and $[c, a]$, and the region enclosed by them.
V_α	the plane with the half-line H_α removed; in other words, the set $\mathbb{C} \setminus H_\alpha \equiv \{re^{it} \mid r > 0 \text{ and } \alpha < t < \alpha + 2\pi\}$. See subsection 1.3B.

$W_\Gamma(a)$	The winding number of the contour Γ with respect to the point a; the net number of times Γ goes around a counterclockwise. See subsection 2.5G.
ω_f	the oscillation of the function f; see p. 401.
$\mathbb{Y}, \mathbb{Y}_-$	The set $\{z = x + iy \mid y \geq 0\}$, the closed upper half-plane, is denoted by $\mathbb{Y}$. The set $\{z = x + iy \mid y \leq 0\}$, the closed lower half-plane, is denoted by $\mathbb{Y}_-$.
$\mathbb{Y}^o, \mathbb{Y}^o_-$	The set $\{z = x + iy \mid y > 0\}$, the open upper half-plane, is denoted by $\mathbb{Y}^o$. The set $\{z = x + iy \mid y < 0\}$, the open lower half-plane, is denoted by $\mathbb{Y}^o_-$.
$\mathbb{Z}$	the set of integers.
$\mathbb{Z}^+$	the set of nonnegative integers.

1

Preliminaries

1.1 PREVIEW

Complex analysis is a coherent and beautiful mathematical theory. As it unfolds, you will encounter theorems that are surprising and fascinating for both their simplicity and their power. We hope that you will enjoy learning the theory for its own sake. Complex analysis provides fundamental ways to understand and solve problems that arise in nature—for example, in fluid flow, electrostatics, magnetostatics, elasticity, and the conduction of heat. Starting in this very Section, we will identify physical problems of the kind that motivate, enliven, and illuminate the theory.

To acquire the powers of the theory, to compute integrals or to solve physical problems, you must dwell for a time in the realm of rigor, precision, and abstraction, and you must master the concepts. We have undertaken to choose an efficient path through the subject matter, and to state all the results in an easy-to-use form. As best we can, we give proofs which are conceptually short and to the point. With few exceptions, reading the proofs will give you exercise in handling the ideas that you most need for understanding and applying the results. In our treatment of the theory, we try to help you gain a ready grasp of the connections among ideas and conditions—the Why of things.

We offer you now a preview of some of what we will achieve in this book. We will briefly describe several boundary value problems; identify and discuss their solutions; and introduce key concepts. Do not expect the methods of solution to be explained in this Section. Chapters 2 and 3 will prepare the way, and in Chapter 5, you will see how to solve these problems and many more. We will often state that the solution to a problem is the only possible one. This book will not prove all of those assertions, but in Section 5.6, we will discuss what is known about uniqueness questions and give references.

We begin by discussing a problem in heat conduction, or, as it may equally well be called, a temperature distribution problem. Consider, in the plane, a set consisting of the inside of a circle, together with the circle itself. Such a set is called a disk. Consider the set as a mathematical model for a thin, flat disk of heat-conducting material. Suppose that by means of some mechanism for heating and cooling, half of the circle is kept at temperature 100°, and the other half at 0°–with perfect insulation at the two points where the hot and cold semicircles meet.[1] Time passes until, whatever the temperatures may have been initially at points inside the disk, they are now constant with respect to time. The temperatures have reached the "steady state," influenced only by the conditions on the circle. What is the temperature at the center of the disk?

For a second specific example, suppose that one-fourth of the circle is maintained at 100° and the rest at 0°. Then what is the eventual temperature at the center?

Figure 1.1-1 shows the two problems just described.

A more general problem: Given arbitrary fixed temperatures at the points of the circle, what will be the temperature at the center? Experiment and theory confirm this plausible answer: Every boundary point has an equal influence on the center, so the temperature at the center must be the average temperature on the circle.

In Chapter 5 we will show you methods for finding the temperature at every point of the disk, not just at the center. Still more generally, we will show how to solve such a problem not just for disks, but for two-dimensional heat-conducting objects of other shapes as well. Consider such an object as a set in the plane, with the temperature held fixed at each point of its boundary, and ask: What are the eventual temperatures at the points inside the set? Since the shape may be much less regular and symmetric than a disk, it is not so easy to obtain an answer by elementary reasoning. Nevertheless, we can apply the conclusion we reached about disks, as follows: *Draw a disk anywhere within the set, and the steady-state temperature at the center will equal the average temperature on the circular boundary.*

That leads us to a definition: Let u be a real-valued function whose domain is a set in the plane. Then u has the **mean value property** if, for every disk contained in the set, the value of u at the center equals the average value of u on the rim. In Section 1.6 we will define a harmonic function as one that satisfies Laplace's equation. It turns out that having the mean value property is the same thing as being harmonic. The Dirichlet problem, which we will treat in Chapter 5, is the general problem of finding a harmonic function whose boundary values are given. Techniques for

---

[1]So that our 2-dimensional mathematical model is understandable in terms of the physics of 3 spatial dimensions, we assume that there is perfect insulation on both sides of the plane of the disk, so that there is no heat flow in the direction perpendicular to that plane. Alternatively, we might assume that the 2-dimensional model is a cross-section of a 3-dimensional model in which all quantities are independent of the third variable. In our discussions of temperature distribution problems, we will not repeat those assumptions.

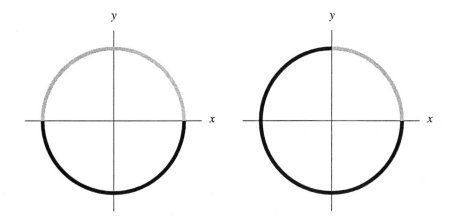

Figure 1.1-1 Two boundary value problems for a heat-conducting disk. In each case, the dark part of the boundary is maintained at 0°, the light gray part at 100°. In the steady state, what is the temperature distribution within the disk? At the center, it will be the average of the temperatures on the boundary: 50° in the first case, 25° in the other.

solving Dirichlet and other boundary value problems are a major showpiece of our book, and we will say more about them later in this Section.

A   It Takes Two Harmonic Functions

The words *map, mapping,* and *function* are synonyms and will be used interchangeably.

We will often find ourselves thinking about a mapping, given by

$$f(x, y) = (u(x, y), v(x, y)),$$

whose domain is all or part of the x, y-plane, and whose range lies in the u, v-plane. A theme of complex analysis is the effort to understand such mappings. We try to visualize them in various ways. We can draw separately the graph of u and the graph of v, in perspective; each is a two-dimensional object in 3-space. For some examples, see Figure 1.2-4. But the graph of f is an object in 4-space; it is not so easy to picture all at once.

One technique is as follows. Draw an x, y-plane and a u, v-plane side-by-side. In the x, y-plane, within the domain of f, select and sketch one or more objects (points, line segments, curves, rectangles, and such). Then sketch their respective images in the u, v-plane. If you are skillful or lucky in your choice of objects, this method can make it easy to visualize the behavior of the mapping; to tell how it preserves or distorts areas, shapes, and proportions. In the first example, we will use sets of curves for "objects." First, some definitions.

An **orthogonal grid** consists of two sets of curves such that wherever a curve in one set intersects a curve in the other set, the tangent lines to the curves at that point

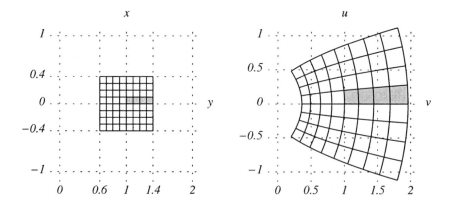

Figure 1.1-2   A rectangular grid in the x, y-plane and its image in the u, v-plane under the mapping $(x, y) \mapsto (x^2 - y^2, 2xy)$ (which we shall later denote as $z \mapsto z^2$).

are well defined and perpendicular. An orthogonal grid is a **rectangular grid** if each of the two sets of curves is made up of parallel line segments. An orthogonal grid is a **polar grid** if one set consists of concentric circles and the other consists of segments of lines that pass through their common center.

Consider the mapping f given by

$$f(x, y) = (x^2 - y^2, 2xy);$$

in other words, the function that maps (x, y) to $(u(x, y), v(x, y))$, where the two coordinate functions are given by $u(x, y) = x^2 - y^2$ and $v(x, y) = 2xy$. In the first picture in Figure 1.1-2, in the vicinity of the point $(1, 0)$ in the x, y-plane, you see a rectangular grid. It consists of nine vertical and nine horizontal line segments; they bound 64 squares, of which four are shaded. The second picture in Figure 1.1-2 shows the 18 curves and 64 regions in the u, v-plane which are the images of those line segments and squares under the mapping f. The shaded regions are the images of the four shaded squares. As you can see, the mapping distorts the part of 2-space covered by the grid in a definite but fairly gentle fashion. Near the point $(1, 0)$, the mapping acts approximately just as a magnification by a factor of 2.

Figure 1.1-3 portrays the same mapping as Figure 1.1-2. You may need to look carefully, and think about it a bit, before this becomes plausible to you. The two figures give us different but consistent information about how the mapping acts in a certain region of the plane. Each shows a grid in the x, y- plane centered at $(1, 0)$, and its image in the u, v-plane. Figure 1.1-2 tells us that the mapping bends a rectangular grid into a certain curved grid. Figure 1.1-3 tells us that it takes a certain curved grid and straightens it out into a rectangular grid.

To obtain Figure 1.1-3, we first drew a rectangular grid in the u, v-plane, consisting of line segments which are parts of the lines whose equations are $u = u_0$ or $v = v_0$ for selected sets of constants u_0, v_0. Then we drew a grid in the (x, y)- plane whose

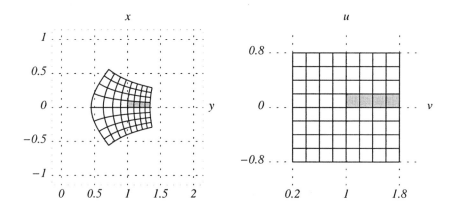

Figure 1.1-3 A certain grid in the x, y-plane and its image in the u, v-plane under the mapping $(x, y) \mapsto (x^2 - y^2, 2xy)$. The image is a rectangular grid.

image is the rectangular grid. In other words, the left-hand picture in Figure 1.1-3 shows a set of level curves for each of the two functions u and v; that is, each curve is part of the locus of an equation $u(x, y) = u_0$ or $v(x, y) = v_0$. If u were temperature, for example, each curve on which u is constant would be called an isotherm.

If you insist on getting a precise grasp of what's going on in Figures 1.1-2 and 1.1-3, you may wish to do Exercises 5 and 6 at the end of the Section. We are happy that we had a computer to help us draw those pictures. Doing such tedious work by hand is not the point. The point is to understand the behavior of certain mappings.

We do not claim to have described how the mapping f behaves "globally," on the entirety of its domain. We have merely taken a look at its behavior "locally," on a small region, near a selected point. Compared to other mappings that we could have considered, its behavior is simple and nice. For one thing, the image of each orthogonal grid is another orthogonal grid. You will be happy to hear that nearly all the mappings that we study in this course are mostly like that. −Which, as you will see, means that we study mappings $(x, y) \mapsto (u(x, y), v(x, y))$ where u and v are non-constant harmonic functions satisfying the **Cauchy-Riemann equations**:

$$\frac{\partial u}{\partial x} = \frac{\partial v}{\partial y} \quad \text{and} \quad \frac{\partial u}{\partial y} = -\frac{\partial v}{\partial x}. \tag{1}$$

If u and v satisfy (1), then v is a **harmonic conjugate** of u, and the two are a **conjugate pair**. We will discuss these ideas further in Section 1.6.

In each example of subsection B, we will be looking for the harmonic function u such that at each point (x, y) of its domain in the plane, $u(x, y)$ is the temperature at that point. As frequently happens, the physical problem seems to ask just for u; as we shall explain later, the mathematics gives us two harmonic functions, u and v, where v is a harmonic conjugate of u. But then v turns out to be relevant to the physical problem also.

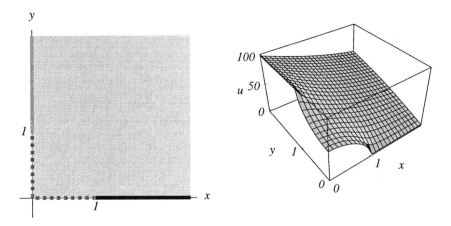

Figure 1.1-4 Pictures for Example 1.1.1. On the left, a flat, heat-conducting object is insulated near a corner, and the rest of the boundary is maintained at 100° or at 0°. On the right, the graph of the temperature function $u(x, y)$.

Imagine, if you were watching a man and woman do the tango, how difficult it would be to appreciate the performance if one of the dancers were invisible to your eyes. Real-valued harmonic functions are like tango dancers; they are best observed in well matched pairs. Complex analysis, one may say, is the study of such pairs.

Let's put it another way. Complex analysis is the study of holomorphic functions, also called analytic functions. They are usually defined after complex numbers and notation are presented. But one may define them in "real" terms, as follows. Let f be given by $f(x, y) = (u(x, y), v(x, y))$, such that the domain of f is an open set (as defined in Section 1.4) in the plane. If u and v are harmonic, and v is a harmonic conjugate of u, then f is **holomorphic** on its domain. If, in addition, it is true that at every point at least one of the partials $\dfrac{\partial u}{\partial x}$ and $\dfrac{\partial u}{\partial y}$ is nonzero, then f is **conformal** on its domain.

B   Heat Flow

We return now to our examples of heat flow problems. Boundary conditions may consist of fixed, known temperature values at the boundary, or else of fixed, known rates of heat flow across the boundary. The two resulting types of problem are called **Dirichlet** and **Neumann problem**, respectively. The next example is a mixture of the two, because part of the boundary is insulated; that is, there is zero heat flow across it.

1.1.1. Example. *Consider the positive quadrant of the x, y-plane, as pictured on the left in Figure 1.1-4, as a mathematical model of a large thin metal sheet with two edges meeting at a right angle. Suppose that over a long period of time:*

- *All points of the interval $[1, \infty)$ on the x-axis are kept at temperature $0°$ Celsius.*

- *All points of the interval $[1, \infty)$ on the y-axis are kept at temperature $100°$ Celsius.*

- *The interval $[0, 1]$ on each axis is insulated, so that no heat can flow across that part of the sheet's boundary.*

- *There are no sources of heat within the metal sheet.*

What is the temperature at each point of the sheet, and how exactly does heat flow from hot edge to cold edge?

Preview of the Solution. The physics of the situation dictates that there is only one possible answer. There is a uniquely determined harmonic function u whose value at each point (x, y) is the temperature there. In Chapter 5 we will use the method of conformal mapping to find u, which turns out to be given by

$$u(x, y) =$$
$$50 - \frac{100}{\pi} \sin^{-1} \left(\frac{\sqrt{4x^2y^2 + (1 + x^2 - y^2)^2} - \sqrt{4x^2y^2 + (-1 + x^2 - y^2)^2}}{2} \right).$$

$$(2)$$

On the right in Figure 1.1-4 appears the graph of u over the square $[0, 2] \times [0, 2]$. Notice the parts of the boundary that are held at constant temperatures, $100°$ and $0°$. It's interesting how the temperature behaves along the insulated part of the boundary, and in the interior near the corner.

Section 5.6 offers a discussion as to why the solution is unique.

The methods that give us the temperature distribution u will give us also a harmonic conjugate of u:

$$v(x, y) =$$
$$- \frac{100}{\pi} \cosh^{-1} \left(\frac{\sqrt{4x^2y^2 + (1 + x^2 - y^2)^2} + \sqrt{4x^2y^2 + (-1 + x^2 - y^2)^2}}{2} \right).$$

$$(3)$$

The graph of v appears on the left in Figure 1.1-5. The numerical value of v at a point has no straightforward significance, but v is pertinent to the physical situation nevertheless, because its level curves are the paths of heat flow.

We have set the messy expressions for u and v before your eyes in order to advertise another merit of complex analysis. As you will see in Chapter 5, there is a briefer, neater way to write down expressions for those functions. It will become easy to understand where they come from and how they behave.

For each constant u_0 between 0 and 100, the condition $u(x, y) = u_0$ determines a curve which, being the set where the temperature equals u_0, is called an isotherm. For each constant v_0 in a certain interval, the condition $v(x, y) = v_0$ determines a

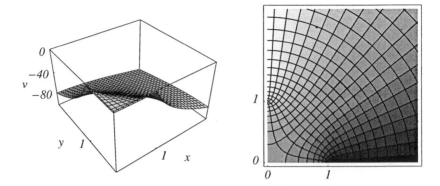

Figure 1.1-5 More pictures for Example 1.1.1. On the left, the graph of v, a harmonic conjugate of the temperature function u. On the right, some of the isotherms and paths of heat flow, which are the level curves of u and v, respectively.

curve which is a path of heat flow. The picture on the right in Figure 1.1-5 shows the isotherms and paths of heat flow for selected sets of values of u_0 and v_0. Each isotherm starts at a point of the insulated boundary and runs off toward infinity. Each path of heat flow, of course, runs from the hot boundary to the cold boundary. As the picture indicates, the isotherms and paths of heat flow form an orthogonal grid. That is forced by the physics; it is also a geometric consequence of the Cauchy-Riemann equations (1). □

Consider the general steady-state temperature problem for the unit disk. A temperature distribution u is given on the boundary; that is, $u(w)$ is known for $|w| = 1$. How does one find the temperature $u(z)$ for each point z inside the disk? Obviously, or at least plausibly, it equals some kind of weighted average of the temperatures on the rim, with the temperatures at closer points having the greater weight. In subsection 5.5.C we will derive the Poisson integral formula, which computes the desired weighted average at every point; and the conjugate Poisson integral formula, which computes a harmonic conjugate of u.

1.1.2. Example. *Consider again the Dirichlet problem indicated in the second picture in Figure 1.1-1. The part of the circle that lies in the first quadrant is maintained at $100°$ and the rest at $0°$. The problem is to find the harmonic function u giving us the steady-state temperature at each point within the disk. (As we stated earlier, $u(0,0) = 25°$.)*

Preview of the Solution. In Chapter 5 we will show how to solve this problem in two ways. Figure 1.1-6 shows the graph of u, generated by numerical evaluation of the Poisson integral over the boundary, a method to be explained in Section 5.5. Figure 1.1-7 was generated from precise expressions for the temperature function u and its harmonic conjugate v, obtained by the conformal mapping method to be explained in Section 5.1. It shows the level curves given by conditions $u(x,y) = c$, where c

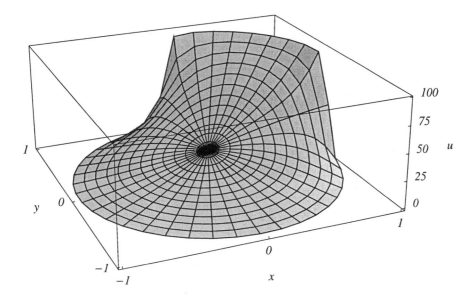

Figure 1.1-6 Graph of the temperature function $u(x, y)$, the solution to Example 1.1.2.

has values $90, 80$, and so on down to 10. Those curves are called isotherms. Also shown are several level curves given by conditions $v(x, y) = c$. Those curves are the paths of heat flow. □

C A Geometric Rule

In 1890, H. A. Schwarz ([64, p. 360]) gave an interesting interpretation of the Poisson integral formula. Schwarz pointed out that the formula is equivalent to an easily stated geometric rule for finding the solution's value at each point. In some cases, the procedure is both easy and amusing to carry out by hand. Therefore we will describe it for you.

For z inside the disk, we may find $u(z)$ as follows. For a point w on the boundary, the line through z and w crosses the rim at one other point; call it w^*. The mapping $w \mapsto w^*$ is a one-to-one mapping from the boundary onto itself. Define a new temperature distribution u^* on the boundary by letting $u^*(w) := u(w^*)$. In other words, we move the temperature at w to the point w^*. Then $u(z)$ will be the average value of u^* on the rim. (Of course, if z is the origin, then u^* and u have the same average.)

1.1.3. Example. *With the boundary values on the unit disk as shown in the first picture of Figure 1.1-1, find $u(z)$ when $z := (0, \frac{1}{2})$, using the geometric rule.*

Solution. Make a sketch. First, find w^* when $w = (-1, 0)$. The line through w and z has equation $y = \frac{1}{2}(x + 1)$. So w^* is the point other than w that satisfies the

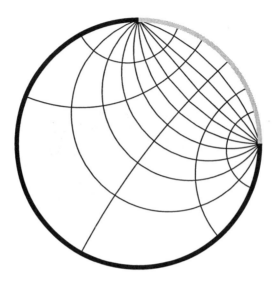

Figure 1.1-7 This graphic depicts the solution to Example 1.1.2 by showing some of the isotherms and paths of heat flow.

equations

$$y = \frac{1}{2}(x+1) \quad \text{and} \quad x^2 + y^2 = 1.$$

Thus $w^* = \left(\frac{3}{5}, \frac{4}{5}\right)$. Similarly (or by symmetry), if $w = (1,0)$, then $w^* = \left(-\frac{3}{5}, \frac{4}{5}\right)$. Now it is clear that under the mapping $w \mapsto w^*$, the upper half of the circle, which is at 100°, maps to the arc that runs from $\left(\frac{3}{5}, \frac{4}{5}\right)$ clockwise around to $\left(-\frac{3}{5}, \frac{4}{5}\right)$, which subtends an angle of approximately 4.99 radians, or about 79% of the circle. The lower half of the circle, which is at 0°, maps to the remaining arc. Therefore the average of the new boundary values, and the temperature at z, is approximately 79°. □

D Electrostatics

By means of several examples, as a preview of Section 5.3, we will now briefly say what an electric field is, and introduce the problem of describing such a field. Once again, the solution is provided by a conjugate pair of harmonic functions, and represented pictorially by an orthogonal grid.

The discussion here is intended to be familiar to students of physics, and merely intriguing to others.

In 3-dimensional x, y, z-space, consider the z-axis as a mathematical model for a long wire which carries a stationary electric charge. Such a charge consists of some fixed number of coulombs (units of charge) per unit length. Let the condition $x^2 + y^2 = 1$ define a cylindrical surface which encloses the wire at its center. The

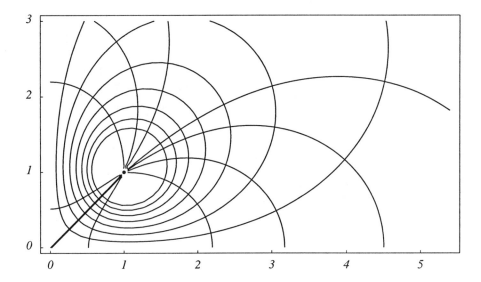

Figure 1.1-8 Problem: In 3-dimensional x, y, z-space, a uniform static charge resides on the line given by $x = 1, y = 1$ inside a vacuum in the region given by $x > 0, y > 0$. Its boundary, consisting of the positive axes, is grounded. The picture shows the 2-dimensional cross section $z = 0$. The curves that surround the point $(1, 1)$ are equipotentials. The paths of force are curves that emanate from that point and go toward points on the boundary.

surface is charge-free; it is grounded. Inside the cylinder, except for the wire, there is a vacuum. We seek to describe the electric field inside the cylinder that is caused by the electric charges on the wire. The value of the field at each point is a vector that gives the direction and magnitude of the force that would be exerted upon a unit of charge if it were located at that point. The force always acts toward or away from the wire, depending on the sign of the charge, with no component in the z-direction. So we may regard the problem as 2-dimensional.

Notice that we are using a line and a cylinder of infinite length. They provide a reasonable mathematical model for a wire and cylinder that are merely very long, where we are concerned with what happens in a cross-section that is far from their ends.

We seek a harmonic function $u(x, y)$ and a conjugate $v(x, y)$ defined for $0 < x^2 + y^2 < 1$. The function u is known as the Green's function of this region. The level curves given by $u(x, y) = c$, for various values of the constant c, are called the equipotential curves. The magnitude of the force per unit charge is constant on each of those curves. It is quite plausible, perhaps obvious, that they are circles centered at the origin. On the circle of radius r, the force per unit charge is proportional to $1/r$. The level curves given by $v(x, y) = c$ are the curves along which the force acts, called paths of flux. If there were a particle inside the region, it would move along one of those paths. They are of course lines through the origin. The picture on the left in Figure 5.2-4, page 337, shows an orthogonal grid formed by level curves from

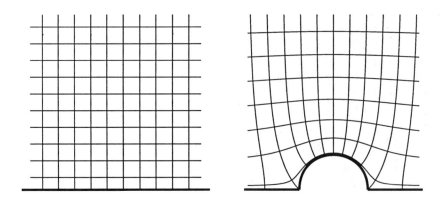

Figure 1.1-9 Streamlines and equipotentials for two fluid flows. On the left, the river bottom is flat; on the right, it has a bump whose cross-section is a half disk.

those two families. This problem will be visited again, and u found explicitly, in Example 5.3.2, page 343.

Now consider a variant on the above problem, in which the cylinder is the same but the wire is off-center, going through $\left(\frac{1}{2}, 0\right)$ instead of $(0,0)$. Again, we seek an orthogonal grid that represents the solution. The conformal mapping method, to be explained later, allows this problem to be reduced to the one above, as will be explained in Example 5.3.3, page 344. The solution is indicated in the picture on the right in Figure 5.2-4, page 337.

Here is another problem like those above. Replace the cylinder by an object whose cross section is the first quadrant, defined by the inequalities $x > 0, y > 0$. The cross section of the grounded surface is the boundary of the first quadrant, consisting of the positive coordinate axes. Let the wire pass through the point $(1,1)$. This problem is also reducible to the first one by the conformal mapping method, as we will show in Example 5.3.5. The orthogonal grid in Figure 1.1-8 represents the solution.

Quite often, one finds that two or more different physical problems and their solutions can be represented by the same mathematical model. A model can represent an electrostatics problem and its solution; or, equally well, a temperature distribution problem and its solution. To be sure, it may be that one of the two physical problems will make sense and will actually arise in practice, while the other will not.

The analogue of temperature is electrostatic potential u. It gives potential energy per unit charge, and may be measured in volts.

Consider the following variant on the first problem above. Remove the wire, so that the inside of the cylinder is a charge-free vacuum. Maintain the part of the cylinder that lies in the first quadrant of 3-space $x > 0, y > 0$ at 100 volts. It is a conductor connected to a battery. Ground the rest of the cylinder's surface, so that it is maintained at 0 volts. The problem is to describe the electric field, which amounts to finding a harmonic function $u(x, y)$ satisfying precisely the boundary conditions represented on the left in Figure 1.1-1. As you see, we have the same mathematical

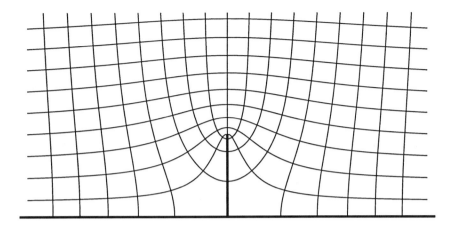

Figure 1.1-10 Streamlines and equipotentials for a fluid flow disrupted by a dam.

model and solution for this problem as for the problem of Example 1.1.2, but with different physical interpretations. The graph of u is shown in Figure 1.1-6. The solution is represented in a different way in Figure 1.1-7, which shows equipotential curves and paths of flux—level curves of u and of its conjugate v, respectively.

E Fluid Flow

The steady flow of a perfect fluid, to be defined later, can be described using a conjugate pair of harmonic functions $u(x,y)$ and $v(x,y)$, the **velocity potential** and the **stream function**, respectively. The fluid flows along the streamlines, which are the level curves given by $v(x,y) = c$. The level curves given by $u(x,y) = c$ are the equipotential curves.

In this preview, we will simply show you some pictures that represent solutions. We will refer to these pictures when we tell the story with further detail in subsection 2.5.I and Section 5.4.

Consider first a fluid occupying the half of x, y, z-space given by $y > 0$. The upper half of the x, y-plane is then a 2-dimensional cross section of the fluid. We assume there is no component of flow in the other dimension. The x-axis is then an impenetrable wall, the "bottom of the river." There are two boundary conditions. One states that at the impenetrable wall, there is no component of flow perpendicular to that wall. The other states that the limit of the flow velocity vector, as y tends to $+\infty$, is α meters per second in a horizontal direction toward the right. The solution is that the flow velocity vector has that same value everywhere in the fluid. The harmonic functions that describe the solution are then given by $u(x,y) := \alpha x$ and $v(x,y) := \alpha y$. The left hand picture in Figure 1.1-9 shows the streamlines, which are parallel horizontal lines; and the equipotential curves, which are vertical lines.

For a second example, consider the same situation but with a cylindrical bump, its cross-section a half disk, on the river bottom. The method of conformal mapping

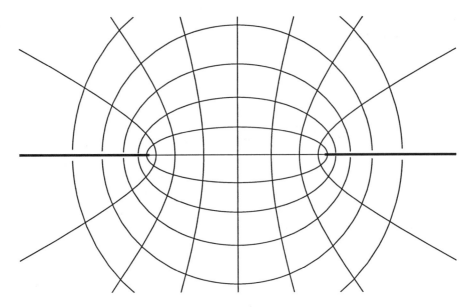

Figure 1.1-11 The orthogonal grid in this picture consists of hyperbolas and ellipses.

allows this problem to be reduced to the first one. The right-hand picture in Figure 1.1-9 describes the solution, showing how the fluid flows over the bump. Notice that the equipotentials meet the boundary at right angles except at the corners, where this would be an impossible requirement. We will present the details in Example 5.4.2, page 352. When we obtain explicit expressions for u and v, we can determine precisely the velocity of the flow at an arbitrary point of the fluid, and in particular in the region near the bump where it varies the most from α. Qualitative information about velocity is present in the picture. Our equipotentials $u = c$ are drawn by computer for equally spaced values of c. Thus where the equipotentials appear closer together, the fluid flows faster.

In our third example, there is an abruptly vertical dam on the river bottom, which is apparently rather disruptive to the flow. This problem also reduces to the first one. Figure 1.1-10 shows a sample of streamlines and equipotentials. The picture leaves open the question, Just how great is the velocity near the top point of the dam? In practice, would a fluid flow in this situation follow the behavior predicted by this mathematical model and become turbulent? We will discuss the problem further in Example 5.4.3, page 354.

F   One Model, Many Applications

You may have realized by now that quite similar mathematical models appear in discussions of temperature distributions, electric fields, and fluid flows. To underline this observation, we offer one model and point out how it fits several physical situations.

Consider two infinite plates in x, y, z-space. One plate consists of all points (x, y, z) with $x \geq 1$ and $y = 0$. The other consists of all points with $x \leq -1$ and $y = 0$. In the cross-section of 3-space with the x, y-plane, the two plates appear as the intervals $(-\infty, -1]$ and $[1, \infty]$ on the real axis. In the plane with those two intervals removed, consider this conjugate pair of harmonic functions:

$$u_1(x, y) = 50 - \frac{100}{\pi} \sin^{-1} \left(\frac{\sqrt{(x+1)^2 + y^2} - \sqrt{(x-1)^2 + y^2}}{2} \right); \quad (4)$$

$$v_1(x, y) = -\frac{100}{\pi} \cosh^{-1} \left(\frac{\sqrt{(x+1)^2 + y^2} - \sqrt{(x-1)^2 + y^2}}{2} \right). \quad (5)$$

As Section 2.2 will explain, the level curves of u_1 and v_1 are hyperbolas and ellipses, respectively, all with focii $(-1, 0)$ and $(+1, 0)$. The ellipses meet the plates at right angles. See Figure 1.1-11. This mathematical model has at least the following three physical interpretations.

- Suppose that the plate on the left were maintained at 100°, and the plate on the right at 0°. Then u_1 and v_1 provide the solution to the temperature distribution problem. The hyperbolas are the isotherms. The ones shown in the Figure are, from left to right, the isotherms of 87.5°, 75°, and so forth, down to 12.5°. The ellipses are the paths of heat flow.

- Suppose that the plate on the left were electrically grounded, maintained at 0 volts; and that the plate on the right were connected to a battery and maintained at 100 volts. Then u_1 and v_1 serve to describe the electric field. The hyperbolas are the equipotentials, and the ellipses are the paths of force.

- Suppose that the two plates are the impenetrable walls for the steady flow of a perfect fluid in space, moving through the aperture between them. Then u_1 and v_1 serve to describe the flow. Suppose that the velocity of the flow at infinity is $\frac{100}{\pi}$ meters per second, downward. Then the functions (4) and (5) are the velocity potential and stream function, respectively. The hyperbolas are the paths of flow, and the ellipses are the equipotentials.

Exercises

1. In each case, show that the given functions u and v satisfy the Cauchy-Riemann equations (1). In other words, show that v is a harmonic conjugate of u. (Review partial differentiation as necessary!)

   (a) $u(x, y) := x$, $v(x, y) := y$.

   (b) $u(x, y) := x$, $v(x, y) := y + 100$.

   (c) $u(x, y) := y$, $v(x, y) := -x$.

   (d) $u(x, y) := x^2 - y^2$, $v(x, y) := 2xy$.

   (e) $u(x, y) := x^2 - y^2$, $v(x, y) := 2xy - 17$.

(f) $u(x, y) := 2xy$, $v(x, y) := y^2 - x^2$.

2. Prove that if v is a harmonic conjugate of u, then $-u$ is a harmonic conjugate of v.

3. Let $u(x, y) = ax + by$, $v(x, y) = cx + dy$, where the four numbers a, b, c, d are constants, not all zero. Consider these three conditions. Prove that (i) and (ii) are equivalent, and that (ii) implies (iii). Does (iii) imply (ii)? Explain.

   (i) v is a harmonic conjugate of u.

   (ii) $a = d$ and $b = -c$.

   (iii) The two lines whose equations are $ax + by = 0$ and $cx + dy = 0$ are perpendicular.

4. Twelve expressions for functions of x and y are given. Arrange them in six ordered pairs such that the second member of each pair is a harmonic conjugate of the first member.

   (a) $e^x \cos y$.   (e) $\sin x \cosh y$.   (i) $y^2 - x^2$.

   (b) $-\cos x \sinh y$.   (f) $3x + 2y$.   (j) $\arctan \dfrac{y}{x}$.

   (c) $e^x \sin y$.   (g) $2xy$.   (k) $\ln \sqrt{x^2 + y^2}$.

   (d) $2x - 3y$.   (h) $-2x - 3y$.   (l) $3x - 2y$.

5. When p and q are two points in the plane, the symbol $[p, q]$ stands for the line segment from p to q. Using that notation, we may identify the nine horizontal line segments in Figure 1.1-2, in the grid on the left, as follows:

$$[(0.6, 0.1k), (1.4, 0.1k)] \qquad (k = -4, -3, \cdots, 3, 4)$$

The nine vertical line segments are:

$$[(1 + 0.1k, -0.4), (1 + 0.1k, +0.4)] \qquad (k = -4, -3, \cdots, 3, 4).$$

Show that the image of each of those 18 line segments is part of a parabola, and find its equation.

6. In Figure 1.1-3, in the grid on the right, there are nine horizontal line segments:

$$[(0.2, 0.2k), (1.8, 0.2k)] \qquad (k = -4, -3, \cdots, 3, 4)$$

and nine vertical line segments:

$$[(1 + 0.2k, -0.8), (1 + 0.2k, 0.8)] \qquad (k = -4, -3, \cdots, 3, 4).$$

Show that each of the 18 curves in the grid on the left, except one, is part of a certain nondegenerate hyperbola, and give its equation.

7. Explain Example 1.1.3 in detail. Use a diagram.

8. With the boundary values on the unit disk as shown in the first picture of Figure 1.1-1, use the geometric rule to show that the horizontal diameter is the 50° isotherm.

9. Consider again the Dirichlet problem discussed in Example 1.1.2. Use the geometric rule to show that the chord from $(0, 1)$ to $(1, 0)$ is the 75° isotherm.

10. Use the geometric rule to find, for Example 1.1.2, the temperature at each of the points $(\frac{1}{2}, 0)$, $(0, \frac{1}{2})$, $(-\frac{1}{2}, 0)$, and $(0, -\frac{1}{2})$.

11. Consider the problem represented by the picture on the left in Figure 1.1-1. Make a sketch representing the solution, showing several isotherms and paths of heat flow on the disk. Aim for plausibility and neatness, not precision. Speculate as necessary.

12. Explain this remark: "The picture on the right in Figure 1.1-5 is like the picture on the left in Figure 1.1-3 in its method of depicting a mapping $(x, y) \mapsto (u, v)$".

13. In the fluid flow problem described in subsection F, where do you think the velocity is the greatest?

Help on Selected Exercises

4. a,c; b,d; d,f; g,i; h,l; k,j.

5. Each of the horizontal segments consists of the points $(x, 0.1k)$ with $0.6 \leq x \leq 1.4$. The image of such a point is (u, v), where $u = x^2 - .01k^2$ and $v = 0.2kx$ and hence $u = \dfrac{v^2}{(0.2k)^2} - .01k^2$.

8. If z is any point on the diameter, the mapping $w \mapsto w^*$ takes the upper semi-circle to the lower semicircle and *vice versa*. So u^* assigns the temperatures 0° and 100^* each to half of the circle. Thus the average of u^*, and hence the temperature at z, is 50°.

10. The approximate answers are, respectively, $40^o, 40^o, 10^o$, and 10°.

11. The isotherms are arcs of circles passing through the points where the hot and cold arcs meet. The paths of heat flow are also circular arcs.

12. In each case, the level curves of u and v form an orthogonal grid because they are harmonic conjugates. Precisely because they are level curves, the image of the grid is a rectangular grid.

1.2 SETS, FUNCTIONS, AND VISUALIZATION

A Terminology and Notation for Sets

It is often helpful to have a short way to write something, provided the reader has become familiar and comfortable with the terms and symbols used . Our purpose here is to state some definitions which will give us brief and precise ways to write things about sets. First we identify the symbols for several familiar sets so that we can use them to give examples.

$$\emptyset \quad := \quad \text{the empty set; the set that contains no elements.}$$
$$\mathbb{R} \quad := \quad \text{the set of real numbers.}$$
$$\mathbb{Z} \quad := \quad \text{the set of integers.}$$
$$\mathbb{N} \quad := \quad \text{the set of positive integers.}$$
$$\mathbb{R}^+ \quad := \quad \text{the set of nonnegative real numbers.}$$
$$\mathbb{Z}^+ \quad := \quad \text{the set of nonnegative integers.}$$
$$\mathbb{R}^2 \quad := \quad \text{the set of ordered pairs } (a_1, a_2) \text{ of real numbers.}$$
$$\mathbb{R}^n \quad := \quad \text{the set of ordered } n\text{-tuples } (a_1, a_2, \cdots, a_n) \text{ of real numbers.}$$

The symbol ":=" signifies that what is on the left is defined as equal to what is on the right.

Let P be a set. Each of the notations $p \in P$ and $p \ni P$ means that p is an element of the set P; alternatively, one may say that p belongs to P, or that p lies in P, or that P contains p. The notation $p \notin P$ means that p is not an element of P. Examples of true statements:

$$\pi \in \mathbb{R}. \qquad \pi \notin \mathbb{Z}. \qquad 5 \in \mathbb{Z}. \qquad -3 \in \mathbb{Z}. \qquad -3 \notin \mathbb{N}. \qquad (0, \pi, -3/2) \in \mathbb{R}^3.$$

Those are complete English sentences, even though each is written entirely in mathematical symbols. Notice that "Let x be a real number" means the same thing as the slightly briefer "Let $x \in \mathbb{R}$." Variants on the use of the symbol: "For $n \in \mathbb{Z}$, we know that $n \in \mathbb{R}$" means "For n in $\mathbb{Z}$, we know that n belongs to $\mathbb{R}$."

Let P and Q be sets. Each of the notations $Q \subseteq P$ and $P \supseteq Q$ means that every element of Q is an element of P; we say then that Q is a subset of P or that P contains Q.

Each of the notations $Q \subset P$ and $P \supset Q$ means that $Q \subseteq P$ and $Q \neq P$; in other words, Q is a subset of P other than P. We say then that Q is a **proper** subset of P or that P contains Q **properly.** Examples of true statements:

$$\mathbb{Z} \subseteq \mathbb{R}. \qquad \mathbb{Z} \subset \mathbb{R}. \qquad \mathbb{N} \subset \mathbb{Z}^+ \subset \mathbb{Z}. \qquad \emptyset \subset \mathbb{R}+. \qquad \mathbb{R} \subseteq \mathbb{R}.$$

Whenever S is a set and C is a statement, the symbol $\{x \in S \mid C\}$ denotes the set of all $x \in S$ such that C is true. For example, $\{x \in \mathbb{R} \mid 0 < x < 1\}$ stands for the set of all real numbers strictly between 0 and 1, the interval customarily written as

$(0, 1)$. The following statements will remind you of the usual notation for intervals of the real line:

$$(a, b) := \{x \in \mathbb{R} \mid a < x < b\}.$$
$$(a, b] := \{x \in \mathbb{R} \mid a < x \leq b\}.$$
$$[a, b) := \{x \in \mathbb{R} \mid a \leq x < b\}.$$
$$[a, b] := \{x \in \mathbb{R} \mid a \leq x \leq b\}.$$

Notice that the symbol "(a, b)" stands for both the interval (a, b) and the ordered pair (a, b). We are obliged to make it clear which is intended.

If we can expect the reader to know from the context what the set S is, then we may write $\{x \mid C\}$ instead of $\{x \in S \mid C\}$. For example, one would usually understand $\{x \mid 0 < x < 1\}$ to mean the same set as $\{x \in \mathbb{R} \mid 0 < x < 1\}$.

Another way to use the braces, "$\{$" and "$\}$," to identify specific sets: The symbol $\{1, 2, 4\}$ stands for the set whose elements are $1, 2$, and 4. Thus the symbol $\{y\}$ denotes the set whose one and only element is y. Such a set is a **singleton**. Notice that $\{y\}$ and y do not mean the same thing; but $x \in \{y\}$ means the same thing as $x = y$.

The symbol $Q \setminus P$ stands for the set of all points p such that $p \in Q$ and $p \notin P$.

Examples of true statements:

$\mathbb{Z}^+ \setminus \mathbb{N} = \{0\}$. $\mathbb{N} \setminus \mathbb{Z}^+ = \emptyset$.

$(1, 3.5) \setminus (3, 5) = (1, 3]$. $(1, 3] \setminus \{3\} = (1, 3)$.

$\mathbb{R}^+ = \{x \in \mathbb{R} \mid x \geq 0\}$. $\mathbb{R} \setminus \mathbb{R}^+ = \{x \in \mathbb{R} \mid x < 0\}$.

$\{p \in Q \mid p \notin P\} = Q \setminus P$. $\{1, 2, 3, 4, 5\} \setminus \{3, 4, 5, 6\} = \{1, 2\}$.

The **union** of two sets P and Q, denoted by $P \cup Q$, is defined as follows:

$$P \cup Q := \{x \mid x \in P \text{ or } x \in Q\}.$$

The "or" is the inclusive or and signifies that at least one of the conditions $x \in P$ and $x \in Q$ must hold. The union of P and Q may be defined equivalently as the smallest set that both contains P and contains Q. Examples:

$\mathbb{R} \cup \mathbb{Z} = \mathbb{R}$. $\mathbb{Z}^+ \cup \mathbb{R}^+ = \mathbb{R}^+$.

$\{0\} \cup \mathbb{N} = \mathbb{Z}^+$. $(1, 4] \cup (4, 7] = (1, 7]$.

$(-\infty, 1) \cup (-1, \infty) = \mathbb{R}$. $\{1, 3, 5\} \cup \{2, 3, 4\} = \{1, 2, 3, 4, 5\}$.

The **intersection** of two sets P and Q is the set

$$P \cap Q := \{x \mid x \in P \text{ and } x \in Q\}.$$

The "and" signifies that both of the conditions $x \in P$ and $x \in Q$ must hold. The intersection of P and Q may be defined equivalently as the largest set that is both a subset of P and a subset of Q. For example: Let A be the set of all positive even integers and B the set of all even integers which are less than 9; that is,

$$A := \{2n \mid n \in \mathbb{Z} \text{ and } 2n > 0\}; \qquad B := \{2n \mid n \in \mathbb{Z} \text{ and } 2n < 9\}.$$

Then $A \cup B$ is the set of all even integers, and $A \cap B = \{2, 4, 6, 8\}$. We must mention that a comma often replaces the word "and" in definitions like those of A and B just given; thus

$$A := \{2n \mid n \in \mathbb{Z}, \, 2n > 0\}; \qquad B := \{2n \mid n \in \mathbb{Z}, \, 2n < 9\}.$$

The **Cartesian product** of two sets A and B is the set of ordered pairs,

$$A \times B := \{(a, b) \mid a \in A, b \in B\}.$$

Similarly, the **Cartesian product** of three sets $A, B,$ and C is the set of ordered triples,

$$A \times B \times C := \{(a, b, c) \mid a \in A, b \in B, c \in C\};$$

and so forth. The Cartesian product of a set A with itself n times, $A \times A \times \cdots \times A$, is denoted by A^n. Thus for example, $\mathbb{R}^2$ is short for $\mathbb{R} \times \mathbb{R}$, $\mathbb{R}^3$ is short for $\mathbb{R} \times \mathbb{R} \times \mathbb{R}$, and so forth. For each $n \in \mathbb{Z}^+$, $\mathbb{R}^n$ is the set of ordered n-tuples of real numbers, and is called the n-**dimensional real Euclidean space**. Notice that $\mathbb{R}^1$ means $\mathbb{R}$.

Let A be a set. A function that maps $A \times A$ into A is a **binary operation** on A. For example, addition and multiplication of real numbers may be thought of as functions defined on $\mathbb{R} \times \mathbb{R}$ with values in $\mathbb{R}$, namely $(x, y) \mapsto x + y$ and $(x, y) \mapsto x \cdot y$, respectively.

B Terminology and Notation for Functions

We will now present some definitions which will give us precise and concise language for the discussion of functions. As examples, we will use familiar real-valued functions of a real variable, including the exponential, logarithm, sine, cosine, hyperbolic sine, and hyperbolic cosine. In Chapter 2 we will extend all of those functions to the appropriate complex domains.

Especially important are the concepts of the image and the pre-image of a set; as you read, watch for the symbols $f[X]$ and $f^{-1}[Y]$ and their definitions.

The words "map" and "mapping" are synonyms for "function," and we will use the three words interchangeably.

The notation $f : A \to B$ means that f is a function with domain A, and that $f(x) \in B$ for every $x \in A$. It means the same thing to say, "f is a function on A with values in B," or "f maps A into B." The element $f(x)$ is **the value of f at x**.

There is nothing in the definition of "function" that specifies what kinds of sets A and B can be; they might be, for example, sets of apples, sets of books, sets of functions, or sets of sets.

So long as B is some set that contains the range of f, it makes sense to speak of "the function $f : A \to B$," and to consider f as a mapping into B.

Beginning calculus deals mostly with the case of a real-valued function of one real variable, typically a function $f : I \to \mathbb{R}$, with one or more continuous derivatives, where the domain is some interval $I \subseteq \mathbb{R}$. You are probably familiar with functions

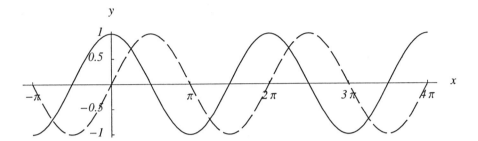

Figure 1.2-1 The graphs of the sine (dashed curve) and cosine (solid curve) over the interval $[-\pi, 4\pi]$.

like those represented in Figures 1.2-1, 1.2-2, and 1.2-3. If not, review their properties as necessary; see, for example, [27, Chapter 1] or [67, Appendix D].

Let $f : A \to B$. Let X be a subset of the domain A. The **image of X under f**, denoted by $f[X]$, is the set of all values $f(x)$ such that $x \in X$:

$$f[X] := \{f(x) \mid x \in X\}.$$

The set $f[A]$ is the **range** of f.

The use of brackets instead of parentheses in "$f[X]$" is meant to help avoid confusion. Notice that $f[X]$ is not "the value of f at X." That phrase does not make sense, X being a set rather than an element of the domain of f. Rather, $f[X]$ is, we may say, "the set of values taken on by f on X."

Consider the case when $f(x) = \cos x$. Then $f : \mathbb{R} \to \mathbb{R}$, though it would also be correct to write $f : \mathbb{R} \to [-10, 10]$ or $f : \mathbb{R} \to [-1, 1]$. Working from your knowledge of the cosine, and with a graph to look at (Figure 1.2-1), you should be able to verify the following statements, which use the $f[\cdots]$ notation:

$f[\mathbb{R}] = [-1, 1].$ $\qquad\qquad$ $f[(0, 2\pi]] = [-1, 1].$

$f[(-\pi, 0]] = (-1, 1].$ $\qquad\qquad$ $f\left[\left[\frac{\pi}{6}, \frac{\pi}{3}\right]\right] = \left[\frac{1}{2}, \frac{\sqrt{3}}{2}\right].$

$f\left[\left[\frac{2\pi}{3}, \frac{5\pi}{6}\right]\right] = \left[-\frac{\sqrt{3}}{2}, -\frac{1}{2}\right].$ $\qquad$ $f\left[\left\{x \mid \frac{\pi}{6} \le |x| \le \frac{\pi}{3}\right\}\right] = \left[\frac{1}{2}, \frac{\sqrt{3}}{2}\right].$

Exercise 3 asks you to complete similar statements, but for the sine instead of the cosine.

Let $f : A \to B$. Let Y be a set. The **pre-image of** of Y **under** f, denoted by $f^{-1}[Y]$, is the set of $x \in A$ such that $f(x) \in Y$:

$$f^{-1}[Y] := \{x \in A \mid f(x) \in Y\}.$$

Notice that the definition does not specify the set Y in any way, but gives a meaning to $f^{-1}[Y]$ for every set Y. Of course, only the points of Y that are in the range of f are relevant. Notice the special case when Y is a singleton: $f^{-1}[\{y\}]$ is the set of all $x \in A$ such that $f(x) = y$, a set which may be empty or may contain one or more elements.

For some examples, consider again the case when $f(x) = \cos x$. Then

- $f^{-1}[\mathbb{R}] = \mathbb{R}$.

- $f^{-1}[[-1, 1]] = \mathbb{R}$.

- $f^{-1}[(-1, 1]]$ is the set of all real numbers which are not equal to an odd integer times π.

- The set $f^{-1}[[\frac{1}{2}, \frac{\sqrt{3}}{2}]]$ is the union of an infinite set of intervals, namely, all the intervals $[2\pi n + \frac{\pi}{6}, 2\pi n + \frac{\pi}{3}]$ and $[2\pi n - \frac{\pi}{3}, 2\pi n - \frac{\pi}{6}]$ for $n \in \mathbb{Z}$; in other notation,

$$f^{-1}\left[\left[\tfrac{1}{2}, \tfrac{\sqrt{3}}{2}\right]\right] = \bigcup_{n \in \mathbb{Z}} \left[2\pi n + \tfrac{\pi}{6}, 2\pi n + \tfrac{\pi}{3}\right] \cup \left[2\pi n - \tfrac{\pi}{3}, 2\pi n - \tfrac{\pi}{6}\right].$$

Exercise 4 is similar, but for the sine instead of the cosine.

Let $f : A \to B$ and $g : X \to B$, where X is a subset of A. If $g(x) = f(x)$ for $x \in X$, then g is the **restriction** of f to X; and then f **extends,** or is an **extension** of, g to A.

Here are some examples. Let f be the cosine, as above, with domain $\mathbb{R}$. Let

$$g(x) = \cos x \quad \text{for } 0 \le x < 2\pi; \qquad h(x) = \cos x \quad \text{for } 0 \le x \le \pi.$$

Then g and h are the restrictions of the cosine to the intervals $[0, 2\pi)$ and $[0, \pi]$; and f extends each of the two functions to $\mathbb{R}$. Notice that the function h is one-to-one. Here are some statements about the pre-images of various sets under g and h:

- $g^{-1}[\mathbb{R}] = [0, 2\pi); \ h^{-1}[\mathbb{R}] = [0, \pi]$.

- $g^{-1}[[-1, 1]] = [0, 2\pi); \ h^{-1}[[-1, 1]] = [0, \pi]$.

- $g^{-1}[(-1, 1]] = [0, \pi) \cup (\pi, 2\pi); \ h^{-1}[(-1, 1]] = [0, \pi)$.

- $g^{-1}[[\frac{1}{2}, \frac{\sqrt{3}}{2}]] = [\frac{\pi}{6}, \frac{\pi}{3}] \cup [\frac{5\pi}{3}, \frac{11\pi}{6}]; \ h^{-1}[[\frac{1}{2}, \frac{\sqrt{3}}{2}]] = [\frac{\pi}{6}, \frac{\pi}{3}]$.

The point x is a **zero** of f if $f(x) = 0$. The set of zeros of f is, of course, $f^{-1}[\{0\}]$.

The function $f : A \to B$ is **one-to-one**[2] if (for $x_1, x_2 \in A$) $f(x_1) = f(x_2)$ only when $x_1 = x_2$. In other words, f is **one-to-one** if for each $b \in f[A]$ there is exactly one $x \in A$ such that $f(x) = b$.

The function $f : A \to B$ is **onto**[3] if $f[A] = B$; that is, if for every $b \in B$ there is an $x \in A$ such that $f(x) = b$. We say then that f maps A onto B.

Let the function f map A onto B. Then f is **m-to-one** from A onto B if for each point $b \in B$, there are exactly m distinct points $a \in A$ such that $f(a) = b$.

---

[2]It is equivalent to say that f is **univalent**, or **injective**, or **an injection**.
[3]It is equivalent to say f is **surjective**, or **a surjection**. If f is both injective and surjective, it is **bijective** or **a bijection**.

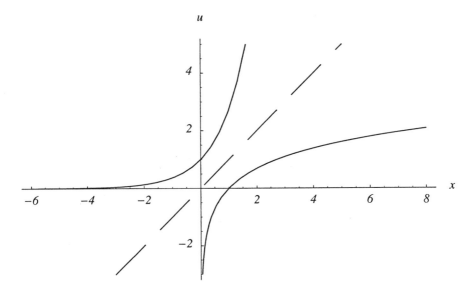

Figure 1.2-2 Graphs of the exponential function $x \mapsto e^x$ and its inverse, the natural logarithm $x \mapsto \ln x$. The two graphs are reflections of each other about the line $u = x$.

We offer the following examples. The logarithm and exponential functions, whose graphs are indicated in Figure 1.2-2, are one-to-one. The logarithm is one-to-one from $(0, \infty)$ onto $\mathbb{R}$. The exponential is one-to-one from $\mathbb{R}$ onto $(0, \infty)$. As for the sine and cosine, each of them maps $\mathbb{R}$ onto $[-1, 1]$; neither function is one-to-one. The hyperbolic sine, or sinh (Figure 1.2-3, second picture), is a one-to-one mapping from $\mathbb{R}$ onto $\mathbb{R}$. The hyperbolic cosine, or cosh, maps $\mathbb{R}$ onto $[1, \infty)$ and is not one-to-one. However, its restriction to $\mathbb{R}^+$ is one-to-one. The

For a one-to-one function $f : A \to B$, the **inverse** of f is the function

$$f^{-1} : f[A] \to A$$

defined as follows: For $y \in f[A]$, let $f^{-1}(y)$ be the unique element $x \in A$ such that $f(x) = y$. Observe that the point (p, q) is on the graph of f if and only if (q, p) is on the graph of f^{-1}. In other words, one graph is the reflection of the other in the line $y = x$. Notice that if f has an inverse, then f^{-1} also has an inverse and $(f^{-1})^{-1} = f$. Examples:

- The function $x \mapsto x^2$, restricted to $[0, \infty)$, has an inverse, called the positive square root.

- The sine maps the interval $[-\frac{\pi}{2}, \frac{\pi}{2}]$ one-to-one onto $[-1, 1]$. Therefore the restriction of the sine to that interval has an inverse, which is denoted by $\sin^{-1}$ or by arcsine.

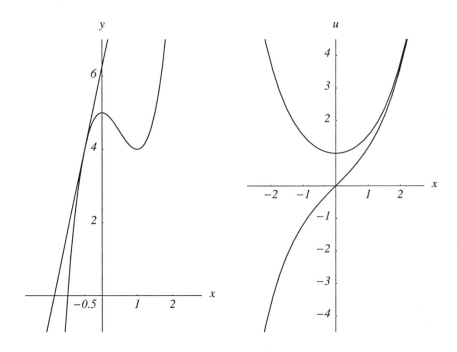

Figure 1.2-3 Let $P(x) = 2x^3 - 3x^2 + 5$. The first picture shows the graph of P and the line tangent to this graph at the point $(-0.5, 4)$. The line has slope $P'(-0.5) = 4.5$. The second picture shows graphs of the hyperbolic cosine, which is an even function, and of the hyperbolic sine, which is odd.

- The cosine maps the interval $[0, \pi]$ one-to-one onto $[-1, 1]$. Therefore the restriction of the cosine to that interval has an inverse, which is denoted by $\cos^{-1}$ or by arccos.

- The function $x \mapsto e^x$ is one-to-one; for each $u > 0$ there exists exactly one real x such that $e^x = u$. Therefore there is a well defined inverse function, the **natural logarithm,** denoted by ln or $\log_e$. The graphs of $x \mapsto e^x$ and $x \mapsto \ln x$ appear in Figure 1.2-2.

Let $f : A \to B$. The set of ordered pairs $\{(x, f(x)) \mid x \in A\}$ is the **graph** of f. A function determines its graph, and the graph of a function determines the function.

If f_1 and f_2 are two functions, then "$f_1 = f_2$" means that they have the same domain, say A, and that $f_1(x) = f_2(x)$ for every $x \in A$.

The **maps-to arrow** "$\mapsto$" may be used to identify a function without necessarily giving it a name. Thus we may write "the function $t \mapsto t^2$" to mean the function whose value at t is t^2.

Let $f : A \to B$. Let $g : B \to C$. The **composition** of g and f is the function $g \circ f$ given by

$$g \circ f(x) := g(f(x)).$$

Notice that the domain of $g \circ f$ is the set of all x in the domain of f such that $f(x)$ is in the domain of g.

Some examples: If $p(x) := \sqrt{x}$ and $q(x) = x^4 + 1$, then:

- $p \circ q(x) = \sqrt{x^4 + 1}.$

- $q \circ p(x) = x^2 + 1.$

- $p \circ p(x) = x^{1/4}.$

- $q \circ q(x) = (x^4 + 1)^4 + 1.$

C Functions from $\mathbb{R}$ to $\mathbb{R}$

This book will emphasize and encourage visualization. In this Section we deliberately belabor this simple-minded remark: To understand and remember a mathematical statement, it often helps to draw a picture. We will point to pictures in various Figures while chatting about some ideas from real analysis. The functions mentioned are chosen for their importance; they are not just examples to be forgotten.

Let $f : I \to \mathbb{R}$ where I is an interval in $\mathbb{R}$ and f has one or more continuous derivatives. The mapping $x \mapsto (x, f(x))$ is one-to-one continuous from I into $\mathbb{R}^2$. Its range is the graph of f. Being in a function, one-to-oneone-to-one correspondence with I by means of that mapping, the graph is a one-dimensional object which is often easy to visualize.

Figure 1.2-1 shows the graphs of the functions $x \mapsto \cos x$ and $x \mapsto \sin x$ over the interval $[-\pi, 4\pi]$. As the picture reminds us, both functions have period 2π, that is,

$$\cos(x + 2\pi) = \cos x \qquad \sin(x + 2\pi) = \sin x \quad \text{for all } x;$$

and antiperiod π, that is,

$$\cos(x + \pi) = -\cos x \qquad \sin(x + \pi) = -\sin x \quad \text{for all } x.$$

Looking at the picture also helps us get the signs correct when we write down the following well known identities:

$$\sin x = \cos\left(x - \frac{\pi}{2}\right), \qquad \cos x = -\sin\left(x - \frac{\pi}{2}\right); \qquad (1)$$

$$\cos x = \sin\left(x + \frac{\pi}{2}\right), \qquad \sin x = -\cos\left(x + \frac{\pi}{2}\right). \qquad (2)$$

The general rule is, Going from $f(x)$ to $f(x + c)$ corresponds to shifting its graph c units to the left.

$$u(x,y) = x \qquad\qquad v(x,y) = y$$

$$u(x,y) = x^2 - y^2 \qquad\qquad v(x,y) = 2xy$$

$$u(x,y) = x^3 - 3xy^2 \qquad\qquad v(x,y) = 3x^2y - y^3$$

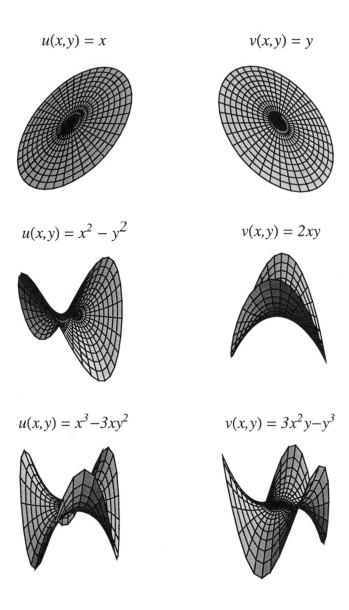

Figure 1.2-4 The graphs of six real-valued functions are shown, over the disk of center $(0,0)$ and radius 1.25. On each row, v is a harmonic conjugate of u. (A different vertical scale is used on each row, to give all the pictures similar proportions.)

Let $P(x) := 2x^3 - 3x^2 + 5$. Figure 1.2-3 shows the graph of P on the interval $[-1.1, 1.7]$, and a line tangent to the graph at $(-0.5, 4)$. The analytic assertions that "P has only one real zero, occurring when $x = -1$" and "$P''(x)$ is positive for $x > 1/2$" become more vivid when we point to the picture and say, "the graph of P crosses the x-axis only at -1" and "the graph is concave up on the interval $(1/2, \infty)$." The fact that $P'(-\frac{1}{2})$ exists and equals $\frac{9}{2}$ may be stated as follows:

$$P(x) = 4 + \left(\frac{9}{2} + \eta(x) \right) \left(x + \frac{1}{2} \right), \quad \text{where} \quad \lim_{x \to -\frac{1}{2}} \eta(x) = 0.$$

But the fact also has a geometric and visual expression: The line given by $x \mapsto 4 + \frac{9}{2}(x + \frac{1}{2})$ is tangent to the graph of P at the point $(-\frac{1}{2}, 4)$ and approximates the graph of P nicely near the point.

Figure 1.2-3 also shows the graphs of the hyperbolic cosine and the hyperbolic sine, defined by:

$$\cosh x := \frac{e^x + e^{-x}}{2} \quad \text{and} \quad \sinh x := \frac{e^x - e^{-x}}{2}.$$

The cosh is an even function and is always at least 1. The sinh is an odd function. They satisfy this identity, easily provable from the definitions just given:

$$\cosh^2 x - \sinh^2 x = 1. \tag{3}$$

D   Functions from $\mathbb{R}^2$ to $\mathbb{R}$

For a real-valued function of two real variables, it is still true that a picture of the graph can do a great deal to clarify the behavior of the function. The graph of a function $u : V \to \mathbb{R}$, where $V \subseteq \mathbb{R}^2$, is the image of the one-to-one mapping $(x, y) \mapsto (x, y, u(x, y))$. Being in one-to-one correspondence with V, by means of that mapping, the graph is a two-dimensional object. Since the graph is a surface lying in 3-space, we can usually draw a satisfactory partial picture, in perspective, upon a two-dimensional surface. We are about to show you some examples. You have probably seen many such pictures, and you may think we're getting carried away, but there is a point to displaying these particular functions. They are all harmonic, and we want you to begin looking for the special features that harmonic functions have in common. We will refer you back to these graphs more than once.

Figure 1.2-4 shows the graphs of six functions, restricted to a disk centered at the origin, which are three conjugate pairs u, v of harmonic functions.

$$\begin{aligned} u &= x, & v &= y; \\ u &= x^2 - y^2, & v &= 2xy; \\ u &= x^3 - 3xy^2, & v &= 3x^2y - y^3. \end{aligned}$$

Each of the six pictures illustrates the mean value property of harmonic functions: The average value of the function on the boundary of the disk equals its value at the

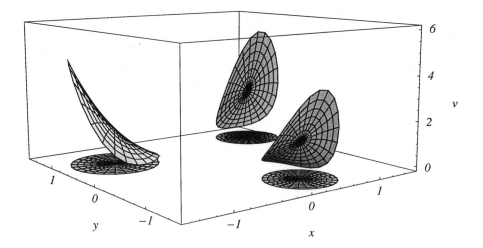

Figure 1.2-5   Three disks in the x, y-plane, and the graphs over those disks of the function given by $v(x, y) = 3x^2 y - y^3$ (compare Figure 1.2-4). Each graph illustrates the mean value property for v.

center. Another property, closely related to that one, is that each function attains its maximum value on the disk at one or more points on the boundary of the disk, but nowhere on the interior; and likewise for the mininum value.

Figure 1.2-5 shows you the graph of $(x, y) \mapsto 3x^2 y - y^3$ restricted to each of three different disks; it comprises three subsets of the graph on the lower right in Figure 1.2-4.

The six functions represented in Figures 1.2-6, 1.2-7, and 1.2-8 comprise another three conjugate pairs of harmonic functions:

$$u = e^x \cos y, \qquad v = e^x \sin y;$$
$$u = \sin x \cosh y, \qquad v = \cos x \sinh y;$$
$$u = \ln \sqrt{x^2 + y^2}, \qquad v = \text{Arctan} \frac{y}{x}.$$

These functions are not polynomials in x and y, and thus in one sense they are inherently more complicated than the six functions of Figure 1.2-4. Nevertheless, each of these functions is simple in certain ways, and when we spot their simplicities, it becomes easier to understand how they behave. Each of the first four is the product of one function of x and another function of y (the graphs of which appear in previous figures). In the case of $(x, y) \mapsto e^x \cos y$, for example, when you intersect the graph of u with a plane on which y or x is constant, you see the graph of $x \mapsto ke^x$ for some constant $k \in [-1, 1]$, or the function $y \mapsto c \cos y$ for some constant $c > 0$, respectively. Each function in the third pair, if we change to polar coordinates in the domain, is a function of one variable: $u = \ln r$, $v = \theta$ (restricted to $r > 0, -\pi < \theta \le \pi$ to make it well defined).

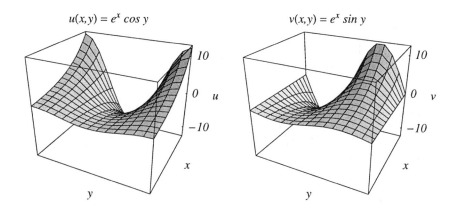

$u(x,y) = e^x \cos y$ $v(x,y) = e^x \sin y$

Figure 1.2-6 The graphs of two functions over the rectangle given by $0 \leq x \leq 2.5$, $0 \leq y \leq 2\pi$. Both functions are periodic-2π in y, and v is a harmonic conjugate of u.

E Functions from $\mathbb{R}^2$ to $\mathbb{R}^2$

You have now seen six examples of conjugate pairs u, v. Each pair gives rise to a mapping from the plane to the plane,

$$(x, y) \mapsto (u(x, y), v(x, y)).$$

One may try to understand the behavior of that mapping merely by studying u and v separately, using the pictures and observations above. But we now return to the method pointed out in Section 1.1, whereby we draw a grid in the x, y-plane and the image of that grid in the u, v-plane. We demonstrated that method in Figure 1.1-2 for the mapping $(x, y) \mapsto (x^2 - y^2, 2xy)$. Let's now use the method for a partial study of another mapping:

$$(x, y) \mapsto (x^3 - 3xy^2, 3x^2y - y^3).$$

Figure 1.2-9 shows a rectangular grid on the left and its image on the right. Figure 1.2-10 shows on the left a polar grid and on the right its image. Each of the grids on the left is centered at the point $p := (0.4, 0.4)$. The image of that point is $q := (-.128, .128)$. Either of the two figures gives a good idea of how the mapping acts on a certain neighborhood of p. It acts by rotating and distorting the grid (and, of course, moving it over to the u, v-plane). By **distortion,** we mean that some parts of the region within the grid are shrunk or magnified more than others; in other words, the shape of the grid changes, not just its size or position. We see a considerable amount of distortion.

In either figure, look at how the mapping acts on the part of the plane that is very close to p. As you see, there is less distortion there. In fact, the mapping acts on the region very close to p approximately as follows: Holding p as the fixed center, it rotates the region counterclockwise 90° and magnifies it by a factor of 0.96 (and then, of course, moves it over to the u, v-plane, with p going to q.)

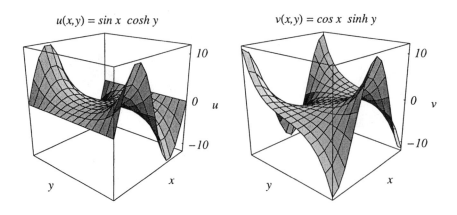

$u(x,y) = \sin x \; \cosh y$ $v(x,y) = \cos x \; \sinh y$

Figure 1.2-7 The graphs over the rectangle given by $0 \leq x \leq 2\pi, -\pi \leq y \leq \pi$ of functions u and v, both periodic-2π in x; v is a harmonic conjugate of u.

This visualization method does not immediately provide a clear idea of how the mapping acts globally; had we used a much larger grid in the domain, the image-grid would overlap itself and be hard to interpret.

Exercises

1. Let $P : \mathbb{R} \to \mathbb{R}, P(x) := 2x^3 - 3x^2 + 5$. In each case below, you are given a set $X \subseteq \mathbb{R}$. Identify the set $P[X]$, the image of X under the mapping P.

   (a) $X = \mathbb{R}$.   (c) $X = (0, 1.2)$.   (e) $X = [0, 2]$.

   (b) $X = [-1, 0]$.   (d) $X = (-0.5, 1]$.   (f) $X = \{5\}$.

2. Let P be as in Exercise 1. In each case below, you are given a set $Y \subseteq \mathbb{R}$. Identify the set $P^{-1}[Y]$, the pre-image of Y under P.

   (a) $Y = \mathbb{R}$.   (c) $Y = [0, 4]$.   (e) $Y = [4, 9)$.

   (b) $Y = \{0\}$.   (d) $Y = (0, 4)$.   (f) $Y = \{4\}$.

3. Let $f : \mathbb{R} \to \mathbb{R}, f(x) := \sin x$. In each case below, you are given a set $X \subseteq \mathbb{R}$. Identify the set $f[X]$. For example, if X is the closed interval $[\frac{\pi}{6}, \frac{\pi}{2})$, then $f[X]$ is the closed interval $[\frac{1}{2}, 1]$.

   (a) $X = \mathbb{R}$.   (d) $X = (-\pi, 0]$.

   (b) $X = (-2\pi, 0]$.   (e) $\left[\frac{\pi}{6}, \frac{\pi}{3}\right] \cup \left[\frac{2\pi}{3}, \frac{5\pi}{6}\right]$.

   (c) $\left[\frac{\pi}{6}, \frac{\pi}{3}\right]$.   (f) $X = \left\{x \middle| \frac{\pi}{6} \leq |x| \leq \frac{\pi}{3}\right\}$.

$u(x,y) = \ln\sqrt{x^2 + y^2} = \ln r$

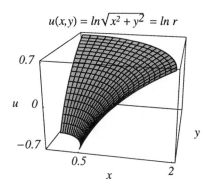

$v(x,y) = \arctan \dfrac{y}{x} = \theta$

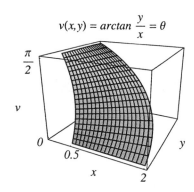

Figure 1.2-8   The graphs of two functions over the domain described in terms of polar coordinates by the inequalities $.5 \le r \le 2$ and $0 \le \theta \le \frac{\pi}{2}$.

4. Let $f : \mathbb{R} \to \mathbb{R}$, $f(x) := \sin x$. In each case below, you are given a set $Y \subseteq \mathbb{R}$. Identify the set $f^{-1}[Y]$.

(a) $Y = \mathbb{R}$.

(b) $Y = [-1, 1]$.

(c) $Y = (-1, 1]$.

(d) $Y = [\frac{1}{2}, \frac{\sqrt{3}}{2}]$.

5. Let $g : \mathbb{R} \to \mathbb{R}^2$, $g(t) = (\cos t, \sin t)$. For each of the six sets $X \subseteq \mathbb{R}$ given in Exercise 3, identify and sketch the set $g[X]$. For example, if X is the closed interval $[0, 2\pi]$, then $g[X]$ is the circle in the plane of center $(0, 0)$ and radius 1.

6. Let $u : \mathbb{R}^2 \to \mathbb{R}$, $u(x, y) = e^x \cos y$. In each case, you are given a set $X \subseteq \mathbb{R}^2$. Identify the set $u[X]$. It may help to look at the graph of u in Figure 1.2-6.

(a) $X = \{(x, y) \mid x > 0\}$.

(b) $X = \{(x, y) \mid x > 0 \text{ and } 0 \le y \le \frac{\pi}{2}\}$.

(c) $X = \{(x, y) \mid x > 0 \text{ and } y = 0\}$.

(d) $X = \{(x, y) \mid \frac{\pi}{2} < y \le \pi\}$.

(e) $X = \{(x, y) \mid x \le 0 \text{ and } y = \frac{\pi}{2}\}$.

(f) $X = \{(x, y) \mid -1 < x < 1 \text{ and } -\pi \le y \le \pi\}$.

7. Read carefully the definition of the symbol $f[X]$ to see how it applies when X is a singleton. Is it true that $f[\{x\}] = f(x)$, or that $f[\{x\}] = \{f(x)\}$?

8. Explain this statement: When f is one-to-one, the symbol f^{-1} has two meanings. It stands for a function whose domain is the range of f, and also for a function whose domain is a set of sets. The two senses of the symbol are related by the equality $f^{-1}[\{y\}] = \{f^{-1}(y)\}$.

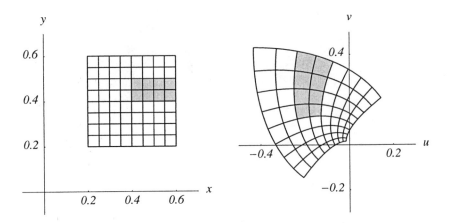

Figure 1.2-9   A rectangular grid in the x, y-plane and its image in the u, v-plane under the mapping $(x, y) \mapsto (x^3 - 3xy^2, 3x^2y - y^3)$ (which we will later denote as $z \mapsto z^3$). A selected set and its image are shaded.

9. In each case, a function f and a set Y are given; identify and sketch the set $f^{-1}[Y]$.

   (a) $f : \mathbb{R} \to \mathbb{R}$, $f(x) = \sin x$, $Y = [\frac{3}{2}, 3]$.

   (b) $f : \mathbb{R} \to \mathbb{R}$, $f(x) = e^x$, $Y = [-1, 1]$.

   (c) $u : \mathbb{R}^2 \to \mathbb{R}$, $u(x, y) = e^x \cos y$, $Y = \{0\}$.

10. In each case, a function f and a set Y are given; identify and sketch the set $f^{-1}[Y]$.

   (a) $f : \mathbb{R} \to \mathbb{R}, f(x) = \sin x$, $Y = [\frac{1}{2}, 2]$.

   (b) $f : \mathbb{R} \to \mathbb{R}$, $f(x) = e^x$, $Y = [-\frac{1}{2}, 2]$.

   (c) $f : \mathbb{R} \to \mathbb{R}^2$, $f(x) = (\cos x, \sin x)$, $Y = \{(u, v) \in \mathbb{R}^2 \mid u \geq 0\}$.

   (d) $u : \mathbb{R}^2 \to \mathbb{R}$, $u(x, y) = e^x \cos y$, $Y = \{u \in \mathbb{R} \mid u \geq 1\}$.

11. In each case, the given function g is a restriction of a function in Exercise 10. Identify and sketch the set $g^{-1}[Y]$:

   (a) $g : [-\frac{\pi}{2}, \frac{\pi}{2}] \to \mathbb{R}$, $g(x) = \sin x$, $Y = [\frac{1}{2}, 2]$.

   (b) $g : [0, \infty) \to \mathbb{R}$, $g(x) = e^x$, $Y = [-1, 1]$.

   (c) $g : [0, 2\pi] \to \mathbb{R}^2$, $g(t) = (\cos t, \sin t)$, $Y = \{(u, v) \in \mathbb{R}^2 \mid u \geq 0\}$.

   (d) $g : \{(x, y) \mid 0 \leq y < \pi\} \to \mathbb{R}$, $g(x, y) = e^x \cos y$, $Y = \{0\}$. Here, the domain is a half-open horizontal strip of height π.

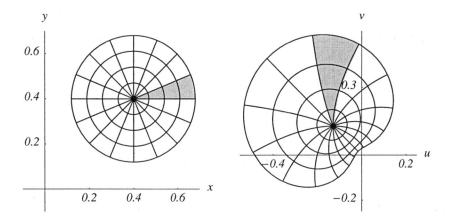

Figure 1.2-10 A polar grid in the x, y-plane and its image in the u, v-plane under the mapping $(x, y) \mapsto (x^3 - 3xy^2, 3x^2 y - y^3)$.

12. Explain whether the following statement is true or false. Let $f(x) = \sin x$. Then f is 4-to-one from $[0, 4\pi)$ onto $[-1, 1]$.

13. Consider the function $u(x, y) = x^2 - y^2$ and its conjugate $v(x, y) = 2xy$. Let $Y = \{1, 2, 4\}$. Identify and sketch the sets $u^{-1}[Y]$ and $v^{-1}[Y]$. You should get an orthogonal grid.

14. Consider the mapping $(x, y) \mapsto (u, v)$, where $u = e^x \cos y$ and $v = e^x \sin y$. In the x, y-plane, sketch the four horizontal line segments on which x runs from $\ln(1/2)$ to $\ln 2$ and y is constant, taking the value 0, $\pi/4$, $y = \pi/2$, or $3\pi/4$. Sketch also the three vertical line segments on which y runs from 0 to $3\pi/4$ and x is constant and equal to $\ln(1/2)$, 0, or $x = \ln 2$. It's an orthogonal grid (See the definition in subsection A). Sketch the image in the u, v-plane of each line segment in that grid. Label your sketch to make clear what maps to what.

15. Let $p(x) = \sin x$ and $q(x) = x^2$. Sketch accurately the graphs of the compositions $p \circ q$ and $q \circ p$ for $-2\pi \leq x \leq 2\pi$.

Help on Selected Exercises

1. (a) $\mathbb{R}$. (b) $[0, 5]$. (c) $[4, 5)$. (d) $[4, 5]$. (e) $[4, 9]$. (f) $\{180\}$.

2. (a) $\mathbb{R}$. (b) $\{-1\}$. (c) $[-1, -0.5] \cup \{1\}$. (d) $(-1, -0.5)$. (e) $[-0.5, 2)$. (f) $\{-0.5, 1\}$.

4. (a) $\mathbb{R}$. (b) $\mathbb{R}$. (c) $\mathbb{R} \setminus \{-\frac{\pi}{2} + 2\pi n \mid n \in \mathbb{Z}\}$.
   (d) $\bigcup_{n \in \mathbb{Z}} [\frac{\pi}{6} + 2\pi n, \frac{\pi}{3} + 2\pi n] \cup [\frac{2\pi}{3} + 2\pi n, \frac{5\pi}{6} + 2\pi n]$.

6. (a) $\mathbb{R}$ (b) $\{u \in \mathbb{R} \mid u \geq 0\}$

9. (a) $\emptyset$, the empty set.   (b) $(-\infty, 0]$.   (c) The union of all the horizontal lines given by $y = (2k+1)\pi/2$, where k is an integer.

11. (a) $[\frac{\pi}{6}, \frac{\pi}{2}]$.   (b) $\{0\}$.   (c) $[0, \frac{\pi}{2}] \cup [\frac{3\pi}{2}, 2\pi]$.   (d) $\{(x, y) \mid y = \frac{\pi}{2}\}$.

12. It's false, because (for example) $f^{-1}[\{1\}] = \{0, 2\pi\}$.

1.3 STRUCTURES ON $\mathbb{R}^2$, AND LINEAR MAPS FROM $\mathbb{R}^2$ TO $\mathbb{R}^2$

Much of this Section is to remind you of things you have seen before, including vector addition, scalar multiplication, the dot product, polar coordinates, and determinants of 2×2 matrices. We will also discuss the linear functions from $\mathbb{R}^2$ to $\mathbb{R}^2$, which is to say, all mappings

$$(x, y) \mapsto (ax + by, cx + dy),$$

where a, b, c, and d are real numbers. For every such mapping, each of the two coordinate functions $(x, y) \mapsto ax + by$ and $(x, y) \mapsto cx + dy$ is harmonic. Especially important to us is the case when those two functions are harmonic conjugates. We will characterize that case in both algebraic and geometric terms. It is a simple matter, but important to understand.

A   The Real Line and the Plane

You are acquainted with the set $\mathbb{R}$ of real numbers. You are familiar also with $\mathbb{R}^2$ or $\mathbb{R} \times \mathbb{R}$, by which we mean the set of ordered pairs (x, y) of real numbers. Each ordered pair is called a point or a vector. When you picture $\mathbb{R}^2$ as a plane in the usual way, you are identifying each point of the plane with the ordered pair (x, y), where x and y are its Cartesian coordinates. Two points of the plane are the same, $(x_1, y_1) = (x_2, y_2)$, if and only if $x_1 = x_2$ and $y_1 = y_2$.

You are also acquainted with addition in $\mathbb{R}$; for every pair x, y of real numbers, their sum $x + y$ is defined, and $x + y$ is also a real number. Addition is thus a binary operation,

$$(x, y) \mapsto x + y,$$

with domain $\mathbb{R} \times \mathbb{R}$ and with values in $\mathbb{R}$. The same can be said for multiplication,

$$(x, y) \mapsto x \cdot y.$$

We will usually omit the dot, writing xy for $x \cdot y$. Those two operations obey certain familiar rules, which we summarize by saying that the system $(\mathbb{R}, +, \cdot)$, or $\mathbb{R}$ for short, is an example of a field. The definition of a field will be stated in Section 2.1.

We will now describe three familiar bits of algebraic structure which are associated with $\mathbb{R}^2$: vector addition, scalar multiplication, and the dot product. The sum of two points of the plane $\mathbb{R}^2$, the usual vector sum, is defined as follows:

$$(x_1, y_1) + (x_2, y_2) := (x_1 + x_2, y_1 + y_2). \tag{1}$$

The symbol "+" on the left-hand side is being defined; the "+" on the right-hand side refers to the addition in $\mathbb{R}$. It is useful to have a geometric idea of this operation. If two points are located for you in a picture, you can locate their sum because you know that the four points $(0,0), (x_1, y_1), (x_2, y_2)$, and $(x_1 + x_2, y_1 + y_2)$ form a parallelogram. If they all lie on some one line, then the parallelogram is **degenerate** and its area is zero.

It is useful to regard $\mathbb{R}$ as a subset of the plane $\mathbb{R}^2$, as follows: For each real number x, we declare "$(x, 0)$" to be another way to write "x." The set of all points $(x, 0)$ in the plane is **the real axis.** Notice that if the two summands on the left-hand side of (1) are both on the real axis, that is, if $y_1 = y_2 = 0$, then their sum, which is the same thing as $x_1 + x_2$, is also on the real axis. Thus equation (1) extends the definition of addition from $\mathbb{R} \times \mathbb{R}$ to all of $\mathbb{R}^2 \times \mathbb{R}^2$.

Scalar multiplication lets you multiply a real number a by a point $(x, y) \in \mathbb{R}^2$:

$$a \cdot (x, y) := (a \cdot x, a \cdot y) \tag{2}$$

The symbol "$\cdot$" in the left-hand expression is being defined; on the right, the symbol "$\cdot$" stands for multiplication in $\mathbb{R}$.

The system $(\mathbb{R}^2, +, \cdot)$ is an example of a real linear space or vector space.

Occasionally we will use the vector notation for a point:

$$(x, y) = x\vec{i} + y\vec{j},$$

which is clear enough if we understand $\vec{i}$ and $\vec{j}$ to stand for the points $(1, 0)$ and $(0, 1)$, respectively.

The **dot product** or **inner product** of two elements $z_1 := (x_1, y_1)$ and $z_2 := (x_2, y_2)$ of $\mathbb{R}^2$ is defined by

$$z_1 \cdot z_2 := x_1 \cdot x_2 + y_1 \cdot y_2. \tag{3}$$

It follows from the definition of the dot product that

$$z_1 \cdot z_2 = z_2 \cdot z_1;$$
$$a(z_1 \cdot z_2) = (az_1 \cdot z_2); \quad \text{and} \tag{4}$$
$$(z_1 + z_2) \cdot z_3 = (z_1 \cdot z_3) + (z_2 \cdot z_3)$$

whenever $z_1, z_2, z_3 \in \mathbb{R}^2$ and $a \in \mathbb{R}$. The system $(\mathbb{R}^2, +, \cdot, \cdot)$ is an example of an inner product space.

A **true multiplication** on a set X is one that maps $X \times X$ onto X. Neither (2) nor (3) is a true multiplication on $\mathbb{R}^2$.

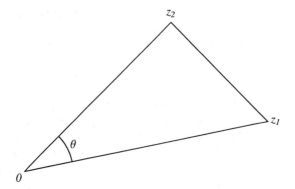

Figure 1.3-1   The triangle $T(0, z_1, z_2)$. The Law of Cosines: $|z_2 - z_1|^2 = |z_2|^2 + |z_1|^2 - 2|z_1||z_2|\cos\theta$.

Let $z := (x, y) \in \mathbb{R}^2$. The **modulus** (or **length**, or **norm**) of z is defined by

$$|z| := (z \cdot z)^{1/2} = (x^2 + y^2)^{1/2}.$$

By the definition and the properties (4),

$$|z_2 - z_1|^2 = (z_2 - z_1) \cdot (z_2 - z_1) = |z_2|^2 - 2z_1 \cdot z_2 + |z_1|^2.$$

for every pair $z_1, z_2 \in \mathbb{R}^2$. It follows then from the Law of Cosines, illustrated in Figure 1.3-1, that

$$z_1 \cdot z_2 = |z_1| \cdot |z_2| \cdot \cos\theta, \tag{5}$$

where θ is the angle at 0 in the triangle $T(0, z_1, z_2)$. Two vectors in the plane are **orthogonal** or **perpendicular** if their dot product $z_1 \cdot z_2$ equals 0, or, equivalently, if $\cos\theta = 0$. The following result is very simple, but worth stating. It follows from (5).

1.3.1. Proposition. *Let z_1 be a fixed nonzero vector in $\mathbb{R}^2$. Then as z_2 varies among all vectors of length 1, the quantity $z_1 \cdot z_2$ varies from $-|z_1|$ to $+|z_1|$ and attains its maximum only when z_2 is a positive multiple of z_1.*

B   Polar Coordinates in the Plane

The Cartesian coordinates of a point (x, y) tell you how to find it: "Starting at the origin, go x units east and then y units north." The polar coordinates tell you how to find it another way: "Starting at the origin, face the direction which is angle θ from the positive x-axis. Go distance r." If $r < 0$, this means that you go distance $-r$ in direction $\theta + \pi$.

An ordered pair of real numbers r, θ is a set of **polar coordinates** for a point $(x, y) \in \mathbb{R}^2$ if these relations hold:

$$x = r\cos\theta \qquad \text{and} \qquad y = r\sin\theta. \tag{6}$$

We will temporarily use the symbol $((r, \theta))$ to denote the point in the plane for which r and θ are polar coordinates. Thus

$$((r, \theta)) = (x, y) \quad \Longleftrightarrow \quad x = r\cos\theta \quad \text{and} \quad y = r\sin\theta.$$

Given x and y, those equations do not determine r and θ uniquely. The ambiguities are precisely as follows: The pairs r, θ and s, η are polar coordinates of the same point, that is, $((r, \theta)) = ((s, \eta))$, if and only if

> $r = s$ and $\eta - \theta = 2\pi n$, where n is an integer; or
>
> $r = -s$ and $\eta - \theta = \pi k$, where k is an odd integer; or
>
> $r = s = 0$.

Given x and y, therefore, there are many ways to choose r and θ satisfying (6). For r we may choose $\sqrt{x^2 + y^2}$ or $-\sqrt{x^2 + y^2}$. Let us agree always to choose the non-negative value. There remain many choices for θ. Given a subset S of the plane, we may try to find a continuous function arg defined on S such that for each point $(x, y) \in S$, the pair $\sqrt{x^2 + y^2}$, $\arg(x, y)$ is a set of polar coordinates for (x, y). Such a continuous function is called an **argument** on S, or sometimes "a branch of the argument" on S.

There is no way to define an argument on a set containing $(0, 0)$.

There is no way to define an argument on the set of all nonzero points of the plane. In fact, whenever the set S contains a circle that goes around the origin, there is no way to define an argument on S. We may justify that assertion as follows. Suppose that there were such an arg. Let (x_0, y_0) be a point on the circle, and let $\theta_0 = \arg(x_0, y_0)$. If a point starts at (x_0, y_0) and moves counterclockwise around the circle, then the value of arg at the point would necessarily increase as the point moves. The value would increase by 2π when the point goes around once and returns to (x_0, y_0). It would follow that $\arg(x_0, y_0) = \theta_0 + 2\pi$, which cannot be the case; we have reached a contradiction.

This interesting little feature of arguments turns out to be important; it is sometimes described by saying, "The origin is a branch point of the argument." We will describe some conventions for dealing with the situation.

One alternative is to adopt the following standard rule, given x and y not both zero, for choosing θ. Define the **principal argument** of the point (x, y) to be the unique number θ in the interval $(-\pi, \pi]$ satisfying (6), and denote it by $\operatorname{Arg}(x, y)$. Thus

$$r = \sqrt{x^2 + y^2} \quad \text{and} \quad \theta = \operatorname{Arg}(x, y) \quad \text{for } (x, y) \neq (0, 0).$$

The function Arg is defined and continuous on the plane except at the origin and on the points of the negative real axis. It is satisfactory provided one does not need to work near those exceptions.

Or we may be more flexible in our choice of domain and argument, as follows. Let $H_\alpha = \{((r, \alpha)) \mid r \geq 0\}$. Such a set is called a half-line; for example, H_π is the non-positive real axis. Let $V_\alpha = \mathbb{R}^2 \setminus H_\alpha$, the plane with the half-line H_α removed.

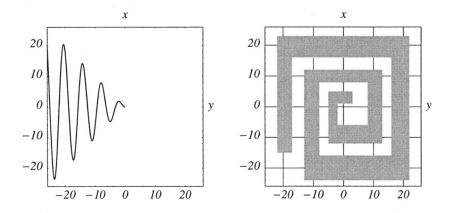

Figure 1.3-2 Two sets on the plane on which an argument can be defined. The first is the complement of the graph of $x \mapsto x \sin x$ $(x \leq 0)$. The second is a polygonal region, indicated by shading.

Then one may define a continuous function arg on V_α (in fact, we have infinitely many choices) such that for every point $(x, y) \in V_\alpha$,

$$((\sqrt{x^2 + y^2}, \arg(x, y))) = (x, y).$$

Notice that if the value of an argument is specified at one point of V_α, it is determined at all points by the requirement of continuity. Often we will write simply "arg" and hold in mind which of the possible arguments we are using. At other times we will need our notation for the argument to carry the information for us, and for that reason we will indicate in a subscript its value at some one point. Thus for example, $\arg_{1 \mapsto 0}$ is the argument on V_π whose value at 1 is 0. The other possibilities on a set containing 1 are, of course, $\arg_{1 \mapsto 2\pi n}$ for $n \in \mathbb{Z}$.

An argument may be definable on other sets besides the sets V_α. In fact, there will be an argument defined on S so long as the complement of S contains some barrier to circumnavigation of the origin. Figure 1.3-2 shows two examples of a set on which there is an argument. You will notice that the range of an argument on a set can be an interval of length greater than 2π.

C   When Is a Mapping $M : \mathbb{R}^2 \to \mathbb{R}^2$ Linear?

We begin with a definition which is formal and algebraic in nature. Then we will prove a Proposition leading to a concrete and geometric idea of linearity.

1.3.2. Definition. A mapping $M : \mathbb{R}^2 \to \mathbb{R}^2$ is **linear** if

$$M(kz) = kM(z) \quad \text{for all } z \in \mathbb{R}^2 \text{ and for all } k \in \mathbb{R}; \tag{7}$$

and

$$M(z_1 + z_2) = M(z_1) + M(z_2) \quad \text{for all} \quad z_1, z_2 \in \mathbb{R}^2. \tag{8}$$

We could write (7) and (8) together as one statement:

$$M(k_1 z_1 + k_2 z_2) = k_1 M(z_1) + k_2 M(z_2)$$
$$\text{for all } z_1, z_2 \in \mathbb{R}^2 \text{ and for all } k_1, k_2 \in \mathbb{R}. \tag{9}$$

A pair of points z_1, z_2 constitutes a **basis** for $\mathbb{R}^2$ if every $z \in \mathbb{R}^2$ can be written in the form $k_1 z_1 + k_2 z_2$. If $\{z_1, z_2\}$ is a basis, and if the values $M(z_1)$ and $M(z_2)$ are known, then $M(z)$ is clearly determined for every $z \in \mathbb{R}^2$, in view of (9).

1.3.3. Proposition. *A mapping $M : \mathbb{R}^2 \to \mathbb{R}^2$ is linear if and only if, for some four real numbers $a, b, c,$ and d, it is given by*

$$M(x, y) := (ax + by, cx + dy) \quad \text{for } (x, y) \in \mathbb{R}^2; \tag{10}$$

If we write elements of $\mathbb{R}^2$ as column vectors and use matrix notation, then (10) becomes

$$M \begin{pmatrix} x \\ y \end{pmatrix} := \begin{pmatrix} ax + by \\ cx + dy \end{pmatrix} = \begin{pmatrix} a & b \\ c & d \end{pmatrix} \begin{pmatrix} x \\ y \end{pmatrix} \quad \text{for} \begin{pmatrix} x \\ y \end{pmatrix} \in \mathbb{R}^2. \tag{11}$$

Proof. It is easy to show that if M is given by (10), then M satisfies (7) and (8). The calculation

$$M(k(x, y)) \equiv M(kx, ky) = (akx + bky, ckx + dky)$$
$$= k(ax + by, cx + dy) = kM(x, y);$$

gives (7). A similar calculation gives (8).

To prove the "only if" part of the Proposition, we first assign names to the images under M of the two canonical basis vectors. Let

$$\begin{pmatrix} a \\ c \end{pmatrix} := M \begin{pmatrix} 1 \\ 0 \end{pmatrix} \quad \text{and} \quad \begin{pmatrix} b \\ d \end{pmatrix} := M \begin{pmatrix} 0 \\ 1 \end{pmatrix}.$$

Then for every $\begin{pmatrix} x \\ y \end{pmatrix} \in \mathbb{R}^2$,

$$M\begin{pmatrix} x \\ y \end{pmatrix} = M\begin{pmatrix} x \\ 0 \end{pmatrix} + M\begin{pmatrix} 0 \\ y \end{pmatrix} \qquad \text{(by (8))}$$

$$= xM\begin{pmatrix} 1 \\ 0 \end{pmatrix} + yM\begin{pmatrix} 0 \\ 1 \end{pmatrix} \qquad \text{(by (7))}$$

$$= x\begin{pmatrix} a \\ c \end{pmatrix} + y\begin{pmatrix} b \\ d \end{pmatrix}$$

$$= \begin{pmatrix} ax + by \\ cx + dy \end{pmatrix}.$$

Thus we may represent the action of M on $\begin{pmatrix} x \\ y \end{pmatrix}$ as left-multiplication by a matrix, as in (11).   □

The proof shows a one-to-one correspondence between the linear mappings $M :$ $\mathbb{R}^2 \to \mathbb{R}^2$ and the 2×2 real matrices. Consider two linear mappings, M_j for $j = 1, 2$, and their corresponding matrices $A_j := \begin{pmatrix} a_j & b_j \\ c_j & d_j \end{pmatrix}$. Then the linear mapping which is the composition $M_1 \circ M_2$ acts on a vector $\begin{pmatrix} x \\ y \end{pmatrix}$ by means of left-multiplication, first by A_2 and then by A_1. Matrix multiplication is associative, so this means that $M_1 \circ M_2$ acts on $\begin{pmatrix} x \\ y \end{pmatrix}$ by means of left-multiplication by the matrix $A_1 A_2$.

As you probably know, matrix multiplication is not commutative; in other words, $A_1 A_2$ does not always equals $A_2 A_1$. It is the same thing to say that two linear mappings do not always commute; $M_1 \circ M_2$ does not always equal $M_2 \circ M_1$.

D   Visualizing Nonsingular Linear Mappings

There are three mutually exclusive possibilities for the map M given by (11). First, if $a = b = c = d = 0$, then M sends all of $\mathbb{R}^2$ to the single point $\begin{pmatrix} 0 \\ 0 \end{pmatrix}$. Second, if $ad - bc = 0$ but the four numbers are not all zero, then the three points $\begin{pmatrix} a \\ c \end{pmatrix}$, $\begin{pmatrix} b \\ d \end{pmatrix}$, and $\begin{pmatrix} 0 \\ 0 \end{pmatrix}$ are collinear, which means that they all lie on one straight line, and M sends all of $\mathbb{R}^2$ onto that line. We are most interested in the third case, when M is

nonsingular, which means $ad - bc \neq 0$. The following Proposition sums up what we need:

1.3.4. Proposition. *For the linear mapping* $M : \mathbb{R}^2 \to \mathbb{R}^2$ *given by* (11), *the following five conditions are equivalent:*

(i) $ad - bc \neq 0$.

(ii) $\begin{pmatrix} a \\ c \end{pmatrix}$ *and* $\begin{pmatrix} b \\ d \end{pmatrix}$ *are independent vectors, which means that neither is a scalar multiple of the other.*

(iii) M *is one-to-one on* $\mathbb{R}^2$.

(iv) M *maps* $\mathbb{R}^2$ *onto* $\mathbb{R}^2$.

(v) *The matrix* $\begin{pmatrix} a & b \\ c & d \end{pmatrix}$ *has an inverse, namely,* $\dfrac{1}{ad - bc} \begin{pmatrix} d & -b \\ -c & a \end{pmatrix}$.

If M is a nonsingular mapping given by (11), it maps the unit square with vertices

$$\begin{pmatrix} 0 \\ 0 \end{pmatrix}, \begin{pmatrix} 1 \\ 0 \end{pmatrix}, \begin{pmatrix} 1 \\ 1 \end{pmatrix}, \text{ and } \begin{pmatrix} 0 \\ 1 \end{pmatrix} \tag{12}$$

onto the parallelogram with vertices

$$\begin{pmatrix} 0 \\ 0 \end{pmatrix}, \begin{pmatrix} a \\ c \end{pmatrix}, \begin{pmatrix} b \\ d \end{pmatrix}, \text{ and } \begin{pmatrix} a+b \\ c+d \end{pmatrix}. \tag{13}$$

As we will prove below, the area of this parallelogram equals $|ad - bc|$. The sign of $ad - bc$ reveals whether the mapping M preserves or reverses the orientation of $\begin{pmatrix} 1 \\ 0 \end{pmatrix}$ and $\begin{pmatrix} 0 \\ 1 \end{pmatrix}$; that is, if you stand at the origin facing $\begin{pmatrix} a \\ c \end{pmatrix}$, and you turn the nearest way to face $\begin{pmatrix} b \\ d \end{pmatrix}$, then you are turning counterclockwise (left) if $ad - bc > 0$, clockwise (right) if $ad - bc < 0$; and M is said to be **orientation-preserving** or **orientation-reversing**, respectively.

By the same token, M maps the unit square with vertices

$$\begin{pmatrix} 0 \\ 0 \end{pmatrix}, \begin{pmatrix} 0 \\ 1 \end{pmatrix}, \begin{pmatrix} -1 \\ 1 \end{pmatrix}, \text{ and } \begin{pmatrix} -1 \\ 0 \end{pmatrix}, \tag{14}$$

which is to the left of the one with vertices (12), onto the parallelogram with vertices

$$\begin{pmatrix} 0 \\ 0 \end{pmatrix}, \begin{pmatrix} b \\ d \end{pmatrix}, \begin{pmatrix} -a+b \\ -c+d \end{pmatrix}, \text{ and } \begin{pmatrix} -a \\ -c \end{pmatrix},$$

which is to the left of the one with vertices (13) if $ad - bc > 0$, but to the right of it if $ad - bc < 0$.

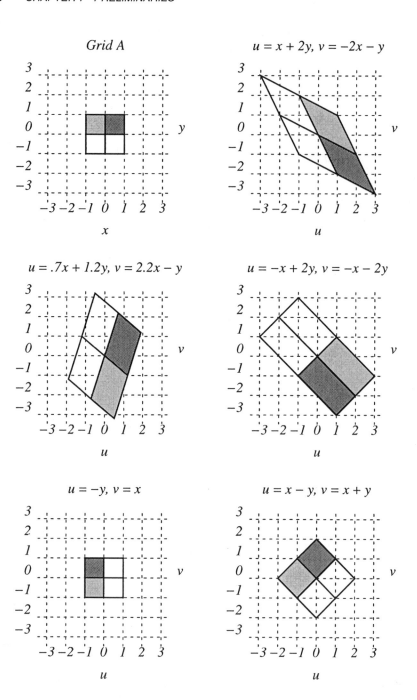

Figure 1.3-3   The picture at the upper left shows an orthogonal grid in the x, y-plane, called Grid A. Each of the other pictures shows the image in the u, v-plane of Grid A under a given linear map $(x, y) \mapsto (ax + by, cx + dy)$.

At the upper left in Figure 1.3-3 is shown a rectangular grid, drawn with solid lines, which marks off four unit squares in the x, y-plane. We will call it Grid A. Each of the other five grids in the Figure is the image in the u, v-plane of Grid A under a different nonsingular linear mapping M. Notice that two of the unit squares of Grid A are shaded, dark and light, and their images in each of the other grids are shaded correspondingly. In each case, once you see what M does to Grid A, it is easy to understand what it does to the plane as a whole.

Observe how each of the five mappings acts on line segments and on angles. The mapping given by

$$u = x + 2y, \quad v = -2x - y$$

takes $\begin{pmatrix} 1 \\ 0 \end{pmatrix}$ to $\begin{pmatrix} 1 \\ -2 \end{pmatrix}$ and $\begin{pmatrix} 0 \\ 1 \end{pmatrix}$ to $\begin{pmatrix} 2 \\ -1 \end{pmatrix}$. Notice that it changes the magnitudes of most angles, increasing some and decreasing others; and it stretches line segments by differing factors. One can say the same about the mapping given by

$$u = .7x + 1.2y, \quad v = 2.2x - y.$$

Notice that it reverses orientation; counterclockwise angles map to clockwise angles; and the light-shaded region is to the right of the dark-shaded region instead of to the left.

The mapping

$$u = -x + 2y, \quad v = -x - 2y$$

also stretches the line segments of Grid A by differing factors, clearly. Does it preserve the magnitudes of angles? One might think so at first glance, since it sends the right angles of the grid to right angles. But in fact, it changes the magnitude of most angles. For example, consider in Grid A the line segment from $(0, 0)$ to $(1, 1)$, which makes angle $\pi/4$ with each of the two unit vectors along the axes; its image is the line segment from $(0, 0)$ to $(1, -3)$, which makes different angles with the images of the two unit vectors.

The mapping

$$u = -y, \quad v = x$$

simply rotates the plane ninety degrees counterclockwise. It preserves both the orientation and magnitude of every angle, and the length of every line segment. The mapping

$$u = x - y, \quad v = x + y$$

rotates the plane $\pi/4$ radians counterclockwise and changes the length of every line segment by a factor of $\sqrt{2}$. It preserves both the orientation and magnitude of every angle.

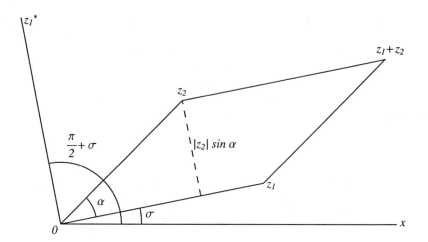

Figure 1.3-4   The area of the parallelogram equals the product of $|z_1|$, the length of its base, and $|z_2| \sin \alpha$, which is its perpendicular height over that base.

E   The Determinant of a Two-by-Two Matrix

Figure 1.3-4 illustrates the next result.

1.3.5. Proposition. *Let* $z_1 := \begin{pmatrix} a \\ c \end{pmatrix}$ *and* $z_2 := \begin{pmatrix} b \\ d \end{pmatrix}$ *be two nonzero points in* $\mathbb{R}^2$.
Let P *be the parallelogram that has for two of its sides the line segments* $[0, z_1]$ *and* $[0, z_2]$. *Let* α *be the angle (with vertex at* 0*) from the first segment toward the second, such that* $-\pi < \alpha \leq \pi$. *Let* D *equal the determinant of the matrix whose column vectors are* z_1 *and* z_2 :

$$D := \det \begin{pmatrix} a & b \\ c & d \end{pmatrix} = ad - bc.$$

Then $|D|$ *equals the area of* P. *If* D *is nonzero, then the sign of* D *is the same as that of* α.

Proof. Write the Cartesian coordinates of z_1 in terms of its polar coordinates:

$$\begin{pmatrix} a \\ c \end{pmatrix} = \begin{pmatrix} |z_1| \cos \sigma \\ |z_1| \sin \sigma \end{pmatrix}.$$

When the line segment $[0, z_1]$ is turned $\frac{\pi}{2}$ radians—this means counterclockwise, with the 0 end fixed—the resulting line segment is $[0, z_1^*]$, where

$$z_1^* := \begin{pmatrix} |z_1| \cos \left(\sigma + \frac{\pi}{2} \right) \\ |z_1| \sin \left(\sigma + \frac{\pi}{2} \right) \end{pmatrix} = \begin{pmatrix} -c \\ a \end{pmatrix}.$$

The last equality follows by the trigonometric identities (2) in subsection 1.2.C.

Observe that D equals the dot product of $z_2 = \begin{pmatrix} b \\ d \end{pmatrix}$ with $z_1^* = \begin{pmatrix} -c \\ a \end{pmatrix}$. When we apply (5) with z_1^*, z_2, and $\frac{\pi}{2} - \alpha$ in the roles of z_1, z_2, and θ, respectively, noting that $|z_1^*| = |z_1|$, we find that

$$D = |z_1||z_2| \cos \left(\frac{\pi}{2} - \alpha \right) = |z_1||z_2| \sin \alpha.$$

Therefore $|D|$ equals the area of P, because $|z_1 \sin \alpha|$ is the height of the parallelogram above the base whose length is $|z_2|$. If $D \neq 0$, it has the same sign as $\sin \alpha$, which has the same sign as α. $\square$

F   Pure Magnifications, Rotations, and Conjugation

A **pure magnification** is a linear mapping given by

$$\begin{pmatrix} k & 0 \\ 0 & k \end{pmatrix} \begin{pmatrix} x \\ y \end{pmatrix} = \begin{pmatrix} kx \\ ky \end{pmatrix},$$

where $k > 0$. It simply multiplies every vector by k. Notice that the matrix is kI, where I is the identity matrix.

The **rotation** R_λ is the linear mapping given by the matrix

$$\begin{pmatrix} \cos \lambda & -\sin \lambda \\ \sin \lambda & \cos \lambda \end{pmatrix}$$

where $\lambda \in \mathbb{R}$. To clarify how R_λ acts, we will write the domain point in terms of its polar coordinates, thus: $\begin{pmatrix} x \\ y \end{pmatrix} = \begin{pmatrix} r \cos \theta \\ r \sin \theta \end{pmatrix}$. Then

$$\begin{aligned} R_\lambda \begin{pmatrix} x \\ y \end{pmatrix} &= \begin{pmatrix} \cos \lambda & -\sin \lambda \\ \sin \lambda & \cos \lambda \end{pmatrix} \begin{pmatrix} r \cos \theta \\ r \sin \theta \end{pmatrix} \\ &= \begin{pmatrix} r \cos \lambda \cos \theta - r \sin \lambda \sin \theta \\ r \sin \lambda \cos \theta + r \cos \lambda \sin \theta \end{pmatrix} = \begin{pmatrix} r \cos(\theta + \lambda) \\ r \sin(\theta + \lambda) \end{pmatrix}. \end{aligned}$$

Thus R_λ simply rotates the plane λ radians counterclockwise. Notice that $R_0 = I$, and $R_\pi = -I$; and that all pure magnifications and rotations, and their compositions, are orientation-preserving.

Conjugation, which is a simple orientation-reversal of the plane, is given by

$$\begin{pmatrix} 1 & 0 \\ 0 & -1 \end{pmatrix} \begin{pmatrix} x \\ y \end{pmatrix} = \begin{pmatrix} x \\ -y \end{pmatrix}.$$

Some observations:

- Pure magnifications and rotations commute; $(kI)R_\lambda = R_\lambda(kI)$ for all λ and k. See Exercises 12 and 13.

- Every composition of pure magnifications and rotations is orientation-preserving.

G   Conformal Linear Mappings

The nonsingular linear mapping $M : \mathbb{R}^2 \to \mathbb{R}^2$ is **isogonal** if it preserves the magnitude (not necessarily the orientation) of angles. It is a **magnification** if there exists a constant $k > 0$ such that

$$|M(z)| = k|z| \qquad \text{for every } z \in \mathbb{R}^2.$$

Those two properties are equivalent.

1.3.6. Proposition. *The nonsingular linear mapping $M : \mathbb{R}^2 \to \mathbb{R}^2$ is isogonal if and only if it is a magnification.*

Proof. Clearly M is isogonal if and only if, whenever A, B, and C are three non-collinear points in the plane, the corresponding angles of the triangle $T(A, B, C)$ and the triangle $T(M(A), M(B), M(C))$ have equal magnitudes. Clearly M is a magnification if and only if, whenever A, B, and C are three noncollinear points in the plane, the lengths of corresponding sides have equal ratios, that is,

$$\frac{|M(B - A)|}{|B - A|} = \frac{|M(C - B)|}{|C - B|} = \frac{|M(A - C)|}{|A - C|};$$

and as we all know, that happens if and only if the corresponding angles of the two triangles have equal magnitudes.   □

   The nonsingular linear mapping $M : \mathbb{R}^2 \to \mathbb{R}^2$ is **conformal** (or **directly conformal**) if it is isogonal and orientation-preserving. It is **inversely conformal** if it is isogonal and orientation-reversing. Some observations and questions:

- All magnifications, all rotations, and therefore all compositions of such maps, are conformal. Are there any others?

- All magnifications, all rotations, the conjugation mapping, and therefore all compositions of such maps, are isogonal. Are there any others?

1.3.7. Proposition. *Let $M : \mathbb{R}^2 \to \mathbb{R}^2$ be a nonsingular linear map given by*

$$M\begin{pmatrix} x \\ y \end{pmatrix} = \begin{pmatrix} a & b \\ c & d \end{pmatrix} \begin{pmatrix} x \\ y \end{pmatrix}. \tag{15}$$

Then

(i) *M is conformal if and only if $a = d$ and $b = -c$.*

(ii) *M is inversely conformal if and only if $a = -d$ and $b = c$.*

Proof. If M is conformal or inversely conformal, then it is both isogonal and a magnification. Therefore the images of the vectors $\begin{pmatrix} 1 \\ 0 \end{pmatrix}$ and $\begin{pmatrix} 0 \\ 1 \end{pmatrix}$—which are

$\begin{pmatrix} a \\ c \end{pmatrix}$ and $\begin{pmatrix} b \\ d \end{pmatrix}$, respectively—are perpendicular and have the same length, $k = (a^2 + c^2)^{1/2} = (b^2 + d^2)^{1/2}$. Write (a, c) in terms of polar coordinates:

$$\begin{pmatrix} a \\ c \end{pmatrix} = \begin{pmatrix} k \cos \lambda \\ k \sin \lambda \end{pmatrix}.$$

If M is orientation-preserving, then

$$\begin{pmatrix} b \\ d \end{pmatrix} = \begin{pmatrix} k \cos \left(\lambda + \frac{\pi}{2} \right) \\ k \sin \left(\lambda + \frac{\pi}{2} \right) \end{pmatrix}, \quad \text{which equals} \quad \begin{pmatrix} -c \\ a \end{pmatrix}$$

(see (2) in subsection 1.2.C). If M is orientation-reversing, then

$$\begin{pmatrix} b \\ d \end{pmatrix} = \begin{pmatrix} k \cos \left(\lambda - \frac{\pi}{2} \right) \\ k \sin \left(\lambda - \frac{\pi}{2} \right) \end{pmatrix}, \quad \text{which equals} \quad \begin{pmatrix} c \\ -a \end{pmatrix}$$

(see (1) in subsection 1.2.C). We have proved the "only if" parts of both (i) and (ii).

Now suppose that $a = d$ and $b = -c$. The point $\begin{pmatrix} a \\ c \end{pmatrix} = \begin{pmatrix} d \\ -b \end{pmatrix}$ is on the circle of radius $k := (a^2 + c^2)^{1/2}$, and therefore $a = k \cos \lambda$ and $c = k \sin \lambda$ for some angle λ. The matrix of the mapping (15) then factors as follows:

$$\begin{pmatrix} a & b \\ c & d \end{pmatrix} = \begin{pmatrix} a & -c \\ c & a \end{pmatrix} = \begin{pmatrix} k \cos \lambda & -k \sin \lambda \\ k \sin \lambda & k \cos \lambda \end{pmatrix} = \begin{pmatrix} k & 0 \\ 0 & k \end{pmatrix} \cdot \begin{pmatrix} \cos \lambda & -\sin \lambda \\ \sin \lambda & \cos \lambda \end{pmatrix}.$$

Therefore the mapping M consists of a rotation followed by a pure magnification (though the order does not matter), and hence M is isogonal. Since the composition of two conformal maps is conformal, we have proved the "if" part of (i).

If $a = -d$ and $b = c$, then the matrix of the mapping (15) is

$$\begin{pmatrix} a & b \\ c & d \end{pmatrix} = \begin{pmatrix} a & c \\ c & -a \end{pmatrix} = \begin{pmatrix} a & -c \\ c & a \end{pmatrix} \cdot \begin{pmatrix} 1 & 0 \\ 0 & -1 \end{pmatrix}.$$

The mapping M therefore consists of conjugation, followed by a pure magnification and a rotation. Since the composition of conjugation with a conformal map is inversely conformal, we have proved the "if" part of (ii). □

1.3.8. Proposition. *Let M, P, and Q be nonsingular linear maps from $\mathbb{R}^2$ to R^2. Let M be conformal, and let P and Q be inversely conformal. Then:*

(i) The compositions $M \circ P$ and $P \circ M$ are inversely conformal.

(ii) The composition $P \circ Q$ is conformal.

(iii) The composition $P \circ M \circ Q$ is conformal.

Proof. We may represent the three maps using matrices. Let

$$M\begin{pmatrix} x \\ y \end{pmatrix} = \begin{pmatrix} a & -c \\ c & a \end{pmatrix}\begin{pmatrix} x \\ y \end{pmatrix},$$

$$P\begin{pmatrix} x \\ y \end{pmatrix} = \begin{pmatrix} s & t \\ t & -s \end{pmatrix}\begin{pmatrix} x \\ y \end{pmatrix}, \quad \text{and} \quad Q\begin{pmatrix} x \\ y \end{pmatrix} = \begin{pmatrix} b & d \\ d & -b \end{pmatrix}\begin{pmatrix} x \\ y \end{pmatrix}.$$

Statements (i) and (ii) are clear from 1.3.7 as soon as we inspect the matrix products:

$$\begin{pmatrix} a & -c \\ c & a \end{pmatrix}\begin{pmatrix} s & t \\ t & -s \end{pmatrix} = \begin{pmatrix} as - ct & at + cs \\ cs + at & ct - as \end{pmatrix}, \text{ so } M \circ P \text{ is inversely conformal;}$$

$$\begin{pmatrix} s & t \\ t & -s \end{pmatrix}\begin{pmatrix} a & -c \\ c & a \end{pmatrix} = \begin{pmatrix} sa + tc & -sc + ta \\ ta - sc & -tc - sa \end{pmatrix}, \text{ so } P \circ M \text{ is inversely conformal;}$$

$$\begin{pmatrix} s & t \\ t & -s \end{pmatrix}\begin{pmatrix} b & d \\ d & -b \end{pmatrix} = \begin{pmatrix} sb + td & sd - tb \\ tb - sd & td + sb \end{pmatrix}, \text{ so } P \circ Q \text{ is conformal.}$$

Statement (iii) can also be proved also by taking the product of the three matrices. It also follows logically from statements (i) and (ii). $\square$

Exercises

1. Explain: If there is one argument defined on a set, then there are infinitely many.

2. For the set shown on the right in Figure 1.3-2, if an argument is defined such that $\arg(20, 0) = 0$, what must be the value of $\arg(0, 2)$? –of $\arg(0, -20)$?

3. Consider the argument defined on the set shown on the left in Figure 1.3-2 that equals 0 on the positive real axis. What is its range?

4. Make a table to provide the answers, for each of the five linear mappings featured in Figure 1.3-3, to these questions: Is it nonsingular? Is it a magnification? Is it a pure magnification? Is it isogonal? Is it conformal? Is it inversely conformal?

5. Find the inverse of each of the five mappings in Figure 1.3-3.

6. For each of the five mappings in Figure 1.3-3, sketch the image of the unit circle $\{(x, y) \mid x^2 + y^2 = 1\}$. For the mapping in the upper right, find the equation of the image.

7. For each of the following linear mappings, sketch by hand the image of Grid A, shown in Figure 1.3-3, labelling the images of the four squares of Grid A. Make a table to provide the answers, for each of the five linear mappings, to these questions: Is it nonsingular? Is it a magnification? Is it a pure magnification? Is it isogonal? Is it conformal? Is it inversely conformal?

    (a) $M_1(x, y) = (x + 2y, 2x + y)$.

(b) $M_2(x,y) = (x + 2y, -2x - y)$.

(c) $M_3(x,y) = (x + 3y, -3x + y)$.

(d) $M_4(x,y) = (x + y, 5x + 3y)$.

(e) $M_5(x,y) = (3x - 3y, -3x + 3y)$.

8. For each of the five mappings in Exercise 7, sketch the image of the unit circle $x^2 + y^2 = 1$.

9. Give an example of a linear mapping $M : \mathbb{R}^2 \to \mathbb{R}^2$ such that the image of the unit circle is the ellipse $u^2 + \dfrac{v^2}{4} = 1$. Are there more than one?

10. Find the 2×2 matrix such that the linear mapping given by it is a counter-clockwise rotation through angle $\frac{\pi}{2}$ followed by a pure magnification by factor 0.27.

11. The proof of Proposition 1.3.7 establishes that if the linear mapping $M : \mathbb{R}^2 \to \mathbb{R}^2$ is conformal, then it is the composition of a pure magnification with a rotation. Also, if it is inversely conformal, then it is the composition of conjugation, a pure magnification, and a rotation. Carry out the factorization in each case.

(a) $M(x,y) = (2x + \sqrt{3}y, -\sqrt{3}x + 2y)$.

(b) $M(x,y) = (-x - y, y - x)$.

(c) $M(x,y) = (y, -x)$.

(d) $M(x,y) = (\sqrt{3}x + 2y, 2x + \sqrt{3}y)$.

12. Find two linear mappings among those illustrated in Figure 1.3-3 which do not commute. Find two that do commute.

13. Do conformal mappings commute? Do inversely conformal mappings commute? For what values of λ does R_λ commute with conjugation?

14. Consider the nonsingular linear map $(x, y) \mapsto (u, v)$, where $u = ax + by, v = cx + dy$. Let $r > 0$, and let C_r be the circle whose equation is $x^2 + y^2 = r^2$.

(a) Show that the image $M[C_r]$ has equation

$$Au^2 + Buv + Cv^2 = (ad - bc)^2 r^2,$$

where $A := c^2 + d^2$, $B := -2(ac + bd)$, and $C := a^2 + b^2$. Suggestion: Look at the inverse of the matrix $\begin{pmatrix} a & b \\ c & d \end{pmatrix}$. It lets you write x and y in terms of u and v. In the equation $x^2 + y^2 = r^2$, replace x and y with the appropriate combinations of u and v, and then simplify.

(b) Show that $B^2 - 4AC < 0$, so that $M[C_r]$ must be an ellipse.

(c) What conditions on the numbers a, b, c, d are equivalent to $M[C_r]$ being a circle?

15. One can also study the circle C_r of Exercise 14 as a parametrized curve. As t moves along the interval $[0, 2\pi]$, the point $z(t) := (r \cos t, r \sin t)$ goes around C_r counterclockwise; and the image of that point under M,

$$M(z(t)) = (ar \cos t + br \sin t, cr \cos t + dr \sin t),$$

travels around an ellipse. Notice that $M(z(0)) = (ar, cr)$ and $M(z(\frac{\pi}{2})) = (br, dr)$. Let $f(t) := |M(z(t))|^2$, the square of the length of $M(z(t))$.

(a) Show that

$$f'(t) = r^2[(b^2 + d^2 - a^2 - c^2) \sin 2t + 2(ab + cd) \cos 2t].$$

(b) If $b^2 + d^2 = a^2 + c^2$, what can you conclude about the ellipse?

(c) If $ab + cd = 0$, what can you conclude about the ellipse?

(d) What condition on the numbers a, b, c, d determines whether $M(z(t))$ goes around the ellipse clockwise or counterclockwise?

(e) What can you say about the cases when M is isogonal?

Help on Selected Exercises

2. $\arg(0, 2) = -\frac{7\pi}{2}$ and $\arg(0, -20) = -\frac{\pi}{2}$.

3. The range is the interval $(-\frac{5\pi}{4}, \frac{5\pi}{4})$.

5. Make use of Proposition 1.3.4, part (v). The inverse of the mapping in the upper right is given by $x = -\frac{1}{3}u - \frac{2}{3}v$, $y = \frac{2}{3}u + \frac{1}{3}v$.

6. One may make the sketches by a careful use of the pictures. To find the equation of the circle's image under a mapping, find the inverse mapping and thus get expressions for x and y in terms of u and v. Then substitute those expressions for x and y in the equation $x^2 + y^2 = 1$ to find the equation for the image of the circle.

7. All are nonsingular; only M_3 has any of the other properties.

10. $\begin{pmatrix} .27 & -.27 \\ .27 & .27 \end{pmatrix}$

15. Before using analytical reasoning, you may wish to experiment. Let a computer graph f for you for several choices of the four constants.

1.4   OPEN SETS, OPEN MAPPINGS, CONNECTED SETS

In subsection A we will define open sets and some other, related concepts. In B we will remind you of the "δ-ϵ" ("delta-epsilon") definition of continuity which is presented in most calculus texts. We will prove its equivalence to another definition, which is stated in terms of open sets and which is often much nicer to work with.

In C we will explain what an open mapping is. With a few exceptions, harmonic and holomorphic functions are open mappings. When that fact is established, their most important properties become easy to understand.

In D, we will give a definition of connectedness which is in terms of open sets, and which is also very pleasant to work with.

A   Distance, Interior, Boundary, Openness

In what follows, we will speak of subsets of $\mathbb{R}^n$; the distance $|a - b|$ between two points a and b of $\mathbb{R}^n$; and functions with domain in $\mathbb{R}^n$ and range in $\mathbb{R}^m$. What we will say will make sense whenever n and m are positive integers. But you may, if you wish, think of n and m as always equal to either 1 or 2, because the action in this book usually takes place in $\mathbb{R}^1$ and $\mathbb{R}^2$.

Let $a := (a_1, \ldots, a_n)$ be an element of $\mathbb{R}^n$. The symbol $|a|$ stands for the **length** or **norm** of a, given by

$$|a| := \left(\sum_{k=1}^{n} a_k^2 \right)^{1/2}.$$

Of course, if $a \in \mathbb{R}^1$, then $|a|$ is the absolute value of a. If $a := (a_1, a_2)$ is a point in the plane $\mathbb{R}^2$, then $|a|$ is called the modulus of a.

If $a, b \in \mathbb{R}^n$, then $|a - b|$ is the **distance** from a to b. The following conditions are satisfied:[4]

- $|a - b| > 0$ whenever $a, b \in \mathbb{R}^n$ and $a \neq b$; and $|a - a| = 0$ for all a.

- $|a - b| = |b - a|$ whenever $a, b \in \mathbb{R}^n$.

- $|a - c| \leq |a - b| + |b - c|$ whenever $a, b, c \in \mathbb{R}^n$ (The Triangle Inequality).

---

[4]What we are saying is that $\mathbb{R}^n$, with the distance function $(a, b) \mapsto |a - b|$, is an instance of a metric space. A **metric space** is a set M together with a function $\rho : M \times M \to \mathbb{R}^+$ such that these conditions are satisfied:

- $\rho(a, b) > 0$ whenever $a, b \in M$.

- $\rho(a, b) = \rho(b, a)$ whenever $a, b \in M$.

- $\rho(a, c) \leq \rho(a, b) + \rho(b, c)$ whenever $a, b, c \in M$.

Such a function is a **distance function** or **metric**. See [8, § 1.3, 6.6]. There are more ways than one to define a metric on $\mathbb{R}^n$; the one we are using is the familiar **Euclidean** metric, treated in calculus texts (for example, see [67, p. 667] or [27, p. 573]).

For $a_0 \in \mathbb{R}^n$ and $r > 0$, the **disk of center a_0 and radius r** is the set of all points of $\mathbb{R}^n$ whose distance to a_0 is less than r:

$$D(a_0, r) := \{a \in \mathbb{R}^n \mid |a - a_0| < r\}.$$

We use the term because we are interested mainly in the case when $n = 2$, and in common usage "disk" suggests a flat, two-dimensional object. Notice that when $n = 1$, $D(a_0, r)$ is the interval $(a_0 - r, a_0 + r)$. In $\mathbb{R}^3$, it would make sense to call it a "ball." A good term for an arbitrary dimension n might be "r-neighborhood."

The **punctured disk of center a_0 and radius r** is the set

$$D'(a_0, r) := \{a \in \mathbb{R}^n \mid 0 < |a - a_0| < r\} \equiv D(a_0, r) \setminus \{a_0\}.$$

The **closed disk of center a_0 and radius r** is the set

$$\bar{D}(a_0, r) := \{a \in \mathbb{R}^n \mid |a - a_0| \leq r\},$$

which is the union of the disk and the set of points a with $|a - a_0| = r$.

Let A be a subset of $\mathbb{R}^n$. Some definitions:

- The point $z_0 \in \mathbb{R}^n$ is a **boundary point** of A if for every $r > 0$, the set $D(z_0, r)$ contains at least one point that is in A and at least one point that is not in A.

- The **boundary** of A, denoted by bA, is the set of boundary points of A.

- The point $z_0 \in \mathbb{R}^n$ is an **interior point** of A if there exists $r > 0$ such that $D(z_0, r) \subseteq A$.

- The **interior** of A, denoted by A^o, is the set of interior points of A.

As we present examples in $\mathbb{R}^2$, you may wish to make sketches to help you picture what is being described.

Since the disk of center 0 and radius 1 in the plane is so frequently used, there are specific names and notations for the sets related to it. Thus the **unit disk** is $D := D(0, 1)$; the **punctured unit disk** is $D' := D'(0, 1)$; the **closed unit disk** is $\bar{D} := \bar{D}(0, 1)$.

1.4.1. Example. *For each of the sets D, D', $\bar{D}$, and $\mathbb{R}^2 \setminus \bar{D}$ in $\mathbb{R}^2$, determine whether the set is open, and identify the boundary and the interior of the set.*

Solution. The disk D is open because for each $z_0 \in D$, the disk $D(z_0, 1 - |z_0|)$ is contained in D. The boundary of D consists of the points z with $|z| = 1$. The interior of D is D.

The punctured disk D' is open because for each $z_0 \in D'$, if r is the smaller of $1 - |z_0|$ and $|z_0|$, the disk $D(z_0, r)$ is contained in D'. The boundary of D' consist of the points z with $|z| = 1$ and the point 0. The interior of D' is D'.

The closed disk $\bar{D}$ is not open, because for $|z_0| = 1$, no disk centered at z_0 is contained in $\bar{D}$. The boundary of $\bar{D}$ consists of the points z with $|z| = 1$. The interior of $\bar{D}$ is D.

The set $B := \{z \mid |z - a_0| > 1\}$ is open, since for each $z_0 \in B$, the set $D(z_0, |z_0| - 1)$ is a subset of B. The boundary of B consists of the points z with $|z| = 1$. The interior of B is B. □

We return now to the case of an arbitrary set $A \subseteq \mathbb{R}^n$. Here are some general observations.

The point z_0 is a boundary point of A if there are both points that are in A, and points that are not in A, which are arbitrarily close to z_0. You should notice that z_0 is a boundary point of A if and only if it is a boundary point of the complement of A; thus $bA = b(\mathbb{R}^n \setminus A)$.

The point z_0 is an interior point of the set A if all points sufficiently close to z_0 are in A. Informally, one might say: A point in A is an interior point if, when you stand at the point, you can move at least a certain distance in any direction without leaving A.

Observe that every point in A is either an interior point of A or a boundary point of A but cannot be both. A point is an interior point of A if and only if it belongs to A and is not a boundary point of A. Thus $A^o = A \setminus bA$.

1.4.2. Example. *For each of the following sets in $\mathbb{R}^2$, identify the boundary and the interior. Let S be a 1×1 square not including its edges:*

$$S = \{(x,y) \mid 0 < x < 1, \ 0 < y < 1\}.$$

Let S_1 be the union of the square S with the set of its four corner-points:

$$S_1 := \{(x,y) \mid 0 < x < 1, 0 < y < 1\} \cup \{(0,0), (1,0), (1,1), (0,1)\}.$$

Let S_2 be the union of S with its bottom edge, excluding the two bottom corners:

$$S_2 := \{(x,y) \mid 0 < x < 1, 0 \leq y < 1\}$$

Let S_3 be the whole square, including the edges:

$$S_3 := \{(x,y) \mid 0 \leq x \leq 1, 0 \leq y \leq 1\}.$$

Let S_4 be S_2 with one point removed:

$$S_4 := S_2 \setminus \left\{\left(\tfrac{1}{4}, \tfrac{1}{4}\right)\right\}$$

Let S_5 be S_2 with one point adjoined:

$$S_5 = S_2 \cup \{(7,7)\}$$

Let $L = \{(x,y) \mid x + y = 2\}$, a line in the plane. Let

$$S_6 = S_3 \cup L.$$

Solution. Let E be the union of the four edges of S, which are of course four line segments of length 1. The interior and the boundary are the same for S, S_1, S_2, and S_3:

$$S_1^o = S_2^o = S_3^o = S^o = S; \qquad bS_1 = bS_2 = bS_3 = bS = E.$$

Notice that the set S and its boundary are disjoint: $S \cap bS = \emptyset$; S_3 contains its boundary; S_2 and S_1 each contains some of the boundary points but not all.

As for the remaining sets:

$$S_4^o = S \setminus \left\{\left(\tfrac{1}{4}, \tfrac{1}{4}\right)\right\} \quad \text{and} \quad bS_4 = E \cup \left\{\left(\tfrac{1}{4}, \tfrac{1}{4}\right)\right\};$$
$$S_5^o = S \quad \text{and} \quad bS_5 = E \cup \{(7, 7)\};$$
$$S_6^o = S \quad \text{and} \quad bS = E \cup L.$$

$\square$

You may have noticed that the definitions we are talking about have a clear meaning only when the dimension n has been stated, or is understood from the context. Change the dimension, and the meaning may change. For example, consider the interval $(0, 1)$. If it is considered as a subset of $\mathbb{R}^1$, it is an open set; its interior is $(0, 1)$, and its boundary is the two-point set $\{0, 1\}$. But if it is considered as a subset of the plane—an interval on the x-axis—then it is not open; its interior is empty; and its boundary is itself. Thus each set of points is considered as a subset of some "universe" of points which must be identified.

Now we are ready to define "open set," along with three other useful terms. Let A be a subset of $\mathbb{R}^n$. Then:

- A is **open** if $A = A^o$; equivalently, if it contains none of its boundary points.

- A is **closed** if $bA \subseteq A$; in other words, if A contains all of its boundary points.

- A is **bounded** if, for some $R > 0$, A is contained in $D(0, R)$.

- A is **compact** if it is both closed and bounded.

Notice that A is closed if and only if its complement $\mathbb{R}^n \setminus A$ is open.

Among the sets S and S_k identified above, only S is open. Only S_3 and S_6 are closed. All the sets are bounded except S_6. Only S_3 is compact.

The ordinary usage of the terms "open" and "closed" may lead one to think that they are opposites—or that every set must be one or the other. A careful reading of the definitions indicates otherwise. If a set contains at least one boundary point, but does not contain all its boundary points, then it is neither open nor closed; examples are S_1, S_2, and S_5. And if a set has an empty boundary, it is both open and closed; for example, the empty set $\emptyset$ and the set $\mathbb{R}^n$.

The following results are simple exercises in using the definitions.

1.4.3. Proposition. *The intersection of two open sets is an open set.*

Proof. Let A and B be two open sets, and let $z_0 \in A \cap B$. All that we need to show is that there exists $r > 0$ such that $D(z_0, r) \subseteq A \cap B$. Since A is open, there exists $r_1 > 0$ such that $D(z_0, r_1) \subseteq A$. Since B is open, there exists $r_2 > 0$ such that $D(z_0, r_2) \subseteq B$. Take r to be the minimum of r_1 and r_2. $\square$

1.4.4. Proposition. *The union of a collection of open sets is an open set.*

Proof. Let $\mathcal{F}$ be a collection of open sets, and let z_0 belong to their union U. Then there is an open set A in the collection $\mathcal{F}$ such that $z_0 \in A$. So there exists $r > 0$ such that $D(z_0, r) \subseteq A$. Therefore $D(z_0, r) \subseteq U$. $\square$

B Continuity in Terms of Open Sets

Next we will state the familiar "δ-ϵ" definition of continuity for a function defined on a subset of one Euclidean space $\mathbb{R}^n$, with values in another, $\mathbb{R}^m$. For calculus-book versions, see [67, p. 71, p. 80] or [27, p. 128, p. 132].

Let $X \subseteq \mathbb{R}^n$. Let $f : X \to \mathbb{R}^m$. Let $a_0 \in X$. Then f is **continuous at** a_0 if for every $\epsilon > 0$ there exists $\delta > 0$ such that

$$|f(a) - f(a_0)| < \epsilon \text{ whenever } a \in X \text{ and } |a - a_0| < \delta. \tag{1}$$

If f is continuous at every point of X, then f is **continuous on** X.

Notice that (1) can be restated as follows:

$$f[D(a_0, \delta) \cap X] \subseteq D(f(a_0), \epsilon);$$

or, equivalently,

$$D(a_0, \delta) \cap X \subseteq f^{-1}[D(f(a_0), \epsilon)]. \tag{2}$$

So we may state the continuity definition this way: f is continuous at a_0 if the pre-image of every disk of center $f(a_0)$ contains the intersection with X of a disk of center a_0.

The set X is mentioned frequently in the preceding discussion, just because f is defined only on X, and we need to avoid speaking of $f(a)$ when $a \notin X$. The set X has taken the place of the space $\mathbb{R}^n$ as our "universe," so far as the domain of f is concerned. To take care of this kind of situation, we need to introduce a variant on open sets, which we previously defined with reference to one of the spaces $\mathbb{R}^n$ as the universe. Let $A \subseteq X \subseteq \mathbb{R}^n$. Then A is **open in** X, or **open relative to** X, if for every point $z_0 \in A$, there exists $r > 0$ such that $D(z_0, r) \cap X \subseteq A$. The statement "$A$ is open," previously defined, means the same thing as "A is open in $\mathbb{R}^n$."

Notice that if X is open, then "open in X" means the same thing as "open." The new definition makes a difference in the case when X is non-open. An example: The interval $(\frac{1}{2}, 1]$ is open in the interval $[0, 1]$.

Some worthwhile and fairly easy observations (Exercise 17):

- If B is open, then $X \cap B$ is open in X.

- Conversely, if A is open in X, then there exists an open set B such that $A = X \cap B$.

1.4.5. Proposition. *Let $X \subseteq \mathbb{R}^n$. Let $f : X \to \mathbb{R}^m$. The following two conditions are equivalent:*

(i) f is continuous on X in the sense of the δ-ϵ definition.

(ii) For every open set $W \subseteq \mathbb{R}^m$, the set $f^{-1}[W]$ is open in X.

Proof. Suppose (i) holds. Let W be an open set in $\mathbb{R}^m$. Let $a_0 \in f^{-1}[W]$. To prove (ii), it suffices to show that there exists $\delta > 0$ such that

$$D(a_0, \delta) \cap X \subset f^{-1}[W]. \tag{3}$$

Since W is open, there exists $\epsilon > 0$ such that $D(f(a_0), \epsilon) \subset W$ and hence, of course,

$$f^{-1}[D(f(a_0), \epsilon)] \subset f^{-1}[W].$$

By (i), there exists $\delta > 0$ such that (1) holds, hence (2) holds, hence (3) holds. It follows that (i) $\implies$ (ii).

Suppose (ii) holds. Let $a_0 \in X$. To prove (i), it suffices to show that f is continuous at a_0. Let $\epsilon > 0$. Then $D(f(a_0), \epsilon)$ is open. By (ii), $f^{-1}[D(f(a_0), \epsilon)]$ is open in X; therefore there exists $\delta > 0$ such that (2) holds and hence (1) holds. Thus (ii) $\implies$ (i). $\square$

C Open Mappings

Let $f : X \to \mathbb{R}^m$, where X is an open subset of $\mathbb{R}^n$. Then f is an **open mapping** if $f[V]$ is open whenever V is open.

Thus if there exists an open set whose image under f is not open, then f is not an open mapping. It is easy to find examples of mappings that are not open:

- Let $f(x) = x^2$. Then (for example) $f[\mathbb{R}] = [0, \infty)$, so f is not open.

- Let $f(x) = \sin x$, shown in Figure 1.2-1. Then $f[(0, \pi)] = (0, 1]$.

- Let $P(x) = 2x^3 - 3x^2 + 5$, shown in Figure 1.2-3. Then $P[(-1, 1)] = (0, 5]$.

- Let $f(x) = \cosh x$, also shown in Figure 1.2-3. Then $f[\mathbb{R}] = [1, \infty)$.

- Let $f(x, y) = \cos \sqrt{x^2 + y^2}$. Then $f[D((0,0), 4)] = [-1, 1]$. See Figure 1.4-1.

Openness is a powerful property, as the next result indicates.

1.4.6. Proposition. *If X is open and $f : X \to \mathbb{R}^m$ is an open mapping, then $z \mapsto |f(z)|$ cannot attain a local maximum at a point $z_0 \in B$. It cannot attain a local minimum at a point $z_0 \in B$ unless $f(z_0) = 0$.*

Proof. Let $r > 0$. Since $f[D(z_0, r)]$ is open, it must contain $D(f(z_0), \epsilon)$ for some $\epsilon > 0$. Therefore it must contain some points $f(z)$ with $|f(z)| > |f(z_0)|$; and unless $f(z_0) = 0$, it must contain some points $f(z)$ with $|f(z)| < |f(z_0)|$. $\square$

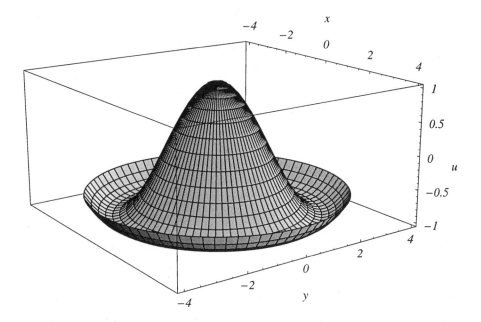

Figure 1.4-1   Let $u(x, y) = \cos \sqrt{x^2 + y^2}$. The graph of u over the disk $D((0,0), 4)$ is shown.

D   Connected Sets

Question: If A is a subset of $\mathbb{R}$ and $u : A \to \mathbb{R}$, and if $u' = 0$ everywhere on A, must u be constant? The answer is No; consider this example, in which A is the union of the intervals $(-2, -1)$ and $(1, 2)$:

$$u(x) := \begin{cases} -1, & \text{for } -2 < x < -1; \\ +1, & \text{for } +1 < x < +2. \end{cases}$$

If we change the question, including the condition on A that it should consist of one interval, then the answer is Yes: $u' = 0$ on A *does* imply that u is constant. For the analogous result in the more general case, when A is a subset of $\mathbb{R}^n$, the appropriate condition to place on A is that it should be **connected** (see Proposition 1.5.2 below).

Let $A \subseteq \mathbb{R}^n$. Let V and W be nonempty subsets of A which are both open in A, such that

$$V \cup W = A \quad \text{and} \quad V \cap W = \emptyset.$$

Then V and W **disconnect** A. If no such sets V and W exist, then A is **connected**.

1.4.7. Proposition. *The continuous image of a connected set is a connected set.*

Proof. If it were the case that some V and W disconnect $f[A]$, then $f^{-1}[V]$ and $f^{-1}[W]$ would disconnect A. $\square$

1.4.8. Proposition. *Every interval $I \subseteq \mathbb{R}$ is connected.*

Proof. Starting with the supposition that some two open sets V and W disconnect I, we will arrive at a contradiction.

There exist points $a \in V \cap I$ and $b \in W \cap I$. We may suppose that $a < b$. Because I is an interval, we know that $[a, b] \subseteq I$. Since V and W are open, there exist numbers a_1 and b_1 such that

$$a < a_1 < b_1 < b; \quad [a, a_1) \subset V; \quad \text{and} \quad (b_1, b] \subset W.$$

Let X be the set of numbers x in $[a, b]$ such that $[a, x) \subset V$. Let s be the smallest number such that $x \leq s$ for all $x \in X$; in other words, s is the least upper bound of the set X. Then $a < s < b$, and s must belong either to V or to W.

If $s \in V$, then because V is open, some open interval $(s - \epsilon, s + \epsilon)$ is contained in V, so that $[a, s + \epsilon) \subseteq V$; which contradicts the choice of s because then s would not be an upper bound of X. If $s \in W$, then because W is open, some open interval $(s - \epsilon, s + \epsilon)$ is contained in W; which also contradicts the choice of s because then it could not be the least upper bound of X. Either way, we have arrived at a contradiction. Therefore I is connected. $\square$

We offer the following result and its proof to show how the concepts of this Section lend themselves to very brief and efficient proofs.

1.4.9. Proposition. *A continuous integer-valued function on an open connected set is constant.*

Proof. If the function f were non-constant, it would take on at least two different integer values. Let m be one of them. The interval $I := (m - \frac{1}{3}, m + \frac{1}{3})$ and the set $J := \{y \in \mathbb{R} \mid |y - m| > \frac{2}{3}\}$ are open and disjoint, and their union contains the range of f. It would follow that the two sets $f^{-1}[I]$ and $f^{-1}[J]$ are open and disjoint and nonempty, so that they would disconnect the domain of f, which cannot be. $\square$

1.4.10. Remark. An alternative approach to defining connectedness begins with the idea that for A to be connected, one should be able, whenever p and q are points in A, to draw a curve that starts at p, lies entirely in A, and ends at q—where of course we might make various decisions as to what we mean by "curve." This approach leads to difficulties. For example, you probably agree that the set

$$B := \{(x, y) \mid x = 0 \text{ and } -1 \leq y \leq 1 \,; \text{ or } x > 0 \text{ and } y = \sin \tfrac{1}{x}\}$$

is one that we ought to regard as connected. But one cannot draw a curve within B that runs from, say, $(0, 1)$ to $(1/2\pi, 1)$. However, the idea of defining connectedness with curves works well when restricted to open sets, as we shall show in 2.5C.

Exercises

1. Identify the interior and the boundary of each of the following sets.

(a) $S_7 = \{(x,y) \mid 0 < x < 1 \text{ or } 1 < x < 2; \text{ and } 0 < y < 1\}$.

(b) $S_8 = S_7 \cup [(1,-1),(1,3)]$.

(c) $S_9 = \{(x,y) \mid x + y = 1\}$.

2. Identify the interior and the boundary of each of the following sets.

(a) the empty set $\emptyset$.

(b) $\mathbb{R}^2$.

(c) $A := \{(0,0)\}$.

(d) $B := \mathbb{R}^2 \setminus \{(0,0)\}$.

3. The **closure** of the set A is $\bar{A} := A \cup bA$. (Thus A is always contained in $\bar{A}$, and A is closed if and only if $A = \bar{A}$.) Identify the closure of the sets S_1 through S_9 defined in subsection A and Exercise 1.

4. Identify the closure of the sets listed in Exercise 2.

5. Give an example of a set that is closed; a set that is not closed; and a set that is neither open nor closed.

6. The point z_0 is an **exterior point** of the set A if there is a neighborhood of z_0 that contains no points of A. The **exterior** of A is the set of exterior points of A. Prove that a point is a boundary point of A if and only if it is neither an interior point nor an exterior point of A.

7. The point z_0 is an **accumulation point** of the set A if every neighborhood of z_0 contains at least one point in A that is not equal to z_0. Identify the set of accumulation points for each of the sets S_1 through S_5 in subsection A. Explain the difference between a boundary point and an accumulation point.

8. Prove that z_0 is an accumulation point of A if and only if every neighborhood of z_0 contains an infinite number of points in A.

9. The point z_0 is an **isolated point** of the set A if there is a neighborhood of z_0 whose intersection with A contains no point of A other than z_0. Identify the set of isolated points for each of the sets S_1 through S_5 in subsection A. Prove that a boundary point of a set A is either an accumulation point of A or an isolated point of A.

10. Consider the subset of $\mathbb{R}^2$,

$$X := \{(x,y) \mid 0 < x < 1, 0 \le y \le e^x\} \cup \{(1,3)\} \cup \{(0,4 + \tfrac{1}{k}) \mid k \in \mathbb{Z}\}.$$

Identify the interior, boundary, isolated points, and accumulation points of X.

11. Give an example of two different sets in $\mathbb{R}^2$ that have the same interior and the same closure.

12. Prove that the intersection of any finite number of open sets is also an open set.

13. Give an example of a collection of open sets whose intersection is not open.

14. Prove that the union of two closed sets is also a closed set. Is the same true for any finite number of closed sets?

15. Give an example of a collection of closed sets whose union is not closed.

16. Prove that the intersection of two closed sets is also a closed set. Is the same true for any finite number of closed sets? Is it true that the intersection of an arbitrary collection of closed sets is also a closed set?

17. Let $A \subseteq X \subseteq \mathbb{R}^n$. Prove that A is open in X if and only if there is an open set B such that $A = B \cap X$.

18. Let $S \subseteq X \subseteq \mathbb{R}^n$. Let X be an open set. Prove that S is open in X if and only if S is open.

19. Give an example of a closed set that is not compact, and an example of a bounded set that is not compact.

20. Let f be a function whose domain is an open set. Prove that if f is one-to-one, so that the function f^{-1} is defined on the range of f, then f is open if and only if f^{-1} is continuous.

21. In this Exercise, we give you a definition and ask you to prove several statements. Let $z_0 \in \mathbb{R}^n$. The set N is a **neighborhood** of z_0 if there exists $r > 0$ such that $D(z_0, r) \subseteq N$. Let $X \subseteq \mathbb{R}^n$. Then N is a **neighborhood of z_0 in** X, or **relative to** X, if there exists $r > 0$ such that $D(z_0, r) \cap X \subseteq N$. You should now be able to prove the following results.

    (a) z_0 is a boundary point of A if and only if every neighborhood of z_0 intersects both A and the complement of A.

    (b) z_0 is an interior point of A if and only if A is a neighborhood of z_0.

    (c) A is open if and only if A is a neighborhood of each of its points.

    (d) Let $A \subseteq \mathbb{R}^n$. Let $f : A \to \mathbb{R}^m$. Let $a_0 \in A$. Then f is continuous at a_0 if and only if for every neighborhood N of $f(a_0)$ in $\mathbb{R}^m$, the set $f^{-1}[N]$ is a neighborhood of a_0 in A.

Help on Selected Exercises

1. Recall that $[p, q]$ denotes the line segment from p to q.

    (a) $S_7^o = S_7$; $bS_7 = [(0,0),(2,0)] \cup [(2,0),(2,1)] \cup [(2,1),(0,1)] \cup [(0,1),(0,0)] \cup [(1,0),(1,1)]$.

    (b) $S_8^o = \{(x, y) \mid 0 < x < 1 \text{ and } 0 < y < 1\}$.; $bS_8 = [(0,0),(2,0)] \cup [(2,0),(2,1)] \cup [(2,1),(0,1)] \cup [(0,1),(0,0)] \cup [(1,-1),(1,3)]$.

    (c) $S_9^o = \emptyset$; $bS_9 = S_9$.

2. (a) $\emptyset^o = \emptyset = b\emptyset$.
   (b) $(\mathbb{R}^2)^o = \mathbb{R}^2$; $b\mathbb{R}^2 = \emptyset$.
   (c) $A^o = \emptyset$ and $bA = A$.
   (d) $B^o = B$ and $bB = A$.

4. Their closures are $\emptyset, \mathbb{R}^2, A$, and $\mathbb{R}^2$, respectively.

10. The interior of X is $\{(x,y) \mid 0 < x < 1 \text{ and } 0 < y < e^x\}$. The boundary of X is

$$\{(0,y) \mid 0 \le y \le 1; \text{ or } y = 4 + \tfrac{1}{k} \text{ for some } k \in \mathbb{Z}; \text{ or } y = 4\}$$
$$\cup \{(x,0) \mid 0 \le x \le 1\} \cup \{(x,e^x) \mid 0 \le x \le 1\}$$
$$\cup \{(1,y) \mid 0 \le y \le e; \text{ or } y = 3\}.$$

The isolated points of X are $(1,3)$ and the points $(0, 4 + \tfrac{1}{k})$ for $k \in \mathbb{Z}$. The set of accumulation points of X is

$$\{(x,y) \mid 0 \le x \le 1 \text{ and } 0 \le y \le e^x\} \cup \{(0,4)\}.$$

13. Consider the sets $D(0, \tfrac{1}{n})$.

14. If the sets are in $\mathbb{R}^n$, note that $\mathbb{R}^n \setminus S$ is open if and only if S is closed. You might then use the fact that, for any family of sets in $\mathbb{R}^n$, the union of the complements equals the complement of the intersection.

15. $\bigcup_{n=3}^{\infty} [\tfrac{1}{n}, 1 - \tfrac{1}{n}] = (0,1)$.

1.5 A REVIEW OF SOME CALCULUS

A Integration Theory for Real-Valued Functions

This subsection will summarize almost all the integration theory that you need as background. Except perhaps for the idea of the principal value of an integral, it should all be familiar.

The real-valued function f defined on the interval $[a, b] \subset \mathbb{R}$ is **piecewise continuous** if either (1) it is continuous at every point; or (2) it is continuous at all but a finite number of points t_0, with $a < t_0 < b$, at each of which the (finite) one-sided limits

$$\lim_{t \to t_0^-} f(t) \quad \text{and} \quad \lim_{t \to t_0^+} f(t)$$

exist and are unequal. In what follows, unless we specify otherwise, assume that all the functions mentioned are real-valued and piecewise continuous on the given interval. You are familiar with the Riemann integral

$$\int_a^b f(t)\,dt$$

of such a function. You know, for example, that if the definition of the function value $f(t)$ is changed for a finite number of values of t, the integral is still defined and still has the same value.

The action $f \mapsto \int_a^b f(t)dt$ is linear. That is, if c_1 and c_2 are constants, then

$$\int_a^b [c_1 f(t) + c_2 g(t)]dt = c_1 \int_a^b f(t)dt + c_2 \int_a^b g(t)dt.$$

If we change the orientation of the interval, the effect is to reverse the sign:

$$\int_b^a f(t)dt = -\int_a^b f(t)dt.$$

For every set of three points a, b, c,

$$\int_a^b f(t)dt = \int_a^c f(t)dt + \int_c^b f(t)dt.$$

If the integrand is a product $g(t)h(t)$, then

$$\int_a^b |g(t)h(t)|dt \leq \max_{a \leq t \leq b} |g(t)| \cdot \int_a^b |h(t)|dt. \tag{1}$$

The Schwarz Inequality, which we will use only in Chapter 6, states that

$$\left| \int_a^b g(t)h(t)dt \right| \leq \left(\int_a^b |g(t)|^2 dt \right)^{1/2} \cdot \left(\int_a^b |h(t)|^2 dt \right)^{1/2}.$$

The Fundamental Theorem of Calculus states that for $a \leq x \leq b$,

$$\frac{d}{dx} \int_a^x f(t)dt = f(x)$$

provided f is continuous at x; and that if $F' = f$ everywhere on $[a, b]$, then

$$\int_a^b f(t)dt = F(b) - F(a).$$

The Mean Value Theorem for Integrals states that if f is continuous on $[a, b]$, then there exists t^* such that $a < t^* < b$ and

$$\int_a^b f(t)dt = f(t^*)(b - a).$$

We will also need the fact of approximability by Riemann sums. The following simple formulation will suffice. Let $\epsilon > 0$. Then for all n sufficiently large,

$$\left| \int_a^b f(t)dt - \frac{b - a}{n} \sum_{k=1}^n f(t_k) \right| < \epsilon, \quad \text{where} \quad t_k = a + \frac{k}{n}(b - a).$$

B Improper Integrals, Principal Values

Certain integrals of non-piecewise continuous functions, and integrals over infinite intervals, are called "improper" but are still assigned values. We will summarize here the usual theory of such integrals.

- If f is piecewise continuous on $[c, R]$ for every $R > c$, then

$$\int_c^\infty f(t)dt := \lim_{R \to \infty} \int_c^R f(t)dt, \tag{2}$$

provided the limit exists and is finite. The improper integral on the left is then said to converge; otherwise, to diverge. Examples:

$$\int_1^\infty \frac{dx}{x^2} = 1. \qquad \int_0^\infty \frac{dx}{x^2 + 1} = \frac{\pi}{2}. \qquad \int_1^\infty \frac{dx}{x} \text{ diverges.}$$

- If f is piecewise continuous on $[-R, c]$ whenever $-R < c$, then

$$\int_{-\infty}^c f(t)dt := \lim_{R \to \infty} \int_{-R}^c f(t)dt, \tag{3}$$

provided the limit exists and is finite. The improper integral on the left is then said to converge; otherwise, to diverge. Examples:

$$\int_{-\infty}^{-2} \frac{dx}{x^3} = -\frac{1}{8}. \qquad\qquad \int_{-\infty}^{-1} \frac{dx}{x} \text{ diverges.}$$

$$\int_{-\infty}^0 \frac{dx}{x^2 + 1} = \frac{\pi}{2}. \qquad\qquad \int_{-\infty}^0 x\,dx \text{ diverges.}$$

- If f is piecewise continuous on every finite interval, and if the integrals (2) and (3) both converge for some (and hence for all) c, then the improper integral $\int_{-\infty}^\infty f(t)dt$ is defined as their sum. It is then said to converge; otherwise, to diverge. Equivalently:

$$\int_{-\infty}^\infty f(t)dt = \lim_{K \to \infty, L \to \infty} \int_{-K}^L f(t)dt.$$

Examples:

$$\int_{-\infty}^\infty \frac{dx}{x^2 + 1} = \pi. \qquad \int_{-\infty}^\infty xe^{-x^2}dx = 0. \qquad \int_{-\infty}^\infty x\,dx \text{ diverges.}$$

- If $c < b$ and if f is piecewise continuous on $[c, b - \epsilon]$ whenever $c < b - \epsilon < b$, then

$$\int_c^b f(t)dt := \lim_{\epsilon \to 0^+} \int_c^{b-\epsilon} f(t)dt, \tag{4}$$

provided the limit exists and is finite. The improper integral on the left is then said to converge; otherwise, to diverge. Examples:

$$\int_0^1 \frac{dx}{\sqrt{1-x}} = 2. \qquad \int_{-1}^0 \frac{dx}{x} \text{ diverges.} \qquad \int_0^{\pi/2} \sec x dx \text{ diverges.}$$

- If $a < c$ and if f is piecewise continuous on $[a + \epsilon, c]$ whenever $a < a + \epsilon < c$, then

$$\int_a^c f(t)dt := \lim_{\epsilon \to 0^+} \int_{a+\epsilon}^c f(t)dt, \tag{5}$$

provided the limit exists and is finite. The improper integral on the left is then said to converge; otherwise, to diverge. Examples:

$$\int_0^1 \frac{dx}{\sqrt{x}} = 2. \qquad \int_0^1 \ln x dx = -1. \qquad \int_0^{10} \frac{dx}{x} \text{ diverges.}$$

- If $a < b$ and if f is piecewise continuous on $[a+\epsilon, b-\epsilon]$ whenever $a < a+\epsilon < b - \epsilon < b$, and if the integrals (4) and (5) both converge for some (and hence for all) c with $a < c < b$, then the improper integral $\int_a^b f(t)dt$ is defined as their sum and is then said to converge; otherwise, to diverge. Equivalently:

$$\int_a^b f(t)dt := \lim_{\epsilon \to 0^+, \eta \to 0^+} \int_{a+\epsilon}^{b-\eta} f(t)dt.$$

Examples:

$$\int_0^1 \frac{dx}{\sqrt{x}\sqrt{1-x}} \text{ converges.} \qquad \int_0^1 \frac{dx}{x\sqrt{1-x}} \text{ diverges.}$$

- If $a < R$ and if f is piecewise continuous on $[a+\epsilon, T]$ whenever $a < a+\epsilon < T$, and if the integrals (2) and (5) both converge for some (and hence all) c with $a < c < \infty$, then the improper integral $\int_a^\infty f(t)dt$ is defined as their sum and is then said to converge; otherwise, to diverge. Examples:

$$\int_0^\infty \frac{dx}{\sqrt{x}(x+1)} = \pi. \qquad \int_0^\infty \frac{dx}{x^p} \text{ diverges.}$$

More generally (and speaking somewhat roughly), a proposed integral may be improper for several reasons; we call it convergent and assign it a finite value only if each and every one of the several limit processes used to resolve the "improprieties" yields a finite limit. As mentioned, the integral $\int_{-\infty}^\infty x dx$ diverges. However, if we take the two limits simultaneously and so that the diverging quantities balance each other off, we do obtain a finite limit, the **principal value** or **Cauchy principal value**:

$$\text{P.V.} \int_{-\infty}^{\infty} x\, dx := \lim_{R \to \infty} \int_{-R}^{R} x\, dx = 0.$$

The convergence to 0 holds only in a weak and delicate sense, as you will agree if you apply the same procedure to the function $x \mapsto x - 1$. Surely a meaningful integral on $\mathbb{R}$ should not change when we regard a point other than 0 as the center of the line! For a more elaborate example, consider the integral

$$\int_{-\infty}^{\infty} \frac{x^3 \sin x}{(x+1)(x-1)(x^2+4)}\, dx,$$

which is improper for six reasons—the infinite endpoints and the behavior on each side of ± 1. A graph of this function appears on the right in Figure 4.4-2, page 289. For convergence, we would require all six of the limits in this sum to exist and be finite (letting $f(x)$ denote the integrand):

$$\lim_{R \to \infty} \int_{-R}^{-2} f(x)\, dx + \lim_{\epsilon \to 0^+} \int_{-2}^{-1-\epsilon} f(x)\, dx$$

$$+ \lim_{\epsilon \to 0^+} \int_{-1+\epsilon}^{0} f(x)\, dx + \lim_{\epsilon \to 0^+} \int_{0}^{1-\epsilon} f(x)\, dx$$

$$+ \lim_{\epsilon \to 0^+} \int_{1+\epsilon}^{2} f(x)\, dx + \lim_{R \to \infty} \int_{2}^{R} f(x)\, dx.$$

None of the six is finite. But if take the limits in such a way as to let the diverging quantities balance each other off, we can obtain a finite principal value:

$$\lim_{\epsilon \to 0^+} \lim_{R \to \infty} \left(\int_{-R}^{-1-\epsilon} f(x)\, dx + \int_{-1+\epsilon}^{1-\epsilon} f(x)\, dx + \int_{1+\epsilon}^{R} f(x)\, dx \right).$$

In Example 4.4.6, we will show by the methods of the Residue Calculus that the this limit equals $\frac{\pi}{5}(\cos 1 + 4e^{-2}) \approx 0.6796166$.

We shall sometimes use the Riemann integral of a continuous function f of two real variables, over a rectangle $R := \{(x, y) \mid a \le x \le c,\ b \le y \le d\}$ in the plane:

$$\int_R f(x, y)\, dx\, dy$$

By Fubini's Theorem, the value of such an integral can be computed as an iterated integral in either order:

$$\int_R f(x, y)\, dx\, dy = \int_b^d \left(\int_a^c f(x, y)\, dx \right) dy = \int_a^c \left(\int_b^d f(x, y)\, dy \right) dx.$$

For a discussion and examples, see [67, Section 13.2] or [27, Section 15.2]. For a proof, see [8, pp. 184–189]. For some interesting connections among Fubini's Theorem, Clairaut's Theorem, and differentiation under the integral sign, see [65] and [17].

C   Partial Derivatives

We will state and discuss some results which we will need, and point out some examples which may be helpful.

Let $u : V \to \mathbb{R}$, where V is an open set in $\mathbb{R}^n$, and let n be a positive integer. If all partial derivatives of u of orders 0 through k are defined and continuous throughout V, then u is a C^k **function on** V. If u is C^2 on the subset V of $\mathbb{R}^2$, for example, then u and its partials

$$\frac{\partial u}{\partial x}, \ \frac{\partial u}{\partial y}, \ \frac{\partial^2 u}{\partial x^2}, \ \frac{\partial^2 u}{\partial y^2}, \ \frac{\partial^2 u}{\partial y \partial x}, \quad \text{and} \quad \frac{\partial^2 u}{\partial x \partial y}$$

are all defined and continuous throughout V. In the system of notation, for partial derivatives that we will use most often, one writes, for those six partials respectively,

$$u_x, \ u_y, \ u_{xx}, \ u_{yy}, \ u_{xy}, \quad \text{and} \quad u_{yx}.$$

Look carefully at the logic of the last two symbols. The partial with respect to y of the partial of u with respect to x is

$$\frac{\partial}{\partial y} \frac{\partial u}{\partial x} = \frac{\partial^2 u}{\partial y \partial x} = (u_x)_y = u_{xy};$$

and the partial with respect to x of the partial of u with respect to y is

$$\frac{\partial}{\partial x} \frac{\partial u}{\partial y} = \frac{\partial^2 u}{\partial x \partial y} = (u_y)_x = u_{yx}.$$

These conventions are not the same in all texts.

The next result is stated without proof. Exercise 2 offers some information.

1.5.1. Theorem (Clairaut). *If u is a C^2 function, then $u_{xy} = u_{yx}$.*

It follows that for C^k functions, the mixed partial derivatives of order up to k are independent of order. For one example, if u is C^3, then $u_{yyx} = u_{yxy} = u_{xyy}$.

A differentiable function g of one variable, say $g : [a, b] \to \mathbb{R}$, is constant if $g' = 0$ on $[a, b]$. The following is the analogue of that result for functions of two or more variables.

1.5.2. Proposition. *Let $u : A \to \mathbb{R}$, where A is an open connected subset of $\mathbb{R}^n$. If all the first-order partials of u exist and equal 0 everywhere on A, then u is constant on A.*

Proof. We will carry out the proof for the case when $n = 2$. Suppose that u is not constant. Let ζ be a number in the range of u. Then the two sets

$$U = \{z \in A \mid u(z) = \zeta\}, \qquad V = \{z \in A \mid u(z) \neq \zeta\}$$

are non-empty and disjoint, and their union is A. Therefore if we prove that they are both open, it follows that they disconnect A, which cannot be.

It suffices to show that whenever $D((a,b),r)$ is a disk that lies in A, u is constant on $D((a,b),r)$, so that the whole disk is contained in the same set, U or V, that (a,b) belongs to.

If (c,d) is a point in that disk, we have

$$u(c,d) - u(a,b) \;=\; u(c,d) - u(c,b) + u(c,b) - u(a,b).$$

By the Mean Value Theorem, applied once to the function $y \mapsto u(c,y)$ and once to the function $x \mapsto u(x,b)$, there are points y_0 and x_0 such that

$$u(c,d) - u(a,b) \;=\; u_y(c,y_0)(d-b) + u_x(x_0,b)(c-a),$$

which equals 0 since u_x and u_y are identically 0. It follows that u is constant and equal to $u(a,b)$ on $D((a,b),r)$. $\qquad\square$

1.5.3. Proposition. *Let g and g_y be defined and continuous on some open set containing the rectangle with opposite corners (a,b) and (c,d). Let*

$$G(x) := \int_c^d g(x,y)dy.$$

Then for all $s \in [a,b]$,

$$G'(s) = \int_c^d g_x(s,y)dy.$$

For a proof see [8, p 194].

Consider a continuous function of one real variable, $f : [a,b] \to \mathbb{R}$. Is there a function F such that $F' = f$ on $[a,b]$? Yes. Simply take $F(x) := \int_a^x f(t)dt$. The next result addresses the corresponding question for functions of two variables, which has a less simple answer.

1.5.4. Proposition. *Let $U(x,y)$ and $V(x,y)$ be C^1 functions defined on a set in the plane that contains the open disk $B := D((a,b),r)$. Then there exists a function $\phi(x,y)$ such that*

$$\phi_x = U \quad and \quad \phi_y = V \quad on \quad B \tag{6}$$

if and only if

$$V_x = U_y \quad on \quad B. \tag{7}$$

Proof. Suppose that ϕ exists, satisfying (6). To see that (7) holds, apply Theorem 1.5.1 with ϕ in the role of u.

For the reverse implication, suppose that U and V satisfy (7). Let

$$\phi(x,y) := \int_a^x U(s,b)ds + \int_b^y V(x,t)dt \quad \text{for } (x,y) \in B. \tag{8}$$

Compute ϕ_x and ϕ_y:

$$\phi_x(x,y) = U(x,b) + \int_b^y V_x(x,t)dt$$
$$= U(x,b) + \int_b^y U_y(x,t)dt$$
$$= U(x,b) + U(x,y) - U(x,b); \text{ and}$$
$$\phi_y(x,y) = 0 + V(x,y),$$

so ϕ satisfies (6). Notice that by 1.5.2, ϕ is unique up to an additive constant. □

1.5.5. Example. *Find ϕ if $U(x,y) := e^x \sin y$ and $V(x,y) := e^x \cos y + y$.*

Solution. It is easy to verify that (7) holds on the whole plane. You may be able to think quickly of a function ϕ satisfying (6); if not, you might reason as follows, using the idea of (8) but avoiding definite integrals. From the first equation in (6) it follows that

$$\phi(x,y) = \int e^x \sin y dx = e^x \sin y + h(y),$$

where $h(y)$ is a "constant of integration"—constant with respect to x, that is, but possibly depending on y. Therefore

$$\phi_y(x,y) = e^x \cos y + h'(y).$$

From that, and the second equation in (6), we conclude that $h'(y) = y$ and hence $h(y) = \frac{1}{2}y^2 + C$. Therefore

$$\phi(x,y) = e^x \cos y + \frac{1}{2}y^2 + C.$$

□

Proposition 1.5.4 is still true if the domain is an arbitrary simply connected open set, a concept which we will present later.

D Divergence and Curl

If you have a background in physics, then you are accustomed to thinking about problems in heat flow and electrostatics, for example, in terms of three spatial dimensions and 3-dimensional vector calculus. To help you see the connection with the 2-dimensional mathematical models which are emphasized in this book, it is important to explain the connection between Proposition 1.5.4, a 2-dimensional result which we will use a great deal, and its 3-dimensional analog. We begin with some terminology.

Let ϕ be a C^2 real-valued (scalar-valued) function of three real variables x, y, z. The **gradient** of ϕ is the vector-valued function, of the same domain, whose components are the first-order partials of ϕ:

$$\mathbf{grad}\,\phi \equiv \nabla\phi := \phi_x\vec{\mathbf{i}} + \phi_y\vec{\mathbf{j}} + \phi_z\vec{\mathbf{k}}.$$

The (3-dimensional) **Laplacian** of ϕ is the scalar-valued function given by

$$\nabla^2\phi := \phi_{xx} + \phi_{yy} + \phi_{zz}.$$

Let $\vec{\mathbf{F}}$ be a C^1 vector field, defined on a subset of x, y, z-space, whose components are three C^1 real-valued functions U, V, and W:

$$\vec{\mathbf{F}} = U\vec{\mathbf{i}} + V\vec{\mathbf{j}} + W\vec{\mathbf{k}}. \tag{9}$$

The **divergence** of $\vec{\mathbf{F}}$ is the scalar field given by

$$\mathrm{div}\,\vec{\mathbf{F}} \equiv \nabla \cdot \mathbf{F} := U_x + V_y + W_z.$$

The **curl** of $\vec{\mathbf{F}}$ is the vector field given by

$$\mathbf{curl}\,\vec{\mathbf{F}} \equiv \nabla \times \vec{\mathbf{F}} := (V_z - W_y)\vec{\mathbf{i}} + (W_x - U_z)\vec{\mathbf{j}} + (U_y - V_x)\vec{\mathbf{k}}.$$

There are two easily proved identities which are worth noting. Whenever ϕ is a C^2 scalar field, the divergence of the gradient of ϕ equals the Laplacian of ϕ:

$$\mathrm{div}\,\mathbf{grad}\,\phi = \nabla^2\phi. \tag{10}$$

Whenever $\vec{\mathbf{F}}$ is a C^1 vector field, the divergence of the curl of $\vec{\mathbf{F}}$ is 0:

$$\mathrm{div}\,\mathbf{curl}\,\vec{\mathbf{F}} = 0. \tag{11}$$

If $\vec{\mathbf{F}} = \mathbf{grad}\,\phi$, then ϕ is a **potential** for $\vec{\mathbf{F}}$. A given vector field may or may not have a potential. The next result says when it does have one.

1.5.6. Proposition. *Let U, V, and W be C^1 functions defined on a set in $\mathbb{R}^3$ that contains the open ball $B := D((a,b,c),r)$. Then there exists a function ϕ such that*

$$\phi_x = U, \quad \phi_y = V, \quad \text{and} \quad \phi_z = W \quad \text{on} \quad B \tag{12}$$

if and only if

$$V_z = W_y, \quad W_x = U_z, \quad \text{and} \quad U_y = V_x \quad \text{on} \quad B. \tag{13}$$

In other words, given a C^1 vector field $\vec{\mathbf{F}}$, then locally there exists ϕ such that $\vec{\mathbf{F}} = \mathbf{grad}\,\phi$ if and only if $\mathbf{curl}\,\vec{\mathbf{F}} = 0$.

The proof is analogous to that of 1.5.4.

In this book, when we consider 3-dimensional vector fields (9), it is nearly always in the special case when the third component is 0 and the first two components are independent of the third variable:

$$W \equiv 0, \quad U_z \equiv 0, \quad \text{and} \quad V_z \equiv 0. \tag{14}$$

Then $\mathbf{curl}\,\vec{F} = (U_y - V_x)\vec{k}$. Thus to assert that the curl is zero is to say that $V_x = U_y$. Evidently, then, 1.5.4 is a special case of 1.5.6.

The 3-dimensional vector fields $\vec{F}$ considered in physics often are both irrotational and incompressible:

$$\mathbf{curl}\,\vec{F} = \vec{0} \quad \text{and} \quad \text{div}\,\vec{F} = 0 \tag{15}$$

throughout the domain. What those two conditions say when the conditions (14) hold is that

$$V_x = U_y \quad \text{and} \quad V_y = -U_x,$$

in other words, the pair V, U satisfies the Cauchy-Riemann equations.

Exercises

1. For which values of p does $\displaystyle\int_0^1 \frac{dx}{x^p}$ converge? Answer the same question for $\displaystyle\int_1^\infty \frac{dx}{x^p}$ and for $\displaystyle\int_0^\infty \frac{dx}{x^p}$.

2. Devise, or find and read, a proof of Theorem 1.5.1. One way to prove it is to show that for every rectangle R contained in the domain of u,

$$\iint_R (u_{xy} - u_{yx})dx\,dy = 0;$$

the integrand being continuous, it follows that it must be zero everywhere (see [8, p. 189]). Or one can prove it by working directly from the definition of the partials as limits of differential quotients. See [67, Appendix F].

3. In the proof of 1.5.4, how did we make use of the fact that the domain is a disk?

4. Verify the identities (10) and (11).

5. Prove 1.5.6.

6. Prove that if the conditions (15) hold and $\vec{F} = \mathbf{grad}\,\phi$, then the Laplacian of ϕ is identically 0.

7. Elaborate on the following discussion, which will be used in Exercise 27, Section 2.6, and in Chapter 6. The area of a region R in the x, y-plane

equals the double integral $\iint\limits_{R} dx\,dy$. The area of the image of R under a C^1 mapping $(x, y) \mapsto (u(x, y), v(x, y))$ equals the double integral $\iint\limits_{R} |u_x v_y - u_y v_x|\,dx\,dy$.

Help on Selected Exercises

1. $p < 1; p > 1$; none.

7. This result is closely related to 1.3.5. Try writing Riemann sums; or see, for example, [67, Section 15.9], [27, pp. 794–796], or [8, Section 4.4].

1.6 HARMONIC FUNCTIONS

We promised you that harmonic functions have remarkable properties. Prepare to be impressed as we point out the geometric implications of the equations that describe them.

A   The Geometry of Laplace's Equation

Let u be a real-valued C^2 function defined on an open subset O of the x, y-plane, such that

$$\text{Laplace's equation:} \quad u_{xx} + u_{yy} = 0 \tag{1}$$

is satisfied throughout O. Then u is **harmonic** on O.

The twelve functions whose graphs appear in Figures 1.2-4, 1.2-6, 1.2-7, and 1.2-8 are all harmonic, as you may verify by computing their first- and then their second-order partial derivatives. For example, if $u = x^2 - y^2$, then

$$u_x = 2x, \ u_{xx} = 2; \quad u_y = -2y, \ u_{yy} = -2.$$

For another, let $u = \sin x \cosh y$. Then $u_x = \cos x \cosh y$, $u_y = \sin x \sinh y$, and thus

$$u_{xx} + u_{yy} = -\sin x \cosh y + \sin x \cosh y \equiv 0.$$

Recall the significance of the sign of the second derivative for the shape of the graph of a C^2 function f of one variable x. For example:

- If $f''(x_0) > 0$, then the graph of f is concave up for x near x_0, and if also $f'(x_0) = 0$, then a local minimum occurs at x_0.

- If $f''(x_0) < 0$, then the graph of f is concave down for x near x_0, and if also $f'(x_0) = 0$, then a local maximum occurs at x_0.

Let u be a C^2 function of two variables. Let (a, b) be a point in its domain. Restricting u first to the line $y = b$, then to the line $x = a$, we see two functions of one variable:

$$x \mapsto u(x, b) \quad \text{and} \quad y \mapsto u(a, y). \tag{2}$$

The values $u_x(a, b)$ and $u_{xx}(a, b)$ tell us about the shape of the first one's graph. The values $u_y(a, b)$ and $u_{yy}(a, b)$ tell us about the shape of the second one's graph. Each of those two graphs is a curve lying in the graph of u.

Let u be harmonic and non-constant. Suppose further that u_{xx} (and hence also u_{yy}) is never zero. Then (1) tells us that at each point, the two second-order partials have equal magnitude and opposite signs. That fact has an immediate consequence for the shape of the graph: If the graph-surface bends one way going from west to east, it must bend the other way going from south to north. Thus when $u_{xx}(a, b) > 0$, the function $x \mapsto u(x, b)$ is concave up at a; but then (1) requires $u_{yy}(a, b) < 0$, so that the function $y \mapsto u(a, y)$ is concave down at b.

We invite you to look carefully at the graphs of four harmonic functions in Figures 1.2-6 and 1.2-7. In each case, a number of curves appear, lying in the graph-surface of u (or v). They are curves of constant x or curves of constant y. In other words, they are graphs of functions of one variable, as in (2). Wherever two of them intersect, you see an illustration of our assertion; if one is concave up, the other is concave down.

This observation applies at all points, and it has a special meaning at a critical point (where both of the first-order partials equal zero). At such a point, when $u_{xx}(a, b) > 0$,, then the function $x \mapsto u(x, b)$ has a local maximum at a, and the function $y \mapsto u(a, y)$ has a local minimum at b. In other words, all critical points are saddle points, so extrema cannot occur at interior points of the domain. So long as u is non-constant, this is still true in the somewhat more subtle case when there are points at which $u_{xx} = u_{yy} = 0$, as we shall prove in subsection C.

B The Geometry of the Cauchy-Riemann Equations

Let u and v be real-valued C^2 functions on the open set O in the x, y-plane. If the

$$\text{Cauchy-Riemann equations:} \quad u_x = v_y \quad \text{and} \quad u_y = -v_x \tag{3}$$

hold throughout O, then the functions are necessarily both harmonic, because

$$
\begin{aligned}
u_{xx} &= (u_x)_x \\
&= (v_y)_x && \text{(by (3), first equation)} \\
&= (v_x)_y && \text{(by Theorem 1.5.1)} \\
&= (-u_y)_y && \text{(by (3), second equation)} \\
&= -u_{yy},
\end{aligned}
$$

and similarly for v.

If u and v are real-valued C^2 functions and the Cauchy-Riemann equations hold on the open set O, then v is a **harmonic conjugate** of u on O. Some easy observations:

- v is a harmonic conjugate of u if and only if $-u$ is a harmonic conjugate of v.

- If c is a constant and v is a harmonic conjugate of u, then $v + c$ is also a harmonic conjugate of u.

If u and v satisfy the Cauchy-Riemann equations at a point, then their gradients at that point,

$$\mathbf{grad}\, u = u_x \vec{\mathbf{i}} + u_y \vec{\mathbf{j}} \quad \text{and} \quad \mathbf{grad}\, v = v_x \vec{\mathbf{i}} + v_y \vec{\mathbf{j}}$$

are two perpendicular vectors of the same length (see 1.3.A). In fact, **grad** v equals **grad** u turned 90° to the left. A consequence is that the level curves of u and the level curves of v form an orthogonal grid (see Exercise 22). Examples appear in Figures 1.1-2 and 1.1-5.

In Section 1.1 we indicated that harmonic functions come in such pairs; the next result says it precisely.

1.6.1. Proposition. *If u is harmonic on the open disk D, then u has a harmonic conjugate on D. If v_1 and v_2 are two such conjugates, then $v_1 - v_2$ is constant.*

Proof. Let $U := -u_y$ and $V := u_x$. Then $V_x = U_y$ because u is harmonic. By 1.5.4, there exists v, unique up to an additive constant, such that $v_x = U$ and $v_y = V$, which means that the Cauchy-Riemann equations hold. □

The result remains true if we replace the disk by an arbitrary simply connected set, a concept which we will explain later.

C The Mean Value Property

Let u be a continuous real-valued function on the open set $O \subseteq \mathbb{R}^2$. Suppose that for every point $(a, b) \in O$, it is true for all sufficiently small $r > 0$ that the value of u at (a, b) equals the average value of u on the circle with center (a, b) and radius r:

$$u(a, b) = \frac{1}{2\pi} \int_0^{2\pi} u(a + r\cos\theta, b + r\sin\theta)\, d\theta. \tag{4}$$

Then u has the **mean value property** on O.

1.6.2. Theorem. *Let u be a continuous real-valued function on the open set $O \subseteq \mathbb{R}^2$. Then u is harmonic on O if and only if u has the mean value property on O.*

We postpone proving this result to discuss its implications. Figure 1.2-5 exhibits three disks in the plane and the graphs of the harmonic function $v = 3x^2y - y^3$ over those disks. Each graph is of course a 2-dimensional surface in x, y, v-space, like a round of pizza dough that's been whirled into the air, whose projection on the x, y-plane is a disk. To say that v has the mean value property is to say that if you travel all the way around the edge of the round of dough, then your average v-coordinate during the trip is equal to the v-coordinate of the center. Similarly, one can visualize the property in each picture in Figure 1.2-4. Each one shows the graph of a harmonic function on a disk.

On the other hand, the function shown in Figure 1.4-1 does not have the mean value property, though it is otherwise perfectly nice. That function takes on its maximum at the origin, and for $0 < r < \frac{\pi}{2}$ has a strictly smaller value all around the circle of radius r. Obviously, a function with the mean value property cannot do such a thing. The full strength of this insight is called the extremum principle, and we will deal with it soon (Theorem 1.6.5).

Later, we will obtain results like 1.6.2 by the efficient and beautiful methods of complex analysis. For the *only if* part, which is known as Gauss's Mean Value Theorem, see 3.1.4; for the *if* part, see 5.5.6. In 5.5.A, we will obtain results that are stronger and more general than 1.6.2. It turns out that for every point p inside the circle—not just for the center—the value of $u(p)$ can be obtained by taking a certain weighted average of the values of u on the circle.

We offer you now a calculus proof of the *only if* part of 1.6.2, for the sake of a more elementary perspective. It may help you get a feeling for the result. This proof will depend on a change of variables in the domain. We change from Cartesian coordinates to a polar coordinate system centered at (a, b).

1.6.3. Lemma. *Let* $u = u(x, y)$ *be a* C^2 *function on an open disk* $D(a, R)$ *in the plane. For* $0 < r < R$ *and all real* θ, *let* $U(r, \theta) = u(x, y)$, *where*

$$x = a + r \cos \theta \quad and \quad y = b + r \sin \theta. \tag{5}$$

Then

$$u_{xx} + u_{yy} = U_{rr} + \frac{1}{r} U_r + \frac{1}{r^2} U_{\theta\theta}. \tag{6}$$

The function U is the composition of the unspecified C^2 function u with the specific mapping (5) from (r, θ) to (x, y). One may prefer the less formal practice of using the same symbol "u" for both U and u, finding that the context will always clear up the ambiguity. We would say then that the right-hand side of (6) is the Laplacian of u in terms of the new variables r and θ. The proof of the Lemma is outlined in Exercise 24.

Proof of Gauss's Mean Value Theorem (the only if *part of 1.6.2).* Let u be a real-valued harmonic function on the open set O. Then u is C^2, and we seek to prove (4). Let $\bar{D}((a, b), s) \subset O$. Let

$$U(r, \theta) := u(a + r \cos \theta, b + r \sin \theta),$$

as in 1.6.3. It suffices to show that the function given by

$$m(r) := \frac{1}{2\pi} \int_0^{2\pi} U(r, \theta) d\theta \qquad (0 < r < s) \tag{7}$$

is constant and equal to $u(a, b)$. Since U is C^2, 1.5.3 applies:

$$m'(r) = \frac{1}{2\pi} \int_0^{2\pi} U_r(r, \theta) d\theta \quad \text{and} \quad m''(r) = \frac{1}{2\pi} \int_0^{2\pi} U_{rr}(r, \theta) d\theta.$$

Thus

$$\frac{d}{dr}(rm'(r)) = \left[m''(r) + \frac{1}{r} m'(r) \right] r$$

$$= \left[\frac{1}{2\pi} \int_0^{2\pi} \left(U_{rr}(r, \theta) + \frac{1}{r} U_r(r, \theta) \right) d\theta \right] r$$

$$= - \left[\frac{1}{2\pi} \int_0^{2\pi} \frac{1}{r^2} U_{\theta\theta}(r, \theta) d\theta \right] r,$$

which equals 0 because the function $\theta \mapsto U(r, \theta)$ is 2π-periodic; and therefore the function $\theta \mapsto U_\theta(r, \theta)$ is 2π-periodic; and thus

$$\frac{1}{2\pi} \int_0^{2\pi} U_{\theta\theta}(r, \theta) d\theta = U_\theta(2\pi) - U_\theta(0) = 0.$$

Therefore $rm'(r) = c$, a constant, and hence $m(r) = c \ln r + d$ for $0 < r < s$, where d is another constant. From the definition of $m(r)$ and the continuity of u, we know that $\lim_{r \to 0+} m(r) = u(a, b)$. But the limit of $m(r)$ can be finite only if $c = 0$. Therefore m is constant and must equal $u(a, b)$. □

1.6.4. Example. *Let u be a real-valued harmonic function whose domain contains the closed disk $\bar{D}((a, b), s)$, and let $u(x, y) > 0$ for all $(x, y) \in \bar{D}((a, b), s)$ Consider the 3-dimensional object*

$$C := \{(x, y, u) \mid (x - a)^2 + (y - b)^2 \le s^2 \text{ and } 0 \le u \le u(x, y).\}$$

It is a modified circular cylinder. Its base is a circular disk of radius r lying in the plane $u = 0$, and its top is the graph of u over that disk. Show that the volume of C is $\pi s^2 u(a, b)$.

Solution. Change the variables by (5) and set up an integral for the volume:

$$V = \int_0^s \left(\int_0^{2\pi} u(a + r \cos \theta, b + r \sin \theta) r d\theta \right) dr$$

For each value of r, the inner integral equals $2\pi u(a, b) r$. Therefore

$$V = 2\pi u(a, b) \int_0^s r dr = \pi s^2 u(a, b).$$

□

1.6.5. Theorem (The Extremum Principle). *Let u be a non-constant, continuous real-valued function on an open connected set $O \subseteq \mathbb{R}^2$ which possesses the mean* *value property. Then u cannot attain an extremum in O. That is, there is no point* $(p, q) \in O$ *at which either*

$$u(p, q) \geq u(x, y) \ \textit{for all} \ (x, y) \in O, \ \textit{or}$$
$$u(p, q) \leq u(x, y) \ \textit{for all} \ (x, y) \in O.$$

Proof. We shall show that if such a point (p, q) did exist, then O would be disconnected, which cannot be. (Recall the definitions and results of 1.4D.) Let

$$V := \{(x, y) \in O \mid u(x, y) = u(p, q)\} \quad \text{and}$$
$$W := \{(x, y) \in O \mid u(x, y) \neq u(p, q)\}.$$

Then V and W are non-empty disjoint sets whose union is O. The set W is the pre-image under u of the set $\mathbb{C} \setminus \{u(p, q)\}$, which is open. Therefore W is open, because u is continuous.

It suffices now to show that V is open, which implies that V and W would disconnect O (as defined in subsection 1.4D), which cannot be.

Let $(a, b) \in V$ and choose $s > 0$ such that $D((a, b), s) \subseteq O$. For every r with $0 < r < s$, consider the function

$$\theta \mapsto u(a + r \cos\theta, b + r \sin\theta) \quad (0 \leq \theta \leq 2\pi).$$

By the mean value property, $u(a, b)$ is both an upper bound for this function and its average. Therefore the function must be identically equal to $u(a, b)$. Thus u is constant on $D((a, b), s)$, which therefore is contained in V. So V is open.   □

D   Changing Variables in a Dirichlet or Neumann Problem

You have no doubt seen many a situation in which a problem becomes easier to solve thanks to a well-selected change of variables. One of the nicest methods for solving Dirichlet and Neumann problems is the method of conformal mapping, of which the main step is to change variables skillfully.

Consider the following problem. Assume that every function mentioned is C^2. Let f be a function given by $f(x, y) = (u(x, y), v(x, y))$, defined on an open set O in the x, y-plane, such that $f[O]$ is an open set in the u, v-plane. Let $G(u, v)$ be a real-valued function defined on $f[O]$. The diagram

$$\mathbb{R} \xleftarrow{\quad G \quad} f[O] \xleftarrow{\quad f \quad} O$$

represents the situation. Let H be the composition of G and f:

$$H(x, y) := G(u(x, y), v(x, y)).$$

What is the relation between the Laplacians of G and H? That is, what is $H_{xx} + H_{yy}$ in terms of $G_{uu} + G_{vv}$?

The first step is to apply the multidimensional Chain Rule (see, for example, [27, Section 13.6] or [67, Section 14.5]) or to obtain the relations among the first partials:

$$H_x = G_u u_x + G_v v_x;$$
$$H_y = G_u u_y + G_v v_y. \tag{8}$$

Exercise 20 asks you to supply the details of what follows. Applying the Chain Rule and the Product Rule to the expressions in (8), one computes H_{xx} and H_{yy}. Their sum can be organized as follows:

$$H_{xx} + H_{yy} = G_{uu}(u_x^2 + u_y^2) + G_{vv}(v_x^2 + v_y^2)$$
$$+ G_{uv}(2u_x v_y + 2u_y v_x)$$
$$+ G_u(u_{xx} + u_{yy}) + G_v(v_{xx} + v_{yy}).$$

Suppose now that the functions u and v satisfy the Cauchy-Riemann equations. Then the last three summands equal 0, and

$$H_{xx} + H_{yy} = (G_{uu} + G_{vv})(u_x^2 + u_y^2). \tag{9}$$

Provided $u_x^2 + u_y^2$ never equals 0, it follows in particular that H is harmonic on O if and only if G is harmonic on $f[O]$.

Thus: If the problem is to find a harmonic function H on O (satisfying certain conditions), then changing variables in the domain by means of a conformal mapping f converts the problem into that of finding a harmonic function G on $f[O]$ (satisfying certain conditions). Often it is easy to find a change of variables such that the converted problem can be solved quickly. That gives us an excellent reason to study conformal mappings.

Exercises

1. In each case, determine whether the given function is harmonic on $\mathbb{R}^2$.

   (a) $v(x, y) = 2xy.$ (c) $w(x, y) = x.$

   (b) $u(x, y) = e^x \cos y.$ (d) $u(x, y) = x^3 - 3xy^2.$

2. In each case, determine whether the given function is harmonic. Identify the domain.

   (a) $u(x, y) = \sqrt{x^2 + y^2}.$ (c) $u(x, y) = \ln(\sqrt{x^2 + y^2}).$

   (b) $v(x, y) = \arctan \frac{y}{x}.$ (d) $v(x, y) = \operatorname{arccot} \frac{x}{y}.$

3. Let $u(x, y) = ax^2 + bxy + cy^2$. What conditions on a, b, and c must hold for u to be harmonic?

4. If $f = u + iv$ is holomorphic and $u(x, y) = x^2 + 6xy - y^2$, and if $f(0, 0) = 2i$, identify $v(x, y)$.

5. Find an example of a harmonic function u such that u^2 is not harmonic.

6. Find a harmonic conjugate for each of the functions in Exercise 1.

7. Find a harmonic conjugate for each of the functions in Exercise 2, at least on some part of its domain.

8. Let $u(x, y)$ be harmonic on the open disk D, such that $u_y \equiv 0$ on D. What can you conclude about u?

9. Let $h(r, \theta)$ be harmonic for $0 < r < r_1$ and all θ, where r and θ are polar coordinates in the plane, such that $h_\theta \equiv 0$. Show that for some constants c and d, h is given by $h(r, \theta) = c \ln r + d$.

10. Let $h(r, \theta)$ be harmonic for $r_0 < r < r_1$ and all θ, where r and θ are polar coordinates in the plane, such that $h_r \equiv 0$. What can you conclude about h?

11. Explain this observation: The function $(r, \theta) \mapsto m(r)$ given by (7) is a harmonic function with $m_\theta \equiv 0$.

12. What is the average value of $x^2 - y^2$ on the circle $(x - 2)^2 + (y + 1)^2 = 1$?

13. Find the volume of the object

$$C = \{(x, y, v) \in \mathbb{R}^3 \mid (x - 2)^2 + (y - 2)^2 \le 0.25 \text{ and } 0 \le v \le 3x^2 y - y^3.\}$$

14. Let u be C^3. If u is harmonic, must u_x be harmonic? Either prove it or give a counterexample.

15. The left-hand side of (1) is called the Laplacian of u and is written $\nabla^2 u$. Let $C^2(O)$ denote the space of all C^2 functions on the open set O. Show that the mapping $u \mapsto \nabla^2 u$, called the Laplacian operator, is a linear operator defined on the space $C^2(O)$. In other words, show that

$$\nabla^2(\alpha u) = \alpha \nabla^2(u) \quad \text{whenever } \alpha \in \mathbb{R} \text{ and } u \in C^2(O); \text{ and} \quad (10)$$
$$\nabla^2(u_1 + u_2) = \nabla^2(u_1) + \nabla^2(u_2) \quad \text{whenever } u_1, u_2 \in C^2(V). \quad (11)$$

It follows from (10) that if u is harmonic and $a \in \mathbb{R}$, then au is also harmonic. It follows from (11) that if u_1 and u_2 are harmonic, then $u_1 + u_2$ is also harmonic. In other words, the harmonic functions on O form a real linear space.

16. The function $u(x, y)$ whose graph is shown in Figure 1.4-1 becomes $U(r, \theta) = \cos r$ when we change coordinates in the domain from Cartesian to polar. Compute the Laplacian of the function to see whether it is harmonic. Notice that for this function, is is easier to compute the right-hand side of (6) than the left-hand side. Can you tell whether this function is harmonic by looking at the graph?

17. In each case, determine whether the given function u is harmonic. Each u is expressed in terms of the polar coordinates r, θ, so that the easiest method may be to compute the right-hand side of (6).

(a) $u = r$.

(c) $u = \ln r$.

(e) $u = r^2 \cos 2\theta$.

(b) $u = \theta$.

(d) $u = r^2 \cos \theta$.

(f) $u = r^2 \cos \theta \sin \theta$.

18. What is the relation between Exercise 2 and Exercise 17, parts (a), (b), and (c)?

19. Prove that if p and q are harmonic, then their product pq is harmonic if and only if the gradients of p and q are perpendicular (note that this is a weaker condition than the requirement that p and q obey the Cauchy-Riemann equations). You may do so by computing the Laplacian of pq in terms of p, q, and their partials, thus:

$$\frac{\partial(pq)}{\partial x} = p\frac{\partial q}{\partial x} + \frac{\partial p}{\partial x}q; \quad \text{so} \tag{12}$$

$$\frac{\partial^2(pq)}{\partial x^2} = 2\frac{\partial p}{\partial x}\frac{\partial q}{\partial x} + p\frac{\partial^2 q}{\partial x^2} + \frac{\partial^2 p}{\partial x^2}q; \quad \text{and so forth.} \tag{13}$$

20. Give in detail the derivation of (9).

21. Prove that if both u and u^2 are harmonic on the open disk D, then u must be constant.

22. Let u be a C^2 function, such that u_x and u_y are never both zero. Let γ be a level curve of u, that is, a curve on which u is constant. Show that at each point of γ, the gradient of u is orthogonal to the tangent vector to the curve. Also show that if v is a harmonic conjugate of u, then wherever a level curve of v intersects a level curve of u, their tangent vectors are orthogonal.

23. The Taylor series for a function $f(x, y)$ at a point (a, b) begins like this:

$$
\begin{aligned}
f(x, y) = {}& f(a, b) + f_x(a, b)(x - a) + f_y(a, b)(y - b) \\
& + \frac{1}{2}f_{xx}(a, b)(x - a)^2 + f_{xy}(a, b)(x - a)(y - b) \\
& + \frac{1}{2}f_{yy}(a, b)(y - b)^2 + \cdots
\end{aligned}
$$

If we take only the terms of order 0 or 1, we get the equation of the plane that is tangent to the graph at $(a, b, f(a, b))$:

$$z = f(a, b) + f_x(a, b)(x - a) + f_y(a, b)(y - b);$$

and (a, b) is called a critical point if both of the first-order partials are zero there. In that case, the tangent plane is horizontal. Now let's group together the terms of second degree and call their sum $\frac{1}{2}Q$:

$$\frac{1}{2}Q(x-a, y-b) = \frac{1}{2}f_{xx}(a,b)(x-a)^2 + f_{xy}(a,b)(x-a)(y-b)$$
$$+ \frac{1}{2}f_{yy}(a,b)(y-b)^2$$

By letting dx and dy stand for $(x-a)$ and $(y-b)$ respectively and omitting "(a,b)," we write Q more briefly:

$$Q(dx, dy) = f_{xx}(dx)^2 + 2f_{xy}dxdy + f_{yy}(dy)^2.$$

As you may recall from multidimensional calculus, when you search for local extrema of a C^2 function $f(x,y)$ you examine each of the critical points; at such a point, if the Hessian $d^2 f := f_{xx}f_{yy} - f_{xy}^2$ is nonzero, its sign will tell you whether a relative extremum or a saddle point occurs at the point. Regardless of whether (a,b) is a critical point, the sign and magnitude of Δ are related to the shape of the graphs of Q and hence of f near the point.

(a) Prove that at every point, the Hessian of a real-valued harmonic function is nonpositive.

(b) Prove that at every point, the Hessians of a harmonic function u and its conjugate v are equal.

24. Prove Proposition 1.6.3. Assume $a = b = 0$.

25. Let u and v be C^2 functions defined on the punctured disk $D'((0,0), s)$. As in Proposition 1.6.3, switch the coordinate system in the domain from Cartesian to polar and look at the functions

$$U(r, \theta) = u(r\cos\theta, r\sin\theta) \quad \text{and} \quad V(r,\theta) = v(r\cos\theta, r\sin\theta)).$$

Show that the Cauchy-Riemann equations (3) are equivalent to

$$rU_r = V_\theta \quad \text{and} \quad rV_r = -U_\theta. \tag{14}$$

26. Consider a mapping $(x,y) \mapsto (u(x,y), v(x,y))$ whose range avoids the origin. We may switch the coordinate system in the range from Cartesian to polar, replacing the pair $u(x,y), v(x,y)$ by the pair $R(x,y), \Theta(x,y)$, where the old and new coordinate functions are related by these equations:

$$u(xy) = R(x,y)\cos\Theta(x,y) \quad \text{and} \quad v(x,y) = R(x,y)\sin\Theta(x,y).$$

Show that the Cauchy-Riemann equations (3) are equivalent to

$$R_x = R\Theta_y \quad \text{and} \quad R_y = -R\Theta_x. \tag{15}$$

27. Consider a mapping $(x, y) \mapsto (u(x, y), v(x, y))$ whose domain and range avoid the origin. We may switch the coordinate system from Cartesian to polar in both the domain and the range, replacing the pair $u(x, y), v(x, y)$ by the pair $R(r, \theta), \Theta(r, \theta)$. Show that the Cauchy-Riemann equations (3) are equivalent to

$$rR_r = R\Theta_\theta \quad \text{and} \quad R_\theta = -rR\Theta_r. \tag{16}$$

28. A C^2 function of three real variables $u(x, y, z)$ is defined to be harmonic if the 3-dimensional Laplacian vanishes, that is,

$$u_{xx} + u_{yy} + u_{zz} = 0.$$

Compute this Laplacian of the function given by

$$u(x, y, z) := \frac{1}{(x^2 + y^2 + z^2)^{1/2}} \quad ((x, y, z) \neq (0, 0, 0)).$$

Compute also the (2-dimensional) Laplacian of the function given by

$$u(x, y) := \frac{1}{(x^2 + y^2)^{1/2}}.$$

Help on Selected Problems

4. $v(x, y) = 2xy + 3y^2 - 3x^2 + 2$.

8. For some constants a and b, $u(x, y) = ax + b$.

10. From Laplace's equation we know that $h_{\theta\theta} \equiv 0$, so $h(r, \theta) = a\theta + b$; but since h is periodic in θ, $a = 0$. So h is constant.

12. By (4), it's 3.

14. Yes. What's to be shown is that $(u_x)_{xx} + (u_x)_{yy} \equiv 0$. The left-hand side equals $(u_{xx} + u_{yy})_x$.

22. Let γ be parametrized by $t \mapsto (x(t), y(t))$. Then for all t in the parameter interval, $u(x(t), y(t)) = c$, a constant. Take the derivative with respect to t on both sides of that equation, obtaining $u_x x' + u_y y' = 0$.

24. Here is an outline of one procedure that works. Write r and θ in terms of x and y:

$$r = (x^2 + y^2)^{1/2}, \qquad \theta = \arctan \frac{y}{x}.$$

You will be able to show that

$$r_x = \frac{x}{r}; \quad r_y = \frac{y}{r}; \quad \theta_x = \frac{-y}{r^2}; \quad \text{and} \quad \theta_y = \frac{x}{r^2}.$$

It may seem a bit odd for r to appear on the right-hand sides of those equations, but it's a way to keep the formulas neater. Then compute, using the multidimensional Chain Rule, this special case of (8):

$$
\begin{aligned}
u_x &= U_r \cdot \frac{x}{r} + U_\theta \cdot \frac{-y}{r^2}; \\
u_y &= U_r \cdot \frac{y}{r} + U_\theta \cdot \frac{x}{r^2}
\end{aligned}
\tag{17}
$$

By applying the Chain Rule and Product Rule to those expressions, one can compute u_{xx} and u_{yy} and prove (6). Carry out this computation in detail. We'll give you some help with the first part. Here's how it comes out:

$$
\begin{aligned}
u_{xx} &= \left[U_{rr} \cdot \frac{x}{r} + U_{r\theta} \cdot \frac{-y}{r^2} \right] \cdot \frac{x}{r} + U_r \cdot \left(\frac{1}{r} - \frac{x^2}{r^3} \right) \\
&\quad + \left[U_{\theta r} \cdot \frac{x}{r} + U_{\theta\theta} \cdot \frac{-y}{r^2} \right] \cdot \frac{-y}{r^2} + U_\theta \cdot \frac{2xy}{r^4} \\
&= U_{rr} \cdot \frac{x^2}{r^2} + U_{\theta r} \cdot \frac{-2xy}{r^3} + U_{\theta\theta} \cdot \frac{y^2}{r^4} \\
&\quad + U_r \cdot \left(\frac{1}{r} - \frac{x^2}{r^3} \right) + U_\theta \cdot \frac{2xy}{r^4}.
\end{aligned}
$$

25. View the equations (17) as a system of linear equations in the unknowns U_r and U_θ. Its determinant is $1/r$, which is nonzero. Therefore one can solve for U_r and U_θ in terms of u_x and u_y, obtaining

$$
\begin{aligned}
U_r &= \frac{1}{r}(xu_x + yu_y) \quad \text{and} \\
U_\theta &= xu_y - yu_x.
\end{aligned}
$$

Similarly, one obtains

$$
\begin{aligned}
V_r &= \frac{1}{r}(xv_x + yv_y) \quad \text{and} \\
V_\theta &= xv_y - yv_x.
\end{aligned}
$$

2

Basic Tools

2.1 THE COMPLEX PLANE

A The Definition of a Field

The real number system $\mathbb{R}$, with its familiar binary operations of addition and multiplication, is an example of a field. The requirements in the following definition are the rules of arithmetic that we use every day.

A **field** is a system $(F, +, \cdot)$, where F is a set containing at least two elements, and where $+$ and $\cdot$ are mappings from $F \times F$ to F, **addition** and **multiplication** respectively, such that the following conditions are satisfied.

   I. $(F, +)$ is a commutative group; that is,

      A. $(\alpha + \beta) + \gamma = \alpha + (\beta + \gamma)$ for all $\alpha, \beta, \gamma \in F$;

      B. $\alpha + \beta = \beta + \alpha$ for all $\alpha, \beta \in F$; and

      C. There is an element $0 \in F$, the **identity** of the group $(F, +)$, such that $\alpha + 0 = \alpha$ for every $\alpha \in F$.

      D. If $\alpha \in F$, there exists an element $-\alpha \in F$ such that $\alpha + (-\alpha) = 0$.

   II. $(\alpha \cdot \beta) \cdot \gamma = \alpha \cdot (\beta \cdot \gamma)$ for all $\alpha, \beta, \gamma \in F$.

  III. $\alpha \cdot \beta = \beta \cdot \alpha$ for all $\alpha, \beta \in F$.

  IV. $(\alpha + \beta) \cdot \gamma = \alpha \cdot \gamma + \beta \cdot \gamma$ for all $\alpha, \beta, \gamma \in F$.

   V. There exists a unique element $1 \in F$ such that $1 \cdot \alpha = \alpha$ for all $\alpha \in F$.

VI. If $\alpha \in F$ and $\alpha \neq 0$, there exists an element $\beta \in F$ such that $\alpha \cdot \beta = 1$.

The two-dimensional real vector space $\mathbb{R}^2$, with the operations of vector addition and scalar multiplication, given by equations (1) and (2) in subsection 1.3A, is not a field. The starting point for complex analysis is the remarkable fact that it is possible to define a true multiplication on the set $\mathbb{R}^2$ which, together with vector addition, makes it a field. When $\mathbb{R}^2$ is given this additional structure, it is called the field of complex numbers, or the complex plane, and denoted by $\mathbb{C}$. The complex structure and notation give us new and efficient ways to describe and work with mappings and objects like lines and circles. To master the theory and application of complex analysis, it is important to be handy with complex arithmetic, and to have a feeling for what its operations represent geometrically. The Section ends with a collection of Exercises, so that you can get the necessary practice.

B   Complex Multiplication

For the sake of emphasis, we will temporarily use the distinctive symbol "$\odot$" for the multiplication that we will define on the plane. To convey a geometric idea of how "$\odot$" works, it seems best to write the definition first in terms of polar coordinates (see subsection 1.3B), where we defined the symbol $((r, \theta))$:

$$((r, \theta)) = (x, y) \quad \Longleftrightarrow \quad x = r \cos \theta \quad \text{and} \quad y = r \sin \theta.$$

Given two points in the plane, $(x_1, y_1) = ((r_1, \theta_1))$ and $(x_2, y_2) = ((r_2, \theta_2))$, we define their product by:

$$((r_1, \theta_1)) \odot ((r_2, \theta_2)) := ((r_1 \cdot r_2, \theta_1 + \theta_2)). \tag{1}$$

Four examples:

$$\left(\left(\frac{3\sqrt{2}}{2}, \frac{\pi}{4}\right)\right) \odot \left(\left(2, -\frac{\pi}{6}\right)\right) = \left(\left(3\sqrt{2}, \frac{\pi}{12}\right)\right);$$

$$\left(\left(\frac{3\sqrt{2}}{2}, \frac{\pi}{4}\right)\right) \odot \left(\left(1, \frac{5\pi}{4}\right)\right) = \left(\left(\frac{3\sqrt{2}}{2}, \frac{3\pi}{2}\right)\right);$$

$$\left(\left(2, -\frac{\pi}{6}\right)\right) \odot \left(\left(1, \frac{5\pi}{4}\right)\right) = \left(\left(2, \frac{13\pi}{12}\right)\right);$$

$$\left(\left(\frac{5}{2}, \frac{5\pi}{6}\right)\right) \odot \left(\left(\frac{8}{5}, -\frac{2\pi}{3}\right)\right) = \left(\left(4, \frac{\pi}{6}\right)\right).$$

Make a sketch for each of those equations, using the polar coordinates to locate each of the two numbers being multiplied and their product. Exercises 2–3 offer more examples.

With a bit of practice, thinking about complex numbers in terms of polar coordinates, you should acquire a geometric sense of multiplication. That is, when you see two numbers represented in a picture, you should be able to point to the location of

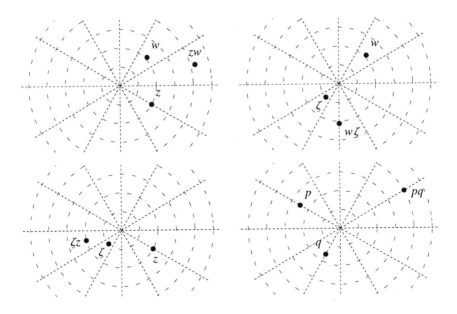

Figure 2.1-1 In each picture, two complex numbers and their product are shown.

their product. See the four pictures in Figure 2.1-1. The polar grid, shown in dashed lines, includes circles of radii 1 to 5 and radial lines at intervals of $\frac{\pi}{6}$ radians. See also Exercise 2.

Now we rewrite (1) giving each point the form (x, y) but with x and y expressed in terms of the polar coordinates of the point:

$$(r_1 \cos\theta_1, \ r_1 \sin\theta_1) \odot (r_2 \cos\theta_2, \ r_2 \sin\theta_2)$$
$$:= \big((r_1 r_2 \cos(\theta_1 + \theta_2), r_1 r_2 \sin(\theta_1 + \theta_2)\big). \tag{2}$$

Using the familiar trigonometric identities,

$$\cos(\theta_1 + \theta_2) = \cos\theta_1 \cos\theta_2 - \sin\theta_1 \sin\theta_2 \quad \text{and}$$
$$\sin(\theta_1 + \theta_2) = \sin\theta_1 \cos\theta_2 + \cos\theta_1 \sin\theta_2,$$

we find that (2) is the same as:

$$(x_1, y_1) \odot (x_2, y_2) := (x_1 x_2 - y_1 y_2, \ x_1 y_2 + y_1 x_2), \tag{3}$$

which would be an equivalent way to define "$\odot$" while making no mention at all of polar coordinates. We are free to use either (3) or (1) when dealing with the product.

We view $\mathbb{R}$ as lying in $\mathbb{R}^2$. That is, we identify the real number x_1 with the point $(x_1, 0)$ in the plane. In fact, we will write x_1 to mean $(x_1, 0)$.

In the case when $(x_1, y_1) \in \mathbb{R}$, that is, when $y_1 = 0$, (3) becomes $x_1 \odot (x_2, y_2) = (x_1 x_2, x_1 y_2)$ and thus complex multiplication agrees with scalar multiplication (see 1.3A). In other words, "$\odot$" extends "$\cdot$" from $\mathbb{R} \times \mathbb{R}^2$ to all of $\mathbb{R}^2 \times \mathbb{R}^2$.

2.1.1. Theorem. *The system* $(\mathbb{C}, +, \odot)$ *is a field.*

Proof. We must check that all the axioms in the definition of a field are satisfied. Axiom I does not involve "$\odot$" but merely lists familiar properties of vector addition. In view of (1), it is clear that Axiom II holds because

$$(r_1 r_2) r_3 = r_1 (r_2 r_3) \quad \text{and} \quad (\theta_1 + \theta_2) + \theta_3 = \theta_1 + (\theta_2 + \theta_3),$$

and Axiom III holds because

$$r_1 r_2 = r_2 r_1 \quad \text{and} \quad \theta_1 + \theta_2 = \theta_2 + \theta_1.$$

The following computation establishes Axiom IV.

$$
\begin{aligned}
&[(x_1, y_1) + (x_2, y_2)] \odot (x_3, y_3) \\
&= (x_1 + x_2, y_1 + y_2) \odot (x_3, y_3) \\
&= [(x_1 + x_2) x_3 - (y_1 + y_2) y_3, \; (x_1 + x_2) y_3 + (y_1 + y_2) x_3] \quad \text{by (3)} \\
&= (x_1 x_3 + x_2 x_3 - y_1 y_3 - y_2 y_3, \; x_1 y_3 + x_2 y_3 + y_1 x_3 + y_2 x_3) \\
&= (x_1 x_3 - y_1 y_3, \; x_1 y_3 + y_1 x_3) + (x_2 x_3 - y_2 y_3, \; x_2 y_3 + y_2 x_3) \\
&= [(x_1, y_1) \odot (x_3, y_3)] + [(x_2, y_2) \odot (x_3, y_3)] \quad \text{by (3)}.
\end{aligned}
$$

Since $(1, 0) \odot (x, y) = (x, y)$ for all (x, y), the element $(1, 0)$ serves as the "1" in Axiom V. To prove Axiom VI, just look at (1) and observe that if $r \neq 0$ then

$$((r, \theta)) \odot ((\tfrac{1}{r}, -\theta)) = ((1, 0)).$$

$\square$

So $(\mathbb{C}, +, \odot)$, or $\mathbb{C}$ for short, is a field. It is the **complex field,** or **complex number system.** Remember: The symbols $\mathbb{R}^2$ and $\mathbb{C}$ designate the same set—the plane—with different algebraic equipment; $\mathbb{C}$ will mean the complex field $(\mathbb{C}, +, \odot)$, while $\mathbb{R}^2$ will stand for the two-dimensional real linear space $(\mathbb{R}^2, +, \cdot)$.

Since "$\odot$" and "$\cdot$" agree wherever they are both defined, from now on we will drop the use of "$\odot$" and use "$\cdot$" for both operations. Most often we will omit "$\cdot$" and write, for example, zw instead of $z \cdot w$.

The two points $1 := (1, 0)$ and

$$i := (0, 1)$$

are a convenient basis for the two-dimensional real linear space $\mathbb{R}^2$. That is, every point $z = (x, y)$ can be written in just one way as a linear combination of those two points:

$$z = (x, y) = x \cdot (1, 0) + y \cdot (0, 1) = x + iy.$$

And then x and y are called the real part and the imaginary part, respectively, of z; the notation is

$$x = \operatorname{Re} z \quad \text{and} \quad y = \operatorname{Im} z.$$

By convention, whenever you see a complex number written in the form $a + ib$, you should assume that a and b are real numbers unless the context indicates otherwise. Note: The symbol j is sometimes used instead of i for $(0, 1)$.

C   Powers and Roots

Observe from (3) that

$$i^2 = -1.$$

Imagine for a moment a time long ago when the real number field $\mathbb{R}$ was understood and accepted. Why would one want a larger field than $\mathbb{R}$? A possible reason would be to include a square root of -1, in other words a solution for the polynomial equation $z^2 + 1 = 0$.[1] Suppose that one starts with the real number system and adjoins some one "imaginary" element, call it i, such that $i^2 = -1$. A field that contains $\mathbb{R}$, and that also contains such an additional element i, must of course also contain $x + iy$ whenever $x, y \in \mathbb{R}$. In such a field, furthermore, the axioms and the fact that $i^2 = -1$ leave us no choice as to how we define multiplication, because the following equalities are forced whenever $x_1, x_2, y_1, y_2 \in \mathbb{R}$:

$$
\begin{aligned}
(x_1 &+ iy_1)(x_2 + iy_2) \\
&= x_1(x_2 + iy_2) + iy_1(x_2 + iy_2) \quad \text{(by Axiom IV)} \\
&= (x_1x_2 + x_1iy_2) + (iy_1x_2 + iy_1iy_2) \quad \text{(by Axioms III and IV)} \\
&= (x_1x_2 - y_1y_2) + i(x_1y_2 + y_1x_2) \quad \text{(by Axioms IA, II, III, and IV)},
\end{aligned}
$$

which is the same as (3).

It turns out that when we adjoin i to $\mathbb{R}$, we achieve much more than the ability to solve $z^2 + 1 = 0$. As we shall prove in 3.1E, every polynomial

$$p(z) := \sum_{k=0}^{n} a_k z^k = a_0 + a_1 z + a_2 z^2 + \cdots + a_n z^n,$$

with each coefficient a_k belonging to $\mathbb{C}$, factors completely over $\mathbb{C}$:

$$p(z) = \prod_{k=0}^{n}(z - z_k) \quad \text{for all } z \in \mathbb{C},$$

where the complex zeros z_k of p are of course not necessarily distinct. We will deal now with the special case when $p(z)$ has the form $z^n - \alpha$.

Thanks to the definition of multiplication in terms of polar coordinates, given by (1), powers of z have an easy formula: If $z = (\!(r, \theta)\!)$, then $z^n = (\!(r^n, n\theta)\!)$. It is easy to understand and visualize the sequence

$$z, z^2, z^3, z^4, \cdots, z^n, \cdots.$$

If $|z| < 1$, then the powers of z move inward, toward the origin. If $|z| > 1$, then the powers move outward, toward infinity. If z is on the unit circle, all its powers are also on the unit circle. Figure 2.1-2 shows some examples.

---

[1] Historically, this is not quite what happened. For the story, see the discussion of Leonhard Euler's work on complex numbers in [16, Chapter 5]. At the time, mathematicians readily accepted the unsolvability of $z^2 + 1 = 0$. But they found the number i unavoidable when they tried to deal with cubic polynomials known to have three real roots, like $x^3 = 6x + 4$.

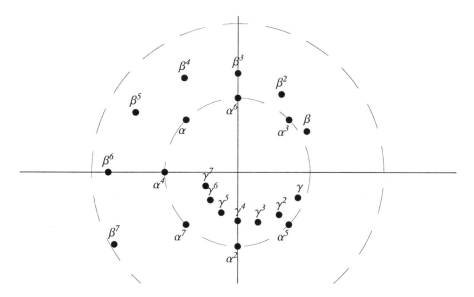

Figure 2.1-2 The first seven powers of each of the complex numbers $\alpha := ((1, \frac{3\pi}{4}))$, $\beta := ((1.1, \frac{\pi}{6}))$, and $\gamma := ((0.9, -\frac{\pi}{8}))$ are shown. The circles of center 0, radii 1 and 2, are shown with dashes.

Consider this problem: Given a complex number $\alpha \neq 0$ and an integer $n > 1$, find all n^{th} roots of α, that is, all z such that $z^n = \alpha$. We can write α as $((|\alpha|, \eta))$, where η is real (and not uniquely determined). Then the problem is to find all $z = ((r, \theta))$ such that $r > 0$ and

$$((r^n, n\theta)) = ((|\alpha|, \eta)),$$

which is equivalent to

$$r^n = |\alpha|, \quad \text{and} \quad n\theta = \eta + 2\pi k \quad \text{for some integer } k.$$

The number r is thereby uniquely determined; the roots are all on the circle of radius $r = |\alpha|^{1/n}$. There are infinitely many choices of the number θ, but only n distinct ones in an interval of length 2π. Therefore there are n distinct n^{th} roots of α :

$$\left(\left(|\alpha|^{1/n}, \frac{\eta + 2k\pi}{n} \right) \right), \qquad k = 0, 1, \ldots, n - 1.$$

2.1.2. Example. *Find the sixth roots of unity, that is, find all z such that $z^6 = 1$.*

Solution. We want to find all $((r, \theta))$ such that $((r^6, 6\theta)) = ((1, 0))$, equivalent to $r = 1$ and $6\theta = 2\pi k$ where k is an integer. Therefore the roots are $((1, \pi k/3))$ for $k = 0, 1, 2, 3, 4, 5$, six evenly spaced points on the unit circle. Since $\cos \dfrac{2\pi}{6} = \dfrac{1}{2}$

and $\sin \dfrac{\pi}{3} = \dfrac{\sqrt{3}}{2}$, the six numbers are, in Cartesian coordinates,

$$1, \quad \frac{1}{2} + i\frac{\sqrt{3}}{2}, \quad -\frac{1}{2} + i\frac{\sqrt{3}}{2}, \quad -1, \quad -\frac{1}{2} - i\frac{\sqrt{3}}{2}, \quad \text{and} \quad \frac{1}{2} - i\frac{\sqrt{3}}{2}.$$

☐

2.1.3. Example. *Find the roots of* $z^4 = 8 + i8\sqrt{3}$. *Illustrate the general fact that if* β *is one of the roots of* $z^n = \alpha$, *then each of the n roots can be represented as* $\sigma\beta$, *where* σ *is an* n^{th} *root of unity.*

Solution. Converted to polar notation, the equation becomes $((r^4, 4\theta)) = ((16, \pi/3))$, equivalent to $r = 2$ and $4\theta = \dfrac{\pi}{3} + 2\pi k$ where k is an integer. Therefore the four roots are $\left(\left(2, \dfrac{\pi}{12} + \dfrac{2\pi k}{4}\right)\right)$ for $k = 0, 1, 2, 3$, four evenly spaced points on the circle of radius 2. The one corresponding to $k = 0$ is $\beta = \dfrac{1 + \sqrt{3}}{2\sqrt{2}} + i\dfrac{\sqrt{3} - 1}{2\sqrt{2}}$, and the four can be represented as $\beta, i\beta, -\beta$, and $-i\beta$.

☐

D   Conjugation

For $z = x + iy \in \mathbb{C}$, let $\bar{z}$ denote $x - iy$, which is the **conjugate**, or **complex conjugate**, of z. We defined conjugation in subsection 1.3F as the mapping from $\mathbb{R}^2$ to $\mathbb{R}^2$ given by

$$\begin{pmatrix} x \\ y \end{pmatrix} \mapsto \begin{pmatrix} 1 & 0 \\ 0 & -1 \end{pmatrix}\begin{pmatrix} x \\ y \end{pmatrix} = \begin{pmatrix} x \\ -y \end{pmatrix}.$$

In complex notation, this is the mapping $z \mapsto \bar{z}$ from $\mathbb{C}$ to $\mathbb{C}$. It may be written, alternatively, as $((r, \theta)) \mapsto ((r, -\theta))$. Notice that

$$\text{Re } z = \frac{z + \bar{z}}{2} \quad \text{and} \quad \text{Im } z = \frac{z - \bar{z}}{2i} \tag{4}$$

and that

$$|z| = (x^2 + y^2)^{1/2} = (z\bar{z})^{1/2}.$$

It is easy to verify the following properties and computational rules:

(i) $\bar{z} = z$ if and only if z is real $(y = 0)$.

(ii) $\bar{z} = -z$ if and only if z is pure imaginary $(x = 0)$.

(iii) $\bar{\bar{z}} = z$ for all $z \in \mathbb{C}$.

(iv) $\overline{z + w} = \bar{z} + \bar{w}$ for all $z, w \in \mathbb{C}$.

(v) $\overline{zw} = \bar{z}\bar{w}$ for all $z, w \in \mathbb{C}$.

(vi) $\overline{\left(\dfrac{z}{w}\right)} = \dfrac{\bar{z}}{\bar{w}}$ for all $z, w \in \mathbb{C}$, provided $w \neq 0$.

E Quotients of Complex Numbers

Given a complex number written as the quotient of two others, say

$$\frac{z}{w} = \frac{x + iy}{u + iv},$$

one prefers to have it in the simpler form $p + iq$. To achieve the simplification, one may multiply it by 1 in the form $\bar{w}/\bar{w}$:

$$\frac{z}{w} = \frac{(x + iy)(u - iv)}{(u + iv)(u - iv)} = \frac{xu + yv}{u^2 + v^2} + i\frac{-xv + yu}{u^2 + v^2}. \tag{5}$$

Some numerical examples:

$$\frac{1 + i}{1 - i} = \frac{(1 + i)(1 + i)}{(1 - i)(1 + i)} = \frac{2i}{2} = i;$$

$$\frac{\sqrt{3} + i}{-3 - 3i} = \frac{(\sqrt{3} + i)(-3 + 3i)}{(-3 - 3i)(-3 + 3i)} = \frac{-1 - \sqrt{3}}{6} + i\frac{-1 + \sqrt{3}}{6};$$

$$\frac{10 - i}{6 + 5i} = \frac{(10 - i)(6 - 5i)}{(6 + 5i)(6 - 5i)} = \frac{55 - 56i}{61} = \frac{55}{61} - \frac{56}{61}i.$$

Alternatively, we may write the two points in terms of their polar coordinates and exploit (1) once more. If $z = ((r, \theta))$ and $w = ((s, \eta))$, with $w \neq 0$, then

$$z \div w = \left(\left(\frac{r}{s}, \theta - \eta\right)\right). \tag{6}$$

Even if we do not know the two points numerically, as soon as we locate z and w in a picture, (6) allows us at least to picture the location of the quotient quickly and easily. The first two examples above work out as follows:

$$\frac{1 + i}{1 - i} = \left(\left(\frac{\sqrt{2}}{\sqrt{2}}, \frac{\pi}{4} - \left(-\frac{\pi}{4}\right)\right)\right) = \left(\left(1, \frac{\pi}{2}\right)\right) = i;$$

$$\frac{\sqrt{3} + i}{-3 - 3i} = \left(\left(\frac{2}{3\sqrt{2}}, \frac{\pi}{6} - \frac{5\pi}{4}\right)\right) = \left(\left(\frac{\sqrt{2}}{3}, -\frac{13\pi}{12}\right)\right).$$

In computing a quotient's second polar coordinate, one might use the trigonometric identity

$$\tan(\theta - \eta) = \frac{\tan\theta - \tan\eta}{1 + \tan\theta\tan\eta};$$

but when evaluating the arctangent one may need to add a multiple of π in order to land in the correct quadrant. The third example above works out as follows:

$$\frac{10 - i}{6 + 5i} = \left(\left(\sqrt{\frac{101}{61}}, \arctan\left(-\frac{1}{10}\right) - \arctan\frac{5}{6}\right)\right)$$

$$= \left(\left(\sqrt{\frac{101}{61}}, \arctan\left(-\frac{56}{55}\right)\right)\right).$$

F   When Is a Mapping $L : \mathbb{C} \to \mathbb{C}$ Linear?

2.1.4. Definition. Let F be a field. A mapping $L : F \to F$ is **linear** if

$$L(\alpha\beta) = \alpha L(\beta) \quad \text{for all} \quad \alpha, \beta \in F; \quad \text{and} \tag{7}$$
$$L(\alpha + \beta) = L(\alpha) + L(\beta) \quad \text{for all} \quad \alpha, \beta \in F. \tag{8}$$

2.1.5. Proposition. *Let F be a field. A mapping $L : F \to F$ is **linear** if and only if, for some $A \in F$, L acts by means of multiplication by A :*

$$L(\alpha) = A\alpha \quad \text{for all} \ \alpha \in F. \tag{9}$$

Proof. Let L be linear. Applying (7) in the case when $\beta = 1$, we find that

$$L(\alpha) = L(\alpha 1) = \alpha L(1) \quad \text{for all } \alpha \in F.$$

Thus the value of L at 1 determines the value of L everywhere; if $A := L(1)$, then (9) holds. Conversely, a mapping L given by (9) satisfies (7) because $L(\alpha\beta) = A\alpha\beta = \alpha A\beta = \alpha L(\beta)$; and satisfies (8) because $L(\alpha + \beta) = A(\alpha + \beta) = A\alpha + A\beta = L(\alpha) + L(\beta)$. $\qquad\square$

As you may have noticed, the proof shows that condition (8) is superfluous in Definition 2.1.4.

Compare carefully Definitions 1.3.2 and 2.1.4. Both are about mappings from the plane to the plane, but in the case of the field $\mathbb{C}$, there is more algebraic structure than in the real linear space $\mathbb{R}^2$ and a more stringent requirement for a mapping to be linear.

Let us now repeat the Proposition for the two fields that interest us, $\mathbb{R}$ and $\mathbb{C}$. A mapping $f : \mathbb{R} \to \mathbb{R}$ is linear if and only if there is a real number A such that $f(x) = Ax$ for all $x \in \mathbb{R}$. In geometric terms, f is linear if and only if the graph of f is a straight line through the origin.

A mapping $L : \mathbb{C} \to \mathbb{C}$ is linear if and only if there is a complex number A such that $L(z) = Az$ for all $z \in \mathbb{C}$. We may write the number in terms of its Cartesian or its polar coordinates:

$$A = ((k, \lambda)) = a + ic$$

and assume $k \geq 0$. The mapping L acts by means of multiplication by $((k, \lambda))$. So what does L do to the plane? It rotates it counterclockwise through angle λ and then multiplies everything by k (or *vice versa;* the order does not matter).

To say it another way: Instead of writing $L(z) = Az$, we may write $L(x + iy) = (a + ic)(x + iy)$; or, dropping the complex notation, $L(x, y) = (ax - cy, cx + ay)$; or, writing the ordered pairs (x, y) and $(ax - cy, cx + ay)$ as column vectors to accommodate matrix notation, we find that L acts by means of multiplication on the left by a matrix:

$$L\begin{pmatrix} x \\ y \end{pmatrix} = \begin{pmatrix} a & -c \\ c & a \end{pmatrix} \begin{pmatrix} x \\ y \end{pmatrix} = \begin{pmatrix} ax - cy \\ cx + ay \end{pmatrix}.$$

Thus $L = kIR_\lambda$, where $k = |A| = (a^2 + c^2)^{1/2}$, as explained in subsection 1.3G.

Thus the linear mappings from $\mathbb{C}$ to $\mathbb{C}$ for which $A \neq 0$ are precisely the conformal linear mappings from $\mathbb{R}^2$ to $\mathbb{R}^2$.

A mapping $N : \mathbb{C} \to \mathbb{C}$ is **conjugate-linear** if

$$N(zw) = \bar{z}N(w) \quad \text{for all } z, w \in \mathbb{C}. \tag{10}$$

2.1.6. Proposition. *A mapping $N : \mathbb{C} \to \mathbb{C}$ is conjugate-linear if and only if it is given by*

$$N(z) = A\bar{z} \quad \text{for all } z \in \mathbb{C}, \tag{11}$$

for some constant $A \in \mathbb{C}$.

Proof. Let N be conjugate-linear. Then with 1 in the role of w in (10), we find that for every $z \in \mathbb{C}$, $N(z) = N(z \cdot 1) = \bar{z} \cdot N(1)$, so (11) holds with $A := N(1)$. Conversely, for every $A \in \mathbb{C}$ the mapping given by (11) clearly satisfies (10). $\square$

If $N : \mathbb{C} \to \mathbb{C}$ is conjugate-linear, given by (11), and we let $A = a + ic = ((k, \theta))$, then we may rewrite (11) as $N(x + iy) = (a + ic)(x - iy)$; or, removing complex notation, $N(x, y) = (ax + cy, cx - ay)$. In matrix notation,

$$N\begin{pmatrix} x \\ y \end{pmatrix} = \begin{pmatrix} a & c \\ c & -a \end{pmatrix} \begin{pmatrix} x \\ y \end{pmatrix} = \begin{pmatrix} ax + cy \\ cx - ay \end{pmatrix}.$$

Thus the conjugate-linear mappings from $\mathbb{C}$ to $\mathbb{C}$ for which $A \neq 0$ are precisely the inversely conformal linear mappings from $\mathbb{R}^2$ to $\mathbb{R}^2$ which we discussed in subsection 1.3G.

G Complex Equations for Lines and Circles

In real notation, the general equation of a line in the plane is

$$ax + by + c = 0 \qquad (a, b, c \in \mathbb{R}; \ a \text{ and } b \text{ not both zero}).$$

In complex notation, the general equation of a line in the plane is

$$\alpha\bar{z} + \bar{\alpha}z + c = 0 \qquad (\alpha \in \mathbb{C}, \ \alpha \text{ nonzero}, c \in \mathbb{R}).$$

To obtain the second from the first, we put $(z + \bar{z})/2$ for x and $(z - \bar{z})/2i$ for y, and let $\alpha = (a + ib)/2$.

In real notation, the general equation of a circle in the plane is

$$(x - x_0)^2 + (y - y_0)^2 = r^2 \qquad (x_0, y_0, r \in \mathbb{R}, \ r > 0). \tag{12}$$

In complex notation, the general equation of a circle in the plane is

$$z\bar{z} - \beta\bar{z} - \bar{\beta}z + k = 0 \qquad (\beta \in \mathbb{C}, \ k \in \mathbb{R}, \ k < |\beta|^2). \tag{13}$$

If $z = x + iy$ and $\beta = x_0 + iy_0$, then (12) becomes

$$|z - \beta|^2 = r^2; \quad \text{or} \quad (z - \beta)(\bar{z} - \bar{\beta}) = r^2,$$

which becomes (13) where $k := |\beta|^2 - r^2$. Notice that if $k = 0$, then the circle passes through the origin; if $k < 0$, then the origin is inside the circle.

H   The Reciprocal Map, and Reflection in the Unit Circle

The **reciprocal map**, illustrated in Figure 2.1-3, is the mapping $G : \hat{\mathbb{C}} \to \hat{\mathbb{C}}$ given by

$$G(z) := \frac{1}{z} = \frac{\bar{z}}{|z|^2}; \qquad \text{or,} \qquad G(((r, \theta))) := \left(\left(\frac{1}{r}, -\theta\right)\right).$$

One purpose of discussing G is to provide further practice and familiarity with complex notation and with complex arithmetic. Another purpose is to establish the properties of this mapping, particularly the fact that the image of every circle and line is a circle or a line. We will then be prepared for the discussion in Section 5.2 of Möbius maps, which are important to the conformal mapping method for solving basic Dirichlet problems.

In much of mathematics, the symbol ∞ appears in statements which are meaningful, while the symbol by itself is not given a meaning. In complex analysis, it is useful to make the symbol stand for a point. The **extended plane** is the complex plane with this one additional point adjoined:

$$\hat{\mathbb{C}} = \mathbb{C} \cup \{\infty\}.$$

We thus make ∞ an honorary member of the complex number system, so to speak, and allow it to participate in complex arithmetic by making these definitions:

$$\frac{1}{0} := \infty; \quad \frac{1}{\infty} := 0; \quad \overline{\infty} := \infty; \quad z + \infty := \infty \text{ whenever } z \in \mathbb{C},$$

This turns out to be very convenient. For one thing, we can avoid having to say repeatedly, "provided the denominator is not zero," and such. Every line may be considered to contain the point ∞.

Reflection in the unit circle is the mapping $J : \hat{\mathbb{C}} \to \hat{\mathbb{C}}$ given by

$$J : z \to \frac{1}{\bar{z}} = \frac{z}{|z|^2}; \qquad \text{or,} \qquad J : ((r, \theta)) \to \left(\left(\frac{1}{r}, \theta\right)\right)$$

We understand from that definition that $J(\infty) = 0$ and $J(0) = \infty$. Note that for $z \in \mathbb{C}'$, $J(z)$ is the point on the ray that runs from 0 through z to ∞ such that $|J(z)| \cdot |z| = 1$.

Some facts to notice:

- G is the composition of reflection in the unit circle and reflection in the real axis (complex conjugation).

- $G(\infty) = 0$ and $G(0) = \infty$.

- G has two fixed points, 1 and -1; everything else "moves."

- G is its own inverse; $G(G(z)) = z$ for all $z \in \hat{\mathbb{C}}$.

- In real notation, $G(x, y) = \left(\dfrac{x}{x^2 + y^2}, \dfrac{-y}{x^2 + y^2}\right)$.

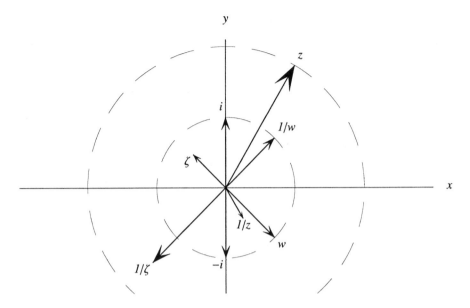

Figure 2.1-3 Four complex numbers and their reciprocals are shown as vectors. The four are i, $z := ((2, \frac{\pi}{3}))$, $w := ((1, -\frac{\pi}{4}))$, and $\zeta := ((\frac{2}{3}, \frac{3\pi}{4}))$.

To illustrate the uses of complex notation for lines and circles, and some properties of the reciprocal map, we pose now the question: If L is a line, what is $G[L]$? We adopt the convention that every line contains the point ∞. It follows, of course, that the image under G of every line contains 0.

Let L be a line. For some $\alpha \in \mathbb{C}$ and $c \in \mathbb{R}$, L has equation

$$\alpha \bar{z} + \bar{\alpha} z + c = 0. \tag{14}$$

Let $w = 1/z, z = 1/w$. Then $z \in L$ if and only if $z = \infty$ or z satisfies (14); which is true if and only if its image w satisfies the equation

$$\frac{\alpha}{\bar{w}} + \frac{\bar{\alpha}}{w} + c = 0;$$

or, equivalently (multiplying both sides by $w\bar{w}$ and rearranging),

$$cw\bar{w} + \bar{\alpha}\bar{w} + \alpha w = 0.$$

Either L contains the origin, or it does not. If it does, then $c = 0$ and $G[L]$ is the line whose equation is

$$\alpha w + \bar{\alpha}\bar{w} = 0,$$

which also contains the origin. If L does not contain the origin, then $c \neq 0$ and $G[L]$ is the circle whose equation is

$$w\bar{w} + \frac{\bar{\alpha}}{c}\bar{w} + \frac{\alpha}{c}w = 0.$$

The circle has center $\bar{\alpha}/c$ and contains the origin.

Consider now the question: If C is a circle, what is $G[C]$?

Let C be the circle of center β and radius r. The equation of the circle is then

$$z\bar{z} - \beta\bar{z} - \bar{\beta}z + k = 0, \tag{15}$$

where $k := |\beta|^2 - r^2$. Let $w = 1/z, z = 1/w$. Then $z \in C$ if and only if z satisfies (15), if and only if its image w satisfies the equation

$$\frac{1}{w\bar{w}} - \frac{\beta}{\bar{w}} - \frac{\bar{\beta}}{w} + k = 0,$$

or, equivalently (multiplying both sides by $w\bar{w}$ and rearranging),

$$kw\bar{w} - \bar{\beta}\bar{w} - \beta w + 1 = 0. \tag{16}$$

If the circle C passes through the origin, then $k = 0$, so the image $G[C]$ is the line whose equation is

$$\bar{\beta}\bar{w} + \beta w - 1 = 0,$$

which does not pass through the origin. If C does not pass through the origin, then $k \neq 0$ and we may rewrite (16) as

$$w\bar{w} - \frac{\bar{\beta}}{k}\bar{w} - \frac{\beta}{k}w + \frac{1}{k} = 0.$$

Therefore in this case $G[C]$ is a circle of center

$$\frac{\bar{\beta}}{k} = \frac{\bar{\beta}}{|\beta|^2 - r^2}$$

and radius

$$R = \sqrt{\left|\frac{\bar{\beta}}{k}\right|^2 - \frac{1}{k}} = \frac{r}{|\beta|^2 - r^2},$$

and it does not pass through the origin.

2.1.7. Example. *Let L be the line $y = \frac{1}{2}$. Identify its image $G[L]$.*

Solution. One method for finding the image is to follow the procedure in the discussion above. First write the equation of L in the form (14): $-iz + i\bar{z} = 1$. Then substitute $1/w$ for z, obtaining $iw - i\bar{w} = w\bar{w}$, which is the equation for the circle through the origin with center $-i$. Make a sketch!

We can arrive at the answer in such a problem somewhat more easily. The discussion above provides the general result that the image of a circle or line is a circle or line. We know that such a set is determined as soon as we know three points on it. So the reciprocals of any three points on L determine $G[L]$. For example, we might pick $i/2, (1 + i)/2$, and ∞; their reciprocals are $-2i, 1 - i$, and 0. So $G[L]$ is the circle with center $-i$ and radius 1.   □

2.1.8. Example. *Let C be the circle with center $1 + i$ and radius $\sqrt{2}$. Identify $G[C]$.*

Solution. Three points on C are 0, 2, and $2i$. Their reciprocals are ∞, $\frac{1}{2}$, and $-\frac{i}{2}$. Therefore $G[C]$ is the line whose equation is $y = -\frac{1}{2} + x$.   □

Exercises 15–24 ask you to draw certain lines and circles, and then find their images under G. Each of Exercises 25 and 31 leads you to a family of circles in which every circle is its own image under G.

I Reflections in Lines and Circles

Let L be a line in the plane. **Reflection in the line L** is the mapping $J_L : \mathbb{C} \to \mathbb{C}$ defined as follows. If z is a point on L, let $J_L(z) = z$. Otherwise, let M be the line perpendicular to L that passes through z, and let $J_L(z)$ be the other point on M such that $J_L(z)$ and z are the same distance from L.

Notice that the mapping $z \mapsto \bar{z}$ is reflection in the x-axis.

We would like to describe J_L using a formula. Let σ be the angle that L makes with the horizontal, and let p be a point on L. Then for every $z \in \mathbb{C}$, there exist $r > 0$, $\theta \in \mathbb{R}$ such that $z = p + re^{i(\sigma+\theta)}$. Make a sketch. Then

$$J_L(z) = p + re^{i(\sigma-\theta)}.$$

In view of the relation $\bar{z} - \bar{p} = re^{-i\theta - i\sigma}$, we may devise an expression without using r and θ:

$$J_L(z) = p + e^{i2\sigma}(\bar{z} - \bar{p}). \tag{17}$$

Notice that if L and M are any two lines, $J_L \circ J_M(z) = C_1 z + C_2$ for some constants $C_1 \neq 0$ and C_2.

One may define reflection in an arbitrary circle as follows. **Reflection in the circle** C of center A and radius R is the mapping $J_C : \hat{\mathbb{C}} \to \hat{\mathbb{C}}$ given by

$$J_C(z) = A + \frac{R^2}{(z - A)}. \tag{18}$$

Thus $J_C(A) = \infty$, and for $z \neq A$, $J_C(z)$ is the point on the ray from A through z to ∞ such that

$$|J_C(z) - A| \cdot |z - A| = R^2. \tag{19}$$

Exercise 32 asks you to do some computations and sketches.

The following interpretation of the mapping J_C, based on results from plane geometry, provides greater insight and will make it easy to understand the important symmetry property of Möbius maps (Proposition 5.2.21). It will also lead to a unification of the ideas of reflection of lines and for circles.

Let C be the circle of center A and radius R. We state now a second definition of J_C which will clearly be equivalent to the one above. Define $J_C(A) := \infty$; $J_C(\infty) := A$; and $J_C z := z$ if z is on C. For other z, let $J_C(z)$ be the point other than z that is common to all circles Γ through z that cross C at right angles. An explanation is required, as to why there is such a point.

Let K be the ray from A through z to ∞. Make a sketch. For every point B on C, there is a unique circle Γ through z that crosses C at right angles at the point B. Thus the segment $[A, B]$, whose length is R, is tangent to Γ. By a plane geometry result (Exercise 35), the point $z_1 \neq z$ at which Γ intersects K satisfies the equation

$$|z_1 - A| \cdot |z - A| = R^2,$$

which tells us that z_1 is independent of Γ.

Reflection in a line L may be defined in the same terms, as follows. Define $J_L(\infty) = \infty$; and for $z \in L$, let $J_L(z) = z$. For other z, let $J_L(z)$ be the point that is common to all circles through z that cross L at right angles.

Exercises

1. In each case, two complex numbers z and w are given. Sketch the parallelogram of which 0, z, and w are three of the vertices, and use the picture to estimate the coordinates of the fourth vertex. Then calculate the sum $z + w$ to check your estimate.

   (a) $z = 5 - \frac{1}{2}i$, $w = 3 + \frac{5}{2}i$.

   (b) $z = 1 + \frac{3}{2}i$, $w = \frac{1}{2} + 2i$.

   (c) $z = -4 - i$, $w = -4 + i$.

   (d) $z = 2 + i$, $w = -2 + i$.

   (e) $z = \frac{1}{10} + i$, $w = i$.

   (f) $z = -\frac{i}{2}$, $w = -1$.

2. Let $z = ((2, -\frac{\pi}{6}))$, $w = ((\frac{3}{2}, \frac{\pi}{4}))$, $\zeta = ((1, -\frac{\pi}{4}))$, $p = ((\frac{5}{2}, \frac{5\pi}{6}))$; and $q = ((\frac{8}{5}, -\frac{2\pi}{3}))$. For each of the ten pairs of complex numbers chosen from that set of five, make a sketch showing the two numbers and also showing their product.

3. In each case, two complex numbers z and w are given. Calculate zw using (1). Sketch z, w, and zw in a picture. Then find the Cartesian coordinates for z and for w, and calculate zw using (3). Check that your results agree.

   (a) $z = ((\frac{3\sqrt{2}}{2}, \frac{\pi}{4}))$, $w = ((2, -\frac{\pi}{6}))$.

   (b) $z = ((1.5, 0))$, $w = ((2, \frac{\pi}{3}))$.

   (c) $z = ((0.8, \frac{\pi}{6}))$, $w = ((2, \frac{\pi}{3}))$.

   (d) $z = ((2, -\frac{\pi}{6}))$, $w = ((1, \frac{5\pi}{4}))$.

4. For each part of Exercise 3, sketch z and w in a picture of the plane, compute $|z|/|w|$, and then locate $z \div w$ in the picture as well, using (6).

5. For the point z in each part of Exercise 1, illustrate the equations (4) by showing z, $\bar{z}$, $z + \bar{z}$, and $z - \bar{z}$ in a sketch.

6. Verify the computational rules (i)–(vi) for complex conjugation in subsection D.

7. In each case you are given a complex number in the form $a + ib$; find $1/(a+ib)$ by the method of (5), which in such cases can be written as follows:

$$\frac{1}{a+ib} = \frac{a-ib}{(a+ib)(a-ib)} = \frac{a}{a^2+b^2} - i\frac{b}{a^2+b^2}.$$

(a) $1 + i\sqrt{3}$.

(b) $\frac{1}{4} - i\frac{\sqrt{3}}{4}$.

(c) $\frac{1}{\sqrt{2}} - i\frac{1}{\sqrt{2}}$.

(d) $\frac{1}{\sqrt{2}} + i\frac{1}{\sqrt{2}}$.

(e) $-\frac{\sqrt{2}}{3} + i\frac{\sqrt{2}}{3}$.

(f) $-\frac{3\sqrt{2}}{4} + i\frac{3\sqrt{2}}{4}$.

Explain the relation between this Exercise and Figure 2.1-3.

8. For each of the pairs z, w in Exercise 1, calculate z/w by the method of writing it as $z\bar{w}/w\bar{w}$, as in (5).

9. Let $g(z) = c\left(z + \dfrac{1}{z}\right)$, where c is a positive constant. Write Re $g(z)$ and Im $g(z)$ in terms of x and y (where $z = x + iy$).

10. Re-do Figure 2.1-2, showing the first ten powers of each of the numbers α, β, and γ, instead of just the first seven.

11. For the number α of Figure 2.1-2, show in a diagram the negative powers α^{-k} for $k = 1, 2, \cdots, 7$. Do the same for β and γ.

12. In each case, consider the values of ζ and its powers in terms of polar coordinates, and show in a picture the points ζ, ζ^2, ζ^3, and ζ^4.

(a) $\zeta = 1$.

(b) $\zeta = i$.

(c) $\zeta = \frac{1+i\sqrt{3}}{2}$.

(d) $\zeta = \frac{9}{10}\frac{1-i\sqrt{3}}{2}$.

(e) $\zeta = 1 + i\sqrt{3}$.

(e) $\zeta = -\frac{i}{2}$.

13. With reference to Examples 2.1.2 and 2.1.3: Make a sketch showing the sixth roots of unity. In the same picture, show the number $32\sqrt{2} - i32\sqrt{2}$ and its sixth roots.

14. In each case, find the roots of the given equation and show them in a diagram.

(a) $z^3 = 1$.

(b) $z^3 = -1$.

(c) $z^3 = -8i$.

(d) $z^3 = 1 + i$.

(e) $z^3 = 1/8$.

(f) $z^4 = 8\sqrt{2} - 8\sqrt{2}i$.

(g) $z^2 = (1+i)^2$.

(h) $z^4 = (1+i)^3$.

(i) $z^7 = \frac{1-i}{2}$.

15. Sketch the lines given by the equations $x = 0$, $x = 1$, and $x = 2$. Identify and sketch the image of each line under the reciprocal map G.

16. Sketch the lines given by the equations $y = 0$, $y = 1$, and $y = 2$. Identify and sketch the image of each line under the reciprocal map G.

17. Since the reciprocal map is conformal, we expect it to map an orthogonal grid (defined in 1.1A) to an orthogonal grid. How is that fact illustrated by Exercises 15 and 16?

18. Let S be the circular arc $\{2e^{it} \mid 0 \le t \le \frac{\pi}{3}\}$. Identify its image under the reciprocal map G. Show in an accurate, well-labelled sketch the two sets S and $G[S]$.

19. Sketch the lines given by the equations $y = 0$, $y = \frac{1}{2}x$, $y = \frac{\sqrt{3}}{2}x$, and $x = 0$. Identify and sketch the image of each line under the reciprocal map G.

20. Sketch the circles with center 0 and radii $1/2, 1$, and 2. Identify and sketch the image of each circle under the reciprocal map G.

21. Since the reciprocal map is conformal, we expect it to map an orthogonal grid to an orthogonal grid. How is that fact illustrated by Exercises 19 and 20?

22. Find the equation of the form $\alpha\bar{z} + \bar{\alpha}z + c = 0$ for each of the lines given by these equations: $x = 0$, $y = 0$, $x = 1$, and $y = 1$.

23. For each of these values of c; $-2, -1, -1/2, 0, 1/2, 1$, and 2, sketch the line whose equation is $x + y = c$; then identify, sketch, and label its image under the reciprocal map G.

24. Sketch the three circles $(x-\frac{1}{2})^2 + y^2 = \frac{1}{4}$, $(x-1)^2 + y^2 = \frac{1}{4}$, $(x-2)^2 + y^2 = \frac{1}{4}$. Identify, sketch, and label their images under the reciprocal map G.

25. Show that if the points $+1$ and -1 both lie on the circle C, then $G[C] = C$. Notice that such a circle has center on the imaginary axis, say at ib, and radius $\sqrt{1 + b^2}$.

26. Make a sketch in the plane of the line L whose equation is $x + y = 1$, and show also the three points $-\frac{1}{2} + i\frac{1}{2}$, $1 + i$, and $2 + i$. For each point z, of the given three, find its reflection $J_L(z)$ and show it in the sketch.

27. In each case, make a sketch showing the given point z and the given line, then compute and show and the reflection of that point in that line.

(a) $z = 1$; $x = 0$.

(d) $z = -4$; $x + y = 4$.

(b) $z = -i$; $y = 0$.

(e) $z = \frac{1-i}{2}$; $x - y = 0$.

(c) $z = -4$; $x + y = 0$.

(f) $z = 0$; $x + y = 0$.

28. If equation (17) defines the mapping J_L, it appears at first glance that the definition may be different depending on the choice of the point $p \in L$. Show that in fact, the mapping is the same regardless of that choice.

29. For what pairs of lines L and M do the mappings J_L and J_M commute?

30. Show that if p and q are two distinct complex numbers, and if $0 < t < 1$, then

$$\left|\frac{z-p}{z-q}\right| = t \tag{20}$$

is the equation of a circle with center $\dfrac{p - t^2 q}{1 - t^2}$. What is the radius of the circle, in terms of p, q, and t? Is p inside the circle?

31. Let $p = 1, q = -1$. Sketch the circles C_t determined by (20) for $t = 1/3, 1/2, 1, 2, 3$. Prove that $G[C_t] = C_t$ for every $t > 0$.

32. In each case, make a sketch showing the given point z and its reflection in the circle of given center A and radius R. (See (18).)

   (a) $z = 2; A = 0, R = 1.$ (d) $z = 1 - i; A = 0, R = 1.$

   (b) $z = -4; A = 4, R = 4.$ (e) $z = -4; A = 0, R = 4.$

   (c) $z = \frac{1-i}{2}; A = 0, R = 1.$ (f) $z = 0; A = 5i, R = 5.$

33. In each case, make a sketch showing $z, \dfrac{1}{\bar{z}},$ and $\dfrac{1}{z}$.

   (a) $z = 2.$ (c) $z = 1 - i.$ (e) $z = -4.$

   (b) $z = -\frac{i}{4}.$ (d) $z = \frac{1-i}{2}.$ (f) $z = -\frac{3+i3}{2\sqrt{2}}.$

34. Prove the following result from plane geometry. Given points z and B, and a line that contains B but not z, there exists a unique circle Γ through z and B that is tangent to the line.

35. Prove the following result from plane geometry, which may be stated by saying, "A tangent squared equals the product of secants." Let A be a point outside the circle Γ. If the line segment $[A, B]$ is tangent to the circle Γ at the point B, and if a line through A intersects Γ in two points z and z_1, then $|z_1 - A| \cdot |z - A| = |A - B|^2$.

36. The following result provides another way of understanding the result in Problem 35. Let C be the circle with center A and radius R. Let K be a an arbitrary half line from A to ∞. Let z and z_1 be two distinct points on K such that $|z - A| > R$ and $|z - A||z_1 - A| = R^2$. Let Γ be an arbitrary circle through z and z_1, and let B be one of the two points in the intersection of C and Γ. Prove that the ratio of $|B - z|$ to $|B - z_1|$ is always the same.

37. Let C be the circle with center A and radius R. Let z and z_1 be any two distinct points such that $|z - A| > R$ and $z_1 = J_C(z)$. Prove that C consists of all the points B such that $\dfrac{|B - z|}{|B - z_1|} = \dfrac{|z_1 - A|}{R}$.

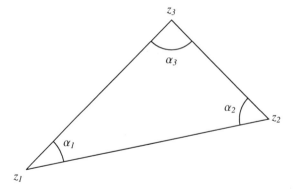

Figure 2.1-4   A picture to accompany Exercise 38. Interpret each angle α_k as counterclockwise, with $0 < \alpha_k < \pi$.

38. The fact that the angles of a triangle add up to π is imbedded in the definition of complex multiplication. Let's use (6) to prove it. Consider the triangle shown in Figure 2.1-4, with angles $\alpha_1, \alpha_2, \alpha_3$ and vertices z_1, z_2, z_3. The angles are all in the interval $(0, \pi)$. Explain why

$$\frac{z_3 - z_1}{z_2 - z_1} = \left(\left(\frac{|z_3 - z_1|}{|z_2 - z_1|}, \alpha_1 \right) \right).$$

Using the corresponding equality for each of the other vertices, and then multiplying the three equations together, we get

$$\frac{z_3 - z_1}{z_2 - z_1} \cdot \frac{z_1 - z_2}{z_3 - z_2} \cdot \frac{z_2 - z_3}{z_1 - z_3}$$
$$= \left(\left(\frac{|z_3 - z_1|}{|z_2 - z_1|} \cdot \frac{|z_1 - z_2|}{|z_3 - z_2|} \cdot \frac{|z_2 - z_3|}{|z_1 - z_3|}, \alpha_1 + \alpha_2 + \alpha_3 \right) \right).$$

It follows that $\alpha_1 + \alpha_2 + \alpha_3 = \pi$. Explain.

39. Do the analogue of Exercise 38 for a quadrilateral.

Help on Selected Exercises

1. (a) $zw = 8 + 2i$.

2. Four of the ten cases are shown in Figure 2.1-1.

3. For parts (a) and (d), see subsection B.

7. (a) The numbèr z in Figure 2.1-3 is $1 + i\sqrt{3}$. Its reciprocal is $\frac{1}{4} - i\frac{\sqrt{3}}{4}$. (c) The number w in the Figure is $\frac{1}{\sqrt{2}} - i\frac{\sqrt{3}}{2}$. Its reciprocal is $\frac{1}{\sqrt{2}} + i\frac{1}{\sqrt{2}}$.

8. (a) $\frac{55}{61} - i\frac{56}{61}$.   (c) $\frac{15}{17} + i\frac{8}{17}$.

9. These are the functions u_1 and v_1 of equation (2), page 353.

12. See Figure 2.1-2 for examples.

18. $G[S] = \{\frac{1}{2}e^{it} \mid -\pi/3 \le t \le 0\}$.

29. The two reflections can be written as $z \mapsto A_1\bar{z} + A_2$ and $z \mapsto B_1\bar{z} + B_2$. If the two maps commute, what does that imply about the constants, and hence about the two lines?

30. Write (20) in the form (13). What does (20) describe if $t > 1$? —if $t = 1$?

35. Show that the triangles $T(A, B, z_1)$ and $T(A, z, B)$ are similar. Let P be the center of the circle Γ. A key result from plane geometry gives $\angle(B, P, z_1) = 2\angle(B, z, z_1)$ and $\angle(B, P, z) = 2\angle(B, z_1, z)$.

36. The common value of those ratios is $\dfrac{R}{|A - z|} \equiv \dfrac{|A - z_1|}{R}$.

2.2   VISUALIZING POWERS, EXPONENTIAL, LOGARITHM, AND SINE

In Sections I.1–I.3, we introduced some ways to understand and visualize the behavior of mappings

$$f(x, y) = (u(x, y), v(x, y)). \tag{1}$$

Now that we have introduced complex structure and complex notation for the plane, we can think of such a mapping as a complex-valued function of a complex variable, writing $z = x + iy$ for (x, y) and $f(z) = u(z) + iv(z)$ instead of (1). We are primarily interested in the case when the domain of f is an open set $O \subseteq \mathbb{C}$ and v is a harmonic conjugate of u, which means that the Cauchy-Riemann equations are satisfied throughout O. In geometric terms, it means that at every point (x, y), if you turn the gradient of u to the left $90°$, then you obtain precisely the gradient of v. Assuming the gradients are everywhere nonzero, it follows that under the mapping $(x, y) \mapsto (u, v)$, the image of every orthogonal grid in O is an orthogonal grid.

   The twelve graphs in Figures 1.2-4, 1.2-6, 1.2-7, and 1.2-8 showed six examples of such a pair u, v. Here they are in complex notation:

$$z \mapsto z = x + i\, y;$$
$$z \mapsto z^2 = (x^2 - y^2) + i\, 2xy;$$
$$z \mapsto z^3 = (x^3 - 3xy^2) + i\, (3x^2y - y^3);$$
$$z \mapsto e^z = e^x \cos y + i\, e^x \sin y;$$
$$z \mapsto \sin z = \sin x \cosh y + i\, \cos x \sinh y;\ \text{and}$$
$$z \mapsto \mathrm{Log}\, z = \ln(x^2 + y^2)^{1/2} + i\, \mathrm{Arg}(x, y)\quad (x^2 + y^2 \ne 0).$$

For each of the first five, $\mathbb{C}$ is the domain. The last one, called the principal logarithm, is defined only for $z \neq 0$, and it suffers from being discontinuous everywhere on the negative real axis (see subsection B).

Those functions are all important and useful, and we shall take pains in this Section to help you understand and visualize them. We will also treat $z \mapsto \cos z$, $z \mapsto \sinh z$, and $z \mapsto \cosh z$.

In the figures just mentioned, the method for picturing each of the mappings $f(x, y) = u + iv$ was to draw separately the graphs of its real and imaginary parts, in x, y, u-space and x, y, v-space, respectively. In Figures 1.1-2, 1.1-3, 1.2-9, and 1.2-10, and in Section 1.3 for linear mappings, we introduced a more promising method, which is to select and sketch grids or other objects in an x, y-plane and their images in a u, v-plane. The trick is to pick objects which are well suited to the function.

A   Powers of z

The mapping $z \mapsto z^n$, where n is a positive integer, is easy enough to understand when we put both z and its image z^n in terms of their polar coordinates:

$$((r, \theta)) \mapsto ((r^n, n\theta)).$$

Two observations to make, and then to hold in mind:

- For each α, the image of the half-line H_α is the half-line $H_{n\alpha}$. The image of a line through the origin is another such line if n is even, a half-line if n is odd.

- Let C_r be the circle of radius r and center 0. Then its image is the circle C_{r^n}. So points inside the unit circle are mapped to points even closer to the origin; points outside the unit circle are mapped to points farther away; and the unit circle is its own image.

If a grid consists of arcs of circles C_r, and segments of half-lines H_α, then the image of that grid also consists of such arcs and segments, and is easy to draw by hand. Exercises 3–5 ask you to try it. To give you some help, a related exercise is done in Figure 2.2-1.

If $n > 1$ and $w \neq 0$, it is ambiguous to speak of the nth root of w, since there are n of them, as we discussed in detail in subsection 2.1C. There is exactly one solution z of the equation $w = z^n$ in every set of the form

$$\left\{ ((r, \theta)) \;\middle|\; r > 0 \text{ and } \theta_0 \leq \theta < \theta_0 + \frac{2\pi}{n}. \right\}$$

The mapping $z \mapsto z^n$ maps that set one-to-one onto $\mathbb{C} \setminus \{0\}$, which we call the punctured plane and denote by $\mathbb{C}'$.

Given an open set $O \subset \mathbb{C}$, there may or may not exist a continuous function $g : O \to \mathbb{C}$ such that $g(w)^n = w$ for each $w \in O$. If there does, then g is **an nth root on** O.

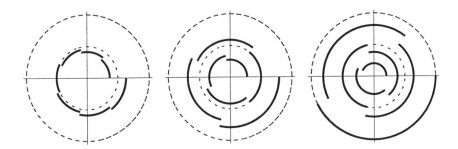

Figure 2.2-1 In each picture, the two dotted circles have center 0 and radii 1 and 2. Each of the seven arcs in the first picture is one-seventh of a circle with center 0 and radius R, where R is approximately $0.74, 0.83, 0.91, 1.00, 1.09, 1.17$, or $1.26 \approx 2^{1/3}$. The second picture shows the images of the seven arcs under the mapping $z \mapsto z^2$, or $((r, \theta)) \mapsto ((r^2, 2\theta))$. The third picture shows their images under the mapping $z \mapsto z^3$, or $((r, \theta)) \mapsto ((r^3, 3\theta))$.

B Exponential and Logarithms

The equation

$$e^{x+iy} := e^x \cos y + i e^x \sin y \tag{2}$$

defines the **exponential function** $z \mapsto e^z$, also written $z \mapsto \exp z$, in terms of familiar real-valued functions of a real variable: the sine, the cosine, and the exponential function $x \mapsto e^x$ (x real), whose graphs were represented in Figures 1.2-1 and 1.2-2. The value of $e^z \equiv e^{x+iy}$ can be written several other ways:

$$e^z = e^{x+iy} = (e^x \cos y, e^x \sin y) = ((e^x, y)).$$

The mapping is easy to understand geometrically when the domain point is written in Cartesian coordinates and its image in polar:

$$(x, y) \mapsto ((e^x, y)).$$

With x held fixed, equal to x_0 say, and letting y run from $-\infty$ to $+\infty$, the mapping is

$$(x_0, y) \mapsto ((e^{x_0}, y)).$$

Thus the exponential function maps the vertical line $x = x_0$ many-to-one onto (winds it around and around!) the circle with center 0 and radius e^{x_0}. More precisely, for every $b \in \mathbb{R}$, it maps the vertical segment $\{(x_0, y) \mid b \le y < b + 2\pi\}$ one-to-one onto that circle. For example, every interval on the y-axis of the form $\{(0, y) \mid b \le y < b + 2\pi\}$ is mapped one-to-one onto the unit circle.

Let $H'_\alpha = \{((r, \alpha)) \mid r > 0\}$, which is the half-line H_α with the origin removed. We will call such a set a **ray** from the origin.

With y held fixed, equal to y_0 say, and letting x run from $-\infty$ to $+\infty$, the mapping is

$$(x, y_0) \mapsto ((e^x, y_0)).$$

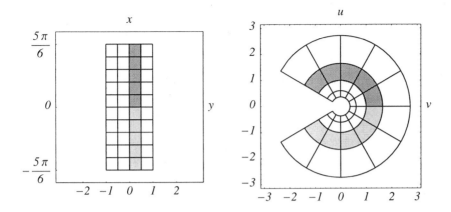

Figure 2.2-2 A grid in the x, y-plane and its image in the u, v-plane under the mapping $z \mapsto e^z$.

Thus the exponential function maps the horizontal line $y = y_0$ one-to-one onto the ray H'_{y_0}. The part of the line $y = y_0$ that lies in the open left half-plane is mapped onto the part of H'_{y_0} that lies within the unit circle, since $x < 0 \Rightarrow e^x < 1$. The part of the line $y = y_0$ that lies in the right half-plane is mapped onto the part of H'_{y_0} that lies outside the unit circle, since $x > 0 \Rightarrow e^x > 1$.

Figure 2.2-2 shows a grid of selected vertical and horizontal line segments in the x, y-plane and their images under the exponential map. Their images form an orthogonal polar grid in the u, v-plane. As we proceed, you may wish to draw some pictures for yourself.

The origin is not in the range because $e^x > 0$ for all x. Thus the exponential function maps $\mathbb{C}$ many-to-one onto $\mathbb{C}'$, by which we mean that it takes infinitely many z to each $w \neq 0$. More precisely, it maps every horizontal strip of height 2π—that is, every set of the form

$$\{x + iy \mid y_0 \leq y < y_0 + 2\pi\}, \tag{3}$$

one-to-one onto $\mathbb{C}'$. It maps the open left half of that strip (where $x < 0$) one-to-one onto the punctured disk $D'(0, 1)$—and the open right half (where $x > 0$) one-to-one onto the set of all $z \in \mathbb{C}$ with $|z| > 1$.

For comparison's sake, note that for each integer $n \geq 1$, the function $z \mapsto z^n$, or $((r, \theta)) \mapsto ((r^n, n\theta))$, is n-to-1 from $\mathbb{C}'$ onto itself (while also taking 0 to 0).

We will say "z is a logarithm of w" if $e^z = w$, which is equivalent to saying that $e^x = |w|$ (equivalently, $x = \ln |w|$) and $e^{iy} = w/|w|$ (equivalently, $y = \arg w$).

For a given $w = u + iv \neq 0$, let $L(w)$ denote the set of all logarithms of w. As we have seen, every nonzero complex number w has exactly one logarithm in every set of the form (3). The one that occurs in the strip $-\pi < y \leq \pi$ is **the principal logarithm** of w, for which we write $\mathrm{Log}\, w$. Note that $\mathrm{Log}\, w = \ln |w| + \mathrm{Arg}\, w$. The set of all logarithms of w is

$$L(w) = \big\{\ln |w| + i(\text{Arg } w + 2\pi in) \,\big|\, n \text{ is an integer}\big\}.$$

We have been speaking of logarithms of one complex number w. In what follows, we speak of logarithms as functions on a set.

If $g : O \to \mathbb{C}$ is continuous, where O is an open set in the plane, and if $e^{g(w)} = w$ for all $w \in O$, then g is **a logarithm** (or, one sometimes says, "a branch of the logarithm") on O, and we may write $\log z$ instead of $g(z)$. Recall the discussion in 1.3B. The ambiguity in "$\log z$" is due precisely to the need to specify the argument. There is a logarithm on the set O if and only if there is a argument on O. Thus if there is one logarithm on O, there are many. Here are some easy but important observations.

- If there is a logarithm on O, then O cannot contain the origin, because 0 is not in the range of the exponential function.

- If O contains a circle C, and if 0 is inside C, then there is no logarithm on O, because there is no argument on O.

- If there is a logarithm $z \mapsto \log z$ on an open set O, then for every integer n there is an nth root on O, namely, $z \mapsto e^{(\log z)/n}$.

- On each of the open sets in Figure 1.3-2, there is a logarithm (and hence infinitely many logarithms).

If there is a logarithm on the connected set O, then its value at any one point $p \in O$ determines which logarithm it is. Often we will write simply "log" and hold in mind which of the possible logarithms we are using. At other times we will need our notation for the logarithm to carry the information for us, and for that purpose we will indicate in a subscript its value at some one point. Thus for example, $\log_{1 \mapsto 0}$ is the logarithm on V_π whose value at 1 is 0. The other possibilities on a set containing 1 are, of course, $\log_{1 \mapsto i2\pi n}$ for $n \in \mathbb{Z}$.

At this point we will abandon the notation $((r, \theta))$ for a point in the plane with polar coordinates r and θ, since $re^{i\theta}$ is now defined and will do nicely.

C   Sin z

We assume that these four real-valued functions of one real variable are well understood:

$$\sin : \mathbb{R} \to [-1, 1], \quad \cos : \mathbb{R} \to [-1, 1], \quad \cosh : \mathbb{R} \to [1, \infty), \quad \sinh : \mathbb{R} \to \mathbb{R}.$$

They were discussed in subsection 1.2C and pictured in Figures 1.2-1 and 1.2-3. In what follows, we will define the extension of the sine to a complex-valued function from $\mathbb{C}$ to $\mathbb{C}$ and discuss its behavior in detail. In the next subsection, we will describe the extensions of the other three and discuss the relations among the four.

We define $\sin z$ for all $z \in \mathbb{C}$ by:

$$\sin(x + iy) := \sin x \cosh y + i \cos x \sinh y. \tag{4}$$

Since the sine and cosine are 2π-periodic and π-antiperiodic on $\mathbb{R}$, so is the sine on $\mathbb{C}$: For all x, y,

$$\sin(x + 2\pi + iy) = \sin(x + iy) \quad \text{and} \quad \sin(x + \pi + iy) = -\sin(x + iy).$$

The second identity, which implies the first, might be restated as follows:

$$\sin(x + iy) = u + iv \quad \Longrightarrow \quad \sin(x + \pi + iy) = -u - iv. \tag{5}$$

Since the cosh and the cosine are even, and the sinh and the sine are odd, it also follows from (4) that

$$\sin(x + iy) = u + iv \quad \Longrightarrow \quad \begin{aligned} \sin(x - iy) &= u - iv \quad \text{and} \\ \sin(-x + iy) &= -u + iv. \end{aligned} \tag{6}$$

Thus the sine commutes with complex conjugation, also known as reflection in the real axis; and it also commutes with reflection in the imaginary axis. If we knew the definition of the sine only on the vertical half-strip

$$V^+ := \left\{ x + iy \mid 0 \le x \le \frac{\pi}{2} \text{ and } y \ge 0 \right\}, \tag{7}$$

and if we also knew that it satisfied (5) and (6), then we could recover the definition everywhere in the plane.

Figure 2.2-3 shows a certain grid in the x, y-plane and its image in the u, v-plane under the sine. As the Figure suggests, and as we shall explain, the sine maps each horizontal line segment of width 2π onto an ellipse, except that it maps the real axis onto the interval $[-1, 1]$ (which might be regarded as a degenerate ellipse). With some exceptions, each vertical line one-to-one onto one half of a hyperbola. We need to remind you of certain facts about ellipses and hyperbolas.

The equation of an ellipse in standard position in the u, v-plane, with the u-axis as its major axis, is

$$\frac{u^2}{a^2} + \frac{v^2}{b^2} = 1, \tag{8}$$

where $a > b > 0$. The ellipse crosses the u-axis at $(\pm a, 0)$ and the v-axis at $(0, \pm b)$. Its focii are $(\pm c, 0)$, where $c^2 = a^2 - b^2$. A point is on the ellipse if and only if the sum of its distances to the two focii equals $2a$, that is,

$$2a = \sqrt{(u + c)^2 + v^2} + \sqrt{(u - c)^2 + v^2}. \tag{9}$$

The equation of a hyperbola in standard position, with the u-axis as its major axis, is

$$\frac{u^2}{a^2} - \frac{v^2}{b^2} = 1. \tag{10}$$

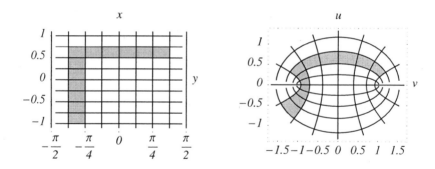

Figure 2.2-3   A rectangular grid and its image under the mapping $z \mapsto \sin z$

The hyperbola crosses the u-axis at $(\pm a, 0)$. Its focii are $(\pm c, 0)$, where $c^2 = b^2 + a^2$. and it has asymptotic lines $v = \pm \dfrac{b}{a} u$. A point is on the hyperbola if and only if the difference of its distances to the two focii equals $2a$, that is,

$$\pm 2a = \sqrt{(u+c)^2 + v^2} - \sqrt{(u-c)^2 + v^2}. \tag{11}$$

We will investigate the action of the sine mapping (4) first on vertical lines, then on horizontal lines.

Consider an arbitrary vertical line L, given by $x = x_0$. Let

$$a = \sin x_0, \qquad b = \cos x_0.$$

Restricted to L, the mapping is given by

$$y \mapsto (u, v) \quad \text{where} \quad u = a \cosh y, \quad v = b \sinh y \quad (y \in \mathbb{R}). \tag{12}$$

Consider three cases.

- If x_0 is not an integer times $\pi/2$, then a and b are both nonzero, and every point $(u, v) \in \sin[L]$ satisfies equation (10) (because of the identity $\cosh^2 y - \sinh^2 y = 1$). In fact, (12) gives a parametrization of one half of the hyperbola—the right half if $a > 0$, the left half if $a < 0$. If $b > 0$, the lower part of the line (where $y < 0$) maps to the lower part of the hyperbola (where $v > 0$) and the upper part to the upper part; and *vice versa* if $b < 0$.

- If $x_0 = n\pi$ $(n \in \mathbb{Z})$, then $a = 0$ and $b = (-1)^n$. Therefore the mapping is $(n\pi, y) \mapsto (0, (-1)^n \sinh y)$ and maps L one-to-one onto the v-axis. If n is even, it maps the lower part of L (where $y < 0$) to the lower part of the v-axis (where $v < 0$) and the upper part to the upper part; and *vice versa* if n is odd.

- If x_0 is an odd multiple of $\pi/2$, so that $x_0 = (2n+1)\pi/2$ for some $n \in \mathbb{Z}$, then $a = (-1)^n$ and $b = 0$. Therefore the mapping is $((2n+1)\pi/2, y) \mapsto ((-1)^n \cosh y, 0)$ and maps L two-to-one onto an infinite interval on the u-axis: $[1, \infty)$ if n is even, $(-\infty, -1]$ if n is odd.

Figure 2.2-3 illustrates the vertical-line analysis; note the vertical line segments in the domain and their hyperbola images in the range.

The discussion so far provides a rather complete understanding of how the mapping behaves. We may conclude, for example, that the vertical half-strip V^+ defined by (7) is mapped one-to-one onto the closed positive quadrant (where $u \geq 0$ and $v \geq 0$). The vertical half-strip

$$V := \left\{ x + iy \ \middle| \ -\frac{\pi}{2} \leq x \leq \frac{\pi}{2} \text{ and } y \geq 0 \right\}, \tag{13}$$

which is the union of V^+ with its reflection in the y-axis, is mapped one-to-one onto the closed upper half-plane (where $v \geq 0$).

Now consider an arbitrary horizontal line K, given by $y = y_0$. Let

$$a = \cosh y_0, \qquad b = \sinh x_0.$$

Restricted to K, the mapping is given by

$$x \mapsto (u, v) \quad \text{where} \quad u = a \sin x, \quad v = b \cos x \quad (x \in \mathbb{R}). \tag{14}$$

Consider two cases.

- If $y_0 = 0$ (so that K is the x-axis), then $a = 1$ and $b = 0$. Therefore $\sin[K]$ is the interval $[-1, 1]$ on the u-axis. In fact, the sine maps each interval $((2n-1)\pi/2, (2n+1)\pi/2]$ on the x-axis one-to-one onto $[-1, 1]$, left-to-right or right-to-left according to whether n is even or odd.

- If $y_0 \neq 0$, then $a > 1$, b is nonzero, and b has the same sign as y_0. Every point $(u, v) \in \sin[L]$ satisfies equation (8) because of the identity $\sin^2 x + \cos^2 x \equiv 1$. In fact, (14) gives a many-to-one parametrization of the ellipse described by (8), going clockwise if $y_0 > 0$, counterclockwise if $y_0 < 0$. Several horizontal segments $\{(x, y_0) \mid -\pi/2 < x < \pi/2\}$ and their elliptical-arc images are shown in Figure 2.2-3.

If you follow the analysis and allow the figure (or other pictures you may draw) to serve as a pictorial shorthand in your mind for the action of the function $z \mapsto \sin z$, then you should be able to use the mapping correctly, and to answer questions about it successfully.

Question: Where does $\sin(x + iy) = 0$? Answer: Precisely where $y = 0$ and x is an integer times π. If you do not "see" the answer immediately, then you can get it straightforwardly by this reasoning: $\sin(x + iy) = 0$ if and only if $\sin x \cosh y = 0$ and $\cos x \sinh y = 0$. Since the cosh is never zero, the first equation implies that $\sin x = 0$ and hence $x = n\pi$. But then $\cos x = \pm 1$, hence by the second equation $\sinh y = 0$ and hence $y = 0$.

Question: Where does $\sin(x + iy) = 2$? You may be able to visualize the answer-set; but one may proceed as follows: $\sin(x + iy) = 2$ if and only if $\sin x \cosh y = 2$ and $\cos x \sinh y = 0$. The second equation implies that either (1) $y = 0$ or (2) $x = n\pi/2$. If (1), then the first equation gives $\sin x = 2$, which never happens. So

(2) holds, and by the first equation, since $\cosh y \geq 1$ for all y, $\sin x > 0$ and hence $x = (4n + 1)\pi/2$, $\sin x = +1$, so $\cosh y = 2$. So the answer-set is

$$\left\{ \frac{(4n + 1)\pi}{2} + i\ln(2 \pm \sqrt{3}) \,\middle|\, n \in \mathbb{Z} \right\}. \tag{15}$$

Exercise 22 asks you to verify the "$\ln(2 \pm \sqrt{3})$."

D The Cosine and Sine, and the Hyperbolic Cosine and Sine

These four functions are closely related, and each can be written in terms of any one of the others, as a composition of it with shifts and/or rotations, so that if you understand one, you understand them all.

The four functions are extended from $\mathbb{R}$ to all of $\mathbb{C}$ as follows.

$$\cos(x + iy) := \cos x \cosh y - i \sin x \sinh y;$$
$$\sin(x + iy) := \sin x \cosh y + i \cos x \sinh y;$$
$$\cosh(x + iy) := \cosh x \cos y + i \sinh x \sin y;$$
$$\sinh(x + iy) := \sinh x \cos y + i \cosh x \sin y.$$

Upon inspecting those definitions, one notices certain relationships.

- It follows from familiar trigonometric identities ((2) in Section 1.2) that:

$$\cos(x + iy) = \sin(x + \frac{\pi}{2} + iy) \quad \text{for all} \quad x, y; \quad \text{equivalently,}$$

$$\cos z = \sin(z + \frac{\pi}{2}) \quad \text{for all} \quad z \in \mathbb{C}.$$

Thus if you move to the right $\pi/2$ units and then apply the sine, you have the cosine.

- Since $i(x + iy) = -y + ix$,

$$\sinh z = -i \sin iz \quad \text{for all} \quad z \in \mathbb{C}.$$

Thus the hyperbolic sine is the composition of (1) rotation by $\pi/2$ radians, (2) the sine, and (3) rotation by $-\pi/2$ radians.

- Again because $i(x + iy) = -y + ix$,

$$\cosh z = \cos iz \quad \text{for all} \quad z \in \mathbb{C},$$

Thus the hyperbolic cosine is the composition of (1) rotation by $\pi/2$ radians and (2) the cosine.

- One might also notice that

$$\cosh z = \sin\left(iz + \frac{\pi}{2}\right)$$

Thus the hyperbolic cosine is the composition of (1) rotation by $\pi/2$ radians, (2) a shift to the right $\pi/2$ units, and (3) the sine.

An alternative, equivalent scheme for defining the four functions is to write them all in terms of the exponential function:

$$\sin z = \frac{e^{iz} - e^{-iz}}{2i}, \qquad \sinh z = \frac{e^z - e^{-z}}{2},$$

$$\cos z = \frac{e^{iz} + e^{-iz}}{2}, \qquad \cosh z = \frac{e^z + e^{-z}}{2}.$$

Exercises

1. Prove from the definition (2) that $e^{z+w} = e^z e^w$ whenever z and w are complex numbers. Use trigonometric identities, and the well-known property of the real exponential according to which $e^{s+t} = e^s e^t$ whenever s and t are real.

2. When $b > 0$ and s is real, we may define $b^s := e^{s \ln b}$. Thus we know that $e^{st} = (e^s)^t$ whenever s and t are real. But this identity does not extend to complex numbers. Give an example of complex numbers z and w such that $e^{zw} \neq (e^z)^w$.

3. Sketch carefully the image of the polar grid shown in Figure 2.2-4 under the mapping $z \mapsto z^2$. You will find this problem easy to do by hand, much more so than the drawing of Figures 1.1-2 and 1.1-3, which illustrate the action of the same mapping on the same region in the plane. In those computer-generated figures, the grids were chosen in order to emphasize the behavior of the mapping near a central point, $(1, 0)$.

4. Sketch carefully another polar grid which has the same image as the one in Figure 2.2-4 under the mapping $z \mapsto z^2$.

5. Sketch carefully the image of the polar grid shown in Figure 2.2-4 under the mapping $z \mapsto z^3$.

6. Sketch carefully the image of the polar grid shown in Figure 2.2-4 under the mapping $z \mapsto z^8$.

7. The mapping $z \mapsto z^2$ is a two-to-one mapping from $\mathbb{C}'$ to $\mathbb{C}'$. That is, every point except 0 is the image of two distinct nonzero points. In Figure 2.2-1, one arc in the first picture consists of all points $((0.83, \theta))$ with $\frac{2\pi}{7} \leq \theta \leq \frac{4\pi}{7}$. Its image consists of all points $((0.68, \theta))$ with $\frac{4\pi}{7} \leq \theta \leq \frac{8\pi}{7}$. This arc, shown in the second picture, is also the image of another, different arc which is not shown. Describe and sketch carefully this arc.

8. In the third picture of Figure 2.2-1 is the arc $\{((.41, \theta)) \mid 0 \leq \theta \leq \frac{6\pi}{7}\}$. It is the image under $z \mapsto z^3$ of each of three arcs, only one of which is shown in the first picture. Identify the three arcs, and show them in an accurate sketch.

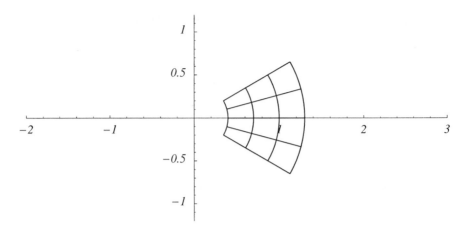

Figure 2.2-4 This polar grid, referred to in Exercises 3–5, consists of arcs with radii 0.4, 0.7, 1.0, and 1.3; and line segments belonging to lines making angles to the horizontal axis $\pm\frac{\pi}{6}, \pm\frac{\pi}{12}$, and 0.

9. Referring again to Figure 2.2-1, sketch as accurately as you can by hand a diagram that shows the images of the original arcs 1 through 7 under the mapping $z \mapsto z^4$.

10. Referring again to Figure 2.2-1, justify and explain this statement: The images of the original arcs 1 through 7 under the mapping $z \mapsto z^7$ are all complete circles. For what other powers, besides 7, will the statement be true?

11. Identify and show in a careful, well-labelled sketch the image under the exponential function of the horizontal half-strip

$$H := \{x + iy \mid x > 0 \text{ and } 0 < y < \frac{\pi}{4}\}.$$

12. Describe and sketch the image under the exponential map of the line $y = x$.

13. Let a, b, and c be three distinct points on a circle with center 0. Let $\tau = \angle(b, 0, c)$, $\sigma = \angle(b, a, c)$. Make a sketch! A theorem of elementary geometry states that $\tau = 2\sigma$. Prove this result using these facts from complex arithmetic:

$$\frac{b - a}{c - a} = \frac{|b - a|}{|c - a|} e^{i\sigma} \quad \text{and} \quad \frac{b}{c} = e^{i\tau}.$$

14. If the function given by $f(z) = e^{e^z}$ is written in the form $u(x, y) + iv(x, y)$, identify $u(x, y)$ and $v(x, y)$.

15. Explain: If there is one logarithm defined on a set, then there are infinitely many.

16. For the set shown on the right in Figure 1.3-2, page 38, if $\log(20)$ is real, what must be the value of $\log(10i)$? —of $\log(-10i)$?

17. Describe thoroughly how the mapping $z \mapsto e^{iz}$ acts on the plane. In particular, describe the images of horizontal and vertical lines.

18. Let O be an open connected subset of the plane that does not contain the origin. Show that there is a logarithm on O if and only if there is an antiderivative of the function $z \to 1/z$ on O.

19. Consider the rectangle $R := [-\pi, \pi] \times [1, 2]$ in the z-plane. Identify its image under the sine mapping. Is the mapping one-to-one on R? Is it one-to-one on the interior of R?

20. Move the rectangular grid in Figure 2.2-3 to the right $\pi/2$ units, obtaining a new rectangular grid. Sketch the image of the new grid under the mapping $z \mapsto \sin z$. Indicate which region is the image of which.

21. Identify and sketch the set of all $z \in \mathbb{C}$ such that $\sin z = i$.

22. Find both real solutions y of the equation $\cosh y = 2$.

23. Identify and sketch the set of all $z \in \mathbb{C}$ such that $\cos z = 2$.

24. Sketch the image of the rectangular grid in Figure 2.2-3 under the mapping $z \mapsto \cos z$, and indicate which region is the image of which.

25. This exercise and the next produce a set of expressions for the real and imaginary parts of $\sin^{-1}$ on the upper half-plane. The sine function maps the vertical half-strip V^+, defined by (7), one-to-one onto the closed positive quadrant $\bar{Q}_I := \{u + iv \mid u \geq 0 \text{ and } v \geq 0\}$. Written without benefit of complex notation, the sine is given by

$$u = \sin x \cosh y, \quad v = \cos x \sinh y. \qquad (16)$$

Given $(u, v) \in Q_I$, solve (16) for $(x, y) \in V^+$. You should obtain:

$$x = \sin^{-1}\left(\frac{\sqrt{(u+1)^2 + v^2} - \sqrt{(u-1)^2 + v^2}}{2}\right); \qquad (17)$$

$$y = \cosh^{-1}\left(\frac{\sqrt{(u+1)^2 + v^2} + \sqrt{(u-1)^2 + v^2}}{2}\right) \qquad (18)$$

26. The sine function maps the vertical half-strip V (see (13)) one-to-one onto the closed upper half-plane $\mathbb{Y} := \{u + iv \mid v \geq 0\}$. Given $(u, v) \in H$, show that the solution of (16) for $(x, y) \in V$ is still given by equations (17) and (18).

27. Consider the function given by $f(z) := z + e^z$ restricted to the horizontal strip $A := \{x + iy \mid -\pi < y < \pi\}$. Either by hand or with a computer, sketch the images of a number of horizontal lines and vertical line segments that lie in A. Describe $f[A]$.

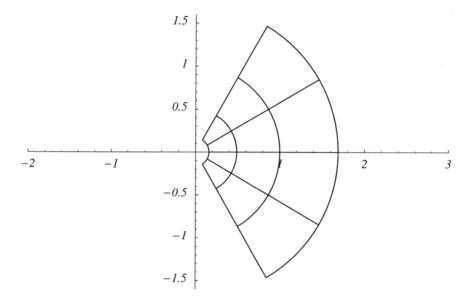

Figure 2.2-5 This polar grid consists of arcs with radii 0.16, 0.49, 1.0, and 1.69; and line segments belonging to lines making angles to the horizontal axis $\pm\frac{\pi}{3}, \pm\frac{\pi}{6}$, and 0.

Help on Selected Exercises

2. $e^{i2\pi} = 1; e^{(i2\pi)^2} = e^{-4\pi^2} \neq 1$. What does $(e^{i2\pi})^{i2\pi}$ equal? When z and w are complex numbers, we define $z^w := e^{w \log z}$, but in so doing we must specify which logarithm is intended.

3. See Figure 2.2-5.

7. The other arc consists of the points $((0.827, \theta))$ such that $\frac{-5\pi}{7} \leq \theta \leq \frac{-3\pi}{7}$.

11. $e^H = \{((r, \theta)) \mid r > 1 \text{ and } 0 < y < \frac{\pi}{4}\}$.

13. Let r be the radius of the circle. Then

$$e^{i2\sigma} = \frac{(b-a)^2}{(c-a)^2} \frac{|c-a|^2}{|b-a|^2} = \frac{b-a}{c-a} \frac{\bar{c}-\bar{a}}{\bar{b}-\bar{a}}$$
$$= \frac{(b-a)(\frac{1}{c}-\frac{1}{a})}{(c-a)(\frac{1}{b}-\frac{1}{a})} = \frac{(b-a)\frac{a-c}{ac}}{(c-a)\frac{a-b}{ab}} = \frac{b}{c} = e^{i\tau}.$$

14. $e^{e^{x+iy}} = e^{e^x \cos y + ie^x \sin y} = e^{e^x \cos y} \cos(e^x \sin y) + ie^{e^x \cos y} \sin(e^x \sin y).$

16. $\ln 10 - i\frac{3\pi}{2}$ and $\ln 10 - i\frac{5\pi}{2}$, respectively.

18. If $e^{g(z)} = z$ for $z \in O$, then taking the derivative gives $e^{g(z)}g'(z) = 1$, which implies that $g'(z) = 1/z$. Conversely, if $h'(z) = 1/z$ for $z \in O$, then

$$\frac{d}{dz}\frac{e^{h(z)}}{z} = \frac{e^{h(z)}h'(z)z - e^{h(z)}}{z^2} \equiv 0$$

and therefore $e^{h(z)} = cz$ for $z \in O$, for some nonzero constant c. The function g given by $g(z) := h(z) - \log c$ is a logarithm on O.

21. It is the set of all numbers $n\pi + i\ln(1 + \sqrt{2})$ where n is an even integer, or $n\pi + i\ln(-1 + \sqrt{2})$ where n is an odd integer. *Note:* $\ln(1 + \sqrt{2}) \equiv -\ln(-1 + \sqrt{2})$.

22. Notice that the hyperbolic cosine is an even function which is two-to-one from $\mathbb{R} \setminus \{0\}$ to the interval $(1, \infty)$. Accordingly, $\cosh y = r$ has two real solutions for each real $r > 1$. Suggestion: Rewrite the equation as $e^y + e^{-y} = 4$, then multiply it by e^y, obtaining a quadratic in e^y. See equation (15) for the answer.

25. Hints for obtaining (17):

- Let $0 < x < \frac{\pi}{2}, a = \sin x, b = \cos x$. Then u and v satisfy (10), and (17) follows from (11).

- To verify that (17) holds also when $x = 0$ or $\pi/2$, begin by observing that $x = 0$ if and only if $u = 0$; and that $x = \frac{\pi}{2}$ if and only if $v = 0$ and $u \geq 1$.

Hints for obtaining (18):

- Let $t \geq 0, a = \cosh t, b = \sinh t$. Then u and v satisfy (8), and (18) follows from (9).

- To verify that (18) holds also when $t = 0$, begin by observing that $t = 0$ if and only if $v = 0$ and $0 \leq u \leq 1$.

26. Note that the right-hand sides of (17) and (18) are, respectively, odd and even in the variable u; and $\sin(x + iy) = u + iv$ if and only if $\sin(-x + iy) = -u + iv$.

27. The function f maps A one-to-one onto the plane with these two half-lines removed: $\{x + i\pi \mid x \leq -1\}$ and $\{x - i\pi \mid x \leq -1\}$.

2.3 DIFFERENTIABILITY

A Differentiability at a Point

Now that we have introduced the complex number system $\mathbb{C}$, we are prepared to define differentiability for functions from $\mathbb{C}$ to $\mathbb{C}$. First, so that you can see how they are related, we will review and re-state the basics of differentiability for the familiar case of functions of one or two real variables. Please notice that in every case, differentiability implies continuity.

The essence of differentiability at a point is always *approximability by a linear mapping*. In subsections 1.3C and 2.1F we took pains to identify linear mappings. We may summarize as follows. The linear mappings from $\mathbb{R}$ to $\mathbb{R}$ are those of the form

$$x \mapsto ax,$$

where a is a real constant. The linear mappings from $\mathbb{R}^2$ to $\mathbb{R}$ are those of the form

$$\begin{pmatrix} x \\ y \end{pmatrix} \mapsto \begin{pmatrix} a & b \end{pmatrix} \begin{pmatrix} x \\ y \end{pmatrix} = ax + by,$$

where a and b are real constants. The linear mappings from $\mathbb{R}^2$ to $\mathbb{R}^2$ are those of the form

$$\begin{pmatrix} x \\ y \end{pmatrix} \mapsto \begin{pmatrix} a & b \\ c & d \end{pmatrix} \begin{pmatrix} x \\ y \end{pmatrix} = \begin{pmatrix} ax + by \\ cx + dy \end{pmatrix},$$

where a, b, c, and d are real constants. The linear mappings from $\mathbb{C}$ to $\mathbb{C}$ are those of the form

$$z \mapsto Az,$$

where $A = a + ic$ is a complex constant, or, without the complex notation,

$$\begin{pmatrix} x \\ y \end{pmatrix} \mapsto \begin{pmatrix} a & -c \\ c & a \end{pmatrix} \begin{pmatrix} x \\ y \end{pmatrix} = \begin{pmatrix} ax - cy \\ cx + ay \end{pmatrix}.$$

Consider first the case of a real-valued function of one real variable. Introductory calculus courses (see, for example, [67, Section 2.8] or [27, Section 2.2]) usually give this definition:

2.3.1. Definition. Let $u : I \to \mathbb{R}$, where I is an open subset of $\mathbb{R}$, and let $x_0 \in I$. The function u is **differentiable at** x_0 if there is a real number a such that

$$a = \lim_{x \to x_0} \frac{u(x) - u(x_0)}{x - x_0}. \tag{1}$$

The number a is then **the derivative of** u **at** x_0, and is denoted by $u'(x_0)$. If u is differentiable ($u'(x_0)$ is defined) for every $x_0 \in I$, then u is **differentiable on** I and the function $u' : I \to \mathbb{R}$ is **the derivative of** u **on** I.

The limit statement (1) can be rewritten in other, equivalent forms. Here are three:

$$\lim_{x \to x_0} \eta(x) = 0, \quad \text{where } \eta(x) := \frac{u(x) - u(x_0)}{(x - x_0)} - a. \tag{2}$$

$$u(x) - u(x_0) = (a + \eta(x)) \cdot (x - x_0) \quad \text{and} \quad \lim_{x \to x_0} \eta(x) = 0. \tag{3}$$

$$u(x) - u(x_0) = a \cdot (x - x_0) + \rho(x) \quad \text{and} \quad \lim_{x \to x_0} \frac{\rho(x)}{|x - x_0|} = 0, \tag{4}$$

$\rho(x)$ being the same as $\eta(x) \cdot (x - x_0)$.

The difference quotient on the right-hand side of (1) is useful sometimes for computing a, but (4) asserts more plainly and succinctly the idea of linear approximability; it says that the function $x - x_0 \mapsto u(x) - u(x_0)$ is approximated by the linear function $x - x_0 \mapsto a \cdot (x - x_0)$; and that the error we make when we use that approximation tends to 0 faster than $x - x_0$, in the sense that it tends to 0 even when divided by $|x - x_0|$. In geometric terms: the tangent line approximates the function's graph nicely near the point. See, for example, the first graph of Figure 1.2-3, and its description in subsection 1.2C

Next we state the standard modern definition of differentiability in the case of a real-valued function of two real variables. As you will see, it is formulated in the spirit of (4).

2.3.2. Definition. Let $u : O \to \mathbb{R}$, where O is an open set in $\mathbb{R}^2$, and let $(x_0, y_0) \in O$. Then u is **differentiable at** (x_0, y_0) if there exist real numbers a, b such that

$$u(x, y) - u(x_0, y_0) = a \cdot (x - x_0) + b \cdot (y - y_0) + \rho(x, y), \tag{5}$$

$$\text{where} \quad \lim_{(x,y) \to (x_0, y_0)} \frac{\rho(x, y)}{|(x - x_0, y - y_0)|} = 0.$$

In case it's not already clear: Equation (5) *defines* $\rho(x, y)$ as the error made when we approximate the function $(x - x_0, y - y_0) \mapsto u(x, y) - u(x_0, y_0)$ by the linear function $(x - x_0, y - y_0) \mapsto a \cdot (x - x_0) + b \cdot (y - y_0)$; and the statement after it says that the ratio of the error to the distance between (x, y) and (x_0, y_0) tends to 0.

What does Definition 2.3.2 have to do with the existence of the partial derivatives of u at the point? Starting with the definition of the partial derivatives of u with respect to x and y at the point (x_0, y_0) (see, for example, [27, Section 13.1] or [67, Section 14.3]) and then using (5), we may compute as follows:

$$u_x(x_0, y_0) = \lim_{x \to x_0} \frac{u(x, y_0) - u(x_0, y_0)}{x - x_0} \tag{6}$$

$$= \lim_{x \to x_0} \frac{a \cdot (x - x_0) + \rho(x, y_0)}{x - x_0} = a$$

and

$$u_y(x_0, y_0) = \lim_{y \to y_0} \frac{u(x_0, y) - u(x_0, y_0)}{y - y_0} \tag{7}$$

$$= \lim_{y \to y_0} \frac{b \cdot (y - y_0) + \rho(x_0, y)}{y - y_0} = b$$

Thus condition (5) implies the existence of the two partial derivatives of u at (x_0, y_0), and they equal a and b, respectively.

The converse is not true; the mere existence of the two partial derivatives at the one point (x_0, y_0) does not imply (5). To understand, notice that the existence of the two limits (6) and (7) depends only on the definition and behavior of $u(x, y)$ at points on the vertical and horizontal lines $y = y_0$ and $x = x_0$, respectively. For an example, see Exercise 1. However, if u and its partials are defined and continuous at every point of an open set, then at every point u is differentiable in the sense of Definition 2.3.2. For a proof, see, for example, [8, Theorem 8, p. 131]. The two partials are then the coordinate functions of a continuous vector-valued function on O, the **gradient** of u:

$$\mathbf{grad}\ u = (u_x, u_y).$$

An alternative notation for **grad** u is ∇u. You will find discussions of the gradient in the multidimensional chapters of calculus books, for example [67, Section 12.6] and [27, Section 13.4]. At each point (x_0, y_0), the dot product of the gradient with $(x - x_0, y - y_0)$ gives us the linear approximation:

$$
\begin{aligned}
u(x,y) - u(x_0, y_0) &\approx (u_x(x_0, y_0), u_y(x_0, y_0)) \cdot (x - x_0, y - y_0) \\
&= a \cdot (x - x_0) + b \cdot (y - y_0),
\end{aligned}
\tag{8}
$$

where $a := u_x(x_0, y_0)$ and $b := u_y(x_0, y_0)$. Among unit vectors $((x - x_0), (y - y_0))$, which one maximizes (8)? In other words, in which direction does the value of u increase most rapidly? By Proposition 1.3.1 the answer is: Precisely in the direction of the gradient (a, b). In the directions perpendicular to the gradient (the direction of $(a, -b)$ or $(-a, b)$), (8) equals 0.

In matrix notation, the value of the approximating linear function in (5) is:

$$
\begin{pmatrix} a & b \end{pmatrix} \cdot \begin{pmatrix} x - x_0 \\ y - y_0 \end{pmatrix}.
$$

In the next case, we will use matrix notation from the beginning. Note that there is still no mention of $\mathbb{C}$, there is still no complex structure, and we are still talking about differentiability in the sense of real-variables calculus.

2.3.3. Definition. Let $f : O \to \mathbb{R}^2$, where O is an open set in $\mathbb{R}^2$, and let $(x_0, y_0) \in O$. Let u, v be the coordinate functions of f, so that

$$f(x,y) = (u(x,y), v(x,y)).$$

Then f is **differentiable at** (x_0, y_0) if there exist real numbers a, b, c, d such that

$$
\begin{pmatrix} u(x,y) - u(x_0, y_0) \\ v(x,y) - v(x_0, y_0) \end{pmatrix} = \begin{pmatrix} a & b \\ c & d \end{pmatrix} \cdot \begin{pmatrix} x - x_0 \\ y - y_0 \end{pmatrix} + \rho(x,y),
\tag{9}
$$

$$
\text{and} \quad \lim_{(x,y) \to (x_0, y_0)} \frac{\rho(x,y)}{|(x - x_0, y - y_0)|} = 0.
$$

This time, the error ρ is vector-valued ($\mathbb{R}^2$-valued); again, it tends to 0 even when divided by the length of $(x - x_0, y - y_0)$.

Condition (9) says precisely that both u and v satisfy Definition 2.3.2. If (9) holds, then a, b and c, d are, respectively, the partial derivatives of u and v; in other words, the 2×2 matrix that appears in (9) is the Jacobian matrix of the mapping f at the point (x_0, y_0) :

$$
\begin{pmatrix} a & b \\ c & d \end{pmatrix} = \begin{pmatrix} u_x(x_0, y_0) & u_y(x_0, y_0) \\ v_x(x_0, y_0) & v_y(x_0, y_0) \end{pmatrix}.
\tag{10}
$$

Let $f : O \to \mathbb{R}^2$, where O is an open set in $\mathbb{R}^2$, and let f be differentiable in the sense of Definition 2.3.3 at each point of O. Then f is **conformal on** O if at each point of O the linear transformation given by the matrix (10) is conformal; and f is **inversely conformal on** O if it is inversely conformal at each point of O.

As you see, we have defined conformality for f on O by reference to conformality for the linear map given by the Jacobian matrix at each point of O. Section 1.3G discussed conformality for linear maps in detail.

Notice that f is conformal on O if at each point of O,

$$u_x = v_y, \; u_y = -v_x, \;\; \text{and} \; u_x^2 + u_y^2 \neq 0;$$

and f is inversely conformal on O if at each point of O,

$$u_x = -v_y, \; u_y = v_x, \;\; \text{and} \; u_x^2 + u_y^2 \neq 0.$$

B Differentiability in the Complex Sense: Holomorphy

Next is a repetition of Definition 2.3.1, but with $\mathbb{C}$ in place of $\mathbb{R}$. We will use complex notation, for example writing z_0 for $x_0 + iy_0$. Differentiability still means approximability by a linear mapping, but linearity from $\mathbb{C}$ to $\mathbb{C}$ is a stronger condition than linearity from $\mathbb{R}^2$ to $\mathbb{R}^2$; it forces the linear mapping given by the Jacobian matrix to be conformal if it is nonsingular.

2.3.4. Definition. Let $f : O \to \mathbb{C}$, where O is an open subset of $\mathbb{C}$, and let $z_0 \in O$. Then f is **differentiable in the complex sense at** z_0, or $\mathbb{C}$-**differentiable at** z_0, if there is a complex number A such that

$$A = \lim_{z \to z_0} \frac{f(z) - f(z_0)}{z - z_0}. \tag{11}$$

The complex number A is then **the derivative of** f **at** z_0, and is denoted by $f'(z_0)$.

The limit statement (11) can be rewritten in other, equivalent forms; here are three.

$$\lim_{x \to x_0} \eta(x) = 0, \;\; \text{where} \; \eta(x) := \frac{f(z) - f(z_0)}{(z - z_0)} - A. \tag{12}$$

$$f(z) - f(z_0) = (A + \eta(z))(z - z_0) \;\; \text{and} \;\; \lim_{z \to z_0} \eta(z) = 0. \tag{13}$$

$$f(z) - f(z_0) = A(z - z_0) + \rho(z) \;\; \text{and} \;\; \lim_{z \to z_0} \frac{\rho(z)}{z - z_0} = 0. \tag{14}$$

Let $A := s + it$. Written without complex notation, as a column vector, the product $A \cdot (z - z_0)$ is

$$\begin{pmatrix} s(x - x_0) - t(y - y_0) \\ t(x - x_0) + s(y - y_0) \end{pmatrix},$$

so if we let $f(x + iy) =: \begin{pmatrix} u(x, y) \\ v(x, y) \end{pmatrix}$ and write (14) without the complex notation, it becomes

$$\begin{pmatrix} u(x, y) - u(x_0, y_0) \\ v(x, y) - v(x_0, y_0) \end{pmatrix} = \begin{pmatrix} s & -t \\ t & s \end{pmatrix} \cdot \begin{pmatrix} x - x_0 \\ y - y_0 \end{pmatrix} + \rho(x, y) \qquad (15)$$

$$\text{where} \qquad \lim_{(x, y) \to (x_0, y_0)} \frac{\rho(x, y)}{|(x - x_0, y - y_0)|} = 0.$$

As we remarked already, if $f(x, y) = (u(x, y), v(x, y))$, then Definition 2.3.3 means no more and no less than that u and v both satisfy Definition 2.3.2 at the point. But if $f(x, y) = (u(x, y), v(x, y))$ satisfies Definition 2.3.4, it means more, namely that the partials of u and v at the point are related by the Cauchy-Riemann equations 1.(3). Provided $f'(z_0) \neq 0$, that makes f conformal at z_0. And then we may write $f'(z_0) = ke^{i\theta}$ with $k > 0, \theta \in \mathbb{R}$. Then the mapping $z - z_0 \mapsto f(z) - f(z_0)$ is approximated by the conformal linear mapping $z - z_0 \mapsto ke^{i\theta}(z - z_0)$. Therefore it is the combination of (1) a rotation (multiplication by $e^{i\theta}$), (2) a magnification (multiplication by k), and (3) a quantity whose ratio to $|z - z_0|$ tends to 0 as z approaches z_0. Of course, θ may be 0; and k may be 1.

Comparing (10) with (15), we notice that s and $-t$ are the coordinates of **grad** u, which means that $u_x = s$ and $u_y = -t$. Since $A = f'(z_0) = s + it$, this means we can write the complex derivative of f as $u_x - iu_y$. In fact, in view of the Cauchy-Riemann equations, there are four ways we can write f' as a combination of partials of u and v :

$$
\begin{aligned}
f' &= u_x - iu_y && \text{(which involves } u \text{ but not } v) \\
&= v_y + iv_x && \text{(which involves } v \text{ but not } u) \\
&= u_x + iv_x && \left(\text{which involves } \frac{\partial}{\partial x} \text{ but not } \frac{\partial}{\partial y} \right) \qquad (16) \\
&= v_y - iu_y && \left(\text{which involves } \frac{\partial}{\partial y} \text{ but not } \frac{\partial}{\partial x} \right).
\end{aligned}
$$

One may see the proof of the last two lines another way, as follows. Rewrite the difference quotient in (11) in terms of $x, y, u,$ and v :

$$\frac{f(z) - f(z_0)}{z - z_0} = \frac{u(x, y) - u(x_0, y_0)}{(x - x_0) + i(y - y_0)} + i \frac{v(x, y) - v(x_0, y_0)}{(x - x_0) + i(y - y_0)}. \qquad (17)$$

As (x, y) approaches (x_0, y_0) along the horizontal line $y = y_0$, by the definition of partial derivatives, the right-hand side of (17) tends to $u_x + iv_x$. As (x, y) approaches (x_0, y_0) along the vertical line $x = x_0$, it tends to $-iu_y + v_y$. But the limit of (17) is $f'(z_0) = s + it$ as (x, y) approaches (x_0, y_0)—no matter by what path the approach is made. Hence $u_x + iv_x = -iu_y + v_y$, hence $u_x = v_y$ and $v_x = -u_y$.

We will use the term "$\mathbb{C}$-differentiable" only when, as above, we consider the significance of $f'(z_0)$ for just one point z_0. Complex analysis is concerned mostly with functions that are $\mathbb{C}$-differentiable at every point of some open set, that is, holomorphic. The following definition is deliberately phrased so that a function enjoys the property of holomorphy *on a set,* never merely *at a point.*

2.3.5. Definition. A **holomorphic function**[2] is a function $f : O \to \mathbb{C}$ for some open set $O \subseteq \mathbb{C}$, such that f is $\mathbb{C}$-differentiable at every $z_0 \in O$. Then f is **holomorphic** on O; the function $f' : O \to \mathbb{C}$ is **the derivative of** f, and f is an **antiderivative** of f' on O.

Let g be a complex-valued function, and let $z_0 \in \mathbb{C}$. Then g is **holomorphic at** z_0 if there is an open set $O \subseteq \mathbb{C}$ such that $z_0 \in O$ and g is holomorphic on O.

A **conjugate-holomorphic** function is a function $\overline{f} = u - iv$, where $f = u + iv$ is holomorphic.

Notice the relation between holomorphy and conformality. The function g is conformal on O if and only if it is holomorphic on O and g' is nonzero at every point of O. The function g is inversely conformal on O if and only if it is conjugate-holomorphic on O and $(\overline{g})'$ is nonzero at every point of O.

2.3.6. Proposition. *Let f be holomorphic, and let p and q be conjugate-holomorphic. Then:*

(i) *The compositions $f \circ p$ and $p \circ f$ are conjugate-holomorphic.*

(ii) *The composition $p \circ q$ is holomorphic.*

(iii) *The composition $p \circ f \circ q$ is holomorphic.*

Proof. If $f = u(x, y) + iv(x, y)$ and $p = x(\sigma, \tau) + iy(\sigma, \tau)$, then

$$f \circ p(\sigma, \tau) = U(\sigma, \tau) + iV(\sigma, \tau),$$

where

$$U(\sigma, \tau) = u(x(\sigma, \tau), y(\sigma, \tau)) \quad \text{and} \quad V(\sigma, \tau) = v(x(\sigma, \tau), y(\sigma, \tau)).$$

One may compute the first partials of U and V as follows:

$$
\begin{aligned}
U_\sigma &= u_x x_\sigma + u_y y_\sigma, \\
U_\tau &= u_x x_\tau + u_y y_\tau; \\
V_\sigma &= v_x x_\sigma + v_y y_\sigma, \\
V_\tau &= v_x x_\tau + v_y y_\tau.
\end{aligned}
$$

It follows that $U + iV$ is conjugate holomorphic, because $u + iv$ is holomorphic, so that $u_x = v_y$ and $u_y = -v_x$; and because $x + iy$ is conjugate-holomorphic, so that $x_\sigma = -y_\tau$ and $x_\tau = y_\sigma$. The other statements can be proved similarly.

You may notice that the results follow essentially from Proposition 1.3.8. $\qquad \square$

---

[2]*Analytic* is often used to mean holomorphic. *Analytic function* is used in Riemann surface theory to mean a holomorphic function whose domain is, in a certain sense, the maximal natural domain for that function.

C Finding Derivatives

Applying the definition, we can show easily that if k is a nonnegative integer, the derivative of $z \mapsto z^k$ is $z \mapsto k z^{k-1}$. To do so, we write down the difference quotient as in (11), simplify it algebraically, and identify the limit. Here it is for the cases $k = 2$ and $k = 3$. As $z \to z_0$,

$$\frac{z^2 - z_0^2}{z - z_0} = z + z_0 \longrightarrow 2z_0,$$

$$\frac{z^3 - z_0^3}{z - z_0} = z^2 + z z_0 + z_0^2 \longrightarrow 3z_0^2;$$

and for the general case:

$$\frac{z^k - z_0^k}{z - z_0} = z^{k-1} + z^{k-2} z_0 + z^{k-3} z_0^2 + \cdots + z_0^{k-1} \longrightarrow k z_0^{k-1}.$$

We will show that on the punctured plane $\mathbb{C} \setminus \{0\}$, the derivative of $z \to z^{-1}$ is $z \to -z^{-2}$, again by taking the limit of the difference quotient, as follows. Let $z_0 \neq 0$. Then

$$\frac{\frac{1}{z} - \frac{1}{z_0}}{z - z_0} = \frac{\frac{z_0 - z}{z z_0}}{z - z_0} \longrightarrow \frac{-1}{z_0^2}.$$

We shall now prove the more general results, that the sums, constant multiples, products, quotients (assuming a zero-free denominator), and compositions of holomorphic functions are also holomorphic, and establish the desirable computational rules, working with hypotheses of pointwise $\mathbb{C}$-differentiability. We begin with compositions.

2.3.7. Theorem (The Chain Rule). *Let* $h(z) := g(f(z)), w_0 := f(z_0)$. *If both* $f'(w_0)$ *and* $g'(z_0)$ *exist, then* $h'(z_0)$ *exists and equals* $g'(w_0) \cdot f'(z_0)$.

Proof. Let $A := f'(w_0), B := g'(z_0)$. Using the form (13) of the differentiability condition, we may state our hypotheses as follows:

$$g(w) - g(w_0) = (B + \eta_g(w))(w - w_0) \quad \text{and} \quad \lim_{w \to w_0} \eta_g(w) = 0 \quad \text{and}$$

$$f(z) - f(z_0) = (A + \eta_f(z))(z - z_0) \quad \text{and} \quad \lim_{z \to z_0} \eta_f(z) = 0.$$

It follows that

$$
\begin{aligned}
h(z) - h(z_0) &= g(f(z)) - g(f(z_0)) \\
&= (B + \eta_g(f(z))(f(z) - f(z_0)) \\
&= (B + \eta_g(f(z))(A + \eta_f(z))(z - z_0).
\end{aligned}
$$

Let $\eta(z) := B \eta_f(z) + A \eta_g(f(z)) + \eta_g(f(z)) \eta_f(z)$. Then

$$h(z) - h(z_0) = (BA + \eta(z))(z - z_0) \quad \text{and} \quad \lim_{z \to z_0} \eta(z) = 0,$$

which is the condition (13) for $h'(z_0)$ to exist and equal BA.   □

2.3.8. Proposition. *If $f'(z_0)$ and $g'(z_0)$ exist, then*

(i) $(f + g)'(z_0) = f'(z_0) + g'(z_0)$.

(ii) $(cf)'(z_0) = cf'(z_0)$.

(iii) $(fg)'(z_0) = f'(z_0)g(z_0) + f(z_0)g'(z_0)$.

(iv) $\left(\dfrac{f}{g}\right)'(z_0) = \dfrac{f'(z_0)g(z_0) - f(z_0)g'(z_0)}{(g(z_0))^2}$ *provided $g(z_0) \neq 0$.*

Proof. The hypotheses, in the form (13), are as follows:

$$f(z) - f(z_0) = (A + \eta_f(z))(z - z_0) \quad \text{and} \quad \lim_{z \to z_0} \eta_f(z) = 0, \qquad (18)$$

$$g(z) - g(z_0) = (B + \eta_g(z))(z - z_0) \quad \text{and} \quad \lim_{z \to z_0} \eta_g(z) = 0. \qquad (19)$$

Now one can obtain the differentiability of each of the four functions at z_0 in the form (13). For the sum:

$$(f + g)(z) - (f + g)(z_0) = (A + B + \eta_f(z) + \eta_g(z)) \cdot (z - z_0)$$
$$\text{and} \quad \lim_{z \to z_0} [\eta_f(z) + \eta_g(z)] = 0.$$

Therefore $(f + g)'(z_0) = f'(z_0) + g'(z_0)$. Item (ii) follows from:

$$(cf)(z) - (cf)(z_0) = (cA + c\eta_f(z))(z - z_0) \quad \text{and} \quad \lim_{z \to z_0} c\eta_f(z) = 0.$$

Item (ii) also follows from the more general (iii). To prove (iii), use (18) again and compute as follows:

$$f(z)g(z) - f(z_0)g(z_0)$$
$$= [f(z_0) + (A + \eta_f(z))(z - z_0)] \cdot [g(z_0) + (B + \eta_g(z))(z - z_0)] - f(z_0)g(z_0)$$
$$= [(Ag(z_0) + Bf(z_0)) + (A\eta_g(z) + B\eta_f(z) + \eta_f(z)\eta_g(z))] \cdot (z - z_0), \quad \text{and}$$
$$\lim_{z \to z_0} (A\eta_g(z) + B\eta_f(z) + \eta_f(z)\eta_g(z)) = 0.$$

To prove (iv), first apply the Chain Rule to the composition of $z \mapsto g(z)$ with $w \mapsto w^{-1}$, obtaining that the derivative of $z \mapsto \dfrac{1}{g(z)}$ at z_0 is $\dfrac{-g'(z_0)}{g(z_0)^2}$. Then apply the Product Rule to the mapping $z \to f(z) \cdot \dfrac{1}{g(z)}$.   □

All polynomials are linear combinations of the power mappings $z \mapsto z^k$, so it follows from items (i) and (ii) of Proposition 2.3.8 that

$$\text{if} \quad p(z) := c_0 + c_1 z + c_2 z^2 + \cdots + c_n z^n,$$
$$\text{then} \quad p'(z) = c_1 + 2c_2 z + \cdots + nc_n z^{n-1}.$$

2.3.9. Definition. Let $f : \mathbb{C} \to \mathbb{C}$. Then f is a **rational** function if $f(z) \equiv p(z)/q(z)$, where p and q are polynomials with no common zeros, and either may be constant, but q is not identically 0.

The quotient rule implies the holomorphy of every rational function p/q on the set where q is nonzero (which is an open set), and allows us to write down an expression for the derivative.

So far in the book, have we given you any examples of holomorphic functions other than polynomials and rational functions? Maybe not explicitly, but actually we have placed a lot of examples within your reach. If $f = u + iv$, and if u and v are C^1, then the complex derivative of f exists if and only if u and v satisfy the Cauchy-Riemann equations. When that happens, one can use (16) to write down f'. So here is a way to produce many more examples of holomorphic functions: Start with a harmonic function $u(x, y)$. At least on some part of its domain, you will be able to find a harmonic conjugate $v(x, y)$. Then $f = u + iv$ will be holomorphic, and $f' = u_x + iv_x$.

2.3.10. Example. *Apply* (16) *to find* f' *in the case when* f *is the exponential, which we defined in subsection 2.2B to be the function given by*

$$e^z := e^{x+iy} = e^x \cos y + ie^x \sin y \quad \text{for } z \in \mathbb{C}.$$

Solution. Since $(e^x \cos y)_x = e^x \cos y$ and $(e^x \sin y)_x = e^x \sin y$, the function $z \mapsto e^z$ equals its own derivative. Evidently, then, the derivatives of all orders exist, and they are all the same function. Notice that for every $k \geq 0$, $f^{(k)}(0) = 1$. $\qquad\square$

The approach of (16) works also for Log z, which also was defined in 2.2B, with domain V_π. We can write it in terms of x and y as follows:

$$\text{Log } z \equiv \text{Log } (x + iy) = \ln \sqrt{x^2 + y^2} + i\text{Arg } (x + iy),$$

Exercise 10 asks you to find the derivative; you will find that it is $1/z$.

D Picturing the Local Behavior of Holomorphic Mappings

The geometric meaning of differentiability in the sense of Definition 2.3.1 is the approximation to the curve/graph at each point by a straight line through the point whose slope is the derivative at that point. We wish now to explain the analogous geometric interpretation of $\mathbb{C}$-differentiability at a point.

Linear mappings from $\mathbb{R}^2$ to itself are easily identified pictorially by showing the image of an orthogonal grid; see Figure 1.3-3. Of the five examples in that figure, the only conformal linear mappings are those that take grid (a) to grid (e), and grid (a) to grid (f). Conformal linear mappings are those that simply rotate and magnify. They are, in complex notation, the mappings $z \mapsto Az$. If $A = ke^{i\theta}$, then the mapping magnifies by k and rotates by θ. When a function f is $\mathbb{C}$-differentiable at a point z_0 and has derivative A, it means that for z near z_0, the mapping $z - z_0 \mapsto f(z) - f(z_0)$ is approximately the same as the conformal linear mapping $z - z_0 \mapsto A \cdot (z - z_0)$.

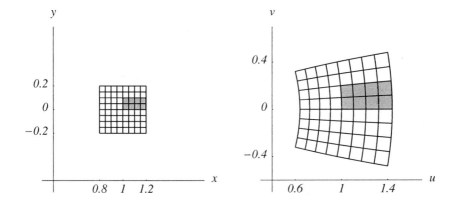

Figure 2.3-1 A rectangular grid centered at $(1, 0)$ in the x, y-plane, and its image in the u, v-plane under the mapping $(x, y) \mapsto (x^2 - y^2, 2xy)$ or $z \mapsto z^2$. The shaded set on the left is mapped to the shaded set on the right.

Figure 1.1-2 shows the action of the mapping $z \mapsto z^2$ on a rectangular grid centered at the point $z_0 = 1$, where the derivative is 2. Figure 1.1-3 shows the same action, but with a different choice of grid. You can see how the mapping acts on a certain neighborhood of $z_0 = 1$. There is "distortion," which means that distances between points are changed by different factors, and angles are changed in size. Figure 2.3-1 shows the same action again, but on a smaller rectangular grid centered at the same point. When the neighborhood of 1 is smaller, there is less distortion, and the mapping $z - 1 \mapsto z^2 - 1$ is more closely approximated by the mapping $z - 1 \mapsto 2(z - 1)$, which is a pure magnification by 2.

Figure 1.2-9 shows the action of the mapping $z \mapsto z^3$ on a rectangular grid centered at the point $z_0 = 0.4 + 0.4i$, which maps to $z_0^3 = -.128 + .128i$. Figure 1.2-10 shows the same action, but on a polar grid; in both of those pictures, there is a good bit of distortion, but there is less near the center of the grid. Figure 2.3-2 shows the same action again, but on a smaller rectangular grid. The derivative at z_0 is $3z_0^2 = .96i = .96e^{i\pi/2}$. On smaller and smaller neighborhoods of z_0, the mapping from $z - (0.4 + 0.4i)$ to $z^3 - (-.128 + .128i)$ is more and more closely approximated by the mapping

$$z - (-0.128 + 0.128i) \mapsto .96i \cdot (z - (0.4 + 0.4i)),$$

which is the combination of a magnification by .96 and a rotation by $\pi/2$.

Those remarks have focused on the character of the mapping in the immediate neighborhood of the center of a grid. Keep in mind that we could draw a grid centered at an arbitrary point in the domain and make similar observations about the action of f, provided f' is nonzero there. The fact that these local features of a holomorphic function hold at every point leads to certain global features of such mappings, which

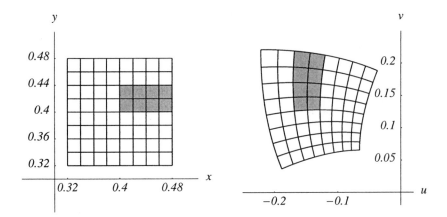

Figure 2.3-2   A rectangular grid in the x, y-plane, center $(.4, .4)$, and its image in the u, v-plane under the mapping $(x, y) \mapsto (x^3 - 3xy^2, 3x^2y - y^3)$, or $z \mapsto z^3$. The shaded set on the left is mapped to the shaded set on the right. Observe that the two pictures have different scales.

have been apparent in many of the figures. For example, orthogonal grids map to orthogonal grids.

Exercises

1. Let

$$\varphi(x, y) := \begin{cases} x^2 & \text{if } y = 0, \\ -7y & \text{if } x = 0, \\ 5 & \text{if } xy \text{ is nonzero and irrational,} \\ -3 & \text{in the remaining cases.} \end{cases}$$

Show that the two partials of φ at $(0, 0)$ exist and equal 2 and -7, respectively, but that φ is not even continuous at the point.

2. Remove parentheses in the expressions $(x + iy)^2$ and $(x + iy)^3$ to show that

$$z^2 = (x^2 - y^2) + i(2xy) \quad \text{and}$$
$$z^3 = (x^3 - 3xy^2) + i(3x^2y - y^3).$$

Use the method of (16) to take their derivatives, and show that the derivatives are $z \mapsto 2z$ and $z \mapsto 3z^2$, respectively.

3. In each case, you are given an expression for $f(z) = f(x + iy)$. Using (16), show that $f'(z)$ exists and find it. Also in each case, identify the sequence of numbers $f(0), f'(0), f''(0), \cdots, f^{(n)}(0), \cdots$.

(a) $\sin(x + iy) = \sin x \cosh y + i \cos x \sinh y$

(b) $\cos(x + iy) = \cos x \cosh y - i \sin x \sinh y$

4. Two ways of writing a complex-valued function of a complex variable appear in the equation $f(z) = u(x, y) + iv(x, y)$. Identify $u(x, y)$ and $v(x, y)$ when $f(z) = e^{e^z}$.

5. Verify that $e^{(1+i)/\sqrt{2}} \approx 1.5 + i1.3$.

6. Explain why conditions (11), (13), and (14) are equivalent.

7. Let f be holomorphic. Prove that $z \mapsto f(\bar{z})$ is conjugate-holomorphic.

8. In each case, you are given an expression for $f(z)$; find $f'(z)$.

(a) z^5

(b) e^{z^2}

(c) $\dfrac{e^z}{z}$ $\quad (z \neq 0)$

(d) $(z^2 + z^4)^2$

(e) ze^z

(f) $\cos z^2$

(g) e^{7z}

(h) $\dfrac{z}{z - 1}$ $\quad (z \neq 1)$

(i) $\cosh^2 z - \sinh^2 z$

9. Given a function f, define the function $\bar{f}$ by: $\bar{f}(z) := \overline{f(z)}$. Show that if both f and $\bar{f}$ are $\mathbb{C}$-differentiable at the point z_0, then $f'(z_0) = 0$.

10. Determine whether Log , the principal logarithm, is holomorphic, and if it is, find the derivative. Recall that

$$\text{Log}\,(x + iy) = \ln \sqrt{x^2 + y^2} + i\text{Arg}\,(x + iy).$$

We will offer you several suggestions for approaching this problem.

- Use (16). You may wish to write the imaginary part as follows:

$$\text{Arg}(x + iy) = \begin{cases} \arctan \frac{y}{x} & \text{if } x \neq 0; \\ \text{arccot} \frac{x}{y} & \text{if } y \neq 0, \end{cases}$$

  where the branches of the inverse trigonometric functions are chosen so that their values lie always in $[-\pi, \pi)$.

- Use the difference quotient (11). Explain the details:

$$\frac{\log z - \log z_0}{z - z_0} = \frac{w - w_0}{e^w - e^{w_0}} \to \frac{1}{e^{w_0}} = \frac{1}{z_0}.$$

- Use the Chain Rule. Since $e^{\log z} = z$, $e^{\log z} \cdot \log' z = 1$, so $\log' z = \frac{1}{z}$. Do you need to assume the existence of $\log'$ to make this argument work?

Help on Selected Exercises

2. $\dfrac{\partial}{\partial x}(x^2 - y^2) + i\dfrac{\partial}{\partial x}2xy = 2x + i2y = 2z.$

3. The derivative of the sine is the cosine, because

$$\frac{\partial}{\partial x}\sin x \cosh y + i\frac{\partial}{\partial x}\cos x \sinh y = \cos x \cosh y - i \sin x \sinh y.$$

Similarly,

$$\frac{\partial}{\partial x}\cos x \cosh y - i\frac{\partial}{\partial x}\sin x \sinh y = -\sin x \cosh y - i \cos x \sinh y.$$

Thus the derivative of the cosine is the negative of the sine. The two sequences asked for are $0, 1, 0, -1, \cdots$ for the sine, and $1, 0, -1, 0, 1, \cdots$ for the cosine. Both sequences have period 4.

8. (a) $5z^4$.   (b) $2ze^{z^2}$.   (d) $2(z^2 + z^4)(2z + 4z^3)$.   (g) $7e^{7z}$.

2.4   SEQUENCES, COMPACTNESS, CONVERGENCE

A   Sequences of Complex Numbers

A **sequence** is a function a whose domain is, for some integer n_0, the set of all integers $n \geq n_0$. We write a_n for the value of a at n. We refer to the sequence a as

$$\text{"the sequence } \{a_n\}_{n=n_0}^{\infty}\text{;"}$$

or, if n_0 is understood or need not be identified, as "the sequence $\{a_n\}$".

We might, for example, speak of "the sequence $\{n^2\}_{n=1}^{\infty}$." This sequence would be adequately identified, perhaps, as "the sequence $\{1, 4, 9, 16, 25, 36, 49, 64, \cdots\}$."

If $\{n_k\}_{k=k_0}^{\infty}$ is a sequence of integers such that $n_0 \leq n_k < n_{k+1}$ for all $k \geq k_0$, then $\{a_{n_k}\}_{k=k_0}^{\infty}$ is a **subsequence** of $\{a_n\}_{n=n_0}^{\infty}$. We may sometimes write $a_{n(k)}$ instead of a_{n_k}.

For example, $\{(2n)^2\}$ is a subsequence of $\{n^2\}$. In other words, $\{4, 16, 36, 64, \cdots\}$ is a subsequence of $\{1, 4, 9, 16, 25, 36, 49, 64, \cdots\}$.

Here are some examples of sequences.

$$
\begin{aligned}
b_n &= (-1)^n & (n = 0, 1, \ldots)\\[4pt]
\zeta_n &= e^{2\pi i n/5} & (n = 0, 1, \ldots)\\[4pt]
w_n &= \frac{1}{n} & (n = 1, 2, \ldots)\\[4pt]
\omega_n &= e^{in} + \frac{(-1)^n(1+i)}{n} & (n = 1, 2, \ldots)\\[4pt]
\beta_n &= 1 + \frac{e^{2\pi i n/5}}{n} & (n = 1, 2, \ldots)
\end{aligned}
\qquad (1)
$$

For each one, make a sketch showing the range of the sequence. In the first case, the range consists of two points, -1 and 1. The range of ζ consists of five points. The ranges of ω and β are represented in Figure 2.4-1.

Let $\{a_n\}_{n=n_0}^{\infty}$ be a sequence, and let S_n be a statement about a_n. We define two phrases:

- **Eventually S_n** means that S_n is true for all but a finite number of values of $n \geq n_0$.

- **Frequently S_n** means that S_n is true for infinitely many values of $n \geq n_0$.

Here are some examples of true statements using the terms "eventually" and "frequently." They refer to the sequences given by (1).

- Frequently $b_n = -1$.

- Frequently $b_n = +1$.

- Frequently $\zeta_n = +1$.

- Eventually w_n lies in the disk $D(0, 0.01)$.

- For every $\epsilon > 0$, eventually $w_n \in D(0, \epsilon)$.

- Frequently $|\omega_n| > 1$.

- Frequently $|\omega_n| < 1$.

- If $|z_0| = 1$ and $\epsilon > 0$, then frequently $|z_0 - w_n| < \epsilon$.

- For every $\epsilon > 0$, eventually $\beta_n \in D(1, \epsilon)$.

- Frequently $\operatorname{Im} \omega_n > 1$.

Let $\{z_n\}$ be a sequence of complex numbers, and let $c \in \mathbb{C}$. The limit statement

$$\lim_{n \to \infty} z_n = c \quad (\text{or } z_n \to c \text{ as } n \to \infty) \tag{2}$$

means that for every $\epsilon > 0$, eventually $|z_n - c| < \epsilon$. If $z_n = x_n + iy_n$ and $c = a + ib$, then of course (2) is equivalent to

$$\lim_{n \to \infty} x_n = a \quad \text{and} \quad \lim_{n \to \infty} y_n = b.$$

It is elementary to prove that if $\lim_{n \to \infty} z_n = c$ and $\lim_{n \to \infty} w_n = d$, then

$$\lim_{n \to \infty} (z_n + w_n) = c + d,$$

$$\lim_{n \to \infty} (z_n \cdot w_n) = cd, \quad \text{and}$$

$$\text{if } \alpha \in \mathbb{C}, \quad \text{then} \quad \lim_{n \to \infty} (\alpha \cdot z_n) = \alpha c.$$

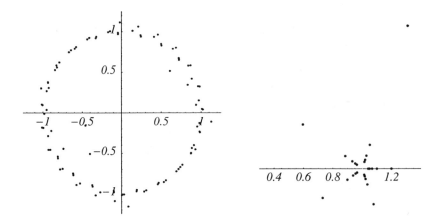

Figure 2.4-1 The first picture shows the range of $n \mapsto e^{in} + (-1)^n \frac{1+i}{n}$ for $1 \leq n \leq 100$. This sequence accumulates at the points of the unit circle. The second picture shows the range of $n \mapsto 1 + \frac{1}{n} e^{2\pi in/5}$ for $1 \leq n \leq 25$. This sequence converges to 1.

When we say that **the limit of the sequence** $\{z_n\} \subset \mathbb{C}$ **exists,** or that **the sequence converges,** we mean that (2) holds for some $c \in \mathbb{C}$. When we say that **the sequence diverges,** or that **the limit of the sequence does not exist,** we mean that (2) is false for every $c \in \mathbb{C}$. The limit statement

$$\lim_{n \to \infty} z_n = \infty, \tag{3}$$

means that for every number $K > 0$, eventually $|z_n| > K$. We say then that $\{z_n\}$ **diverges to infinity.**

There are variations on (3) for real-valued sequences $\{x_n\}$. The limit statement $\lim_{n\to\infty} x_n = \infty$ means that for every number $K > 0$, eventually $x_n > K$; whereas $\lim_{n\to\infty} x_n = -\infty$ means that for every number $K > 0$, eventually $x_n < -K$.

In Section 1.4 we defined a continuous function using the familiar δ-ϵ statement. Exercise 11 asks you to provide a proof of the following result, which characterizes continuity in terms of sequences.

2.4.1. Proposition. *For a function $f : A \to B$, the following two conditions are equivalent.*

(i) *f is continuous on A.*

(ii) *Whenever $\{z_n\}$ is a sequence of points in A such that $\lim_{n\to\infty} z_n = z_0$ and $z_0 \in A$, then $\lim_{n\to\infty} f(z_n) = f(z_0)$.*

In Section 1.4 we defined a closed set to be one that contains all of its boundary points. We can now characterize the property of being closed in terms of sequences.

2.4.2. Proposition. *For a set A, the following two conditions are equivalent.*

(i) *A is closed.*

(ii) *If there is a sequence of points in A that converges to the point z_0, then $z_0 \in A$.*

Proof. First we assume (i) and prove (ii). Let $\{z_n\}$ be a sequence of points in A such that $\lim_{n \to \infty} z_n = z_0$. Suppose that z_0 is not in A; this leads to a contradiction, as follows. For every $r > 0$, eventually z_n belongs to the disk $D(z_0, r)$. Therefore z_0 is a boundary point of A, which must belong to A since A is closed.

Next we assume (ii) and prove (i). It suffices to prove that if z_0 is a boundary point of A, then $z_0 \in A$. For each integer $n > 0$, the disk $D(z_0, \frac{1}{n})$ contains at least one point that is in A; select one and call it z_n. The sequence thus defined converges to z_0, which by (ii) must belong to A. $\qquad\qquad\square$

B   The Limit Superior of a Sequence of Reals

Let S be a set of real numbers. If there exists $b \in \mathbb{R}$ such that

$$b \geq x \quad \text{for all} \quad x \in S,$$

then S is **bounded above** and b is an **upper bound** of S. If there exists $a \in \mathbb{R}$ such that

$$a \leq x \quad \text{for all} \quad x \in S,$$

then S is **bounded below** and a is a **lower bound** of S.

Let B be the set of all upper bounds of the set $S \subseteq \mathbb{R}$. There are exactly three possibilities: Either

- $B = \mathbb{R}$, which holds if S is the empty set $\emptyset$; or

- $B = \emptyset$, for example when $S = \mathbb{Z}$; or

- B is an interval of the form $[b^*, \infty)$.

In the last case, b^* is the **least upper bound** or **supremum** or **sup** of the set S, and we write $b^* = \sup S$. In the case $B = \emptyset$, we define $\sup S := -\infty$.

Let A be the set of all lower bounds of the set $S \subseteq \mathbb{R}$. There are exactly three possibilities: Either

- $A = \mathbb{R}$, which holds if S is the empty set $\emptyset$; or

- $A = \emptyset$, for example when $S = \mathbb{Z}$; or

- A is an interval of the form $(-\infty, a^*]$.

In the last case, a^* is the **greatest lower bound** or **infimum** or **inf** of the set S, and we write $a^* = \inf S$. In the case $A = \emptyset$, we define $\inf S := \infty$.

Let $\{x_n\}$ be a real-valued sequence. Consider the sequence of sets

$$X_m := \{x_n \mid n \geq m\} \quad (m = 1, 2, \dots).$$

and the sequence of their least upper bounds and greatest lower bounds:

$$b_m := \sup X_m, \qquad a_m := \inf X_m.$$

For every m, $X_{m+1} \subseteq X_m$, and therefore

$$b_{m+1} \leq b_m \quad \text{and} \quad a_{m+1} \geq a_m \quad \text{for all} \quad m.$$

The suprema b_m are either all finite or all equal to ∞. If they are all ∞, then $\lim_{m \to \infty} b_m$ of course means ∞. Otherwise they form a nonincreasing sequence; such a sequence must have a limit, or else diverge to $-\infty$.

The infima a_m are either all finite or all equal to $-\infty$. If they are all $-\infty$, then $\lim_{m \to \infty} a_m$ of course means $-\infty$. Otherwise they form a nondecreasing sequence; such a sequence must have a limit, or else diverge to ∞.

We are now prepared to state the definition of the limit superior and the limit inferior of a sequence. Let $\{x_n\}$ be a real-valued sequence. The **limit superior** of the sequence is

$$\limsup_{n \to \infty} x_n := \lim_{m \to \infty} \sup\{x_n \mid n \geq m\}. \tag{4}$$

The **limit inferior** of the sequence is

$$\liminf_{n \to \infty} x_n := \lim_{m \to \infty} \inf\{x_n \mid n \geq m\}. \tag{5}$$

The right-hand sides of (4) and (5) are always meaningful, provided it is understood that the values ∞ and $-\infty$ are allowed. For examples, see Exercise 4.

In the next Proposition, we speak only of the limsup, leaving it to you to realize that there is an analogue of the result for the liminf.

Notice that $\ell = \limsup_{n \to \infty} x_n$ if and only if for every $\epsilon > 0$,

$$\text{frequently} \quad \ell - \epsilon < x_n \quad \text{and} \quad \text{eventually} \quad x_n < \ell + \epsilon.$$

2.4.3. Proposition. *Let $\{x_n\}$ be a real-valued sequence, and let*

$$\ell = \limsup_{n \to \infty} x_n.$$

Then $\{x_n\}$ has a subsequence $\{x_{n(k)}\}$ such that $\lim_{k \to \infty} x_{n(k)} = \ell$.

Proof. For every $\epsilon > 0$, frequently $\ell - \epsilon < x_n \leq \ell + \epsilon$. Applying this fact first with $\epsilon = 1$, we may select $n(1)$ such that

$$\ell - 1 < x_{n(1)} \leq \ell + 1.$$

We proceed by induction. When $n(k)$ has been selected, we may select $n(k+1)$ such that

$$\ell - \frac{1}{k+1} < x_{n(k+1)} \leq \ell + \frac{1}{k+1} \quad \text{and} \quad n(k) < n(k+1).$$

$\square$

C Implications of Compactness

In Section 1.4 we defined a compact set to be one that is closed and bounded. It will be useful to characterize compactness in terms of sequences.

2.4.4. Proposition. *For a set $A \subseteq \mathbb{R}^m$, the following two conditions are equivalent.*

(i) A is compact.

(ii) Every sequence of points in A has a subsequence that converges to a point of A.

Proof. We will give a proof for the case when $m = 2$. The proof for the general case is not much different. We will regard $\mathbb{R}^2$ as $\mathbb{C}$.

First we suppose (i) and prove (ii). Let $\{z_n\}$ be a sequence of points in A. Let $z_n = x_n + iy_n$. Since A is bounded, the real sequences $\{x_n\}$ and $\{y_n\}$ are bounded above and below. Let $x_0 = \limsup_{n \to \infty} x_n$. Then there is a subsequence $x_{n(k)}$ that converges to x_0. The sequence $\{y_{n(k)}\}$ is bounded. Let $y_0 = \limsup_{k \to \infty} y_{n(k)}$. Then there is a subsequence $\{y_{n(k(j))}\}_{j=1}^{\infty}$ that converges to y_0. It follows that $\{z_{n(k(j))}\}_{j=1}^{\infty}$ is a subsequence of $\{z_n\}$ whose limit is $x_0 + iy_0$, which must belong to A because A is closed.

Next we suppose (ii) and prove (i). It suffices to prove that A is closed and bounded.

If A were not closed, then by 2.4.2 there would exist a sequence $\{z_n\}$ of points in A that converges to a point z_0 not in A. Every subsequence of it would also converge to z_0, so no subsequence could converge to a point in A.

If A were not bounded, then for every $n \in \mathbb{Z}^+$ there would be a point $z_n \in A$ such that $|z_n| > n$; and no subsequence of $\{z_n\}$ could converge to a point in A. $\square$

The sequence $\{z_n\}_{n=0}^{\infty} \subset \mathbb{C}$ is a **Cauchy sequence** if for every $\epsilon > 0$, there exists n_0 such that

$$|z_n - z_m| < \epsilon \quad \text{whenever} \quad n \geq m \geq n_0. \tag{6}$$

2.4.5. Proposition. *A sequence of complex numbers $\{z_n\}_{n=0}^{\infty}$ is a Cauchy sequence if and only if there exists $c \in \mathbb{C}$ such that $\lim_{n \to \infty} z_n = c$.*

Exercise 15 asks you to provide a proof.

2.4.6. Proposition. *Let $K \subset O$, where K is a compact set and O is an open set. Then there exists $d > 0$ such that:*

$$\text{If } z \in K \text{ and } w \notin O, \text{ then } |z - w| > d. \tag{7}$$

Proof. Suppose that there were no such d. Then the implication (7) fails for every $d > 0$. For each integer $n > 0$, it would fail for $\frac{1}{n}$, so that there would exist points $z_n \in K$ and $w_n \notin O$ such that $|z_n - w_n| < \frac{1}{n}$. Since K is compact, there would be a subsequence $\{z_{n(k)}\}$ that converges to some point z_0 in K (and hence in O). It would then follow that $w_{n(k)} \to z_0$. So eventually $w_{n(k)}$ would be in O. We have reached a contradiction. $\qquad\square$

It follows, for example, that if an open set contains $\bar{D}(a,r)$, then it contains $D(a, r + d)$ for some $d > 0$.

2.4.7. Proposition. *Continuous functions preserve compactness. That is, if f is continuous and K is a compact set contained in the domain of f, then $f[K]$ is compact.*

Proof. It suffices to show that every sequence of points in $f[K]$ has a subsequence that converges to a point of $f[K]$. Such a sequence may be written as $\{f(z_n)\}$ where $\{z_n\}$ is a sequence of points in K. Then $\{z_n\}$ has a subsequence $\{z_{n(k)}\}$ that converges to a point $z_0 \in K$. Then $\{f(z_{n(k)})\}$ is a subsequence of $\{f(z_n)\}$, and it converges to $f(z_0)$ because f is continuous. $\qquad\square$

2.4.8. Proposition. *If f is continuous on the compact set K, then f is uniformly continuous on K.*

Proof. Let $\epsilon > 0$. It suffices to show that there exists $\delta > 0$ such that:

$$\text{If } z \in K, \ w \in K, \text{ and } |z - w| < \delta, \text{ then } |f(z) - f(w)| < \epsilon. \qquad (8)$$

Suppose that there were no such δ, so that the implication (8) fails for every choice of δ. Then for each n it would fail for $\frac{1}{n}$; there would exist z_n and w_n in K such that

$$|z_n - w_n| < \frac{1}{n} \quad \text{and} \quad |f(z_n) - f(w_n)| \geq \epsilon. \qquad (9)$$

By the compactness of K, there would be a subsequence $\{z_{n(k)}\}$ converging to some point z_0 in K. Then, since f is continuous at z_0, $f(z_{n(k)}) \to f(z_0)$. It would follow that $w_{n(k)} \to z_0$, hence $f(w_{n(k)}) \to f(z_0)$, hence $|f(z_{n(k)}) - f(w_{n(k)})| \to 0$, which would contradict (9). $\qquad\square$

D Sequences of Functions

Consider a sequence, of functions $\{f_n\}_{n=0}^{\infty}$ with a common domain $S \subseteq \mathbb{C}$ and with values in $\mathbb{C}$. For each $z \in S$, then, $\{f_n(z)\}_{n=0}^{\infty}$ is a sequence of complex numbers. If the limit of that sequence exists for every $z \in S$, we may define a function f on S by

$$f(z) := \lim_{n \to \infty} f_n(z) \quad \text{for } z \in S.$$

Then the sequence $\{f_n\}$ **converges**, or **converges pointwise, to** f on S. In this case it is true for each $z \in S$ that for every $\epsilon > 0$, eventually $|f_n(z) - f(z)| < \epsilon$. If the following stronger condition holds: For every $\epsilon > 0$

$$\text{eventually} \quad \sup_{z \in S} |f_n(z) - f(z)| < \epsilon,$$

then f_n converges **uniformly to** f **on** S. If for every compact set $K \subseteq S$, f_n converges uniformly to f on K, then f_n converges **subuniformly to** f **on** S.

The sequence of functions given by $f_n(z) = z^n$ converges pointwise to 0 on the set $D(0,1)$, but it does not converge uniformly to 0 on $D(0,1)$. In fact, for every $n > 0$,

$$\sup_{z \in D(0,1)} |z^n - 0| = 1.$$

However, for each $r < 1$ the sequence does converge uniformly to 0 on $D(0,r)$. So it converges subuniformly to 0 on $D(0,1)$.

2.4.9. Proposition. *Let* $\{f_n\}$ *be a sequence of continuous complex-valued functions whose common domain is the open set* $O \subseteq \mathbb{C}$. *Let* f_n *converge subuniformly to* f *on* O. *Then* f *is continuous on* O.

Proof. Let $z_0 \in O$. It suffices to prove that f is continuous at z_0. Let $\epsilon > 0$. Let $\bar{D}(0,r) \subset O$. Then $f_n \to f$ uniformly on $\bar{D}(0,r)$, which is a compact subset of O. Thus there exists N such that $|f_N(z) - f(z)| < \epsilon/3$ for all $z \in \bar{D}(0,r)$. Since f_N is continuous, there exists $\delta > 0$ such that

$$|z - z_0| < \delta \implies |f_N(z) - f_N(z_0)| < \epsilon/3.$$

It follows that

$$|z - z_0| < \delta \implies |f(z) - f(z_0)| < \epsilon,$$

because

$$|f(z) - f(z_0)| = |f(z) - f_N(z) + f_N(z) - f_N(z_0) + f_N(z_0) - f(z_0)|$$
$$\leq |f(z) - f_N(z)| + |f_N(z) - f_N(z_0)| + |f_N(z_0) - f(z_0)|.$$

$\square$

Exercises

1. Explain these statements, which refer to the sequences defined in (1):

   (a) The constant sequence $\{1\}$ is a subsequence of $\{b_n\}$.

   (b) The constant sequence $\{-1\}$ is a subsequence of $\{b_n\}$.

   (c) The constant sequence $\{1\}$ is a subsequence of $\{\zeta_n\}$.

   (d) The sequence $\left\{\dfrac{1}{k^2}\right\}$ is a subsequence of $\{w_n\}$.

2. Give an example of a sequence that converges to 0, and that is frequently real, frequently pure imaginary, and never equal to 0.

3. Let $\gamma_k = 1 + i + \dfrac{i^k}{k}$. Which of the following statements, if any, are true frequently? Which ones are true eventually?

(a) γ_k is real. (c) $\operatorname{Re} \gamma_k = 0$.

(b) γ_k lies on the line $x = 1$. (d) $|\gamma_k - 1 - i| < 0.001$.

4. In each case, find the limit inferior and the limit superior of the given sequence.

(a) $(-1)^n \left(1 + \dfrac{1}{n}\right)$

(d) $(1 + (-1)^n)n \left(1 + \dfrac{1}{n}\right)$

(b) $(1 + (-1)^n) \left(1 + \dfrac{1}{n}\right)$

(e) $(-1 + (-1)^n)n \left(1 + \dfrac{1}{n}\right)$

(c) $(-1 + (-1)^n) \left(1 + \dfrac{1}{n}\right)$

(f) $\sin \left(\dfrac{\pi}{6} + \dfrac{n\pi}{2}\right)$

5. In each case, find the limit superior of the given sequence.

(a) $n^{1/n}$

(d) $(n^{3^n})^{(1/3^n)}$

(b) $\operatorname{Re} \left(1 + i + \dfrac{i^k}{k}\right)$

(e) $\operatorname{Im} e^{2\pi i n/5}$

(c) $(3^n)^{1/2n}$

(f) $(3^n)^{1/3^n}$

6. Let $\{c_n\}$ be a sequence of complex numbers. Let $c \in \mathbb{C}$. Prove that these two statements are equivalent:

(a) For every $\epsilon > 0$, frequently $c_n \in D(c, \epsilon)$.

(b) There is a subsequence $\{c_{n_k}\}$ of $\{c_n\}$ such that $\lim_{k \to \infty} c_{n_k} = c$.

Notice that for every point z_0 on the unit circle, the sequence $\{\beta_n\}$ of (1) has a subsequence converging to z_0.

7. Explain: Given a sequence $\{x_n\}$, every subsequence of a subsequence of $\{x_n\}$ is also a subsequence of $\{x_n\}$.

8. Explain: Let $\{z_{n(k)}\}_{k=1}^{\infty}$ be a subsequence of $\{z_n\}_{n=1}^{\infty}$. If $\lim_{n \to \infty} z_n = z_0$, then $\lim_{k \to \infty} z_{n(k)} = z_0$.

9. Prove that the point z_0 is a boundary point of the set A if and only if there exist sequences $\{z_k\}$ and $\{\zeta_k\}$ such that for every k, $z_k \in A$ and $\zeta_k \notin A$; and such that $\lim_{k \to \infty} z_k = z_0$ and $\lim_{k \to \infty} \zeta_k = z_0$.

10. Prove the analogue of Proposition 2.4.3 for the limit inferior.

11. Let $A \subseteq \mathbb{C}$ and let $f : A \to \mathbb{C}$. Let $z_0 \in A$. Prove that the following two conditions are equivalent.

(i) f is continuous at z_0 relative to A.

(ii) For every sequence $\{z_n\} \subset A$ such that $\lim_{n \to \infty} z_n = z_0$, it is also true that $\lim_{n \to \infty} f(z_n) = f(z_0)$.

Then prove Proposition 2.4.1.

12. Prove that $\limsup_{n \to \infty}(x_n + y_n) \leq \limsup_{n \to \infty} x_n + \limsup_{n \to \infty} y_n$. Give an example to show that equality does not always hold.

13. We proved 2.4.4 for the case $n = 2$. Write a proof for the general case.

14. Proposition 2.4.4 is equivalent to the Bolzano-Weierstrass Theorem, which is often taught in advanced calculus courses. Find and study a treatment of that theorem, for example [8, Section 1.7].

15. Prove Proposition 2.4.5, using 2.4.4(ii).

16. Let u be a real-valued harmonic function on a bounded connected open set O. Prove that if u is defined and continuous on $\bar{O}$, and if

$$M := \sup\{u(x,y) \mid (x,y) \in \bar{O}\}$$

is finite, then there exists a point (x_0, y_0) on the boundary of O such that $u(x_0, y_0) = M$. Is the statement still true if the boundedness condition on O is omitted?

17. Let p and q be real-valued harmonic functions on a connected open set O. Prove that if O is bounded and if p and q are continuous on $\bar{O}$, and if $p = q$ on the boundary of O, then $p = q$ on O. Is the statement still true if the boundedness of O is omitted?

Help on Selected Exercises

4. (a) -1 and 1. (c) -2 and 0. (f) $-\frac{\sqrt{3}}{2}$ and $\frac{\sqrt{3}}{2}$.

5. (a) 1. (c) $3^{1/2}$. (f) 1.

15. A brief sketch: If the sequence is Cauchy, it is bounded. Since it is bounded, it has a subsequence which converges to some point $c \in \mathbb{C}$; then it follows that the sequence itself converges to c. If $\lim_{n \to \infty} z_n = c$, then there exists n_0 such that $|z_n - c| < \epsilon/2$ if $n \geq n_0$; (6) follows.

16. Establish and use the compactness of O. Recall 1.6.5. Distinguish the case when u is constant. Consider the the function $u(x,y) := e^x \cos y$ and the unbounded strip $O := \{(x,y) \mid -\pi/2 < y < \pi/2\}$.

17. Apply the result of Exercise 16 with $p - q$ in the role of u.

2.5 INTEGRALS OVER CURVES, PATHS, AND CONTOURS

A Integrals of Complex-Valued Functions

Subsection 1.5A summarized the theory of integrals of continuous real-valued functions. We define the integral of a continuous complex-valued function $f = u + iv$ over the interval $[a, b]$ in terms of the integrals of the real-valued u and v, thus:

$$\int_a^b f(t)dt := \int_a^b u(t)dt + i \int_a^b v(t)dt. \tag{1}$$

It follows from the definition that all of the statements in 1.5A are true whenever f and g are continuous complex-valued functions on $[a, b]$, and r and s are arbitrary complex-valued constants. Another result, perhaps more subtle, is that

$$\left| \int_a^b f(t)dt \right| \leq \int_a^b |f(t)|dt \tag{2}$$

whenever f is a continuous complex-valued function. To prove it, consider first the special case when the complex number (1) is real and nonnegative, so that

$$\int_a^b v(t)dt = 0 \quad \text{and} \quad \left| \int_a^b f(t)dt \right| = \int_a^b u(t)dt;$$

in that case, (2) follows because $u(t) \leq |f(t)|$ for every t. Now consider the general case. We can write the complex number (1) as $re^{i\theta}$ for some $r \geq 0$ and some real θ. The value of each side of (2) remains the same if we replace $f(t)$ by $g(t) = e^{-i\theta}f(t)$. Therefore it suffices to prove (2) when g is in place of f; and thus the general case reduces to the special case already proved.

There are several terms to be defined, including "smooth curve," the most basic; "path;" "closed path;" and "contour." Before we give detailed definitions and launch into a thorough discussion, it may be useful to give you a brief indication of the relations among them: A path consists of one or more smooth curves. Thus every smooth curve is also a path. A contour consists of one or more closed paths. Thus every closed path is also a contour.

B Curves

A **curve** in $\mathbb{C}$ is a set γ together with a continuous mapping $z : [a, b] \to \mathbb{C}$, where $[a, b]$ is a compact interval of $\mathbb{R}$, such that the set γ is the range of z. We speak then of "the curve γ given by z."[3] You may find it helpful to think of a curve as the motion

---

[3] A more formal formulation would be as follows: A **curve** is an ordered pair (z, γ), where z is a continuous mapping $z : [a, b] \to \mathbb{C}$ and γ is the range of z. Once we started being more formal, we would notice that z determines γ, and restate the definition as follows: A **curve** is a continuous mapping $z : [a, b] \to \mathbb{C}$. Perhaps we would then name its range γ. As you see, I have chosen a less formal, perhaps less efficient way to define and to speak of a curve; I think it works well.

in the plane of a particle during a finite time-interval $[a, b]$. At each instant t during this interval of time, the particle is located at the point $z(t)$. Thus γ is the set of its locations during the time interval. Please note that the set γ by itself is not the curve. Without knowing z, we do not know the motion; we do not know when, or how fast, or how often, the particle moves through each location on γ. When we speak of "the curve γ," we mean the curve γ given by a specific z. To emphasize that remark, we offer some examples. The curves

$$
\begin{array}{lll}
\gamma_1 \text{ given by } z_1(t) := e^{it} \equiv \cos t + i \sin t & (0 \le t \le 2\pi), \\
\gamma_2 \text{ given by } z_2(t) := e^{-it} \equiv \cos(-t) + i \sin(-t) & (0 \le t \le 2\pi), \\
\gamma_3 \text{ given by } z_3(t) := e^{i2t} \equiv \cos 2t + i \sin 2t & (0 \le t \le \pi), \quad \text{and} \\
\gamma_4 \text{ given by } z_4(t) := e^{it} \equiv \cos t + i \sin t & (0 \le t \le 3\pi),
\end{array}
$$

are different curves, even though each γ_j is the same set, namely, the circle $|z| = 1$. The motions are different; the first curve winds around once counterclockwise, the second winds around once clockwise, the third winds around once counterclockwise in half the time, and the fourth goes around 1.5 times. The standard way to go around a circle is once, counterclockwise, starting at the rightmost point. Thus when we speak of "the circle $|z| = 1$," we mean the curve given by z_1.

More generally, whenever $a \in \mathbb{C}$ and $r > 0$, "the circle $|z - a| = r$" means the curve given by

$$
z(t) = a + re^{it} \qquad (0 \le t \le 2\pi). \tag{3}
$$

And we will say "the circle $|z - a| = r$, traversed clockwise" if we mean the curve given by

$$
z(t) = a + re^{-it} \qquad (0 \le t \le 2\pi).
$$

The following variation on a circle will often be used. The symbol $\mathrm{Arc}_a(r, t_0, t_1)$ will denote the curve given by

$$
z(t) = a + re^{it} \qquad (t_0 \le t \le t_1),
$$

which is an **arc**, or **circular arc**. We will write $\mathrm{Arc}(r, t_0, t_1)$ to mean $\mathrm{Arc}_0(r, t_0, t_1)$.

Let p and q be a fixed pair of complex numbers. The curves

$$
\begin{array}{lll}
\tau_1 \text{ given by } \zeta_1(t) := (1 - t)p + tq & (0 \le t \le 1), & \tag{4} \\
\tau_2 \text{ given by } \zeta_2(t) := (1 - t^2)p + t^2 q & (0 \le t \le 1), \quad \text{and} \\
\tau_3 \text{ given by } \zeta_3(t) := (1 - t)q + tp & (0 \le t \le 1),
\end{array}
$$

are different curves, even though every τ_j equals the line segment between p and q. The motions are different; the first curve moves from p to q at constant speed; the second moves from p to q at a speed that varies in a certain way; and the third moves from q to p at constant speed. When we speak of "the line segment $[p, q]$," we mean the curve τ_1.

We return now to the general case. Consider the curve γ given by $z : [a, b] \to \mathbb{C}$. Some terminology:

- $[a, b]$ is the **parameter interval** of the curve.

- The mapping z is the **parametrization** of the curve.

- $z(a)$ is the **initial point** and $z(b)$ is the **terminal point** of the curve; the curve **runs from** $z(a)$ **to** $z(b)$.

- The curve is **closed** if $z(a) = z(b)$.

- The symbol $-\gamma$ denotes the curve given by

$$w(t) = z(b + a - t) \qquad (a \leq t \leq b).$$

Examples among the curves defined above: $\gamma_2 = -\gamma_1$ and $\tau_3 = -\tau_1$.

- The curve is a C^1 **curve** if z' is defined and continuous throughout $[a, b]$, where $z'(a)$ and $z'(b)$ are understood to be one-sided derivatives.

- For a C^1 curve, for each $t \in [a, b]$ the **tangent vector** at t is $z'(t)$; the **unit tangent**, or **direction**, at t is the vector $\dfrac{z'(t)}{|z'(t)|}$; it is defined and its length equals 1, provided $z'(t) \neq 0$.

- A **smooth** curve is a C^1 curve such that $z'(t)$ is nonzero for every t such that $a < t < b$.

The smoothness property assures that the direction of the curve is well-defined and changes continuously, except perhaps at the initial and terminal points. The C^1 requirement alone is not enough to make the curve free of cusps and kinks, as one might think at first. For example, sketch the C^1 curve given by $z(t) = (t - \sin t) + i(1 - \cos t)$, the well-known cycloid (see, for example, [67, pp. 643-644]). Whenever t equals an integer times 2π, a cusp occurs; as t passes through such a value, the tangent vector passes through the value 0 and reverses direction.

Notice that the definition of a smooth curve does not rule out the possibility that z is constant, as for example when $a = b$.

The **length** of a C^1 curve γ given by $z : [a, b] \to \mathbb{C}$ is

$$\int_\gamma |dz| := \int_a^b |z'(t)| \, dt.$$

Consider two examples. The length of the circle $|z - a| = r$, given of course by (3), equals

$$\int_0^{2\pi} \left| \frac{d}{dt}(a + re^{it}) \right| dt = \int_0^{2\pi} |rie^{it}| \, dt = 2\pi r.$$

The length of the line segment $[p, q]$ given by $t \mapsto (1 - t)p + tq$ $(0 \le t \le 1)$ is

$$\int_0^1 \left| \frac{d}{dt}((1 - t)p + tq) \right| dt = \int_0^1 |q - p| dt = |q - p|.$$

Let γ be a C^1 curve given by $z : [a, b] \to \mathbb{C}$. Let f be a continuous complex-valued function whose domain contains γ. Then **the integral of f over** γ, for which we write

$$\int_\gamma f \quad \text{or} \quad \int_\gamma f(z) dz, \tag{5}$$

is defined to be

$$\int_a^b f(z(t)) z'(t) dt, \tag{6}$$

which is called a **pullback** of (5) to the interval $[a, b]$.

In this section we will write out and/or evaluate a few instances of a complex integral $\int_\gamma f$ "by the definition," that is, by using a pullback. This procedure will break the quantity $\int_\gamma f$ into recognizable parts for you. It will help you appreciate what it really is, and help you understand some of the possible physical interpretations. The integral will be a combination of real integrals of the kind one sees in calculus, some of which are easy to evaluate—but some of which are not! In this and future chapters we will prove a number of theorems that allow us to obtain $\int_\gamma f$ in terms of other quantities, which at first seem unrelated to integrals, and which often are quite easy to evaluate exactly. We will then have a bag of tricks for easily evaluating many real and complex integrals.

For examples, consider again the cases when γ is a circle or a line segment. For an arbitrary continuous function f defined on the circle $|z - a| = r$,

$$\int_{|z-a|=r} f(z) dz = \int_0^{2\pi} f(a + re^{it}) r i e^{it} dt.$$

In the important and useful case when $f(z) = \dfrac{1}{z - a}$, this becomes

$$\int_{|z-a|=r} \frac{dz}{z - a} = \int_0^{2\pi} \frac{1}{re^{it}} r i e^{it} dt = 2\pi i.$$

For an arbitrary continuous function f defined on the line segment $[p, q]$,

$$\int_{[p,q]} f(z) dz = \int_0^1 f((1 - t)p + tq)(q - p) dt. \tag{7}$$

The Fundamental Theorem of Calculus can be lifted from the line into the plane, as follows. If there is a holomorphic function F whose domain contains γ, and if $F' = f$, then $F(z(t))$ is an antiderivative for $f(z(t))z'(t)$ throughout $[a,b]$ and therefore (6) equals $F(z(b)) - F(z(a))$. Restated briefly:

$$F' = f \text{ on a } C^1 \text{ curve } \gamma \quad \Longrightarrow \quad \int_\gamma f = F(z(b)) - F(z(a)). \qquad (8)$$

For example, if n is a nonnegative integer, $F(z) = \dfrac{z^{n+1}}{n+1}$, and $f(z) = z^n$, then F is an antiderivative for f throughout the plane and therefore

$$\int_\gamma z^n dz = \frac{q^{n+1} - p^{n+1}}{n+1}$$

for every γ in the plane that runs from p to q.

2.5.1. Proposition. *For every continuous complex-valued function f whose domain contains the C^1 curve γ,*

$$\int_{-\gamma} f = -\int_\gamma f.$$

Proof. If γ is given by $z : [a,b] \to \mathbb{C}$, then $-\gamma$ is given by $w(t) = z(s(t))$, where $s(t) = b + a - t$. So

$$\int_{-\gamma} f = \int_a^b f(w(t))w'(t)dt = \int_a^b f(z(s(t)))z'(s(t))s'(t)dt$$

$$= \int_b^a f(z(s))z'(s)ds = -\int_\gamma f.$$

$\square$

2.5.2. Proposition. *Let γ be a C^1 curve, and let f be a continuous complex-valued function whose domain contains γ. Then*

$$\left| \int_\gamma f \right| \le \max_{z \in \gamma} |f(z)| \cdot \int_\gamma |dz|.$$

Proof. The left-hand side can be written as a pullback and then estimated, using (2):

$$\left| \int_a^b f(z(t))z'(t)dt \right| \le \int_a^b |f(z(t))z'(t)|dt$$

$$\le \max_{a \le t \le b} |f(z(t))| \cdot \int_a^b |z'(t)|dt.$$

$\square$

We will now define an equivalence relation $\sim$ among curves. Consider two C^1 curves,

$$\gamma_1 \text{ given by } z_1 : [a_1, b_1] \to \mathbb{C}; \quad \text{and} \quad \gamma_2 \text{ given by } z_2 : [a_2, b_2] \to \mathbb{C}.$$

Then γ_1 and γ_2 are **equivalent,** and we write $\gamma_1 \sim \gamma_2$, if there exists a C^1 mapping $\phi : [a_1, b_1] \to [a_2, b_2]$ with $\phi(a_1) = a_2, \phi(b_1) = b_2$, and such that

$$\phi'(t) > 0 \quad \text{and} \quad z_1(t) = z_2(\phi(t)) \quad \text{for each } t \in [a_1, b_1].$$

In other words, two curves are equivalent when they are the same set with different parametrizations, one obtainable from the other by a suitable change of variable. It is often convenient to make such a change of parametrization.

2.5.3. Proposition. *Let $[c, d]$ be an arbitrary compact interval of $\mathbb{R}$. Then every curve is equivalent to one whose parameter interval is $[c, d]$.*

Proof. Let γ be an arbitrary curve, given by, say, $z : [a, b] \to \mathbb{C}$. Define a change of variable $\phi : [c, d] \to [a, b]$ by $\phi(t) = a + \dfrac{t - c}{d - c}(b - a)$. Then $\gamma \sim \gamma_1$ where γ_1 is given by $w(t) = z(\phi(t))$ $(c \leq t \leq d)$. Notice that if γ is C^1, then so is γ_1; and if γ is smooth, then so is γ_1. $\qquad\square$

2.5.4. Proposition. *If two C^1 curves are equivalent, $\gamma_1 \sim \gamma_2$, then*

$$\int_{\gamma_1} f = \int_{\gamma_2} f$$

for every continuous complex-valued function f defined on $\gamma_1 = \gamma_2$; and

$$\int_{\gamma_1} |dz| = \int_{\gamma_2} |dz|.$$

Proof. It is a matter of changing variables, letting $s = \phi(t)$ so that $ds = \phi'(t)dt$ and

$$\begin{aligned}
\int_{\gamma_1} f &= \int_{a_1}^{b_1} f(z_1(t))z_1'(t)dt \\
&= \int_{a_1}^{b_1} f(z_2(\phi(t)))z_2'(\phi(t))\phi'(t)dt \\
&= \int_{a_2}^{b_2} f(z_2(s))z_2'(s)ds = \int_{\gamma_2} f \; ; \text{ and}
\end{aligned}$$

$$\begin{aligned}
\int_{\gamma_1} |dz| &= \int_{a_1}^{b_1} |z_1'(t)|dt \\
&= \int_{a_1}^{b_1} |z_2'(\phi(t))||\phi'(t)|dt \\
&= \int_{a_2}^{b_2} |z_2'(s)|ds = \int_{\gamma_2} |dz|.
\end{aligned}$$

$\square$

Here are some examples with line segments. If $p = a + ib$ and $q = c + ib$, then following (7) we get

$$\int_{[p,q]} f = \int_0^1 f(a(1-t) + ct + ib)(c - a)dt. \tag{9}$$

But since the line segment $[p, q]$ is horizontal, it is easier and more elegant to use x itself as the parameter. The parametrization is then

$$x \mapsto x + ib \ (a \leq x \leq c); \quad \text{and} \quad \int_{[p,q]} f = \int_a^c f(x + ib)dx, \tag{10}$$

which equals (9). For a vertical line segment $[p, q]$, say with $p = c + ib$ and $q = c + id$, we can follow (7) and get

$$\int_{[p,q]} f = \int_0^1 f(c + i((1-t)b + dt)i(d - b)dt, \tag{11}$$

or use the more natural parametrization

$$y \mapsto c + iy \ (b \leq y \leq d); \quad \text{and} \quad \int_{[p,q]} f = \int_b^d f(c + iy)idy, \tag{12}$$

which equals (11).

C   Paths

A **path** γ is the union of a finite number of smooth curves,

$$\gamma_j \ \text{given by} \ z_j : [a_j, b_j] \to \mathbb{C} \quad \text{for} \ 1 \leq j \leq J,$$

such that (if $J > 1$) the terminal point of γ_j equals the initial point of γ_{j+1} for $1 \leq j < J$. The **length** of γ is the sum of their lengths. For a continuous complex-valued function f whose domain contains $\cup_{j=1}^J \gamma_j$, the **integral of** f **over** γ is

$$\int_\gamma f = \sum_{j=1}^J \int_{\gamma_j} f.$$

We may write $\gamma = \gamma_1 + \gamma_2 + \cdots + \gamma_J$.[4]

The **initial point of** γ is $p := z_1(a_1)$, and the **terminal point of** γ is $q := z_J(b_J)$, and we say then that γ **runs from** p **to** q. The path is **closed** if $p = q$. Two paths

---

[4]This may seem a strange new use of the symbol "+," but consider this: If $F(x) = x$ and $G(x) = x^2$, then we commonly write $F + G$ for the function $x \mapsto x + x^2$. Each γ_j identifies a mapping $f \mapsto \int_{\gamma_j} f$, defined on a class of functions, with values in the complex numbers. The value of γ at f is the sum of the values of the γ_j at f.

are **equivalent** if they are sequences of the same number of C^1 curves which are respectively equivalent.

If γ is the union of the curves $\gamma_1, \cdots, \gamma_J$, then the symbol $-\gamma$ denotes the path which is the union of the curves $-\gamma_J, \cdots, -\gamma_1$.

If γ and ρ are two paths, and the terminal point of γ equals the initial point of ρ, then the symbol $\gamma + \rho$ denotes the path which is the union of γ and ρ.

2.5.5. Proposition. *Let γ be a path, and let f be a continuous complex-valued function whose domain contains γ. Then*

$$\left| \int_\gamma f \right| \leq \max_{z \in \gamma} |f(z)| \cdot \int_\gamma |dz|.$$

Proof. This follows easily when we apply 2.5.2 to each of the C^1 curves γ_j that make up γ. □

2.5.6. Proposition. *Let $g_1, g_2, \ldots$ be continuous complex-valued functions such that the domain of each g_n contains the path γ. If $\lim_{n \to \infty} g_n(z) = g(z)$ uniformly for $z \in \gamma$, then*

$$\lim_{n \to \infty} \int_\gamma g_n = \int_\gamma g.$$

Proof. Note that the function g is continuous on γ by the proof of 2.4.9. Let $\epsilon > 0$. Since the convergence is uniform, eventually $\max_{z \in \gamma} |g_n(z) - g(z)| < \epsilon$. Therefore

$$\left| \int_\gamma (g_n(z) - g(z)) dz \right| \leq \epsilon \cdot \int_\gamma |dz|,$$

by Proposition 2.5.5. □

2.5.7. Proposition. *Let f be continuous, and let F be holomorphic, such that $F' = f$ on the open set $O \subset \mathbb{C}$. Let γ be a path in O that runs from p to q. Then*

$$\int_\gamma f = F(q) - F(p).$$

Proof. The result (8) applies to each of the C^1 curves γ_j that make up γ, so that

$$\int_\gamma f = \sum_{j=1}^{J} \int_{\gamma_j} f = \sum_{j=1}^{J} F(z_j(b_j)) - F(z_j(a_j)),$$

a telescoping sum which equals $F(z_J(b_J)) - F(z_1(a_1)) \equiv F(q) - F(p)$. □

The union of line segments $[p_1, p_2], [p_2, p_3], \cdots, [p_{n-1}, p_n]$ is an example of a path, for which we write

$$[p_1, p_2] + [p_2, p_3] + \cdots + [p_{n-1}, p_n],$$

or, for the sake of brevity, $[p_1, p_2, \cdots, p_n]$.

2.5.8. Example. *Let a, b, c, and d be real numbers, with $a < c$ and $b < d$. Consider the rectangle R whose four corners are $a + ib, c + ib, c + id$, and $a + id$; and whose oriented boundary ∂R is the closed path $[a + ib, c + ib, c + id, a + id, a + ib]$. Using convenient parametrizations as in (10) and (12), write the integral of the function f over this path as the sum of four pullbacks.*

Solution.

$$\int_{\partial R} f = \int_a^c f(x + ib)dx + \int_b^d f(c + iy)idy$$
$$+ \int_c^a f(x + id)dx + \int_d^b f(a + iy)idy, \tag{13}$$

which, interestingly, equals

$$\int_a^c (f(x + ib) - f(x + id))dx + i \int_b^d (f(c + iy) - f(a + iy))dy.$$

$\square$

2.5.9. Example. *Let a, b, and c be complex numbers. Let $T = T(a, b, c)$ be the triangle with vertices a, b, c (containing the inside as well as the three edges). Write the integral of the function f over the boundary of T as the sum of three pullbacks to integrals over $[0, 1]$.*

Solution. Provided the closed path $[a, b, c, a]$ goes counterclockwise, as we shall always assume it does, it is the oriented boundary of the triangle and is denoted by ∂T. To bring into view all the pertinent definitions, we will write an integral over this path first as the sum of three integrals over segments, then as the three pullbacks to integrals over $[0, 1]$:

$$\int_{\partial T} f = \int_{[a,b]} f + \int_{[b,c]} f + \int_{[c,a]} f$$
$$= (b - a) \int_0^1 f(a(1 - t) + bt)dt$$
$$+ (c - b) \int_0^1 f(b(1 - t) + ct)dt \tag{14}$$
$$+ (a - c) \int_0^1 f(c(1 - t) + at)dt.$$

$\square$

A **piecewise smooth** curve is one such that z' is piecewise continuous and nonzero at every point where it is continuous. By Proposition 2.5.3, a path $\gamma = \{\gamma_j\}_{j=1}^J$ is always equivalent to one such that for each j, the parameter interval for γ_j is $\left[\frac{j-1}{J}, \frac{j}{J}\right]$.

Thus a path could just as well be defined as a piecewise smooth curve with parameter interval $[0, 1]$. But we prefer to incorporate into our definition of a path the practical convenience, in case we need to do actual computations, of making a separate choice of parametrization and parameter interval for each curve making up the path.

D   Pathwise Connected Sets

Some easy but worthwhile observations:

- If the path γ runs from p to q, then the path $-\gamma$ runs from q to p.

- If the path γ_1 runs from p to q, and the path γ_2 runs from q to r, then the path $\gamma_1 + \gamma_2$ runs from p to r.

- The constant path runs from p to p.

Thus, if S denotes an arbitrary set, the statement "there exists a path γ in S that runs from p to q" defines an equivalence relation among the points of S. The equivalence classes are the **components** of S. Thus for $p \in S$, the component of S to which p belongs is the set of $q \in S$ such that some path γ in S runs from p to q.

For the sake of brevity, henceforth we will say "p and q are connected in S" or "q is connected to p in S" to mean "there exists a path γ in S that runs from p to q"

Let A be a subset of $\mathbb{R}^n$. If p and q are connected in A for all $p, q \in A$, then A is **pathwise connected**.

The set $A \subset \mathbb{R}^n$ is **convex** if for every pair of points $p, q \in O$ the line segment $[p, q]$ is contained in O.

Obviously every convex set is pathwise connected. In particular, every interval of $\mathbb{R}$ is pathwise connected, and every open ball in $\mathbb{R}^n$ is pathwise connected. The complement of a ball is pathwise connected, clearly, though not convex.

2.5.10. Proposition. *Let B be a bounded subset of $\mathbb{R}^n$, and let A denote its complement. Then A has exactly one unbounded component.*

Proof. There exists a number b such that $|z| \le b$ for every $z \in B$. Thus A contains the set $A_1 := \{q \mid |q| > b\}$, which is clearly pathwise connected. The component of A of which A_1 is a subset is unbounded, and every other component of A is bounded by b. $\square$

Recall the definition of a connected set from 1.4D. From henceforth, we will make use of the properties of connectedness and pathwise connectedness only for open sets—for which they are equivalent.

2.5.11. Proposition. *Let O be an open subset of $\mathbb{R}^n$. Then O is connected if and only if it is pathwise connected.*

Proof. First, we will show that connectedness implies pathwise connectedness. Suppose that the open set O is connected. Let $p \in O$, and let V be the set of all $q \in O$ which are connected to p in O. Let W consist of all $z \in O$ which are not connected to p in O. It suffices to prove $W = \emptyset$.

The set V is open. Proof: Let $q \in V$; since O is open there exists $\delta > 0$ such that $D(q, \delta) \subset O$. Every $z \in D(q, \delta)$ is connected to q in O and hence to p. Therefore $D(q, \delta) \subset V$.

The set W is open. Proof: Let $z \in W$. Because O is open, there exists $\epsilon > 0$ such that $D(z, \epsilon) \subset O$. No point $w \in D(z, \epsilon)$ is connected to p, because if it were, then z would be connected to p. Therefore $D(z, \epsilon) \subset W$.

If W were nonempty, then V and W would disconnect O, which cannot be.

To establish the converse, suppose now that O is pathwise connected, and also disconnected by sets V and W; we will arrive at a contradiction. Let $p \in V$, $q \in W$. There exists a path in O that runs from p to q, so there is a continuous function $c : [0, 1] \to O$ such that $c(0) = p$ and $c(1) = q$. Since c is continuous, the two sets $V_0 := c^{-1}[V]$ and $W_0 := c^{-1}[W]$ are non-empty, disjoint open sets in $[0, 1]$ whose union is $[0, 1]$. It would follow that $[0, 1]$ is disconnected, which by 1.4.8 is false. $\square$

E   Independence of Path and Morera's Theorem

Let f be a continuous complex-valued function the an open set $O \subseteq \mathbb{C}$. Then **integrals of f are independent of path in** O if one of these two equivalent conditions holds:

(i) If γ_1 and γ_2 are two paths in O with the same initial point and the same terminal point, then

$$\int_{\gamma_1} f = \int_{\gamma_2} f.$$

(ii) If γ is a closed path in O, then

$$\int_\gamma f = 0.$$

Conditions (i) and (ii) are easily seen to be equivalent, as follows. If (ii) holds, and if γ_1 and γ_2 are as in (i), then the path γ consisting of γ_1 followed by $-\gamma_2$ is a closed path, and

$$0 = \int_\gamma f = \int_{\gamma_1} f - \int_{\gamma_2} f.$$

Therefore (i) follows from (ii); and (ii) follows from (i) because if γ is a closed curve in O, we may apply (i) with $\gamma_1 = \gamma$ and $\gamma_2 = -\gamma$, obtaining

$$\int_\gamma f = -\int_\gamma f \quad \text{and hence} \quad \int_\gamma f = 0.$$

2.5.12. Theorem. *Let f be a continuous complex-valued function on the open connected set $O \subseteq \mathbb{C}$. Then the following two conditions are equivalent.*

(i) f has an antiderivative on O.

(ii) Integrals of f are independent of path in O.

If O is convex, then the following condition is also equivalent:

(iii) There is a point $a \in O$ such that $\int_{\partial T} f = 0$ whenever T is a triangle contained in O and a is a vertex of T.

Proof. If $F' = f$ on O, and if $p, q \in O$, then $\int_\gamma f = F(q) - F(p)$ for every path γ that runs from p to q in O, by 2.5.7. Thus (ii) follows from (i).

The key part of the Theorem, which is that (i) follows from (ii), is due to Giacinto Morera [43]. To prove it, fix a point $a \in O$. For $z \in O$, let γ be a path in O that runs from a to z, and define

$$F(z) := \int_\gamma f.$$

By (ii), this definition does not depend on the choice of γ, so the definition of F is unambiguous. All that remains is to prove that $F' = f$ throughout O. Let $z_0 \in O$. There exists $r > 0$ such that $D(z_0, r) \subseteq O$. If γ_0 is a path in O that runs from a to z_0, then

$$F(z_0) = \int_{\gamma_0} f.$$

If $z \in D(z_0, r)$, let $\gamma = \gamma_0 + [z_0, z]$. Then γ runs from a to z, so

$$F(z) = \int_\gamma f = \int_{\gamma_0} f + \int_{[z_0, z]} f.$$

Therefore

$$F(z) - F(z_0) = \int_{[z_0, z]} f = \int_0^1 f((1-t)z_0 + tz)(z - z_0)dt, \qquad (15)$$

so

$$\frac{F(z) - F(z_0)}{z - z_0} = \int_0^1 f((1-t)z_0 + tz)dt.$$

Since f is continuous at z_0, the values of f on the segment $[z_0, z]$ are all as close as you please to $f(z_0)$, provided z is sufficiently close to z_0. Therefore the limit of the right-hand side as $z \to z_0$ exists and equals $f(z_0)$. So $F'(z_0)$, the limit of the left-hand side, also exists and equals $f(z_0)$.

Thus whenever O is open and connected, (i) and (ii) are equivalent; and (ii) certainly implies (iii) since the boundary of every triangle is a closed path. All that remains is to prove that when O is convex and (iii) holds, then (i) follows. Let a be the point given in (iii). Since O is convex, the segment $[a, z]$ lies in O whenever $z \in O$. Therefore a good candidate for antiderivative is given by

$$F(z) := \int_{[a, z]} f.$$

Let $z_0 \in O$. It suffices to show $F'(z_0)$ is defined and equal to $f(z_0)$. Because O is convex, the triangle $T := T(a, z_0, z)$ is contained in O whenever $z \in O$. By (iii), then, $\int_{\partial T} f = 0$, which may be written as

$$\int_{[a,z]} f - \int_{[a,z_0]} f = \int_{[z_0,z]} f,$$

which gives us (15). We complete the proof as before. $\qquad\square$

2.5.13. Remark. It is important to know other, weaker conditions that suffice to imply that the integrals of a holomorphic function are independent of path. At this early stage of our presentation, the convexity of the domain is merely a convenient and easy condition to use. It is certainly stronger than what's really needed in Theorem 2.5.12 above, or in Corollary 2.5.15 below. If you wish to explore one weaker condition that suffices, see Exercises 19 and 20. But there is no need to dwell on small improvements, since the Cauchy Theorem of Section 3.2 will soon arrive and provide the most satisfying answers.

F   Goursat's Lemma

The proof of Goursat's Lemma is a gem. It is too much fun to pass up. There are three ideas that make it work.

(i) The first is especially easy. If a complex number w equals the sum of four others, $w = w_1 + w_2 + w_3 + w_4$, then $|w|$ is less than or equal to four times the maximum of the four numbers $|w_j|$. Therefore $|w_j| \geq \frac{1}{4}|w|$ for at least one value of j.

(ii) Let f be a continuous function whose domain contains the triangle $T := T(a, b, c)$. Let $I(a, b, c)$ stand for the integral of f over ∂T:

$$I(a,b,c) := \int_{\partial T(a,b,c)} f = \int_{[a,b]} f + \int_{[b,c]} f + \int_{[c,a]} f \qquad (16)$$

(see subsection 2.5C, especially Example 2.5.9). To follow the discussion, draw the triangle and label the three vertices. Let C be a point on the segment $[a, b]$ other than a or b, so that

$$\int_{[a,b]} f = \int_{[a,C]} f + \int_{[C,b]} f.$$

In the same way, let $A \in [b, c]$ and $B \in [c, a]$. If you add to your picture the segments $[A, B]$, $[B, C]$, and $[C, A]$, then you see five triangles—the original one $T(a, b, c)$ and four similar subtriangles. We claim that the integral of f over $\partial T(a, b, c)$ equals the sum of the integrals of f over the boundaries of the four subtriangles:

$$I(a,b,c) = I(a,C,B) + I(C,b,A) + I(A,c,B) + I(A,B,C). \qquad (17)$$

To see why (17) is true, observe that

$$\int_{\partial T(a,b,c)} f = \int_{[a,b]} f \qquad + \int_{[b,c]} f \qquad + \int_{[c,a]} f$$

$$= \int_{[a,C]} f + \int_{[C,b]} f + \int_{[b,A]} f + \int_{[A,c]} f + \int_{[c,B]} f + \int_{[B,a]} f$$

$$+ \int_{[C,B]} f + \int_{[B,C]} f$$

$$+ \int_{[A,C]} f + \int_{[C,A]} f$$

$$+ \int_{[B,A]} f + \int_{[A,B]} f \qquad = \cdots$$

Each of the last three sums-of-two-integrals equals 0 (by 2.5.1). Now we can arrange the sum of those twelve integrals into four sums of three, so as obtain (17):

$$\cdots = \int_{[a,C]} f + \int_{[C,B]} f + \int_{[B,a]} f$$

$$+ \int_{[C,b]} f + \int_{[b,A]} f + \int_{[A,C]} f$$

$$+ \int_{[A,c]} f + \int_{[c,B]} f + \int_{[B,A]} f$$

$$+ \int_{[B,C]} f + \int_{[C,A]} f + \int_{[A,B]} f .$$

(iii) The third idea is for estimating the integral of f over a path that lies very close to a point z_0 such that $f'(z_0)$ exists. For z in the domain of f we may write

$$f(z) = f(z_0) + f'(z_0)(z - z_0) + \eta(z)(z - z_0),$$

where η is continuous and $\lim_{z \to z_0} \eta(z) = 0$ (as pointed out in Section 2.3). The function $z \mapsto f(z_0) + f'(z_0)(z - z_0)$ has an antiderivative, namely, $z \mapsto f(z_0)z + \frac{1}{2}f'(z_0)(z - z_0)^2$. Therefore its integral over every closed path equals 0 (Theorem 2.5.12). Therefore for every closed path γ that lies in the domain of f,

$$\int_\gamma f = \int_\gamma \eta(z)(z - z_0)dz.$$

By 2.5.5, then,

$$\left| \int_\gamma f \right| \leq \max_{z \in \gamma} |\eta(z)(z - z_0)| \cdot \int_\gamma |dz|. \qquad (18)$$

We will put together the three ideas (i), (ii), and (iii) to obtain the result.

2.5.14. Theorem (Goursat's Lemma). *Let f be holomorphic on the open set $O \subseteq \mathbb{C}$. Then for every triangle T contained in O,*

$$\int_{\partial T} f = 0.$$

Proof. Let $T := T(a, b, c)$. In the notation (16), what we wish to show is that $I(a, b, c) = 0$. Use (ii), letting the midpoints of the sides of $T(a, b, c)$ play the roles of A, B, and C.

$$I(a, b, c) = I\left(a, \frac{a+b}{2}, \frac{c+a}{2}\right) + I\left(\frac{a+b}{2}, b, \frac{b+c}{2}\right)$$
$$+ I\left(c, \frac{c+a}{2}, \frac{b+c}{2}\right) + I\left(\frac{a+b}{2}, \frac{b+c}{2}, \frac{c+a}{2}\right).$$

Notice that the perimeter of each of the four smaller triangles is half the perimeter of the original triangle $T(a, b, c)$; likewise for the diameter (the length of the longest side). Use (i); the modulus of at least one of those four integrals on the right is at least one-fourth the modulus of $I(a, b, c)$.

The process just described, passing from a triangle to a smaller one within it, may be repeated an arbitrary number of times. First, we restate the first iteration using different notation: Let a_0, b_0, and c_0 denote a, b, and c, respectively. Let d denote the diameter of the triangle $T(a_0, b_0, c_0)$; in other words,

$$d := \max(|b_0 - a_0|, |c_0 - b_0|, |a_0 - c_0|).$$

The diameter of each of the four subtriangles is $d/2$. Let a_1, b_1, c_1 denote the vertices of one of the four, selected to satisfy

$$|I(a_1, b_1, c_1)| \geq \frac{1}{4}|I(a_0, b_0, c_0)|.$$

Proceeding inductively, we obtain a sequence $\{(a_n, b_n, c_n)\}_{n=0}^{\infty}$ such that for each n, the diameter of the triangle $T(a_n, b_n, c_n)$ is $d2^{-n}$, and

$$|I(a_n, b_n, c_n)| \geq \frac{1}{4}|I(a_{n-1}, b_{n-1}, c_{n-1})|$$
$$\geq \frac{1}{4^2}|I(a_{n-2}, b_{n-2}, c_{n-2})|$$
$$\geq \quad \cdots \quad \geq \frac{1}{4^n}|I(a_0, b_0, c_0)|. \tag{19}$$

There is a unique point z_0 in the intersection of all the triangles (see Exercise 21). Since $f'(z_0)$ exists, we may use (iii) and the estimate (18):

$$|I(a_n, b_n, c_n)| \leq \max_{z \in \partial T(a_n, b_n, c_n)} |\eta(z)||z - z_0| \cdot \int_{\partial T(a_n, b_n, c_n)} |dz|. \tag{20}$$

Let m_n denote the maximum value of $|\eta(z)|$ for $z \in \partial T(a_n, b_n, c_n)$. Then for z on that path, $|\eta(z)||z - z_0| \leq m_n d2^{-n}$. The length of the path is no greater than $3d2^{-n}$. Therefore

$$|I(a_n, b_n, c_n)| \leq \frac{3d^2 m_n}{4^n}. \tag{21}$$

Putting (19) and (21) together, we find that

$$|I(a_0, b_0, c_0)| \leq d^2 m_n \quad \text{for every } n.$$

Since $m_n \to 0$ as $n \to \infty$, $I(a_0, b_0, c_0) = 0$.   □

2.5.15. Corollary. *Let f be holomorphic on the open convex set $O \subseteq \mathbb{C}$. Then integrals of f are independent of path in O.*

Proof. Condition (iii) of Theorem 2.5.12 holds. In fact, by Goursat's Lemma, the integral of f around every triangle in O equals 0. By that Theorem, the conclusion follows.   □

2.5.16. Theorem (Goursat's Lemma, Slightly Improved). *Let O be an open subset of $\mathbb{C}$. Let $a \in O$. Let $q : O \to \mathbb{C}$ such that q is continuous on O and holomorphic on $O \setminus \{a\}$. If T is a triangle contained in O, and if a is a vertex of T, then*

$$\int_{\partial T} q = 0.$$

Proof. Notice that the hypothesis does not imply that q is *not* holomorphic at a. In fact, as we shall show later, it turns out that it must be holomorphic at that point also.

Let $T = T(a, b, c)$. The result is immediate if a, b, and c are collinear, so we may suppose they are not. Make a sketch. Fix an $s > 0$ such that $D(a, s) \subset O$, and let $B := \max\{|q(w)| : |w - a| \leq s\}$. Let $0 < \epsilon < s$. Let a_1, a_2 be points on the segments $[a, b]$ and $[a, c]$, respectively, which lie in $D(a, \epsilon)$. Then

$$I(a, b, c) = I(a, a_1, a_2) + I(a_1, b, a_2) + I(a_2, b, c).$$

Each of the last two integrals on the right equals 0, by Goursat's Lemma 2.5.14. The first one is bounded by $4B\epsilon$, since the perimeter of that triangle is no more than 4ϵ. Since ϵ is an arbitrary positive number, $I(a, b, c) = 0$.   □

G The Winding Number

Let γ be a closed path given by $w : [0, 1] \to \mathbb{C}$, and let z be a point in $\mathbb{C}$ that does not lie on γ. Then the **winding number** of γ with respect to z is

$$W_\gamma(z) := \frac{1}{2\pi i} \int_\gamma \frac{dw}{w - z}. \tag{22}$$

The **inside** of γ is the set of all points z such that $W_\gamma(z) \neq 0$. The **outside** of γ is the set of all points z such that $W_\gamma(z) = 0$.

For a fixed γ and a fixed $z \notin \gamma$, there is a condition which is obviously suffi-cient (though it is not necessary) for z to be outside γ. That is, if there exists an antiderivative of the function $w \mapsto \dfrac{1}{w - z}$ on γ, then of course $W_\gamma(z) = 0$. As Exercise 13 asks you to prove, the existence of such an antiderivative is equivalent to the existence of an argument $w \mapsto \arg(w - z)$, and thus also to the existence of a logarithm $w \mapsto \log(w - z) = \ln|w - z| + i \arg(w - z)$.

To understand what the winding number means in the general case, recall the fact that a logarithm $w \mapsto \log(w - z)$ always exists locally on $\mathbb{C} \setminus \{z\}$. The pullback of the integral (22),

$$\int_0^1 \frac{w'(t)}{w(t) - z} dt, \tag{23}$$

can be written as the sum of integrals over small subintervals of $[0, 1]$, like

$$\int_{t_0}^{t_1} \frac{w'(t)}{w(t) - z} dt,$$

where $0 \le t_0 < t_1 < 1$. Let r be the distance from z to γ. If we take $t_1 - t_0$ sufficiently small, then the disk with center $w(t_0)$ and radius r will contain $w(t)$ for all t with $t_0 \le t \le t_1$. On that disk, there is a logarithm $w \mapsto \log(w - z)$, so

$$\int_{t_0}^{t_1} \frac{w'(t)}{w(t) - z} dt \;=\; \log(w(t) - z)\big|_{w(t_0)}^{w(t_1)} \;=\; \arg(w(t_1) - z) - \arg(w(t_0) - z).$$

Thus the integral computes the net change that takes place in $\arg(w(t) - z)$ as t increases from t_0 to t_1. That change is the same regardless of the choice of argument. (For details, see Exercise 22.)

It follows that the integral (22) computes the net number of times $w(t)$ goes counterclockwise around the point z as the parameter t increases from 0 to 1. More briefly: The winding number counts the number of times γ winds around the point z.

We have discussed the number $W_\gamma(z)$ for a fixed γ and z. The next result considers W_γ as a function.

2.5.17. Proposition. *Let γ be a closed path given by $w : [0, 1] \to \mathbb{C}$. Then the function W_γ is constant on each component of $\mathbb{C} \setminus \gamma$ and is always integer-valued. It equals 0 on the unbounded component.*

Proof. The derivative of W_γ can be written as an integral:

$$\frac{d}{dz} W_\gamma(z) = \frac{1}{2\pi i} \int_\gamma \frac{dw}{(w - z)^2} \quad \text{for all } z \notin \gamma. \tag{24}$$

This is "differentiation under the integral sign;" see Exercise 17. For each z not on γ, the function $w \mapsto (w - z)^{-2}$ has an antiderivative on $\mathbb{C} \setminus \{z\}$, namely, $w \mapsto -(w - z)^{-1}$. Therefore its integral (24) over the closed path γ must equal 0.

Therefore W_γ is constant on every component of the complement of γ (see Exercise 18). It remains to show that it is integer-valued.

Let γ be given by $w : [0, 1] \to \mathbb{C}$. Let

$$p(s) := c + \int_0^s \frac{w'(t)}{w(t) - z} dt \qquad \text{for } 0 \le s \le 1,$$

where c is chosen so that $e^c = w(0) - z$. Then

$$p(1) - p(0) = 2\pi i \, W_\gamma(z)$$

and

$$p'(s) = \frac{w'(s)}{w(s) - z}.$$

A computation shows that

$$\frac{d}{ds}\left(\frac{e^{p(s)}}{w(s) - z}\right) = 0,$$

so that the function

$$s \mapsto \frac{e^{p(s)}}{w(s) - z}$$

must be constant. By our choice of c, it equals 1 when $s = 0$. Therefore it equals 1 everywhere, which means that $p(s)$ is a logarithm for $w(s) - z$ for every $s \in [0, 1]$. Since

$$\frac{e^{p(1)}}{w(1) - z} = \frac{e^{p(0)}}{w(0) - z}$$

and $w(1) = w(0)$, we must have $e^{p(1)} = e^{p(0)}$, and it follows that $p(1) - p(0)$ must be $2\pi i$ times an integer. The first statement of the Proposition is proved.

Using Proposition 2.5.5, we obtain the estimate

$$|W_\gamma(z)| \le \frac{\int_\gamma |dz|}{2\pi \min_{w \in \gamma} |w - z|}. \tag{25}$$

As pointed out in the proof of Proposition 2.5.10, the unbounded component of the complement of γ contains, for some b, all z with $|z| > b$. Therefore it contains points z for which the right-hand side of (25) is arbitrarily small. Therefore the constant value of the winding number on that component must be 0. $\qquad\square$

H   Green's Theorem

The time has come to show you the connection between the complex integrals discussed above, and the integrals over curves that are familiar to you from multi-dimensional calculus. In particular, we will state and prove the customary version of Green's Theorem for the special case of a rectangle, and explain its relation to Goursat's Lemma.

A C^1 curve in $\mathbb{C}$ is of course the same thing as a C^1 curve in $\mathbb{R}^2$, but with complex notation. Let the curve γ be given by $t \mapsto z(t)$ $(0 \leq t \leq 1)$. Its parametrization could just as well be written as $t \mapsto (x(t), y(t))$.

Let U and V be continuous real-valued functions defined on γ. Then the integral of $U\,dx + V\,dy$ (called a continuous real-valued 1-form) over γ is defined by

$$\int_\gamma U\,dx + V\,dy := \int_0^1 \left(U(x(t), y(t))x'(t) + V(x(t), y(t))y'(t) \right) dt. \qquad (26)$$

The functions U and V may be considered as the components of a vector field $\vec{\mathbf{F}}$, whose values may be written in the $\vec{\mathbf{i}}, \vec{\mathbf{j}}$ notation for vectors:

$$\vec{\mathbf{F}}(x, y) = U(x, y)\vec{\mathbf{i}} + V(x, y)\vec{\mathbf{j}}.$$

Various physical quantities are computed by integrals like (26). For example, if $\vec{\mathbf{F}}$ is a force field, then the integrand on the right-hand side of (26) is, for each t, the dot product of the force with the tangent vector to the curve. The integral then computes the work done when a particle moves along the curve.

If $f(z) = f(x + iy) = u(x, y) + iv(x, y)$, then the real and imaginary parts of the complex integral $\int_\gamma f$ are integrals of certain 1-forms:

$$\int_\gamma f(z)\,dz = \int_0^1 \left(u(x, y) + iv(x, y) \right) \left(x'(t) + iy'(t) \right) dt \qquad (27)$$

$$= \int_0^1 \left(u(x, y)x'(t) - v(x, y)y'(t) \right) dt + i \int_0^1 \left(v(x, y)x'(t) + u(x, y)y'(t) \right) dt$$

$$= \int_\gamma u\,dx - v\,dy + i \int_\gamma v\,dx + u\,dy.$$

As we shall see, the last two integrals sometimes compute two closely related physical quantities.

2.5.18. Theorem (Green's Theorem). *Let U and V be C^1 functions whose domain is an open set containing the rectangle*

$$R = \{(x, y) \mid a \leq x \leq c \text{ and } b \leq y \leq d\}.$$

Then

$$\int_{\partial R} U\,dx + V\,dy = \iint_R \left(\frac{\partial V}{\partial x} - \frac{\partial U}{\partial y} \right) dx\,dy. \qquad (28)$$

Proof. Our procedure is similiar to that of (13), but without complex structure. The left-hand side of (28) is the sum of integrals over the four edges of R. We begin with the bottom edge. If we use the parametrization $x \mapsto (x, b)$ $(a \leq x \leq c)$, then (26) becomes

$$\int_{[(a,b),(c,b)]} U\,dx + V\,dy = \int_a^c U(x,y)\,dx.$$

Proceeding in this manner counterclockwise, we obtain:

$$\int_{\partial R} U\,dx + V\,dy = \int_a^c U(x,b)\,dx + \int_b^d V(c,y)\,dy$$
$$+ \int_c^a U(x,d)\,dx + \int_d^b V(a,y)\,dy$$
$$= \int_a^c (U(x,b) - U(x,d))\,dx + \int_b^d (V(c,y) - V(a,y))\,dy.$$

The right-hand side of (28) is the difference of two integrals over R; evaluate each as an iterated integral, choosing the order of iteration suitably:

$$\iint_R \frac{\partial V}{\partial x}(x,y)\,dx\,dy \;-\; \iint_R \frac{\partial U}{\partial y}(x,y)\,dx\,dy$$
$$= \int_b^d \left(\int_a^c \frac{\partial V}{\partial x}(x,y)\,dx \right) dy - \int_a^c \left(\int_b^d \frac{\partial U}{\partial y}(x,y)\,dy \right) dx$$
$$= \int_b^d (V(c,y) - V(a,y))\,dy - \int_a^c (U(x,d) - U(x,b))\,dx.$$

Therefore the two sides of (28) are equal. □

2.5.19. Corollary. *Let $f = u + iv$ be a C^1 function whose domain is an open set containing the rectangle*

$$R = \{(x,y) \mid a \leq x \leq c \text{ and } b \leq y \leq d\},$$

and such that u and v satisfy the Cauchy-Riemann equations. Then

$$\int_{\partial R} f = 0.$$

Proof. Applying (27) in the case when $\gamma = \partial R$, we find that

$$\int_{\partial R} f = \int_{\partial R} u\,dx - v\,dy + i \int_{\partial R} v\,dx + u\,dy.$$

Applying Green's Theorem to the real and imaginary parts separately, we obtain

$$\int_{\partial R} u\,dx - v\,dy = -\iint_R \left(\frac{\partial v}{\partial x} + \frac{\partial u}{\partial y} \right) dx\,dy$$

and

$$\int_{\partial R} v\,dx + u\,dy = \iint_R \left(\frac{\partial u}{\partial x} - \frac{\partial v}{\partial y} \right) dx\,dy.$$

$\square$

2.5.20. Remark. We proved Goursat's Lemma 2.5.14 for a triangle; the rectangle version follows easily. We proved Green's Theorem and its Corollary for a rectangle; the triangle version is almost as easy. So the two approaches give the same conclusion. But Goursat's Lemma did not use the hypothesis that f be C^1. Instead, it used the seemingly weaker hypothesis that f' exists everywhere—which implies that the partials exist and satisfy the Cauchy-Riemann equations at each point, but does not obviously imply that they are C^1. We shall show later that it does indeed imply that they must be C^1; and must, in fact, amazingly, be C^∞.

2.5.21. Remark. Using the same ideas that give Goursat's Lemma, one can prove a version of Green's Theorem which does not need the C^1 hypothesis, and which is strong enough to imply Goursat's Lemma in its full strength (see [1] and [35]).

I   Irrotational and Incompressible Fluid Flow

Consider the flow of a fluid. Let U, V, and W be real-valued C^2 functions defined on the convex open set $O \subset \mathbb{R}^3$, giving at each point $(x, y, z) \in O$ the three components of the velocity vector of the fluid flow:

$$\vec{\mathbf{F}}(x, y, z) = U(x, y, z)\vec{\mathbf{i}} + V(x, y, z)\vec{\mathbf{j}} + W(x, y, z)\vec{\mathbf{k}}. \tag{29}$$

This is a **steady** flow, which means that $\vec{\mathbf{F}}$ is independent of time. If the effects of viscosity can be ignored, then we may suppose that the velocity vector field is irrotational and incompressible. That is to say, the curl and the divergence of $\vec{\mathbf{F}}$, discussed in subsection 1.5D, are both identically 0 on O. Under those assumptions, the fluid is **perfect**.

We have begun the discussion by writing down a 3-dimensional fluid flow (29)(and temporarily using "z" to denote a real variable!), because the physical phenomenon inherently requires three spatial dimensions. But the methods of this book deal with the special case when the third component of $\vec{\mathbf{F}}$ vanishes and the first two components depend only on x and y. So we simplify the mathematical model by saying: Consider a 2-dimensional fluid flow velocity vector field

$$\vec{\mathbf{F}}(x, y) = U(x, y)\vec{\mathbf{i}} + V(x, y)\vec{\mathbf{j}} \tag{30}$$

defined on an convex open set $O \subseteq \mathbb{R}^2$.

Consider a rectangle $R \subset O$, with the edges of ∂R parametrized in the usual way. For each edge, for each t in the parameter interval,

$$\vec{\mathbf{T}}(t) := x'(t)\vec{\mathbf{i}} + y'(t)\vec{\mathbf{j}}$$

is the unit vector that is tangent to ∂R, pointing along the path in the counterclockwise direction. The dot product $\vec{\mathbf{F}} \cdot \vec{\mathbf{T}}$ is the component of flow along the path. Its integral over ∂R is the **circulation** of the fluid around the path:

$$\int_{\partial R} \vec{\mathbf{F}} \cdot \vec{\mathbf{T}} = \int_{\partial R} U\,dx + V\,dy = \iint_R (V_x - U_y)\,dx\,dy. \tag{31}$$

The second equality is given by Green's Theorem. A fluid flow is **irrotational** or **circulation-free** if it has zero circulation around the boundary of every rectangle contained in O.[5] It follows that the flow is irrotational if and only if

$$U_y = V_x \tag{32}$$

throughout O, because if $V_x - U_y$ were nonzero at a point, then it would be nonzero and have the same sign throughout some neighborhood of that point; and then the integral on the right in equation (31) would be nonzero for every rectangle contained in that neighborhood.

At each point of ∂R the unit vector

$$\vec{\mathbf{N}}(t) := y'(t)\vec{\mathbf{i}} - x'(t)\vec{\mathbf{j}}$$

is perpendicular to ∂R, and it points to the outside of R. Therefore the dot product $\vec{\mathbf{F}} \cdot \vec{\mathbf{N}}$ is the component of flow out of R. Its integral around ∂R is the **flux**, which is the amount of fluid leaving R per unit time:

$$\int_{\partial R} \vec{\mathbf{F}} \cdot \vec{\mathbf{N}} = \int_{\partial R} U\,dy - V\,dx = \iint_R (U_x + V_y)\,dx\,dy. \tag{33}$$

The second equality is due to Green's Theorem. A fluid flow is **incompressible** if for every rectangle $R \subset O$, the flux out of R, across ∂R, equals 0. It follows that the flow is incompressible if and only if

$$V_x = -U_y \tag{34}$$

throughout O.

Thus: If we associate with the fluid flow $\vec{\mathbf{F}} := U\vec{\mathbf{i}} + V\vec{\mathbf{j}}$ the complex-valued function $F = U + iV$, then what we have found is that $\vec{\mathbf{F}}$ is irrotational and incompressible if and only if both (32) and (34) hold, which means precisely that F is conjugate-holomorphic throughout O.

Notice that the holomorphic function involved is not not $F = U + iV$, but $\bar{F} = U - iV$. In the following restatement of what we have found, U and V are renamed U and $-V$.

Let $F := U + iV$ be a C^1 function on the convex open set $O \subseteq \mathbb{C}$. Associate with F the vector field

$$\vec{\mathbf{F}}(x, y) := U(x, y)\vec{\mathbf{i}} - V(x, y)\vec{\mathbf{j}}.$$

---

[5]It would be equivalent and perhaps more natural to say, if it has zero circulation around every closed path in O whose inside is contained in O. It is simpler to consider just rectangles.

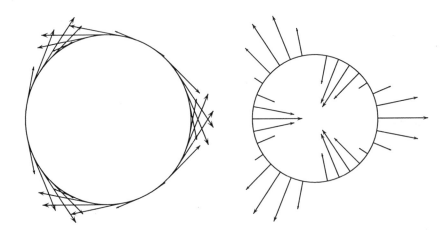

Figure 2.5-1   Pólya diagrams

If γ is a path in O, then the integral of F over γ may be written as the combination of two real integrals, respectively the integrals of the tangential and normal components of $\vec{\mathbf{F}}$ over γ:

$$\int_\gamma F = \int_\gamma U\,dx - V\,dy + i\int_\gamma V\,dx + U\,dy$$

$$= \int_\gamma \vec{\mathbf{F}}\cdot\vec{\mathbf{T}} + i\int_\gamma \vec{\mathbf{F}}\cdot\vec{\mathbf{N}}.$$

If γ is a closed path, then the last two integrals compute the vector field's circulation around γ and its total flux across γ, respectively. Whenever R is a rectangle contained in O and $\gamma = \partial R$, by Green's Theorem

$$\int_\gamma F = \iint_R (-V_x - U_y)\,dx\,dy + \iint_R (U_x - V_y)\,dx\,dy,$$

which equals 0 for all R if and only if the Cauchy-Riemann equations hold throughout the convex open set O. Thus the function F is holomorphic on O precisely when the vector field $\vec{\mathbf{F}}$ is irrotational and incompressible.

Diagrams that illustrate this interpretation of a complex integral are sometimes called Pólya diagrams; see [5] and [4]. Let $F(x,y) = (x^2 - y^2) + i2xy$. The associated vector field is then given by

$$\vec{\mathbf{F}}(x,y) = (x^2 - y^2)\vec{\mathbf{i}} - 2xy\vec{\mathbf{j}}.$$

Figure 2.5-1 shows Pólya diagrams for this case, where the integral is over the unit circle. In the picture on the left, at each of 24 points (x,y) on the circle, the projection of $\vec{\mathbf{F}}(x,y)$ onto the unit tangent is shown. In the picture on the right, at each of those points, the projection onto the unit normal is shown.

We will return to fluid flow problems in Section 5.4.

J Contours

A **contour** is the union of a finite number of closed paths, γ_k for $1 \le k \le n$. Thus of course a contour may consist of exactly one closed path. If f is defined on Γ, then the **integral of f over** Γ is the sum of its integrals over the paths γ_k:

$$\int_\Gamma f := \sum_{k=1}^n \int_{\gamma_k} f.$$

In view of this additive character of the action $f \mapsto \int_\Gamma f$, we may write

$$\Gamma = \gamma_1 + \gamma_2 + \cdots + \gamma_n.$$

The **winding number of Γ with respect to** the point $z \notin \Gamma$ is

$$W_\Gamma(z) = \sum_{k=1}^n W_{\gamma_k}(z).$$

The **inside** of a contour is the set on which the winding number is nonzero; the **outside**, where it equals 0. The **length** of a contour is the sum of the length of the closed paths that make it up:

$$\int_\Gamma |dz| := \sum_{k=1}^n \int_{\gamma_k} |dz|.$$

The contour $-\Gamma$ is defined to consist of the closed paths $-\gamma_k$. Thus for f defined and continuous on Γ,

$$\int_{-\Gamma} f = -\int_\Gamma f.$$

The **sum** of two contours Γ_1, Γ_2 is their union; thus for all f,

$$\int_{\Gamma_1+\Gamma_2} f = \int_{\Gamma_1} f + \int_{\Gamma_2} f.$$

For example, let $\Gamma = C_2 - C_1$, where C_r denotes the circle $|z| = r$. Then the contour Γ consists of the circle $|z| = 2$, traversed in the usual counterclockwise direction, and the circle $|z| = 1$ traversed clockwise.

The closed path $\gamma = C_2 + [2,1] - C_1 + [1,2]$ has nearly the same properties as Γ, but the contour Γ is less complicated. Obtaining the closed path γ requires us to put in and take out the line segment $[1,2]$, which is a somewhat artificial thing to do. Observe that $\int_\Gamma f = \int_\gamma f$ whenever both integrals are defined, by 2.5.1.

Exercises

1. Let $a \in \mathbb{C}$ and let $R > 0$. In each case, verify the given statement by computing a pullback.

   (a) $\displaystyle \int_{|z-a|=R} \frac{dz}{z-a} = 2\pi i.$

   (b) $\displaystyle \int_{|z-a|=R} (z-a)dz = 0.$

   (c) $\displaystyle \int_{\mathrm{Arc}_a(R,t_0,t_1)} \frac{dz}{z-a} = (t_1 - t_0)i.$

   (d) If n is a positive integer, then $\displaystyle \int_{|z-a|=R} (z-a)^n dz = 0.$

2. Evaluate $\displaystyle \int_{[0,1+i]} \bar{z}dz.$

3. Here are three statements about integrals over a semicircle of center $a \in \mathbb{C}$ and radius $R > 0$. In each case, verify the given statement by computing a pullback:

   (a) $\displaystyle \int_{\mathrm{Arc}_a(R,0,\pi)} (z-a)dz = 0.$

   (b) $\displaystyle \int_{\mathrm{Arc}_a(R,0,\pi)} (z-a)^2 dz = -\frac{2}{3}R^3.$

   (c) If $n \in \mathbb{N}$, then $\displaystyle \int_{\mathrm{Arc}_a(R,0,\pi)} (z-a)^n dz = \begin{cases} 0 & \text{if } n \text{ is odd,} \\ -\dfrac{2R^{n+1}}{n+1} & \text{if } n \text{ is even.} \end{cases}$

4. Point out the results in this section which imply that, if $a \in \mathbb{C}$ and $r > 0$,

$$\frac{1}{2\pi i} \int_{|w-a|=r} \frac{dw}{w-z} = \begin{cases} 1 & \text{if } |z-a| < r; \\ 0 & \text{if } |z-a| > r. \end{cases}$$

5. Explain why (9) and (10) are equal, and why (11) and (12) are equal.

6. In each case, verify the statement by applying Proposition 2.5.7.

   (a) If $A = 0$ and $B = i$, then $\displaystyle \int_{[A,B]} z dz = -\frac{1}{2}.$

   (b) If $A = 2$ and $B = i$, then $\displaystyle \int_{[A,B]} z dz = -\frac{5}{2}.$

   (c) If $A = 2$ and $B = i$, then $\displaystyle \int_{[A,B]} z^2 dz = -\frac{8+i}{3}.$

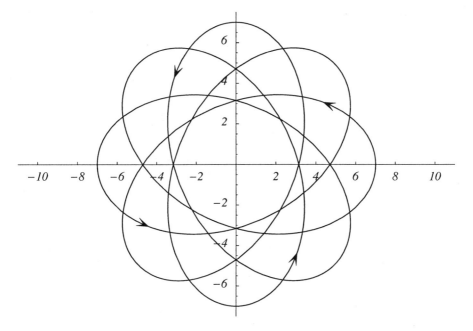

Figure 2.5-2   The closed path shown is given by $t \mapsto 5e^{i5t/4} + 2e^{-i3t/4}$ for $0 \le t \le 8\pi$.

7. Verify the statement in each part of Exercise 6 by using a pullback.

8. Let R be the rectangle with vertices $-2-i, 1-i, 1+i$, and $-2+i$. Explain why
$$\int_{\partial R} z^2 dz = 0.$$ Then evaluate the integrals on each of the four edges separately, by the method of your choice.

9. Evaluate $\dfrac{1}{2\pi i}\displaystyle\int_{|z|=1} e^z dz.$

10. Let Γ be the contour consisting of two circles, $|z - 1| = 2$ counterclockwise and $|z - 1| = 1$ clockwise. Then the complement of Γ has three components. In each component, what is the value of the winding number?

11. Figure 2.5-2 shows a closed path γ whose complement in the plane has 34 components. What is the value of W_γ on each of the 34 components? The caption gives the parametrization, which is one-to-one with only the obvious exceptions.

12. Let γ be the closed polygonal path $[-1 + i, -1 - i, 1 + i, 1 - i, -1 + i, 1 + i, -1 - i, -1 + i, 1 - i, 1 + i, -1 + i]$. The complement of γ has several components. Find the winding number in each one.

13. Recall the discussion of logarithms in subsection B. Let O be a connected open set that does not contain the point z. Then the following three conditions are equivalent.

   (a) There is an antiderivative on O of the function $w \mapsto \dfrac{1}{w - z}$.

   (b) There is a logarithm on O of the function $w \mapsto w - z$, which means a continuous function $w \mapsto \log(w - z)$ such that $e^{\log(w-z)} = w - z$ for $w \in O$.

   (c) There is an argument $w \mapsto \arg(w - z)$ for $w \in O$.

14. Let R be a rectangle as in Green's Theorem. Show that the area of R equals
$$\frac{1}{2i} \int_{\partial R} \bar{z} dz.$$

15. Prove that the relation $\sim$ in the definition of equivalent curves does in fact qualify as an equivalence relation; that is, it is reflexive, symmetric, and transitive:

   (a) For every curve γ, $\gamma \sim \gamma$.

   (b) If γ_1 and γ_2 are curves such that $\gamma_1 \sim \gamma_2$, then $\gamma_2 \sim \gamma_1$.

   (c) If γ_1, γ_2, and γ_3 are curves such that $\gamma_1 \sim \gamma_2$ and $\gamma_2 \sim \gamma_3$, then $\gamma_1 \sim \gamma_3$.

16. Consider the seven specific curves γ_j and τ_k defined near the beginning of subsection B. Prove that $\gamma_1 \sim \gamma_3 \sim -\gamma_2$ and $\tau_1 \sim \tau_2 \sim -\tau_3$. Are there any other equivalences among those curves?

17. Let f be a continuous complex-valued function whose domain includes the C^1 curve γ. Let

$$g(z) = \int_\gamma \frac{f(w)}{w - z} dw \qquad \text{for } z \in \mathbb{C} \setminus \gamma.$$

Prove that g is differentiable and that

$$g'(z) = \int_\gamma \frac{f(w)}{(w - z)^2} dw. \tag{35}$$

18. Prove that if $F : O \to \mathbb{C}$, where O is an open connected set, and if $F' = 0$ on O, then F is constant on O.

19. Here we will define a special type of pathwise connected set. Let O be an open subset of $\mathbb{C}$. Then O is **starlike** if there exists a point $a \in O$ such that for every $z \in O$, the line segment $[a, z]$ is contained in O. Such a point a is a **hub** for O.

   (a) Prove that every convex open set is starlike.

(b) Prove that the set V_α is starlike.

(c) Give an example of a starlike set that has only one hub.

20. Prove Theorem 2.5.12, modified as follows. Replace "convex" with "starlike." In condition (iii), replace "a point $a \in O$" with "a hub a of O."

21. Justify the statement in the proof of Goursat's Lemma, "There is a unique point z_0 in the intersection of all the triangles."

22. Explain why the integral (23) can be written as the sum of a finite number of integrals over small subintervals of $[0, 1]$ of the kind considered.

Help on Selected Exercises

2. By pullback,

$$\int_{[0,1+i]} \bar{z}\,dz = \int_0^1 \underbrace{t(1-i)}_{\overline{z(t)}}\,\underbrace{(1+i)}_{z'(t)}\,dt = 1.$$

8. To evaluate $\displaystyle\int_{[1-i,1+i]} z^2\,dz$ by one method: $\cdots = \left.\dfrac{z^3}{3}\right|_{1-i}^{1+i} = \dfrac{4i}{3}$. By another

method: $\cdots = i\displaystyle\int_{-1}^1 (1+2iy-y^2)dy = i\left(y+y^2-\dfrac{y^3}{3}\right)\bigg|_{-1}^1 = \dfrac{4i}{3}$.

10. For every point z between the two circles, $W_\Gamma(z) = 1$; if $|z-1| < 1$ or $|z-1| > 2$, then $W_\Gamma(z) = 0$.

12. The winding number is 0 on every component.

13. Compare Exercise 13, page 164 in Section 2.2.

17. Here is an outline of a proof. Fix a point $z \in \mathbb{C}\setminus\gamma$. The derivative $g'(z)$ is the limit of the difference quotient, if it exists:

$$\lim_{\zeta\to z}\frac{g(\zeta)-g(z)}{\zeta-z} = \lim_{\zeta\to z}\int_\gamma f(w)\frac{\frac{1}{w-\zeta}-\frac{1}{w-z}}{\zeta-z}\,dw.$$

For each $w \in \gamma$, the difference quotient that appears in the integrand has a limit:

$$\lim_{\zeta\to z}\frac{\frac{1}{w-\zeta}-\frac{1}{w-z}}{\zeta-z} = \lim_{\zeta\to z}\frac{1}{(w-\zeta)(w-z)} = \frac{1}{(w-z)^2}. \qquad (36)$$

First explain why there is a positive number d such that $|w-z| \geq d$ for all $w \in \gamma$. Then, to prove (35), show that the limit (36) is uniform in w. That is, for every $\epsilon > 0$, there exists $\delta > 0$ such that

$$\left|\frac{1}{(w-\zeta)(w-z)} - \frac{1}{(w-z)^2}\right| < \epsilon \text{ for all } w \in \gamma \text{ whenever } |\zeta - z| < \delta.$$

20. Here is a proof that when O is starlike and (iii) holds, then (i) follows. Let a be a hub as in (iii). Since the segment $[a, z]$ lies in O whenever $z \in O$, a good candidate for antiderivative is given by $F(z) := \int_{[a,z]} f$. Let $z_0 \in O$. It suffices to show $F'(z_0)$ is defined and equal to $f(z_0)$. There is a disk $D(z_0, r)$ contained in O. Because O is starlike, the triangle $T := T(a, z_0, z)$ is contained in O for every $z \in D(z_0, r)$. By (iii), then, $\int_{\partial T} f = 0$, which may be written in the form

$$\int_{[a,z]} f - \int_{[a,z_0]} f = \int_{[z_0,z]} f.$$

That gives us (15). We complete the proof as before.

21. Show that $\{a_0, b_0, c_0, a_1, b_1, c_1, a_2, \cdots\}$ is a Cauchy sequence and apply 2.4.5.

22. The function $t \mapsto w(t)$ is continuous and hence uniformly continuous on $[0, 1]$, by 2.4.8. Therefore there exists $\delta > 0$ such that $|w(t) - w(t_0)| < r$ whenever $|t - t_0| < \delta$. If $K > 1/\delta$, then we can use the subintervals $[\frac{k}{K}, \frac{k+1}{K}]$ for $k = 0, 1, \cdots, K - 1$.

2.6 POWER SERIES

A Infinite Series

An **infinite series**, or **series**, consists of two sequences of complex numbers, $\{a_k\}_{k=0}^{\infty}$ and $\{s_n\}_{n=0}^{\infty}$, where

$$s_n = \sum_{k=0}^{n} a_k \quad \text{for} \quad n = 0, 1, 2, \cdots.$$

The numbers s_n are the **partial sums** of the series, and the numbers a_k are the **summands**. We speak of "the series $\sum_{k=0}^{\infty} a_k$." If there is a complex number S such that $\lim_{n \to \infty} s_n = S$, then the series **converges**; if not, the series **diverges**.

2.6.1. Proposition. *If the series* $\sum_{k=0}^{\infty} a_k$ *converges, then* $\lim_{k \to \infty} a_k = 0$.

Proof. Let $S := \lim_{n \to \infty} s_n$. Then $\lim_{n \to \infty} s_{n-1} = S$ also. Therefore of course $\lim_{n \to \infty}(s_n - s_{n-1}) = 0$. $\quad\square$

The converse of the Proposition is not true. For example, by the integral test $\sum_{k=1}^{\infty} \frac{1}{k}$ does not converge, since

$$\sum_{k=1}^{n} \frac{1}{k} > \int_{1}^{n+1} \frac{dx}{x} = \log(n+1) \to \infty \quad \text{as} \quad n \to \infty.$$

For an explanation of the integral test, see [67, Section 11.3].

The series $\sum_{k=0}^{\infty} a_k$ **converges absolutely** if $\sum_{k=0}^{\infty} |a_k|$ converges.

2.6.2. Proposition. *If a series converges absolutely, then it converges.*

Proof. Again by 2.4.5, the series $\sum_{k=0}^{\infty} |a_k|$ converges absolutely if and only if the partial sums form a Cauchy sequence, which means that for every $\epsilon > 0$ there exists n_0 such that

$$\sum_{k=m}^{n} |a_k| < \epsilon \quad \text{whenever} \quad n \geq m \geq n_0.$$

By the Triangle Inequality, it follows that

$$\left| \sum_{k=m}^{n} a_k \right| < \epsilon \quad \text{whenever} \quad n \geq m \geq n_0.$$

Thus the partial sums of the series $\sum_{k=0}^{\infty} a_k$ form a Cauchy sequence, hence it converges. $\square$

The converse of the Proposition is not true. The alternating series $\sum_{n=1}^{\infty} \frac{(-1)^n}{n}$ converges, but not absolutely. Recall the Alternating Series Test (see, for example, [67, Section 11.5] or [27, p. 442]).

B The Geometric Series

It is easy to identify and describe the set of z for which the **geometric series**

$$\sum_{k=0}^{\infty} z^k \tag{1}$$

converges. Figure 2.1-2, page 88, indicates how the sequence $\{z^k\}_{k=0}^{\infty}$ behaves for various $z \in \mathbb{C}$. The powers move outward toward infinity if $|z| > 1$, and inward toward 0 if $|z| < 1$. If $|z| = 1$, the points z^k are all on the unit circle. Since the summands of a series must tend to 0 if it converges, we know therefore that (1) diverges if $|z| \geq 1$. We shall see that it converges when $|z| < 1$, and the sum is easy to compute.

Observe what happens when the nth partial sum is multiplied by $1 - z$ and the product is simplified:

$$\left(\sum_{k=0}^{n} z^k\right)(1 - z) = (1 + z + z^2 + \ldots + z^n)(1 - z)$$

$$= 1 + z + z^2 + \ldots + z^n \tag{2}$$

$$-z - z^2 - \ldots - z^n - z^{n+1}$$

$$= 1 - z^{n+1}.$$

If $z \neq 1$, dividing by $1 - z$ gives a concise expression for the nth partial sum:

$$\sum_{k=0}^{n} z^k \equiv 1 + z + \ldots + z^n = \frac{1 - z^{n+1}}{1 - z} = \frac{1}{1 - z} - \frac{z^{n+1}}{1 - z}, \tag{3}$$

which, if $|z| < 1$, clearly converges to $\dfrac{1}{1 - z}$ as $n \to \infty$; thus

$$\frac{1}{1 - z} = \sum_{k=0}^{\infty} z^k \qquad \text{for } |z| < 1. \tag{4}$$

Consider the partial sums (3) as a sequence of functions defined on $D(0, 1)$. We can show that for each $r < 1$, they converge uniformly on $D(0, r)$, as follows: Given $\epsilon > 0$, choose n_0 sufficiently large so that

$$\frac{r^{n_0+1}}{1 - r} < \epsilon.$$

Note that the closer r is to 1, the larger n_0 must be. Then

$$n \geq n_0 \implies \max_{|z| \leq r} \left| \frac{1}{1 - z} - \sum_{k=0}^{n} z^k \right| = \max_{|z| \leq r} \left| \frac{z^{n+1}}{1 - z} \right| \leq \frac{r^{n_0+1}}{1 - r} < \epsilon.$$

In other words, for every $r < 1$ and $\epsilon > 0$, eventually the distance from the partial sum to $\dfrac{1}{1 - z}$ is less than ϵ uniformly on $D(0, r)$. However, the convergence is not uniform on $D(0, 1)$; the modulus of the difference between partial sum and limit is not even bounded there:

$$\sup_{|z| < 1} \left| \frac{z^{n+1}}{1 - z} \right| = \infty \quad \text{for all } n.$$

If a series can be recognized as a variant of (1), it becomes easy to evaluate. Some examples:

- $\displaystyle\sum_{k=0}^{\infty}(-1)^k 2^k z^k = \frac{1}{1+2z}$ for all $z \in D(0, \tfrac{1}{2})$, because the series is just (1), and we apply (4) again, but with $-2z$ in the role of z.

- $\displaystyle\sum_{k=0}^{\infty} 5^k(z-i)^k = \frac{1}{1-5(z-i)}$ for all $z \in D(i, \tfrac{1}{5})$; just apply (4) with $5(z-i)$ playing the role of z.

- $\displaystyle\sum_{k=0}^{\infty}\left(\frac{1}{z}\right)^k = \frac{1}{1-\frac{1}{z}} = \frac{z}{z-1}$ for $|z| > 1$; just let $1/z$ play the role of z in (4).

- $\displaystyle\sum_{k=m}^{\infty} z^k = z^m \sum_{k=m}^{\infty} z^{k-m} = z^m \sum_{n=0}^{\infty} z^n = \frac{z^m}{1-z}$ for $|z| < 1$; here, we have related the given series to (1) by a factorization and a change of index, $n = k - m$, and used (4) again.

- $0.12121212\cdots = \displaystyle\sum_{k=1}^{\infty} 12 \cdot 10^{-2k} = 12 \sum_{k=1}^{\infty}\left(\frac{1}{100}\right)^k = \frac{\frac{12}{100}}{1-\frac{1}{100}} = \frac{12}{99}.$

- $17.005555555\cdots = 17 + \displaystyle\sum_{k=3}^{\infty} 5 \cdot 10^{-k} = 17 + \frac{\frac{5}{1000}}{1-\frac{1}{10}} = 17 + \frac{5}{900}.$

Whenever b is a nonzero complex number, one can find a series that represents the function $z \mapsto \frac{1}{b-z}$ for every z in the disk $D(0, |b|)$ as follows:

$$\frac{1}{b-z} = \frac{1}{b}\left(\frac{1}{1-\frac{z}{b}}\right) = \frac{1}{b}\sum_{k=0}^{\infty}\left(\frac{z}{b}\right)^k = \sum_{k=0}^{\infty}\frac{z^k}{b^{k+1}}. \tag{5}$$

In the special case when $b = 1$, equation (5) is the same as (4).

The function $z \mapsto \dfrac{1}{b-z}$ is very nice everywhere in the plane except at the point b. It has a series representation (5) on the disk $D(0, |b|)$; so why shouldn't it have one on any other disk that does not include the point b? In fact, as we are about to see, for every $a \neq b$ there is a series in powers of $z - a$ that represents the function on the largest disk you can draw, centered at a, that does not contain the point b.

2.6.3. Proposition. *Let $a, b \in \mathbb{C}$, $a \neq b$, and consider the function $z \mapsto \frac{1}{b-z}$.*

(i) *For $|z - a| < |b - a|$, the function can be written as the sum of a series in powers of $z - a$:*

$$\frac{1}{b-z} = \sum_{k=0}^{\infty}\left(\frac{1}{b-a}\right)^{k+1}(z-a)^k. \tag{6}$$

The convergence is uniform for $|z - a| \leq s$ if $s < |b - a|$.

(ii) *For $|z - a| > |b - a|$, the function can be written as the sum of a series in negative powers of $z - a$:*

$$\frac{1}{b - z} = -\sum_{k=1}^{\infty} (b - a)^{k-1} \left(\frac{1}{z - a} \right)^k \tag{7}$$

The convergence is uniform for $|z - a| \geq s$ if $s > |b - a|$.

Proof. The following elementary computations, which we will use repeatedly, provide the proof:

(i) For $|z - a| < |b - a|$,

$$\frac{1}{b - z} = \frac{1}{b - a - (z - a)} = \frac{1}{b - a} \cdot \frac{1}{1 - \left(\dfrac{z - a}{b - a} \right)}$$

$$= \frac{1}{b - a} \sum_{k=0}^{\infty} \left(\frac{z - a}{b - a} \right)^k , \tag{8}$$

which is (6). Except for the constant factor $\dfrac{1}{b - a}$, the series is just (1) with $\dfrac{z - a}{b - a}$ in the role of z, and thus it converges for $\left| \dfrac{z - a}{b - a} \right| < 1$ and converges uniformly for $\left| \dfrac{z - a}{b - a} \right| < r$ if $r < 1$; take $s = r|b - a|$. Therefore the convergence set is the open disk with center a and radius $|b - a|$.

(ii) For $|z - a| > |b - a|$,

$$\frac{1}{b - z} = \frac{-1}{z - a - (b - a)} = -\frac{1}{z - a} \cdot \frac{1}{1 - \left(\dfrac{b - a}{z - a} \right)}$$

$$= -\frac{1}{z - a} \sum_{k=0}^{\infty} \left(\frac{b - a}{z - a} \right)^k , \tag{9}$$

which is (7). Except for the factor $\dfrac{1}{b - a}$, the series is just (1) with $\dfrac{b - a}{z - a}$ in the role of z, and thus it converges for $\left| \dfrac{b - a}{z - a} \right| < 1$, and converges uniformly for $\left| \dfrac{b - a}{z - a} \right| < r$ if $r < 1$. Take $s = \dfrac{|b - a|}{r}$.   □

2.6.4. Example. *Expand the function $z \mapsto \dfrac{1}{1 + iz}$ as a series in powers of $z - 1$. What is the radius of convergence?*

Solution. We use essentially the procedure (8):

$$\frac{1}{1+iz} = \frac{i}{i-z} = \frac{i}{i-1-(z-1)}$$

$$= \frac{i}{i-1} \cdot \frac{1}{1 - \left(\dfrac{z-1}{i-1}\right)}$$

$$= \frac{1-i}{2} \cdot \sum_{k=0}^{\infty} \left(\frac{z-1}{i-1}\right)^k$$

$$= \frac{1-i}{2} \cdot \sum_{k=0}^{\infty} \left(\frac{1}{i-1}\right)^k (z-1)^k,$$

which has radius of convergence $\sqrt{2}$. The procedure would allow us to expand the function in powers of $z - a$ for arbitrary $a \neq i$. □

C An Improved Root Test

You may recall the root test for the convergence of a series. See, for example, [67, Section 11.6]. In simplest form it is as follows. Let $c_n \geq 0$ for all n and suppose that the limit

$$L := \lim_{n \to \infty} (c_n)^{1/n}$$

exists. If $L < 1$, then the series $\sum_n c_n$ converges. If $L > 1$, it diverges. That version of the root test appears to be useless if the limit is undefined. In fact, though, the limit need not exist; the test can be stated in terms of the limsup, which is always defined (see 2.4B).

2.6.5. Proposition. *Let $\{c_n\}$ be a sequence of complex numbers and let*

$$\ell := \limsup_{n \to \infty} |c_n|^{1/n}.$$

If $\ell < 1$, then the series $\sum_n c_n$ converges absolutely. If $\ell > 1$, then c_n does not tend to 0 and hence the series $\sum_n c_n$ diverges.

Proof. If $\ell < 1$, pick a number r such that $\ell < r < 1$. Then

$$\text{eventually} \quad |c_n|^{1/n} < r \quad \text{and hence} \quad |c_n| < r^n.$$

By comparison with the geometric series, the series $\sum_n |c_n|$ must converge. If $\ell > 1$, then

$$\text{frequently} \quad |c_n|^{1/n} > 1 \quad \text{and hence} \quad |c_n| > 1.$$

It follows that c_n does not tend to 0, so that $\sum_n c_n$ cannot converge. □

When $\ell = 1$, the series may converge or diverge. For example, the series $\sum n^{-p}$ has $\ell = 1$ regardless of p; it converges if $p > 1$, diverges if $p \leq 1$.

D Power Series and the Cauchy-Hadamard Theorem

A **power series** is an infinite series of the form

$$\sum_{n=0}^{\infty} a_n(z-a)^n, \tag{10}$$

where $a \in \mathbb{C}$ is a fixed point, $\{a_n\}_{n=0}^{\infty} \subset \mathbb{C}$ is a fixed sequence, and z varies over $\mathbb{C}$. One may think of a power series as a family of infinite series, a different one for each value of z. The **set of convergence** of the power series (10) is the set of all z for which it converges. For z in that set, let $f(z)$ denote the sum. Then f is a well-defined complex-valued function on the set of convergence. When $z = a$, there is only one nonzero summand, namely, $a_0(z-a)^0 = a_0$. Therefore a always belongs to the set of convergence, and $f(a) = a_0$.

The power series

$$\sum_{n=1}^{\infty} na_n(z-a)^{n-1} \tag{11}$$

is the **derived series** or **first derived series** of (10). The derived series looks as if it has something to do with f'; indeed it does, as we shall explain. More generally, the kth **derived series of** (10) is

$$\sum_{n=k}^{\infty} n(n-1)\cdots(n-k+1)a_n\,(z-a)^{n-k}. \tag{12}$$

2.6.6. Theorem (Cauchy-Hadamard). *Consider the power series* (10). *Let*

$$R := \frac{1}{\limsup_{n\to\infty} |a_n|^{1/n}}. \tag{13}$$

In defining R we follow the conventions $\frac{1}{\infty} := 0$ and $\frac{1}{0} := \infty$, so $0 \leq R \leq \infty$.

(i) *If $R = 0$, then the power series converges only when $z = a$.*

(ii) *If $R = \infty$, then the power series converges for all $z \in \mathbb{C}$.*

(iii) *If $0 < R < \infty$, then the power series converges for $|z - a| < R$ and diverges for $|z - a| > R$.*

(iv) *If $0 < R \leq \infty$ and $0 < r < R$, then the power series converges uniformly for $|z - a| \leq r$.*

Proof. For each $z \neq a$, let

$$\ell(z) := \limsup_{n\to\infty} |a_n(z-a)^n|^{1/n}.$$

Proposition 2.6.5, with $a_n(z-a)^n$ in the role of c_n, tells us that the series converges if $\ell(z) < 1$ and diverges if $\ell(z) > 1$. The four parts of the Theorem now follow easily. First, observe that

$$\ell(z) = |z-a| \cdot (\limsup_{n \to \infty} |a_n|^{1/n}) = \frac{|z-a|}{R}.$$

(i) If $R = 0$, then $\ell(z) = \infty$ for all $z \neq a$.

(ii) If $R = \infty$, then $\ell(z) = 0$ for all $z \in \mathbb{C}$.

(iii) If $0 < R < \infty$, then $\ell(z)$ is less than 1 when $|z-a| < R$ and greater than 1 when $|z-a| > R$.

(iv) If $0 < r < R \leq \infty$, then $\ell(r) = \limsup |a_n(r-a)^n|^{1/n} = \frac{|r-a|}{R}$. Choose s such that $\frac{|r-a|}{R} < s < 1$. Eventually $|a_n r^n|^{1/n} < s$. In other words, for some n_0, $|a_n r^n| < s^n$ for all $n \geq n_0$. Therefore

$$n \geq n_0 \quad \text{and} \quad |z-a| \leq r \quad \Longrightarrow \quad |a_n(z-a)^n| < s^n$$

and hence

$$n \geq n_0 \quad \Longrightarrow \quad \sup_{|z-a|\leq r} \left| \sum_{k=n}^{\infty} a_k(z-a)^k \right| < \sum_{k=n}^{\infty} s^k = \frac{s^{n+1}}{1-s},$$

which tends to 0 as $n \to \infty$ because $s < 1$. $\square$

The number R is the **radius of convergence**, and $D(a, R)$ the **disk of convergence**, of the power series. Notice that the Theorem, in case (iii), is silent about convergence at points of the circle $|z-a| = R$. Some of the Exercises delve into what happens at those points, in various cases. Depending on the series, the set of convergence consists of the disk of convergence together with none, some, or all of the points on the circle.

2.6.7. Corollary. *The sum of a convergent power series is continuous on the disk of convergence.*

Proof. Part (iv) of the Theorem is the relevant result. The partial sums of the power series are polynomials, so they are all continuous on $\mathbb{C}$. They converge subuniformly on $D(a, R)$ to the sum of the power series. Therefore the sum is continuous on $D(a, R)$ by 2.4.9. $\square$

2.6.8. Corollary. *A power series and its derived series have the same radius of convergence.*

Proof. By the Theorem, the radii of convergence of the series (10) and (11) are the reciprocals of $\limsup |a_n|^{1/n}$ and $\limsup |na_n|^{1/n}$, respectively. The two are equal because $n^{1/n} \to 1$ as $n \to \infty$. (See Exercise 12.) $\qquad\square$

Please note that we have not yet established what the derived series converges to.

It follows from the Corollary, by induction on k, that the kth derived series (12) has the same radius of convergence R for every k.

2.6.9. Remark. An elementary criterion for the convergence of a series $\sum_{n=0}^{\infty} c_n$ is as follows. If the limit

$$L := \lim_{n \to \infty} \left| \frac{c_{n+1}}{c_n} \right|$$

exists (allowing the values 0 and ∞), then the series converges if $L < 1$ and diverges if $L > 1$. This result is sometimes convenient for determining the radius of convergence of a power series. For example, consider the power series

$$\sum_{n=0}^{\infty} 2^n z^{2n}.$$

To apply the criterion, regard z as fixed and let $c_n = 2^n z^{2n}$. Then

$$\left| \frac{c_{n+1}}{c_n} \right| = \frac{2^{n+1} |z|^{2n+2}}{2^n |z|^{2n}} = 2|z|^2.$$

The limit exists, of course. It is less than 1 if $|z| < 1/\sqrt{2}$, greater than 1 if $|z| > 1/\sqrt{2}$. Therefore the radius of convergence is $1/\sqrt{2}$.

E   Uniqueness of the Power Series Representation

Let f be a complex-valued function whose domain is a neighborhood of the point a. If the derivatives of f of all orders exist at a, then the **Taylor series** of f at a is defined to be

$$\sum_{n=0}^{\infty} \frac{f^{(n)}(a)}{n!} (z - a)^n.$$

The next result is that if a power series has a positive radius of convergence, and if f is its sum, then the derivatives of f of all orders exist throughout the disk of convergence, and the power series must be the Taylor series of f.

2.6.10. Theorem. *Let*

$$f(z) = \sum_{n=0}^{\infty} a_n (z - a)^n \quad \text{for} \quad z \in D(a, R), \tag{14}$$

where the radius of convergence R is positive. Then f is holomorphic, and the first derived series converges to f':

$$f'(z) = \sum_{n=1}^{\infty} n a_n (z-a)^{n-1} \quad \text{for } z \in D(a,R). \tag{15}$$

Furthermore, the derivatives of all orders $f', f'', \cdots, f^{(k)}, \cdots$ exist throughout $D(a,R)$, and for every positive integer k,

$$f^{(k)}(z) = \sum_{n=k}^{\infty} n(n-1)\cdots(n-k+1) a_n (z-a)^{n-k} \quad \text{for } z \in D(a,R). \tag{16}$$

The coefficients in (14) can be recovered from the values of f and its derivatives at a, thus:

$$a_n = \frac{f^{(n)}(a)}{n!} \quad \text{for } n = 0, 1, \cdots. \tag{17}$$

Proof. By Corollary 2.6.8, the derived series of (14) also has radius of convergence R. Let g be its sum:

$$g(z) := \sum_{n=0}^{\infty} n a_n (z-a)^{n-1} \quad \text{for } z \in D(a,R). \tag{18}$$

We need to show that $f' = g$. For $m = 1, 2, \cdots$, let

$$f_m(z) = \sum_{n=0}^{m} a_n (z-a)^n \quad \text{and} \quad g_m(z) = \sum_{n=0}^{m} n a_n (z-a)^{n-1}.$$

For every m, the polynomials f_m and g_m are of course holomorphic on $\mathbb{C}$, and $f'_m = g_m$. We will now make use of Proposition 2.5.12 and its proof. Since every g_m has a antiderivative, we know that for every closed path γ in $D(a,R)$,

$$\int_\gamma g_m = 0 \quad \text{for all } m.$$

Therefore, since g_m converges uniformly to g on γ,

$$\int_\gamma g = \lim_{m\to\infty} \int_\gamma g_m = 0.$$

Therefore g has an antiderivative on $D(a,R)$. One antiderivative of g is the function $z \mapsto \int_{[a,z]} g$, which equals

$$\lim_{m\to\infty} \int_{[a,z]} g_m = \lim_{m\to\infty} (f_m(z) - f_m(a)) = \lim_{m\to\infty} (f_m(z) - a_0) = f(z) - a_0.$$

Therefore $f' = g$ on $D(a, R)$.

We have proved the part of the Proposition that ends with equation (15). Apply this result with f', then f'', and so forth, in the role of f, to obtain equation (16) for all k. When $z = a$, (16) becomes $f^{(k)}(a) = k!a_k$, and (17) follows.   □

The implication (14) $\implies$ (17) in Theorem 2.6.10 is an important uniqueness result: The sum of a convergent power series determines its coefficients uniquely. Thus if two power series converge on an open disk to the same sum, then they are the same; that is, for each n, the nth coefficients are equal. This uniqueness result is the basis for the power series method of solving differential equations. The two examples below will introduce the method.

In subsection 3.1C we will prove another uniqueness result: To determine the coefficients a_n, and hence to determine f, it actually suffices to know the values of f on a sequence of points accumulating at a.

2.6.11. Example. *Let* $f(z) = \sum_{n=0}^{\infty} a_n z^n$ *for* $z \in D(0, r)$, *where* $r > 0$. *Suppose that* f *satisfies the differential equation* $f' = f$ *and the initial condition* $f(0) = 1$. *Identify the coefficients* a_n.

Solution. The differential equation tells us that

$$\sum_{1}^{\infty} na_n z^{n-1} = \sum_{0}^{\infty} a_n z^n.$$

We may rewrite that equation so that the same symbol appears as the superscript on z on both sides:

$$\sum_{0}^{\infty} (n+1)a_{n+1} z^n = \sum_{0}^{\infty} a_n z^n.$$

By the uniqueness result,

$$a_{n+1} = \frac{a_n}{n+1} \quad \text{for } n = 0, 1, 2, \cdots; \quad \text{so } a_{n+1} = \frac{a_0}{(n+1)!} \quad \text{for } n = 0, 1, 2, \cdots.$$

Since $f(0) = 1$, we know that $a_0 = 1$, and it follows that f is given by

$$f(z) = \sum_{n=0}^{\infty} \frac{z^n}{n!}. \tag{19}$$

The radius of convergence is ∞, as Exercises 8, 16, and 23 ask you to show. Thus we have found a solution of $f' = f$ whose domain is $\mathbb{C}$.   □

2.6.12. Example. *Let* $g(z) = \sum_{n=0}^{\infty} a_n z^n$ *for* $z \in D(0, r)$, *where* $r > 0$, *and suppose that* g *satisfies the differential equation* $g'' + g = 0$. *Identify the coefficients* a_n *to the extent possible. Show that they are all uniquely determined when* $g(0)$ *and* $g'(0)$ *are specified.*

Solution. The differential equation becomes

$$\sum_{n=2}^{\infty} n(n-1)a_n z^{n-2} + \sum_{n=0}^{\infty} a_n z^n = 0$$

or, equivalently,

$$\sum_{n=0}^{\infty} [(n+2)(n+1)a_{n+2} + a_n] z^n = 0 \quad \text{for } |z| < r.$$

By the uniqueness result,

$$a_{n+2} = -\frac{a_n}{(n+2)(n+1)} \quad \text{for } n = 0, 1, \cdots.$$

Thus

$$a_2 = -\frac{a_0}{2 \cdot 1}, \quad a_4 = -\frac{a_2}{4 \cdot 3} = \frac{a_0}{4!},$$

and so forth. Thus all the coefficients on even powers can be written in terms of a_0;

$$a_{2k} = \frac{(-1)^k a_0}{(2k)!} \quad \text{for } k = 1, 2, \ldots.$$

For the coefficients on odd powers, one shows similarly that

$$a_{2k+1} = \frac{(-1)^k a_1}{(2k+1)!} \quad \text{for } k = 1, 2, \ldots.$$

Thus we are free to choose a_0 and a_1, but then the other coefficients are determined; and the general solution is

$$g(z) = a_0 \sum_{k=0}^{\infty} \frac{(-1)^k z^{2k}}{(2k)!} + a_1 \sum_{k=0}^{\infty} \frac{(-1)^k z^{2k+1}}{(2k+1)!}. \tag{20}$$

For each of the two series, the radius of convergence is ∞. The solutions we have found are defined and satisfy the differential equation throughout $\mathbb{C}$. Note that $a_0 = g(0)$ and $a_1 = g'(0)$.   □

2.6.13. Remark. Exercise 16 asks you to confirm that the Taylor series for the exponential, sine, and cosine functions (as defined in 2.2B) are precisely the power series that appear in the two Examples above. Soon (Theorem 3.1.6) we will show that the Taylor series of a holomorphic function converges to the function. Once that is done, we will be justified in re-writing (19) and (20), respectively, as $f(z) = e^z$ and $g(z) = a_0 \cos z + a_1 \sin z$.

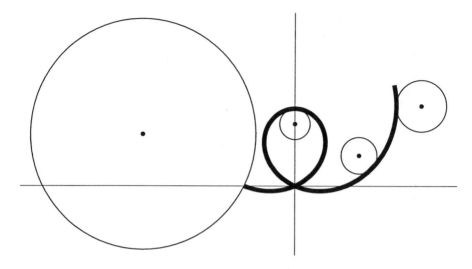

Figure 2.6-1   A pictorial dramatization of 2.6.14. A path is shown in bold. Let h be a function whose domain includes the path. The Cauchy integral of h over the path gives a function g defined everywhere off the path. For every point not on the path, draw the largest disk with that center that does not intersect the path. (Four such disks are shown.) Then g has a power series representation on that disk.

F   Integrals That Give Rise to Power Series

If γ is a curve given by $w : [a, b] \to \mathbb{C}$ and $h : \gamma \to \mathbb{C}$, then h is **piecewise continuous** on γ if $t \mapsto h(w(t))$ is piecewise continuous (see 1.5A).

Let h be a piecewise continuous complex-valued function defined on the path γ. Then the **Cauchy integral** of h over γ is the function given by

$$g(z) := \frac{1}{2\pi i} \int_\gamma \frac{h(w)}{w - z} dw \qquad (z \notin \gamma). \tag{21}$$

The function $(z, w) \mapsto \dfrac{1}{w - z}$ is the **Cauchy kernel**.

The next result states that g is well defined and holomorphic off γ without any further conditions on h. The function h may or may not be continuous, and its domain needs only to contain γ. On the other hand, no relation is asserted between g and h.

2.6.14. Proposition. *Let h be a piecewise continuous complex-valued function defined on the path γ, and let g be defined by (21). Let z_0 be a point that is not on the path γ. Let r be the distance from z_0 to γ; in other words, let $D(z_0, r)$ be the largest open disk with center z_0 that does not intersect the path. Then g has a power series representation on $D(z_0, r)$.*

Proof. Figure 2.6-1 illustrates the situation. By Proposition 2.6.3 (with z_0 and w in the roles of a and b), for each $z \in D(z_0, r)$,

$$\frac{1}{w - z} = \sum_{k=0}^{\infty} \left(\frac{1}{w - z_0} \right)^{k+1} (z - z_0)^k, \quad \text{convergent if } w \in \gamma \text{ and } z \in D(z_0, r).$$

Let $0 < s < r$. The convergence is uniform for $w \in \gamma$ and $|z - z_0| \le s$, since for all such z and w,

$$\left| \frac{z - z_0}{w - z_0} \right| \le \frac{s}{r} < 1.$$

Therefore the order of integration and summation can be reversed after we substitute the power series for $\dfrac{1}{w - z}$ in (21):

$$
\begin{aligned}
g(z) &= \frac{1}{2\pi i} \int_\gamma \frac{h(w)}{w - z} dw \\
&= \frac{1}{2\pi i} \int_\gamma h(w) \sum_{k=0}^{\infty} \left(\frac{1}{w - z_0} \right)^{k+1} (z - z_0)^k dw \\
&= \sum_{k=0}^{\infty} \left(\frac{1}{2\pi i} \int_\gamma \frac{h(w)}{(w - z_0)^{k+1}} dw \right) (z - z_0)^k.
\end{aligned}
\tag{22}
$$

$\square$

2.6.15. Remark. Notice that by the uniqueness of the power series representation (Proposition 2.6.10), the coefficient on $(z - z_0)^k$ in (22) must equal $g^{(k)}(z_0)/k!$. Since z_0 is an arbitrary point not on γ, it follows that there is an integral representation for $g^{(k)}$ everywhere on the complement of the path:

$$g^{(k)}(z_0) = \frac{k!}{2\pi i} \int_\gamma \frac{h(w)}{(w - z_0)^{k+1}} dw \quad \text{for } k = 0, 1, 2, \cdots \quad (z_0 \notin \gamma).$$

Next, we will arrive at the same results by another approach. Consider the following rather general situation. Let $\varphi(z, w)$ be a continuous function of two complex variables defined whenever $z \in O$ and $w \in E$, where O is open and E is a set containing a path γ given by $w : [0, 1] \to E$. Note that $E = \gamma$ is possible. Consider the function

$$g(z) = \int_\gamma \varphi(z, w) dw \equiv \int_0^1 \varphi(z, w(t)) w'(t) dt \quad \text{for } z \in O. \tag{23}$$

In (23) we have "integrated away" the dependence on w, obtaining a function of z only. This is a general type of integral representation of a function g. Consider now the difference quotient

$$\frac{g(z) - g(z_0)}{z - z_0} = \frac{1}{z - z_0} \left(\int_\gamma \varphi(z, w) dw - \int_\gamma \varphi(z_0, w) dw \right)$$

$$= \int_\gamma \frac{\varphi(z, w) - \varphi(z_0, w)}{z - z_0} dw.$$

If the limit

$$\lim_{z \to z_0} \frac{\varphi(z, w) - \varphi(z_0, w)}{z - z_0} = \frac{\partial \varphi}{\partial z}(z_0, w)$$

exists and is uniform for w on the path γ, then we can properly reverse the order of the two limit operations:

$$g'(z) = \int_\gamma \frac{\partial \varphi}{\partial z}(z, w) dw, \tag{24}$$

so that g is holomorphic on O. A Cauchy integral (21) is the special case of (23) in which

$$\varphi(z, w) = \frac{1}{2\pi i} \frac{h(w)}{w - z},$$

$E = \gamma$, and O is the complement of γ. Differentiation works as in (24):

$$g'(z) = \frac{1}{2\pi i} \int_\gamma \frac{h(w)}{(w - z)^2} dw.$$

Thus we have an integral that represents g'. Furthermore, (24) applies equally well with g' in the role of g, so that

$$g''(z) = \frac{2}{2\pi i} \int_\gamma \frac{h(w)}{(w - z)^3} dw;$$

and so forth. Thus the function g defined by (21) is holomorphic. In fact, it is infinitely differentiable at each point off the path:

$$g^{(n)}(z) = \frac{n!}{2\pi i} \int_\gamma \frac{h(w)}{(w - z)^{n+1}} dw \quad \text{for } n = 0, 1, 2, \cdots \quad (z \notin \gamma).$$

Exercises

1. Without referring to any of the results in this section, devise an elementary proof of the following result, which is an important precursor to the uniqueness result

of Theorem 2.6.10: Let p and q be two polynomials:

$$p(z) := \sum_{k=0}^{n} a_k z^k = a_0 + a_1 z + a_2 z^2 + \cdots + a_n z^n,$$

$$q(z) := \sum_{k=0}^{m} b_k z^k = b_0 + b_1 z + b_2 z^2 + \cdots + b_m z^m.$$

Suppose that $p(z) = q(z)$ for all $z \in \mathbb{C}$. Then $m = n$ and $a_k = b_k$ for all k.

2. Find the series in powers of z that converges to $\dfrac{1}{\frac{1}{2} - z}$ and identify the set of convergence.

3. Find the series in powers of $z - 3$ that converges to $\dfrac{1}{1 - z}$ and determine the set of convergence.

4. Find the series in powers of $z - 1 - i$ that converges to $\dfrac{1}{1 - z}$ and determine the set of convergence.

5. In each case, compute the exact sum.

(a) $\displaystyle\sum_{k=0}^{\infty} \left(\frac{1}{2}\right)^k$

(b) $\displaystyle\sum_{k=0}^{\infty} \left(\frac{1+i}{2}\right)^k$

(c) $\displaystyle\sum_{k=-4}^{\infty} \left(\frac{1}{3}\right)^k$

(d) $\displaystyle\sum_{k=-\infty}^{0} \left(\frac{2}{i}\right)^k$

(e) $\displaystyle\sum_{k=0}^{\infty} \left(\frac{i}{2}\right)^k$

(f) $\displaystyle\sum_{k=3}^{\infty} \left(\frac{1}{2}\right)^k$

(g) $\displaystyle\sum_{k=-4}^{\infty} \left(\frac{i}{3}\right)^k$

(h) $\displaystyle\sum_{k=4}^{8} \left(\frac{1}{3}\right)^k$

(i) $\displaystyle\sum_{k=0}^{\infty} \left(\frac{2i}{3}\right)^k$

(j) $\displaystyle\sum_{k=3}^{\infty} \left(\frac{i}{2}\right)^k$

(k) $\displaystyle\sum_{k=-4}^{4} \left(\frac{1}{2}\right)^k$

(l) $\displaystyle\sum_{k=4}^{8} \left(\frac{3}{i}\right)^k$

6. Give an example of two power series, $\sum_{n=0}^{\infty} a_n z^n$ with radius of convergence 2 and $\sum_{n=0}^{\infty} b_n z^n$ with radius of convergence 3, such that the radius of convergence of $\sum_{n=0}^{\infty} a_n b_n z^n$ is *not* 6.

7. Consider the power series $\sum a_n z^n$. Unlike the limit superior in the Cauchy-Hadamard Theorem, the limit $K := \lim\limits_{n \to \infty} \left| \dfrac{a_{n+1}}{a_n} \right|$ may or may not exist. Prove that if it does exist, then the radius of convergence R of is $1/K$.

8. In each case, use either the method of Remark 2.6.9 or that of Exercise 7 to obtain the radius of convergence of the given series.

(a) $\displaystyle\sum_{n=1}^{\infty} n^n z^n$

(c) $\displaystyle\sum_{n=0}^{\infty} \frac{z^n}{n!}$

(e) $\displaystyle\sum_{n=0}^{\infty} 3^n z^n$

(b) $\displaystyle\sum_{n=0}^{\infty} 3^n z^{2n}$

(d) $\displaystyle\sum_{n=0}^{\infty} 3^n z^{3^n}$

(f) $\displaystyle\sum_{n=0}^{\infty} n^2 z^{n^2}$

9. For each series in Exercise 8, show how the Cauchy-Hadamard Theorem can be used to obtain the radius of convergence.

10. In each case, find the radius of convergence of the given series and justify your answer.

(a) $\displaystyle\sum_{n=1}^{\infty} n z^{2n}$

(c) $\displaystyle\sum_{n=0}^{\infty} \frac{2^n z^n}{n!}$

(e) $\displaystyle\sum_{n=0}^{\infty} \frac{3^n z^n}{4^n + 5^n} z^n$

(b) $\displaystyle\sum_{n=0}^{\infty} \frac{2^n z^{2n}}{n^2 + n}$

(d) $\displaystyle\sum_{n=1}^{\infty} \left(1 - \frac{1}{n}\right)^{n^2} z^n$

(f) $\displaystyle\sum_{n=1}^{\infty} n^{(\ln n)^2} z^n$

11. The Cauchy-Hadamard Theorem offers no conclusion about convergence when $|z| = R$ ("on the circle of convergence" or "on the rim"). Give an example of a power series with $R = 1$ that ...

   (a) ... converges uniformly for $|z| \leq 1$.

   (b) ... diverges at every z with $|z| = 1$.

   (c) ... converges at some points on the rim but not all.

   (d) ... converges uniformly for $|z| \leq 1$, but such that its derived series converges nowhere for $|z| = 1$.

12. Prove that $\lim_{n \to \infty} n^{1/n} = 1$.

13. Let $f(z) = \sum_{n=0}^{\infty} a_n z^n$ for $z \in D(0, r)$, where $r > 0$. Suppose that f satisfies the differential equation $f' = cf$, where c is a nonzero constant, and the initial condition $f(0) = 1$. Identify the coefficients a_n.

14. Let $g(z) = \sum_{n=0}^{\infty} a_n z^n$ for $z \in D(0, r)$, where $r > 0$, and suppose that g satisfies the differential equation $g'' + 2g = 0$, $g(0) = 2$, and $g'(0) = 0$. What can you say about g?

15. Find the Taylor series for the function $z \mapsto z^3$ at each of the points 0, 1, and 2.

16. Verify that the Taylor series at 0 of the exponential, sine, and cosine functions are respectively as follows:

$$\sum_{n=0}^{\infty} \frac{z^n}{n!}; \quad \sum_{k=0}^{\infty} \frac{(-1)^k z^{2k}}{(2k)!}; \quad \text{and} \quad \sum_{k=0}^{\infty} \frac{(-1)^k z^{2k+1}}{(2k+1)!}.$$

Verify that the radius of convergence for each is ∞.

17. Find the Taylor series for the function $z \mapsto e^z$ at 1.

18. Find the Taylor series for the function $z \mapsto \cos z$ at $\frac{\pi}{2}$.

19. Prove that the series $\sum_0^{\infty} a_k b_k$ converges if these three conditions hold: (i) the partial sums of the series $\sum a_k$ are bounded; (ii) the numbers b_k tend to 0 as $k \to \infty$; and (iii) $\sum_0^{\infty} |b_k - b_{k+1}| < \infty$. *Note:* Dedekind proved this result in 1863; see [33, p. 137]. The three conditions are easily satisfied if, for example, $a_k = (-1)^k$ and $0 < b_{k+1} < b_k$ for all k; thus the familiar Alternating Series Test is a special case.

20. Show that the result of Exercise 19 applies when $a_k = z^k$, where $|z| = 1$ and $z \neq 1$, and $b_k = \frac{1}{k+1}$. Obtain an example of a power series with radius of convergence 1 that converges at every point of the circle $|z| = 1$ except 1. The series may be written $\sum_{k=1}^{\infty} \frac{e^{ikt}}{k}$. Prove that for each $\alpha \in (0, \pi)$, its convergence is uniform for $\alpha < t < 2\pi - \alpha$. *Note:* Written in that form, this is the Fourier series for the sawtooth wave, and provides an excellent example of Gibbs's phenomenon. See, for example, [31, Section 1.5].

21. Prove that the power series $\sum_1^{\infty} n^2 z^n$ converges to $\frac{z(1+z^2)}{(1-z)^4}$.

22. Prove that if a power series $\sum_0^{\infty} c_n (z-a)^n$ converges uniformly on the open disk $D(a,r)$, then it converges uniformly on the closed disk $\bar{D}(a,r)$.

23. Use the Cauchy-Hadamard Theorem to find the radius of convergence for each of the power series

$$\sum_{n=1}^{\infty} \frac{z^n}{n!}, \quad \sum_{n=1}^{\infty} \frac{n!}{n^n} z^n, \quad \text{and} \quad \sum_{n=1}^{\infty} \frac{(n!)^2}{n^n} z^n.$$

24. The Bessel function of the first kind of order k is given by

$$J_k(z) := \sum_{n=0}^{\infty} \frac{(-1)^n \left(\frac{z}{2}\right)^{k+2n}}{n!(n+k)!}.$$

Show that the series has radius of convergence ∞.

25. Show that the function J_k in the previous Exercise satisfies the differential equation $z^2 f'' + zf' + (z^2 - k^2)f = 0$.

26. Consider the differential equation

$$(1 - z^2)y'' - 2zy' + \lambda y = 0. \tag{25}$$

Suppose that there is a solution $y(z) = \sum_{n=0}^{\infty} a_n z^n$, where the power series has a positive radius of convergence. Show that if the appropriate power series for y, y', and y'' are substituted in (25), then it becomes (if we write the summands given by $n = 0$ and $n = 1$ separately),

$$(2a_2 + \lambda a_0) + (6a_3 + (\lambda - 2)a_1)z$$

$$+ \sum_{n=2}^{\infty} [(n+2)(n+1)a_{n+2} - (n(n+1) - \lambda)a_n] z^n = 0.$$

By the uniqueness theorem, then, the information contained in the differential equation about the function y is equivalent to the information contained in these equations involving the coefficients a_n :

$$2a_2 + \lambda a_0 = 0;$$
$$6a_3 + (\lambda - 2)a_1 = 0;$$
$$(n+1)(n+2)a_{n+2} - (n(n+1) - \lambda)a_n = 0 \quad \text{for } n = 2, 3, \cdots.$$

It follows that in the general solution $y(z)$, a_0 and a_1 may be assigned any values whatever, and then the other coefficients are determined. Show that if $\lambda = N(N + 1)$, where N is a non-negative integer, then one solution must be a polynomial. With appropriate normalization, these are the Legendre polynomials P_N. For the cases $N = 0$ through 5, they are as follows:

$$P_0(z) = 1; \qquad\qquad\qquad P_1(z) = z;$$
$$P_2(z) = \frac{1}{2}(3z^2 - 1); \qquad\qquad P_3(z) = \frac{1}{2}(5z^3 - 3z);$$
$$P_4(z) = \frac{1}{8}(35z^4 - 30z^2 + 3); \qquad P_5(z) = \frac{1}{8}(63z^5 - 70z^3 + 15z)$$

Show that P_0 through P_5 are indeed solutions as claimed.

27. Let f be a one-to-one holomorphic function on D: $f(z) = \sum_{n=0}^{\infty} c_n z^n$. Prove that the area of $f[D]$ equals $\pi \sum_{n=1}^{\infty} n|c_n|^2$.

Help on Selected Exercises

2. $\dfrac{1}{\frac{1}{2} - z} = \dfrac{1}{\frac{1}{2}(1 - 2z)} = 2\sum_{k=0}^{\infty}(2z)^k$, convergent for $|z| < \dfrac{1}{2}$.

3. As in Exercise 2, the method is to follow the procedure (8), but with 3 in the role of a and 1 in the role of b.

5. (a) 2.   (b) $\frac{2}{1-i} \equiv 1 + i$.   (c) $\frac{243}{2}$.   (d) $\frac{2}{2-i} \equiv \frac{4}{5} + i\frac{2}{5}$.

8. (a) 0.   (b) $\frac{1}{\sqrt{3}}$.   (c) ∞.   (d) 1.   (e) $\frac{1}{3}$.   (f) 1.

10. (a) 1.   (b) $\frac{1}{\sqrt{2}}$.   (c) ∞.   (d) e.   (e) $\frac{5}{3}$.   (f) 1.

11. Exercise 19 may provide an idea for part 11c.

12. One would like to show that for $\epsilon > 0$, eventually $n^{1/n} \le 1 + \epsilon$. To do so, consider the following. If $n^{1/n} > 1 + \epsilon$, then

$$n > (1+\epsilon)^n = 1 + n\epsilon + n(n-1)\epsilon^2 + \cdots > n(n-1)\epsilon^2,$$

which implies that $\epsilon^{-2} + 1 > n$.

15. At 0: z^3. At 1: $1 + 3(z-1) + 3(z-1)^2 + (z-1)^3$. At 2: $8 + 12(z-2) + 6(z-2)^2 + (z-3)^3$.

16. See Section 2.3, Example 2.3.10 and Exercise 3. We have shown that the functions are differentiable everywhere, and identified their derivatives. So this is a matter of evaluating the successive derivatives at 0 for each of the functions and using the definition of the Taylor series (subsection E).

19. Let $s_n = \sum_{k=1}^n a_k$. Then for some B, $|s_n| \le B$ for all n. Apply Abel's partial summation method, as follows:

$$
\begin{aligned}
a_1 b_1 &= s_1 b_1 \\
&= s_1(b_1 - b_2) + s_1 b_2; \\
a_1 b_1 + a_2 b_2 &= s_1(b_1 - b_2) + s_2(b_2 - b_3) + s_2 b_3;
\end{aligned}
$$

and by induction,

$$\sum_{k=1}^n a_k b_k = \sum_{k=1}^n s_k(b_k - b_{k+1}) + s_n b_{n+1}.$$

The result follows. One may estimate the error thus:

$$\left| \sum_{k>n} a_k b_k \right| \le B \left(\sum_{k>n} |b_k - b_{k+1}| + |b_{n+1}| \right).$$

21. Make use of the relation between the power series for a function and the one for its derivative. Let $f(z)$ be the sum of the given series on $D(0,1)$. Write down the series for g such that $f(z) = zg'(z)$. Then write down the series for h such that $g(z) = zh'(z)$, and notice that $h(z) = \frac{z}{1-z}$. Then work backward to identify g and f.

23. Sometimes it is useful to know precisely how fast $n!$ tends to ∞ as $n \to \infty$. Stirling's Formula provides this information:

$$n! = \left(\frac{n}{e}\right)^n \sqrt{n}\, C_n, \qquad \text{where} \qquad \lim_{n\to\infty} C_n = \sqrt{2\pi}.$$

For a proof, see [8, p. 252, pp. 299–301].

27. It follows from Exercise 7, Section 1.5, on page 71, that the area equals

$$\iint_D |f'(x+iy)|^2 dxdy. \tag{26}$$

Change variables in the domain from Cartesian to polar, and, since $f'(z) = \sum_{n=0}^{\infty} c_n n z^{n-1}$, (26) becomes

$$\int_0^1 \int_0^{2\pi} \left(\sum_{n=0}^{\infty} c_n n r^{n-1} e^{i(n-1)\theta}\right) \left(\sum_{m=0}^{\infty} \bar{c}_m m r^{m-1} e^{-i(m-1)\theta}\right) r\, dr\, d\theta.$$

3

The Cauchy Theory

Titled in honor of the great French mathematician Baron Augustin Louis Cauchy (1789–1857), this chapter presents, in a modern form, the framework of complex function theory that we owe primarily to his efforts.

The chapter is not an account of exactly what Cauchy contributed. This book is concerned with the logic of the theory, and with a sequence of presentation conducive to learning, and certainly not with the historical order of discoveries. Chapters 1 and 2 include ideas due to Cauchy, as well as ideas that arose after Cauchy's time. Chapter 3 itself includes, for example, a formulation and proof of Cauchy's Theorem that has been developed in recent decades.

In his recent book [66], Frank Smithies of St. John's College, Cambridge, traces in detail the process by which Cauchy developed the theory in the years from 1814 to 1831. After you have completed a first course in complex analysis, you may find the book's last chapter [66, pp. 186–204], "Summary and Conclusions," both understandable and interesting. The brief and accessible "Historical Introduction" in Reinhold Remmert's book [57, pp. 1–7] discusses Cauchy's contributions alongside those of Georg Friedrich Bernhard Riemann (1826–1866) and Karl Theodor Wilhelm Weierstrass (1815–1897). See also the comments on the "Cauchy, Riemann, and Weierstrass points of view," in [57, pp. 239–241]. Short biographies of those and other workers, and a summary of literature, appear in the back of Remmert's book [57, pp. 416–434]. As both Remmert and Smithies mention, Carl Friedrich Gauss (1777–1855) wrote a letter in 1811 which makes clear that Gauss already knew about the Cauchy integral theorem at that time. But the letter was not published until 1880.

Cauchy's work began with the effort to find the exact values of certain definite integrals. Chapter 4 will present such problems and their solutions, using the theory developed in this Chapter. Section 3.2 presents a powerful and easy-to-use version of

Cauchy's Theorem. Section 3.3 explains isolated singularies and residues and gives many examples. Section 3.4 shows how the Residue Theorem follows easily from Cauchy's Theorem, and you will then be prepared to begin Chapter 4. Section 3.5 uses the two big theorems to prove the Argument Principle and related results, and to develop the important mapping properties of holomorphic functions. Section 3.6 introduces the Riemann sphere, providing a fascinating way to visualize the extended plane.

3.1 FUNDAMENTAL PROPERTIES OF HOLOMORPHIC FUNCTIONS

A Integral and Series Representations

To say that f is holomorphic on an open set means that f' is defined at every point. We have established (2.6.10) that the sum f of a power series is holomorphic on the disk of convergence; and, more than that, its derivatives of all orders exist: f', f'', f''', and so forth. We know also (2.6.14) that if a function can be written as a Cauchy integral, then it has a power series representation. Now we will show that every holomorphic function can be written as a Cauchy integral. It then follows that every holomorphic function has a power series representation on every disk that lies in its domain. Thus we arrive at the startling fact that the existence of f' implies the existence of the derivatives of all orders.

3.1.1. Theorem (Cauchy's Theorem for a Convex Domain). *Let f be holomorphic on the convex open set $O \subseteq \mathbb{C}$. Let Γ be a contour in O. Then*

$$\int_\Gamma f = 0. \tag{1}$$

Also,

$$f(z) \cdot W_\Gamma(z) = \frac{1}{2\pi i} \int_\Gamma \frac{f(w)}{w-z} dw \quad \text{for all} \quad z \in O \setminus \Gamma, \tag{2}$$

For every point $z \in O$, we can of course find a contour Γ such that $W_\Gamma(z) = 1$; for example, we can let Γ be a circle centered at z. Then (2) gives f, on a set containing the point, as a Cauchy integral.

The Theorem becomes false if the word "convex" is omitted and no other change is made. An example should come to mind: The function $z \mapsto z^{-1}$ is holomorphic on $\mathbb{C}'$, and $\int_{|z|=1} \frac{dz}{z} = 2\pi i$.

Proof. Corollary 2.5.15 of Goursat's Lemma gives (1). It remains to prove (2). Let $z \in O$. The function given by

$$q(w) := \begin{cases} \dfrac{f(w) - f(z)}{w - z} & \text{for } w \neq z; \\[2mm] f'(z) & \text{for } w = z, \end{cases} \tag{3}$$

is holomorphic at every point in O other than z, because it is the quotient of two holomorphic functions, the denominator being nonzero. It is also continuous at z. By 2.5.16, $\int_{\partial T} q = 0$ for every triangle T contained in O for which z is a vertex. By 2.5.12, q has an antiderivative on O, and hence $\int_\Gamma q = 0$ for every contour Γ in O. Therefore if $z \in O \setminus \Gamma$, then

$$
\begin{aligned}
0 = \int_\Gamma q(w)dw &= \int_\Gamma \frac{f(w) - f(z)}{w - z} dw \\
&= \int_\Gamma \frac{f(w)}{w - z} dw - f(z) \int_\Gamma \frac{dw}{w - z} \\
&= \int_\Gamma \frac{f(w)}{w - z} dw - f(z) \cdot 2\pi i \cdot W_\Gamma(z);
\end{aligned}
\tag{4}
$$

and (2) follows. $\square$

3.1.2. Example. *Evaluate the integrals*

$$
\int_{|z|=1} e^z dz, \quad \int_{|z|=1} \frac{e^z}{z - (1+i)} dz, \quad \text{and} \quad \int_{|z|=1} \frac{e^z}{z + \frac{1}{2}} dz.
$$

Solution. We apply Theorem 3.1.1 with the unit circle in the role of Γ. Each of the first two integrals equals 0, because each of the functions $z \mapsto e^z$ and $z \mapsto \dfrac{e^z}{z - (1+i)}$ is holomorphic on a convex set, $D(0, 1.1)$ for example, that contains Γ. Equation (2), with $f(z) = e^z$, implies that the third integral equals $2\pi i e^{-1/2}$. $\square$

3.1.3. Example. *Let Γ be the contour shown in Figure 2.5-2, page 163. Evaluate the integrals*

$$
\int_\Gamma e^z dz, \quad \int_\Gamma \frac{e^z}{z - (8+8i)} dz, \quad \int_\Gamma \frac{e^z}{z + \frac{1}{2}} dz, \quad \text{and} \quad \int_\Gamma \frac{e^z}{z - 6i} dz.
$$

Solution. Each of the first two integrals equals 0, because each of the functions $z \mapsto e^z$ and $z \mapsto \dfrac{e^z}{z - (8+8i)}$ is holomorphic on a convex set, $D(0, 8)$ for example, that contains Γ. Equation (2), with $f(z) = e^z$, gives the answer for the second two integrals. Since $W_\Gamma(-\frac{1}{2}) = 5$ and $W_\Gamma(6i) = 1$, we obtain that their values are $10\pi i e^{-1/2}$ and $2\pi i e^{6i}$, respectively. $\square$

Theorem 3.1.1 will serve reasonably well for most applications, but it often needs to be combined with *ad hoc* arguments. The next section will reveal a more powerful Cauchy integral representation theorem which is not limited to convex domains, and which is easy to understand and use.

3.1.4. Remark. Theorem 3.1.1 implies the mean value property of harmonic functions (subsection 1.6C), as follows. Given a function u harmonic on $\bar{D}(a, r)$, and hence on $D(a, r + \epsilon)$ for some $\epsilon > 0$, let f be a holomorphic function of which u is the real part. Apply (2) with the circle $|w - a| = r$ in the role of Γ:

$$f(z) = \frac{1}{2\pi i} \int_{|w-a|=r} \frac{f(w)}{w - z} dw \quad \text{for } z \in D(a, r).$$

The circle is of course given by $w(\theta) = a + re^{i\theta}$ $(0 \leq \theta \leq 2\pi)$. Now let $z = a$ and write the integral as a pullback:

$$f(a) = \frac{1}{2\pi i} \int_0^{2\pi} \frac{f(re^{i\theta})}{re^{i\theta}} ire^{i\theta} d\theta = \frac{1}{2\pi} \int_0^{2\pi} f(a + re^{i\theta}) d\theta.$$

Take the real part on both sides:

$$u(a) = \frac{1}{2\pi} \int_0^{2\pi} u(a + re^{i\theta}) d\theta.$$

3.1.5. Remark. This is a reminder of the discussion at the end of 2.6F. Under the hypothesis of Theorem 3.1.1, consider the case when $z \in O \setminus \Gamma$ and $W_\Gamma(z) = 1$. Then (2) becomes

$$f(z) = \frac{1}{2\pi i} \int_\Gamma \frac{f(w)}{w - z} dw.$$

From this it follows that $f'(z)$ exists (as we knew already), and we have an integral representation for it:

$$f'(z) = \frac{1}{2\pi i} \int_\Gamma \frac{f(w)}{(w - z)^2} dw.$$

In fact, the derivatives of all orders exist, and we have an integral representation for each of them:

$$f^{(n)}(z) = \frac{n!}{2\pi i} \int_\Gamma \frac{f(w)}{(w - z)^{n+1}} dw.$$

The existence of all the derivatives follows also, by 2.6.10, from the existence of a power series representation, which is the next result.

3.1.6. Theorem (The Power Series Representation). *Let f be holomorphic on the open set O. Let $D(a, r)$ be a disk contained in O. Then f is the sum of a power series that converges on that disk:*

$$f(z) = \sum_{k=0}^{\infty} c_k(z - a)^k \quad \text{for } |z - a| < r.$$

For every $k \in \mathbb{Z}^+$, and every number s with $0 < s < r$,

$$c_k = \frac{1}{2\pi i} \int_{|w-a|=s} \frac{f(w)}{(w - a)^{k+1}} dw.$$

Also, for every $k \in \mathbb{Z}^+$,

$$c_k = \frac{f^{(k)}(a)}{k!};$$

the series is of course the Taylor series of f *at* a.

Proof. Select a number s with $0 < s < r$ and a point z with $|z - a| < s$. By 3.1.1,

$$f(z) = \frac{1}{2\pi i} \int_{|w-a|=s} \frac{f(w)}{w - z} dw.$$

By the procedure in the proof of 2.6.14, with the circle $|z - a| = s$ in the role of γ and f in the role of h,

$$f(z) = \sum_{k=0}^{\infty} c_{k,s}(z - a)^k \quad \text{for } |z - a| < s, \quad \text{where}$$

$$c_{k,s} = \frac{1}{2\pi i} \int_{|w-a|=s} \frac{f(w)}{(w - a)^{k+1}} dw.$$

(5)

We have written $c_{k,s}$ instead of c_k because it appears, at first, that these coefficients depend on s. But if $0 < s_1 < s_2 < r$, we have, at least for $|z - a| < s_1$,

$$\sum_{k=0}^{\infty} c_{k,s_1}(z - a)^k = \sum_{k=0}^{\infty} c_{k,s_2}(z - a)^k,$$

both series converging to $f(z)$. Therefore $c_{k,s_1} = c_{k,s_2} = \dfrac{f^{(k)}(a)}{k!}$ for all k (by 2.6.10). For each k, therefore, $c_{k,s}$ is the same for all s, so that we may as well write c_k. Thus there is only one series given by (5), and the Proposition is proved. $\square$

Notice that if $O = \mathbb{C}$, then r may be taken arbitrarily large, and the series converges for all z.

3.1.7. Remark. In developing the theory, we have made use of several properties that a function might have: independence of path for its integrals, existence of its derivative, representation as a power series, representation as a Cauchy integral. We have been in the process of tying these properties closely together, showing that they are very nearly equivalent. The Theorem just proved completes the logical circuit, as the next two Corollaries indicate. The first is a kind of converse to 2.5.15.

3.1.8. Corollary. *Let f be a continuous function on the open connected set O. If the integrals of f are locally independent of path, then f is holomorphic on O.*

Proof. For each point $a \in O$, there is a disk $D(a, r)$ on which the integrals of f are independent of path. Therefore f has an antiderivative F on that disk, by 2.5.12. Being holomorphic, F has a power series representation on the disk by 3.1.6.

Therefore the derivatives $F', F'', \cdots$ all exist, so in particular $f' \equiv F''$ exists, so that f is holomorphic on the disk. Therefore f is holomorphic on O.   □

3.1.9. Corollary. *Let f be continuous on the open set O and holomorphic on $O \backslash \{a\}$, where a is a point in O. Then f is holomorphic at a.*

Proof. Let $r > 0$ such that $D(a, r) \subseteq O$. By 2.5.16, condition (iii) of 2.5.12 is satisfied, and hence f has an antiderivative F on $D(a, r)$. The conclusion follows in in the previous proof.   □

3.1.10. Remark. Recall the function q, the difference quotient given by (3). We now know more about q than when we used it in the proof of Theorem 3.1.1. Thanks to Corollary 3.1.9, we know that q is holomorphic at z as well as at the other points where f is holomorphic. We can prove that fact another way, as follows. Since f is holomorphic at z, by 3.1.6 there is a series representation

$$f(w) = f(z) + f'(z)(w - z) + \frac{f''(z)}{2}(w - z)^2 + \cdots$$

for w near z. Therefore q also has a power series representation:

$$q(w) \equiv \frac{f(w) - f(z)}{w - z} = f'(z) + \frac{f''(z)}{2}(w - z) + \cdots.$$

Therefore q is holomorphic, and not merely continuous, at z.

The preceding Remark paves the way for the following Proposition, a technical result which is useful to know. In 3.1.1 we stated and proved (1) and (2) as if they were separate results. In fact, they are equivalent.

3.1.11. Proposition. *Let Γ be a contour contained in the open set $O \subseteq \mathbb{C}$. The following two conditions are equivalent.*

(i) For every f holomorphic on O,

$$\int_\Gamma f = 0. \tag{6}$$

(ii) For every f holomorphic on O,

$$f(z) \cdot W_\Gamma(z) = \frac{1}{2\pi i} \int_\Gamma \frac{f(w)}{w - z} dw, \quad provided\ z \in O \backslash \Gamma. \tag{7}$$

Proof. Suppose that (i) holds. Let f be holomorphic on O, let $z \in O \backslash \Gamma$. We wish to prove (7). Let q be the difference quotient as in (3). We know now q is holomorphic, so by (i), $0 = \int_\Gamma q$. Repeat the computation (4) in the proof of 3.1.1, and you get (7).

Suppose now that (ii) holds. Apply (7) with the holomorphic function given by $F(w) := f(w)(w - z)$ in the role of f, and you get (6).   □

B Eight Ways to Say "Holomorphic"

The next Theorem summarizes and ties together much of the information about holomorphic functions that we have presented so far.

3.1.12. Theorem. *Let $f = u + iv$ be a complex-valued function on the open set O. The following conditions are equivalent.*

I. *f is holomorphic on O (that is, f has a derivative on O).*

II. *At each point of O, u and v are differentiable in the real-variables sense and the Cauchy-Riemann equations are satisfied:*

$$u_x = v_y \quad and \quad u_y = -v_x.$$

III. *f has an antiderivative locally on O.*

IV. *f is continuous on O; and in each disk $D(a,r)$ contained in O, there is a point z such that the integral of f over the boundary of every triangle in the disk equals 0:*

$$\int_{[z,b,c,z]} f = 0 \quad whenever\, b, c \in D(a,r).$$

V. *f is continuous on O; and locally on O, the integrals of f are independent of path.*

VI. *On each disk $D(a,r)$ contained in O, f has a power series representation:*

$$f(z) = \sum_{k=0}^{\infty} c_k (z-a)^k \quad for \quad |z-a| < r.$$

VII. *The derivatives of f of all orders: $f', f'', f''', \cdots$ exist at every point of O.*

VIII. *f is continuous on O and holomorphic on $O \setminus E$, where E is a set containing a finite number of points.*

You will find it rewarding, before moving on, to meditate at length on the eight conditions and the logical connections among them. (Exercise 9 asks you to write up a proof.) Having such a Theorem gives us multiple ways to make use of holomorphy when we know we have it, and multiple ways to test for it when we are in doubt.

C Determinism

A general principle: A small amount of information about a holomorphic function may suffice to determine it completely. For example, since a holomorphic function f on a disk is given by a power series, to identify f it suffices to know just the values of f and its derivatives at one point, by Theorem 2.6.10. The significance of the integral representation formula (2) may be expressed as follows: When the values of f on Γ are known, the values of f inside Γ are determined and, if they have been lost or forgotten, can be recovered by means of the Cauchy integral over Γ. We now derive another instance of the principle from the power series representation. We prove it first for a disk, then for an arbitrary open connected set.

3.1.13. Lemma. *Let f be the sum of a power series on a disk:*

$$f(z) := \sum_{n=0}^{\infty} c_n(z-a)^n = c_0 + c_1(z-a) + \cdots \qquad for\ z \in D(a,r).$$

Suppose that f has a zero at each point of a sequence $\{z_k\}$ in $D'(a,r)$ that converges to a:

$$z_k \neq a \quad and \quad f(z_k) = 0 \ for\ all\ k; \quad and \quad \lim_{k \to \infty} z_k = a.$$

Then $f \equiv 0$.

Proof. We know that $c_0 = 0$ because $c_0 = f(a) = \lim_{k \to \infty} f(z_k)$. It suffices to prove that $c_n = 0$ for all $n > 0$. Suppose not. Then there would be a smallest integer m such that $c_m \neq 0$. Let

$$F(z) := \sum_{n=m}^{\infty} c_n(z-a)^{n-m} = c_m + c_{m+1}(z-a) + \cdots \qquad for\ z \in D(a,r).$$

Then $f(z) = (z-a)^m F(z)$. For every k, since $f(z_k) = 0$ and $z_k \neq a$, it would follow that $F(z_k) = 0$. Since $z_k \to a$, and since F is continuous (by 2.6.7), it would follow that $F(a) = 0$, which would imply that $c_m = 0$, which cannot be true. $\square$

Notice how the proof uses the hypothesis that there are zeros of f which converge to a but are not equal to a. We take this occasion to define a bit of terminology. The point a is an **accumulation point** of the set A if A contains a sequence of points $\{z_k\}$ such that for all k, $z_k \neq a$; and $\lim_{k \to \infty} z_k = a$.

3.1.14. Theorem. *Let f be a holomorphic function whose domain is the connected open set O. If there is a point in O which is an accumulation point of the set of zeros of f, then f is identically zero.*

Proof. Let V be the set of accumulation points in O of $f^{-1}[\{0\}]$. We know that V is non-empty; we need to show that $V = O$. It suffices to show that both V and $O \setminus V$ are open; because then, if $O \setminus V$ were nonempty, the two sets would disconnect O, which cannot happen.

Let $a \in V$. Pick $r > 0$ such that $D(a,r) \subseteq O$. There is a sequence of points $z_k \in D'(a,r)$ such that $z_k \to a$ and $f(z_k) = 0$. By the Lemma, $f \equiv 0$ on $D(a,r)$, so $D(a,r) \subseteq V$. Thus V is open.

Let $b \in O \setminus V$. Then $f(b)$ may or may not be zero, but in either case there exists $r > 0$ such that $D(b,r) \subseteq O$ and $D'(b,r)$ contains no zeros of f. So $D(b,r)$ is contained in $O \setminus V$. Thus $O \setminus V$ is open. $\square$

The two corollaries below are thinly disguised reformulations of Theorem 3.1.14. They are frequently useful.

3.1.15. Corollary. *Let g_1 and g_2 be holomorphic on the open connected set O, such that $g_1 = g_2$ on a set that has an accumulation point within O. Then $g_1 = g_2$ throughout O.*

Proof. Apply the Theorem with $g_1 - g_2$ in the role of f. □

3.1.16. Corollary. *Let f be a non-constant holomorphic function whose domain is the connected open set O. Let $a \in O$, and let $q \in \mathbb{C}$. Then there exists $r > 0$ such that $\bar{D}(a, r) \subset O$ and*

$$f(z) \neq q \quad for \ 0 < |z - a| \leq r.$$

Proof. Suppose not. Then for every $n > 0$ there would exist $z_n \in D'(a, \frac{1}{n})$ such that $f(z_n) = q$. The Theorem, applied with $z \mapsto f(z) - q$ in the role of f, tells us that $f(z) \equiv q$, so that f would be constant, which is false. □

3.1.17. Remark. We now digress to tell you briefly about an advanced result, beyond this book, which you may enjoy knowing about. When contemplating a striking result like Theorem 3.1.14, it is a good practice to try to discover its limitations; and to investigate exactly what is needed in its hypothesis. Asking good questions can lead to further interesting and useful results. We might ask, as Karl Weierstrass asked in the 1870s, the following question: Let $A \subset O \subseteq \mathbb{C}$, where O is open. For what kind of set A can we be sure that there exists a holomorphic function f on O that equals zero at each point of A, but has no other zeros in O? The Theorem tells us that we must rule out a set A which has an accumulation point in O. On the other hand, if A consists of a finite number of points, we can write down a polynomial whose zeros are precisely the points of A. And we can readily think of examples of an infinite set A which is precisely the zero set of a holomorphic function. For example, if A is the set of integers on the real axis, consider $z \mapsto \sin \pi z$. It turns out that the hypothesis of Theorem 3.1.14 is the best possible hypothesis.

3.1.18. Theorem. *Let $A \subset O \subseteq \mathbb{C}$, where O is open and A has no accumulation point in O. Then there exists f holomorphic on O such that f has a zero at each point of A and no other zeros in O.*

For a proof of this advanced theorem, we refer you to [61, Chapter 15], [23, Chapter VIII] or [20, Chapter 8]. Weierstrass proved it in the case $O = \mathbb{C}$ in 1876.

The Theorem has an interesting consequence: For every open set $O \subseteq \mathbb{C}$, there exist holomorphic functions f on O for which O is the **natural domain** of f. That is, f cannot be extended to make it holomorphic on any open set properly containing O. To prove this, choose a set $A \subset O$ whose accumulation points include all boundary points of O but no points of O itself. Weierstrass's Theorem 3.1.18 gives us an f holomorphic on O and zero only on A. Suppose that there were an open set O_1 strictly larger than O, and a function f_1 holomorphic on O_1 agreeing with f on O. Then some boundary points of O would belong to O_1 and would also be accumulation points of zeros of f_1. But then by Theorem 3.1.14, f_1 would be identically zero on O_1, which is not the case.

D   Liouville's Theorem

An **entire** function is one that is holomorphic on $\mathbb{C}$. If F is an entire function, then at every point of the plane, its Taylor series has infinite radius of convergence (by 3.1.6). In particular, consider the Taylor series at the origin:

$$F(z) = \sum_{k=0}^{\infty} c_k z^k \qquad \text{for all } z \in \mathbb{C}, \text{ where}$$

$$c_k = \frac{1}{2\pi i} \int_{|w|=s} \frac{F(w)}{w^{k+1}} dw \qquad \text{for every } s > 0. \tag{8}$$

The next result is known as Liouville's Theorem. It follows easily from results published by Cauchy in 1827 (see [66, p. 138 and p. 145]), and Cauchy published it explicitly in 1844 ([12]). It was treated in lectures by Joseph Liouville in 1847, and that seems to be how it got its name. Gauss may have known the result in 1816.

3.1.19. Theorem. *A bounded entire function is constant.*

Proof. There exists a number $B > 0$ such that $|F(z)| \leq B$ for all $z \in \mathbb{C}$. We use 2.5.2 to estimate the integrals in (8). We find that for every $s > 0$,

$$|c_k| \leq \frac{1}{2\pi} \max_{|w|=s} \left| \frac{F(w)}{w^{k+1}} \right| \left| \int_{|w|=1} |dw| \right| \leq \frac{1}{2\pi} \frac{B}{s^{k+1}} 2\pi s = \frac{B}{s^k}.$$

Provided $k > 0$, the right-hand side tends to zero as $s \to \infty$. Therefore $c_k = 0$. Therefore $F(z) \equiv c_0$. $\qquad\qquad\qquad\square$

E   The Fundamental Theorem of Algebra

The degree of a non-constant polynomial P is a positive integer n. We may write

$$P(z) = \sum_{k=0}^{n} c_k z^k = z^n \left(c_n + \frac{c_{n-1}}{z} + \frac{c_{n-2}}{z^2} + \cdots + \frac{c_0}{z^n} \right),$$

where $c_n \neq 0$. As $|z| \to \infty$, the quantity in parentheses tends to c_n. Therefore $\lim_{|z|\to\infty} P(z) = \infty$; more precisely, $\lim_{|z|\to\infty} \frac{|P(z)|}{|z|^n} = |c_n|$. So at least in some sense, for large z, $P(z)$ behaves like the mapping $z \mapsto z^n$, which is easy to understand. We know in particular that $z \mapsto z^n$ maps the plane onto itself. We will show that the same is true for P.

3.1.20. Lemma. *If P is a non-constant polynomial, then $P[\mathbb{C}] = \mathbb{C}$.*

Proof. If there existed a complex number w not in the range of P, then the function given by

$$g(z) := \frac{1}{P(z) - w}$$

would be entire. If $\epsilon > 0$, then there exists R such that $|g(z)| < \epsilon$ for $|z| > R$. Let

$$m := \min\{|P(z) - w| \mid z \in \bar{D}(0, R)\},$$

which would be positive. Therefore

$$|g(z)| \le \max\{\epsilon, \tfrac{1}{m}\} \quad \text{for all } z \in \mathbb{C}.$$

But then g, being a bounded entire function, would be constant, and hence P would be constant, which is not the case. □

The next result is the Fundamental Theorem of Algebra. Over a hundred different proofs have been given for this result. See [9, pp. 117–118] for a discussion, with references. See [66, pp. 79–80] for a discussion of the several proofs that Cauchy gave.

3.1.21. Theorem. *Let P be a polynomial of degree $n > 0$ with coefficients in $\mathbb{C}$:* $P(z) = \sum_{k=0}^{n} c_k z^k$. *Then there exist n complex numbers $a_1, \cdots, a_n$, not necessarily distinct, such that*

$$P(z) = c_n(z - a_1) \cdot \ldots \cdot (z - a_n).$$

Proof. Let $a \in \mathbb{C}$. Long division of $P(z)$ by $z - a$ gives $P(z) = (z - a)Q(z) + R$, where Q is a polynomial of degree $n - 1$ and R is a constant. Evidently these three conditions are equivalent:

- $R = 0$.

- $P(a) = 0$.

- $(z - a)$ is a factor of $P(z)$.

Since P maps $\mathbb{C}$ onto $\mathbb{C}$, there exists a such that $P(a) = 0$ and hence $z - a$ is a factor of $P(z)$. Proceeding by induction, we obtain the Theorem. □

F   Subuniform Convergence Preserves Holomorphy

3.1.22. Theorem (Weierstrass). *Let $\{f_k\}$ be a sequence of holomorphic functions on the open set O. If $f_k \to f$ subuniformly on O, then f is holomorphic on O.*

Proof. For every contour $\Gamma \subset O$, the convergence of f_k to f is uniform on Γ, and therefore (by 2.5.6)

$$\int_\Gamma f = \lim_{k \to \infty} \int_\Gamma f_k. \tag{9}$$

The function f is holomorphic on O if and only if its integrals are locally independent of path (recall 2.5.15 and 3.1.8. Let $D(a,r) \subseteq O$. For every contour $\Gamma \subset D(a,r)$, each integral on the right-hand side of (9) equals 0 because each f_k is holomorphic. Therefore the integral on the left also equals 0.

Thus the integrals of f are locally independent of path, so f is holomorphic on O.   □

Exercises

1. Evaluate the following integrals, using Theorem 3.1.1.

   (a) $\displaystyle\int_{|z|=4} \cos z \, dz.$

   (b) $\displaystyle\int_{|z|=4} \frac{\cos z}{z - \pi} dz.$

   (c) $\displaystyle\int_{|z|=1} \frac{\cos z}{z - \pi} dz.$

   (d) $\displaystyle\int_{|z|=4} \frac{\cos z}{(z - \pi)(z - 2\pi)} dz.$

   (e) $\displaystyle\int_{|z|=4} \frac{z - 2\pi}{z - \pi} dz.$

   (f) $\displaystyle\int_{|z|=4} \frac{\cos z}{z(z - \pi)} dz.$

2. Let Γ be the contour shown in Figure 2.5-2. Use Theorem 3.1.1 to evaluate each of the following integrals.

   (a) $\displaystyle\int_{\Gamma} \cos z \, dz.$

   (b) $\displaystyle\int_{\Gamma} \frac{e^z \cos z}{z} dz.$

   (c) $\displaystyle\int_{\Gamma} \frac{e^z}{z - 4i} dz.$

   (d) $\displaystyle\int_{\Gamma} \frac{z \cos z}{z - 2\pi} dz.$

   (e) $\displaystyle\int_{\Gamma} \frac{z - 2}{z + 4} dz.$

   (f) $\displaystyle\int_{\Gamma} \frac{z^3}{z - 4} dz.$

3. Explain this remark: If Theorem 3.1.1 were true with the word "convex" omitted, then all winding numbers would equal 0.

4. Prove that if u is harmonic on an open set O, then all the partials of u of all orders exist and are continuous on O.

5. Let f and g be holomorphic on $D(a,r)$, with Taylor series

$$f(z) = \sum_{n=0}^{\infty} a_n(z - a)^n \quad \text{and} \quad g(z) = \sum_{m=0}^{\infty} b_m(z - a)^m. \qquad (10)$$

Consider the power series

$$\sum_{p=0}^{\infty} c_p(z - a)^p, \qquad (11)$$

where $c_0 = a_0 b_0$, $c_1 = a_0 b_1 + a_1 b_0$, and in general

$$c_p = \sum \{a_n b_m \mid n + m = p\} \equiv \sum_{k=0}^{p} a_{p-k} b_k,$$

for each $p = 0, 1, \cdots$. The series (11) is the **Cauchy product** of the first two series. Let $h(z) = f(z)g(z)$. Then of course h is holomorphic on $D(a, r)$. Prove that the Taylor series of h at a is (11).

6. Use the result of Exercise 5 to compute the first few summands in the Taylor series at 0 for the function

$$z \mapsto \cos z \sin z \equiv \left(1 - \frac{z^2}{2!} + \frac{z^4}{4!} - \cdots \right) \left(z - \frac{z^3}{3!} + \cdots \right).$$

7. Carry out the following discussion in detail. Let f and g be holomorphic on $D(a, r)$, given by Taylor series as in (10), such that g has no zeros on that disk. Then the Taylor series for the holomorphic function f/g can be obtained using long division, as follows. Let f_N and g_N be the partial sums of order N, and carry out the long division of the two polynomials. We find that $f_N = p_N g_N + r_N$, where p_N and r_N are polynomials and the degree of r_N is at most $N - 1$. For each integer m, the coefficient c_m on $(z - a)^m$ is the same in all the polynomials p_N for $N > m$, and that will be the coefficient on z^m in the Taylor series for f/g, because for $0 < s < r$,

$$c_m = \frac{1}{2\pi i} \int_{|w-a|=s} \frac{f(w)}{g(w)(w-a)^{m+1}} dw$$

$$= \lim_{N \to \infty} \frac{1}{2\pi i} \int_{|w-a|=s} \frac{f_N(w)}{g_N(w)(w-a)^{m+1}} dw = \cdots.$$

8. Find the Taylor series for $z \mapsto \tan z$ at 0, using the result from Exercise 7.

9. Write a complete proof of Theorem 3.1.12.

10. Let f be a function with a power series representation on a disk, say $D(0, 1)$. Lemma 3.1.13 tells us that surprisingly little information about such a function suffices to identify it completely. For example, if $f\left(\frac{1}{n}\right) = \frac{1}{n}$ for $n = 1, 2, \cdots$, then f is completely determined, and in fact is given by $f(z) = z$. In each case, use the given information to identify the function. Is it unique?

(a) $f\left(\dfrac{1}{n}\right) = 4$ for $n = 1, 2, \cdots$.

(b) $f\left(\dfrac{i}{n}\right) = -\dfrac{1}{n^2}$ for $n = 1, 2, \cdots$.

(c) $f\left(\dfrac{1}{2n+1}\right) = \dfrac{1}{n}$ for $n = 1, 2, \cdots$.

11. This is a proof of the maximum principle due to d'Alembert. Let f be a non-constant holomorphic function on some neighborhood of 0, so that one

may write $f(z) = \sum_{k=0}^{\infty} a_k z^k$. Suppose that $a_0 = 1$ and let m be the smallest value of $k > 0$ such that $a_k \neq 0$. Show that for each real θ,

$$\lim_{t \to 0^+} (|f(te^{i\theta})|^2 - 1)t^{-m} = 2\mathrm{Re}\,(a_m e^{im\theta}).$$

Noting the arbitrariness of θ, show that $|f|$ possesses neither a relative maximum nor a relative minimum at 0.

12. In each case, determine whether there is a function f with a power series representation on $D(0,1)$ that has the given property; and if so, identify it.

   (a) $f\left(\dfrac{1}{n^2}\right) = 4$ for $n = 1, 2, \cdots$; and $f\left(-\dfrac{1}{2}\right) = 3$.

   (b) $f\left(\dfrac{1}{n}\right) = f\left(-\dfrac{1}{n}\right) = \dfrac{1}{n}$ for $n = 1, 2, \cdots$.

   (c) $f\left(\dfrac{i^n}{n}\right) = -\dfrac{1}{n^2}$ for $n = 1, 2, \cdots$.

13. Determine whether there is a function f with a power series representation on $D(0,1)$ such that

$$f\left(\frac{1}{2n}\right) = f\left(\frac{1}{2n+1}\right) = \frac{1}{n} \quad \text{for } n = 1, 2, \cdots.$$

   If so, identify it.

14. Prove that the power series $\sum_{n=1}^{\infty} \dfrac{n^2 z^n}{n!}$ converges to $e^z(z + z^2)$.

15. Let f be holomorphic on the disk $D(a,1)$, and let g be holomorphic on the disk $D(a, 0.1)$. If $f = g$ on $D(a, 0.01)$, what can you say about the radius of convergence of the power series of g at a?

16. Let u be harmonic on the open disk $D(a, r)$. Suppose that $\{z_k\}$ is a sequence of points not equal to a, such that $z_k \to a$ and $u(z_k) = 0$ for all k. Must u be identically zero on the disk? Justify your answer; if it's No, give an example.

17. Prove that if f is an entire function, and if $\mathrm{Re}\, f(z) \leq M$ for all z, where M is finite, then f is constant.

18. This Exercise generalizes Liouville's Theorem. Let f be an entire function. Suppose that f has polynomial growth toward infinity; that is, suppose there are finite positive constants A, B, and K such that

$$|f(z)| \leq B|z|^K \quad \text{whenever} \quad |z| > A. \tag{12}$$

Prove that f is a polynomial with degree at most $[K]$, which means the greatest integer less than or equal to K.

19. Suppose that in Exercise 18 we made no mention of a constant A and wrote, instead of (12),

$$|f(z)| \le B|z|^K \quad \text{for all} \quad z \in \mathbb{C}.$$

What would be the conclusion?

20. Let f and g be entire functions such that $f(z) = g(\frac{1}{z})$ for $0 < |z| < \infty$. Prove that f is constant.

21. An essay question: Let f be a continuous complex-valued function on the open set $O \subseteq \mathbb{C}$, and let γ be a closed path in O. What conditions on f, O, and/or γ suffice to assure that $\int_\gamma f = 0$?

22. Given $f = u + iv$ holomorphic on $\bar{D}(a,r)$, Remark 3.1.4 obtains $u(a)$ as a certain integral of u over the circle $|z - a| = r$. In fact, it equals the average value of u on the circle. For arbitrary $z \in D(a,r)$, obtain $u(z)$ as an integral over the circle.

23. Let $f = u + iv$ be holomorphic on the disk $D(0, R)$, so that we know it has a power series representation thereon:

$$f(z) = u(z) + iv(z) = \sum_{n=0}^{\infty} c_n z^n \qquad (|z| < R).$$

One may separate the real and imaginary parts of the series, as follows. Let $c_n =: a_n + ib_n$, and write z as $re^{i\theta}$, where $0 < r < R$. Then $z^n = r^n e^{in\theta} = r^n(\cos n\theta + i \sin n\theta)$. Show that

$$u(re^{i\theta}) = a_0 + \sum_{n=1}^{\infty} r^n(a_n \cos n\theta - b_n \sin n\theta), \qquad (13)$$

$$v(re^{i\theta}) = b_0 + \sum_{n=1}^{\infty} r^n(b_n \cos n\theta + a_n \sin n\theta).$$

24. Use the result of Exercise 23 to prove that if u is a real-valued harmonic function on $D(0, R)$, then u has a series representation of the type (13). Let $0 < r < R$. Prove these integral formulas, which give the coefficients in (13):

$$a_0 = \frac{1}{2\pi} \int_0^{2\pi} u(re^{i\theta}) d\theta;$$

$$r^n a_n = \frac{1}{\pi} \int_0^{2\pi} u(re^{i\theta}) \cos n\theta d\theta \quad \text{for } n \ne 0;$$

$$r^n b_n = \frac{1}{\pi} \int_0^{2\pi} u(re^{i\theta}) \sin n\theta d\theta \quad \text{for } n \ne 0.$$

25. In this book, we have made full use of your prior knowledge of the real-valued calculus versions of the exponential, sine, and cosine, beginning by using them to define the complex-valued versions in Section 2.2. In Section 2.3 we showed that those functions are holomorphic (at the end of subsection C, page 124, and Exercise 3, page 127). In Section 2.6, Examples 2.6.11 and 2.6.12, we showed that certain differential equations lead us to the Taylor series for these functions. Theorem 3.1.1 assures us that these functions, being holomorphic on the plane, must have power series representations, which are necessarily their respective Taylor series. Thus we are justified in writing:

$$e^z = \sum_{n=0}^{\infty} \frac{z^n}{n!},$$

$$\cos z = \sum_{k=0}^{\infty} \frac{(-1)^k z^{2k}}{(2k)!}, \quad \text{and} \tag{14}$$

$$\sin z = \sum_{k=0}^{\infty} \frac{(-1)^k z^{2k+1}}{(2k+1)!}.$$

An alternative procedure in a complex analysis course is to set aside any and all prior knowledge of the functions and their properties; avoid mentioning them until after power series theory is fully developed; then present them from scratch, starting with the following definitions. The **exponential** is the unique entire function f such that

$$f' = f \quad \text{and} \quad f(0) = 1.$$

The **sine** is the unique entire function g such that

$$g'' + g = 0, \quad g(0) = 0, \quad \text{and} \quad g'(0) = 1.$$

The **cosine** is the unique entire function h such that

$$h'' + h = 0, \quad h(0) = 1, \quad \text{and} \quad h'(0) = 0.$$

From those definitions one may identify completely the power series for the functions, as in (14). You are invited to carry out the rest of the presentation. First, prove the identities (a)–(k) below. Then, with π defined as twice the smallest value of x such that $\cos x = 0$, prove the results (l) through (w).

(a) $\cos z = \cos(-z)$.

(b) $\sin z = -\sin(-z)$.

(c) $\cos' z = -\sin z$.

(d) $\sin' z = \cos z$.

(e) $\exp(iz) = \cos z + i \sin z$.

(f) $\cos z = \frac{1}{2}(\exp(iz) + \exp(-iz))$.

(g) $\sin z = \frac{1}{2i}(\exp(iz) - \exp(-iz))$.

(h) $\cos(z + w) = \cos z \cos w - \sin z \sin w$.

(i) $\sin(z + w) = \sin z \cos w + \cos w \sin z$.

(j) $\cos^2 z + \sin^2 z = 1$.

(k) $\cos 2 < 0$.

(l) $0 < \pi < 4$.

(m) $\cos t > 0$ for $0 \le t < \frac{\pi}{2}$.

(n) $\sin t > 0$ for $0 < t \le \frac{\pi}{2}$.

(o) $\sin(\frac{\pi}{2}) = 1$, $\exp(\frac{i\pi}{2}) = i$.

(p) The mapping $t \mapsto \exp(it)$ maps $[0, \frac{\pi}{2}]$ one-to-one onto $\{z = x + iy :$ $|z| = 1, x \ge 0, y \ge 0\}$.

(q) $\cos(z + \frac{\pi}{2}) = -\sin z$ for all z.

(r) $\sin(z + \frac{\pi}{2}) = \cos z$ for all z.

(s) The mapping $t \mapsto \exp(it)$ maps $[0, 2\pi)$ one-to-one onto $\{z : |z| = 1\}$.

(t) $\exp(2\pi i k) = 1$ if and only if k is an integer.

(u) $\exp(z) = \exp(a)$ if and only if $z - a$ is an integer times $2\pi i$.

(v) $\cos(z) = \cos a$ if and only if $z - a$ is an integer times 2π.

(w) $\sin(z) = \sin a$ if and only if $z - a$ is an integer times 2π.

Help on Selected Exercises

1. (a) 0. (b) $-2\pi i$. (c) 0. (d) $2i$. (e) $-2\pi^2 i$. (f) $-i$.

2. (a) 0. (b) $10e\pi i$. (c) $6\pi i e^{4i}$. (d) $4\pi^2 i$. (e) $-36\pi i$. (f) $384\pi i$.

5. Briefly: Let $0 < s < r$. By Theorem 3.1.6, with h in the role of f,

$$c_p = \frac{1}{2\pi i} \int_{|z-a|=s} \frac{\left(\sum_{n=0}^{\infty} a_n(z-a)^n\right)\left(\sum_{m=0}^{\infty} b_m(z-a)^m\right)}{(z-a)^{p+1}} dz$$

$$= \frac{1}{2\pi i} \sum_{n=0}^{\infty} \sum_{m=0}^{\infty} a_n b_m \int_{|z-a|=s} (z-a)^{n+m-p-1} dz.$$

The last integral is nonzero only when $n + m = p$.

6. You should obtain

$$z - \frac{4}{3!}z^3 + \frac{16}{5!}z^5 - \frac{64}{7!}z^7 + \cdots,$$

To check the result, obtain directly the Taylor series for the function $z \mapsto \frac{1}{2} \sin 2z$.

12. (b) If such an f exists, then the function $z \mapsto f(z) - z$ equals zero at each point of the sequence $\{1/n\}$. By 3.1.13, $f(z) \equiv z$. Hence $f(-1/n) = -1/n$ for all $n \in \mathbb{N}$, which cannot be.

17. Apply Liouville's Theorem to the entire function $z \mapsto e^{f(z)}$.

22. You might just take the real part of the integral that gives $f(z)$. We will arrive at a more satisfactory answer in Section 5.5.

24. *Hints:* (1) We know there exists v on the disk such that $u + iv$ is holomorphic; we know that $u + iv$ has a power series representation. (2) The functions $x \mapsto \sin nx$ and $x \mapsto \cos mx$, where m and n are positive integers, have the following properties:

$$\int_0^{2\pi} \cos mx \sin nx\, dx = 0 \text{ for all } m, n;$$

$$\int_0^{2\pi} \cos mx \cos nx\, dx = \begin{cases} 0 & \text{if } m \neq n, \\ \pi & \text{if } m = n; \end{cases}$$

$$\int_0^{2\pi} \sin mx \sin nx\, dx = \begin{cases} 0 & \text{if } m \neq n, \\ \pi & \text{if } m = n. \end{cases}$$

3.2   CAUCHY'S THEOREM

In many applications, one can do without the Theorem that we are about to present. One can piece together convex sets to make up the domain of the function in question, and then use Theorem 3.1.1; or one can devise other *ad hoc* arguments. This book does not promote such techniques. For greater efficiency in applications, and for the satisfaction of a better understanding, you will find it worthwhile to know and to use the quite general and powerful Cauchy's Theorem 3.2.1.

Let f be holomorphic on the open set O, and let Γ be a contour in O. The question is, What conditions on f, O and/or Γ allow us to assert that $\int_\Gamma f = 0$? We have shown that if O is convex, then $\int_\Gamma f = 0$ for every f and Γ. As you know, if O is arbitrary, then we cannot draw that conclusion. But as we are about to show, it is still true that $\int_\Gamma f = 0$ for every Γ that winds around no points outside O. Such a curve is said to be null-homologous in O; and the next result is known as the homology version of Cauchy's Theorem.

3.2.1. Theorem (Cauchy). *Let Γ be a contour in the open set $O \subseteq \mathbb{C}$ such that*

$$W_\Gamma(z) = 0 \quad \text{for all } z \notin O. \tag{1}$$

Let f be holomorphic on O. Then

$$\int_\Gamma f(w)dw = 0 \tag{2}$$

and

$$f(z) \cdot W_\Gamma(z) = \frac{1}{2\pi i} \int_\Gamma \frac{f(w)}{w - z} dw, \quad \text{provided } z \in O \setminus \Gamma. \tag{3}$$

One may state the Theorem without mentioning the set O: If f is holomorphic on and inside the contour Γ, then $\int_\Gamma f = 0$. The unmentioned O is then simply the domain on which f is holomorphic.

Subsection A presents an elementary proof of the Theorem. You may wish to proceed to subsection B and postpone reading the proof. Sometimes it is best first to study the statement of a theorem and understand its uses. One may then read the proof later, with greater appreciation.

Chapter 6, intended for use in somewhat more advanced courses, includes two additional proofs of Cauchy's Theorem. The elegant proof by John D. Dixon is given in Section 6.1. Section 6.2 presents Runge's Theorem and shows how it implies Cauchy's Theorem.

A Černý's 1976 Proof

This approach to the Theorem seems to have appeared in print first in the 1976 paper [13] by I. Černý, although it is possible that it was known much earlier. See Burckel's discussion [9, p. 264].

What we must show is that *every* contour Γ that satisfies (1) must also satisfy (2) and (3). We begin with a result that asserts the existence of *certain* contours that satisfy (2) and, in a restricted sense, (3).

3.2.2. Lemma. *Let K be a compact subset of the open set $O \subseteq \mathbb{C}$. Then there exists a contour Γ_0, lying within O but not touching K, which winds around each point of K exactly once but does not wind around any point of the complement of O. In briefer, symbolic language: There exists a contour $\Gamma_0 \subset O \setminus K$ such that*

$$W_{\Gamma_0}(z) = \begin{cases} 0 & \text{for } z \notin O, \\ 1 & \text{for } z \in K; \end{cases} \tag{4}$$

and such that if f is holomorphic on O, then

$$f(z) = \frac{1}{2\pi i} \int_{\Gamma_0} \frac{f(w)}{w - z} dw \quad \text{for } z \in K. \tag{5}$$

Proof. You should find the result up through condition (4) quite believable, even obvious. Try this: Sketch any case you can think up, of an open set O with a compact

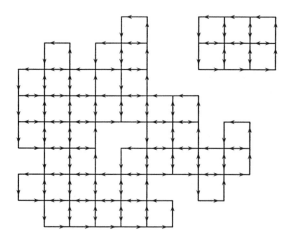

Figure 3.2-1 An illustration for the proof of Lemma 3.2.2. A contour consisting of the boundaries of 46 squares, each boundary traversed counterclockwise. If an edge belongs to two of the squares, then it occurs twice in the contour, traversed once in each direction.

subset K; then describe a Γ_0 as claimed. For example, suppose that $O = \mathbb{C}'$. Every compact subset K is contained in the annulus $s \leq |z| \leq S$ for some s and S. Let Γ_0 consist of the circle $|z| = 2S$, traversed counterclockwise, and the circle $|z| = s/2$, traversed clockwise.

For the general case, we may produce a Γ_0 consisting of horizontal and vertical line segments, as follows. Choose $S > 0$ large enough so that K is contained in the interior of the square whose four vertices are $(\pm S, \pm S)$. For a positive integer n, partition that square into $(2n)^2$ subsquares, each of size $\frac{S}{n}$ by $\frac{S}{n}$, as follows. For integers j and k with $-n \leq j < n$ and $-n \leq k < n$, let

$$Q_{j,k} := \left\{ z = x + iy \,\middle|\, \frac{j}{n}S \leq x \leq \frac{j+1}{n}S \text{ and } \frac{k}{n}S \leq y \leq \frac{k+1}{n}S \right\}.$$

Choose n large enough so that no square $Q_{j,k}$ can touch both K and the complement of O. Let $\{Q_h\}_{h=1}^m$ be a one-to-one enumeration of the squares $Q_{j,k}$ that touch K. Then $K \subseteq \cup_{h=1}^m Q_h \subset O$. The paths $\{\partial Q_h\}$ consist of $4m$ line segments. Some of the segments γ occur twice, that is, both γ and $-\gamma$ occur. Let $\{\gamma_j\}_{j=1}^J$ be an enumeration of the segments that occur just once; notice that none of them touch K. Let $\Gamma_0 := \gamma_1 + \cdots + \gamma_J$. Then $\Gamma_0 \cap K = \emptyset$. One may show that Γ_0 is a contour; we leave this result as Exercise 2.

For every continuous function ϕ on $\cup_{h=1}^m Q_h$,

$$\sum_{h=1}^m \int_{\partial Q_h} \phi(w)dw = \sum_{j=1}^J \int_{\gamma_j} \phi(w)dw = \int_{\Gamma_0} \phi(w)dw. \tag{6}$$

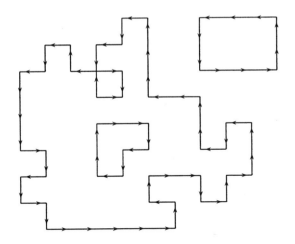

Figure 3.2-2 Figure 3.2-1 again, but with the edges eliminated that occur twice. The edges that remain make up the contour Γ_0. It consists of three (or four) closed paths.

Observe these winding number values:

$$W_{\partial Q_h}(z) \equiv \frac{1}{2\pi i}\int_{\partial Q_h}\frac{dw}{w-z} = \begin{cases} 0 & \text{for } z \notin Q_h, \\ 1 & \text{for } z \in Q_h^o. \end{cases}$$

Also,

$$\frac{1}{2\pi i}\int_{\partial Q_h}\frac{f(w)}{w-z}dw = \begin{cases} 0 & \text{for } z \notin Q_h, \\ f(z) & \text{for } z \in Q_h^o, \end{cases}$$

by 3.1.1, which is the convex-domain case of Cauchy's Theorem. Applying (6) first with $\phi(w) := \dfrac{1}{2\pi i(w-z)}$ and then with $\phi(w) := \dfrac{f(w)}{2\pi i(w-z)}$, we find that

$$1 = \frac{1}{2\pi i}\int_{\Gamma_0}\frac{dw}{w-z} \quad \text{and} \quad f(z) = \frac{1}{2\pi i}\int_{\Gamma_0}\frac{f(w)}{w-z}dw$$

for all $z \in \cup_{h=0}^{m}Q_h^o$, a set whose closure contains K. Hence the same is true for all $z \in K$, by the continuity of the function ϕ in each case. $\qquad\square$

Proof of Cauchy's Theorem. Let f, O, and Γ be as given in the hypothesis. Let K be the compact set consisting of all points on or inside Γ. It suffices to prove that $\int_{\Gamma}f = 0$. Let Γ_0 be the contour given by Lemma 3.2.2, so that for each $z \in K$,

$f(z) = \dfrac{1}{2\pi i} \displaystyle\int_{\Gamma_0} \dfrac{f(w)}{w - z} dw$. Now an application of Fubini's Theorem, allowing a change in the order of integration, gives the desired result:

$$\int_\Gamma f(z)dz = \int_\Gamma \left(\frac{1}{2\pi i} \int_{\Gamma_0} \frac{f(w)}{w - z} dw \right) dz = \int_{\Gamma_0} f(w) \left(\frac{1}{2\pi i} \int_\Gamma \frac{dz}{w - z} \right) dw,$$

which equals zero because the inner integral is $-W_\Gamma(w)$, which equals 0 for every w on Γ_0. $\qquad\square$

B   Simply Connected Sets

The open connected set $O \subseteq \mathbb{C}$ is **simply connected** if every contour Γ contained in O satisfies (1). It would be equivalent to say, if every closed path Γ contained in O satisfies (1)—since a closed path is a contour, and every contour is made up of a finite number of closed paths. Examples: Disks and convex sets are simply connected; annuli and punctured disks are not.

3.2.3. Corollary. *Let O be an open, simply connected subset of $\mathbb{C}$. Let f be holomorphic on O. Then:*

  (i) *f has an antiderivative on O.*

  (ii) *If f has no zeros in O, then f has a logarithm on O; that is, there exists a function g which is holomorphic on O such that $e^{g(z)} = f(z)$ for all $z \in O$.*

  (iii) *If f has no zeros in O, then for each integer $n \geq 2$, f has an nth root on O; that is, there exists a function h such that $h(z)^n = f(z)$ for all $z \in O$.*

Proof. By Cauchy's Theorem, $\int_\Gamma f = 0$ for every closed path Γ contained in O. In other words, integrals of f are independent of path in O, so f has an antiderivative F on O (recall 2.5.12). Thus (i) is proved. To prove (ii), apply (i) to the function f'/f. It has an antiderivative g. Select a point $a \in O$ and adjust g by adding a constant so that $e^{g(a)} = f(a)$. The derivative of $\dfrac{e^g}{f}$ is identically 0, so it is constant. The constant value must be 1 because the value at a is 1. To prove (iii), let g be the function given by (ii), and then let $h = e^{g/n}$. $\qquad\square$

If O does not contain the origin, then (ii) and (iii) apply with the function $z \mapsto z$ in the role of f, and they tell us that there are holomorphic functions $\log z$, $\sqrt{z}$, and so forth, defined on O.

For an interesting list of eleven conditions on a set that are equivalent to its being simply connected, see [61, Theorem 13.11]. The next result points out two of them.

3.2.4. Proposition. *Let O be an open connected subset of $\mathbb{C}$. The following conditions are equivalent.*

  (i) *O is simply connected.*

(ii) Every holomorphic function on O has an antiderivative on O.

(iii) The integrals of every holomorphic function on O are independent of path in O.

Proof. By Corollary 3.2.3, (i) implies (ii). Let $z_0 \notin O$. If (ii) holds, then in particular the function $z \mapsto \dfrac{1}{z - z_0}$ has an antiderivative on O. It follows that the $W_\Gamma(z_0) = 0$ for every contour Γ in O, so that O is simply connected. Thus (ii) implies (i). The equivalence of (ii) and (iii) is Morera's Theorem 2.5.12. $\square$

3.2.5. Proposition. *Let O and O_1 be open connected subsets of $\mathbb{C}$. Suppose that O and O_1 are conformally equivalent, that is, there exists a one-to-one holomorphic mapping g from O onto O_1. Then if O is simply connected, O_1 is also simply connected.*

Proof. Let f be holomorphic on O_1, and let γ be a closed path in O_1, given say by $z : [0, 1] \to O_1$. It suffices to show that $\int_\gamma f = 0$. Observe that

$$\int_\gamma f = \int_0^1 f(z(t))z'(t)dt = \int_0^1 f(g(g^{-1}(z(t)))) \frac{d}{dt}(g(g^{-1}(z(t))))dt,$$

simply because $g(g^{-1}(z(t))) \equiv z(t)$, so

$$\int_\gamma f = \int_\rho f(g(w))g'(w)dw, \tag{7}$$

where ρ is the closed path in O given by

$$w(t) := g^{-1}(z(t)) \quad \text{for } 0 \le t \le 1.$$

Since the function $w \mapsto f(g(w))g'(w)$ is holomorphic on the simply connected set O, the integral (7) equals 0. $\square$

C   Subuniform Boundedness, Subuniform Convergence

These are some interesting results which are now easy to prove, now that Lemma 3.2.2 is available.

3.2.6. Proposition. *Let $\mathcal{F}$ be a family of holomorphic functions on the open set $O \subseteq \mathbb{C}$. Suppose that $\mathcal{F}$ is subuniformly bounded on O; that is, for every compact set $K_0 \subset O$, there is a finite number M_0 such that*

$$|f(z)| \le M_0 \quad \text{for all } z \in K_0 \text{ and for all } f \in \mathcal{F}. \tag{8}$$

Then the family $\{f' \mid f \in \mathcal{F}\}$ is also subuniformly bounded on O.

Proof. Let K be a compact subset of O. It suffices to show that for some number M,

$$|f'(z)| \leq M \quad \text{for all } z \in K \text{ and for all } f \in \mathcal{F}. \tag{9}$$

Let Γ_0 be the contour provided by Lemma 3.2.2, and let O_0 be the inside of Γ_0.

By 2.4.6, there exists a positive number d such that $|w - z| \geq d$ whenever $z \in K$ and $w \notin O_0$. In particular, then,

$$\frac{1}{|w - z|} < \frac{1}{d} \quad \text{if } z \in K \text{ and } w \in \Gamma_0. \tag{10}$$

Let K_0 be the union of Γ_0 and its inside. Then K_0 is compact, so there is a finite number M_0 as in (8).

Recall from 3.1.5 the Cauchy integral representation for derivatives. For every $f \in \mathcal{F}$ and every z inside Γ_0,

$$f'(z) = \frac{1}{2\pi i} \int_{\Gamma_0} \frac{f(w)}{(w - z)^2} dw. \tag{11}$$

If $z \in K$, then,

$$|f'(z)| \leq \frac{M_0}{2\pi d^2} \int_{\Gamma_0} |dw|.$$

Let M be the number on the right, and (9) is satisfied. $\qquad\qquad\square$

3.2.7. Proposition. *Let $\{f_k\}$ be a sequence of holomorphic functions on the open set O. If $f_k \to f$ subuniformly on O, then $f'_k \to f'$ subuniformly on O.*

Proof. We know that f is holomorphic by 3.1.22.

Let K be a compact subset of O. It suffices to show that

$$\max_{z \in K} |f'(z) - f'_k(z)| \to 0 \quad \text{as } k \to \infty.$$

Given K, let Γ_0 be the contour provided by Lemma 3.2.2. Again, there exists $d > 0$ such that (10) holds. Again, we use the representation (11). It follows that for $z \in K$,

$$f'(z) - f'_k(z) = \frac{1}{2\pi i} \int_{\Gamma_0} \frac{f(w) - f_k(w)}{(w - z)^2} dw.$$

Therefore by (2.5.5):

$$\max_{z \in K} |f'(z) - f'_k(z)| \leq \frac{\max_{w \in \Gamma_0} |f(w) - f_k(w)|}{2\pi d^2} \int_{\Gamma_0} |dw|.$$

The right-hand side tends to 0 as $k \to \infty$, since $f_k \to f$ uniformly on Γ_0. $\qquad\square$

Exercises

1. In the proof of Lemma 3.2.2, why is it possible to "choose n large enough so that no square $Q_{j,k}$ can touch both K and the complement of O?

2. Prove that the path Γ_0 devised in the proof of Lemma 3.2.2 is a contour. That is, the segments γ_j can be grouped into a finite number of closed paths.

3. Explain the remark in the caption of Figure 3.2-2, that the contour consists of three (or four) closed paths.

4. For the contour Γ_0 of Figure 3.2-2, what values does the winding number W_{Γ_0} take on, and where?

5. Let $\mathcal{F}$ be a subuniformly bounded family of holomorphic functions on the open set $O \subseteq \mathbb{C}$. Prove that $\mathcal{F}$ is **subuniformly equicontinuous** on O. That is, for each compact set $K \subset O$ and each $\epsilon > 0$, there exists $\delta > 0$ such that

$$|f(z_1) - f(z_2)| < \epsilon \quad \text{whenever } f \in \mathcal{F}; \ z_1, z_2 \in K; \ \text{ and } |z_1 - z_2| < \delta.$$

6. Let a_k be holomorphic on the open set O for $k = 1, 2, \cdots$. Suppose that for each compact set $K \subset O$, if $c_{k,K} = \max\{|a_k(z)| \mid z \in K\}$, then $\sum_{1}^{\infty} c_{k,K}$ is finite. Prove that $f(z) = \sum_{k=1}^{\infty} a_k(z)$ is holomorphic on O and $f'(z) = \sum_{k=1}^{\infty} a'_k(z)$.

7. Let $O := \{x + iy \mid x > 1\}$. Define $w^z := e^{z \, Logw}$. Prove that

$$\zeta(z) = \sum_{n=1}^{\infty} \frac{1}{n^z}$$

defines a holomorphic function ζ on O, and that

$$\zeta'(z) = -\sum_{n=1}^{\infty} \frac{\log n}{n^z}.$$

Help on Selected Exercises

1. Consider Proposition 2.4.6. Does it suffice to have $\frac{\sqrt{2}}{n} < d$?

2. One must show that if $[a, b]$ is one of the segments in Γ_0, then there is exactly one segment $[b, c]$ in Γ_0.

5. A partial sketch: Given K and ϵ, let Γ_0 be the contour provided by Lemma 3.2.2. If f is holomorphic on O, then

$$f(z_1) - f(z_2) = \frac{1}{2\pi i} \int_{\Gamma_0} \frac{f(w)(z_1 - z_2)}{(w - z_1)(w - z_2)} dw \quad \text{whenever } z_1, z_2 \in K.$$

6. A partial sketch: Let $f_n(z) = \sum_{k=1}^{n} a_k(z)$. For each compact set $K \subset O$,

$$\sup_{z \in K} |f(z) - f_n(z)| \leq \sum_{k=n+1}^{\infty} c_{k,K} \to 0 \quad \text{as } n \to \infty.$$

Thus $f_n \to f$ subuniformly on O.

7. This is the **Riemann zeta function**, though this Exercise defines it only on part of its natural domain. For more information, see [68, Chapter 6]. *Hints:* Note that $|n^z| = |e^{z \ln n}| \geq e^{x \ln n} = n^x$ if $z = x + iy$. Let $x_0 > 1$. On the compact set $K = \{x + iy \mid x \geq x_0\}$,

$$\sum_{n=m}^{\infty} n^{-x_0} \to 0 \quad \text{as } m \to \infty.$$

3.3 ISOLATED SINGULARITIES

A The Laurent Series Representation on an Annulus

Let $a \in \mathbb{C}$ and let $0 \leq r_1 < r_2 \leq \infty$. Then the **annulus** with center a, inner radius r_1, and outer radius r_2 is the set[1]

$$\text{Ann}_a(r_1, r_2) := \{z \mid r_1 < |z - a| < r_2\}.$$

We will sometimes refer to it as "the annulus $r_1 < |z - a| < r_2$." We will often write $\text{Ann}(r_1, r_2)$ for $\text{Ann}_0(r_1, r_2)$. As you will see, annuli are of interest because they are the convergence sets for Laurent series.

A **negative-power series** is one of the form

$$\sum_{k=1}^{\infty} c_{-k}(z - a)^{-k} \quad \text{or, equivalently,} \quad \sum_{k=-\infty}^{-1} c_k(z - a)^k.$$

---

[1] We now have two different terms and notations for some sets: $\text{Ann}_a(0, r_2)$ and the punctured disk $D'(a, r_2)$ are the same set; $\text{Ann}(r_1, \infty) \equiv D(\infty, r_1)$; and $\text{Ann}(0, \infty) \equiv \mathbb{C}'$.

Section 2.6 discussed positive-power series in detail. We can use the results from that section here, since a negative-power series can be written as a positive-power series:

$$\sum_{k=1}^{\infty} c_{-k}(z-a)^{-k} = \underbrace{\sum_{k=1}^{\infty} c_{-k}\zeta^{k}}_{=:g(\zeta)}, \quad \text{where} \quad \zeta = \frac{1}{z-a}$$

By the Cauchy-Hadamard Theorem, the positive-power series in ζ has radius of convergence

$$R := \frac{1}{\limsup_{k\to\infty} |c_{-k}|^{1/k}}.$$

So it converges for $|\zeta| < R$ and diverges for $|\zeta| > R$. If $r_1 := 1/R$, then this is what follows about the negative-power series: It converges for $|z-a| > r_1$ and diverges for $|z-a| < r_1$. On the set of convergence, it is the composition of the holomorphic functions $\zeta \mapsto g(\zeta)$ and $z \mapsto \zeta := \dfrac{1}{z-a}$, and hence it is holomorphic.

A **Laurent series** is one of the form

$$\sum_{k=-\infty}^{\infty} c_k(z-a)^k := \underbrace{\sum_{k=-\infty}^{-1} c_k(z-a)^k}_{\text{negative powers}} + \underbrace{\sum_{k=0}^{\infty} c_k(z-a)^k}_{\text{positive powers}}.$$

It **converges** at z if each of the two series on the right converges at z.

For some r_1, with $0 \le r_1 \le \infty$, the negative-power series converges for $|z-a| > r_1$ to a holomorphic function f_1, and diverges for $|z-a| < r_1$. For each $s_1 > r_1$, convergence is uniform for $|z-a| \ge s_1$.

For some r_2, with $0 \le r_2 \le \infty$, the positive-power series converges for $|z-a| < r_2$ to a holomorphic function f_2, and it diverges for $|z-a| > r_2$. For each $s_2 < r_2$, convergence is uniform for $|z-a| \le s_2$.

Assume now that $r_1 < r_2$, so that the two series have overlapping convergence sets. Then the Laurent series converges pointwise on the annulus $\text{Ann}_a(r_1, r_2)$ to the holomorphic function $f := f_1 + f_2$; and for every pair of numbers s_1, s_2 such that $r_1 < s_1 < s_2 < r_2$,

$$\lim_{K\to\infty} \sum_{n=-K}^{K} c_k(z-a)^k = f(z) \quad \text{uniformly for } s_1 \le |z-a| \le s_2.$$

So: On an annulus, if f is the sum of a Laurent series, then f is holomorphic. The next result is that the converse is also true.

3.3.1. Theorem (The Laurent Series Representation). *If f is a holomorphic function on the annulus $A := \text{Ann}_a(r_1, r_2)$, then it has a Laurent series representation:*

$$f(z) = \sum_{k=1}^{\infty} c_{-k}(z-a)^{-k} + \sum_{k=0}^{\infty} c_k(z-a)^k \quad \text{for all } z \in A. \tag{1}$$

The first series converges for $|z - a| > r_1$, and the second converges for $|z - a| < r_2$. The coefficients c_k are uniquely determined. For every integer k and every number s between r_1 and r_2,

$$c_k = \frac{1}{2\pi i} \int_{|w-a|=s} \frac{f(w)}{(w-a)^{k+1}} dw. \tag{2}$$

Proof. Let $r_1 < s_1 < s_2 < r_2$. Let Γ be the contour consisting of two circles: $|z - a| = s_2$ traversed counterclockwise, and $|z - a| = s_1$ traversed clockwise. Then

$$W_\Gamma(w) = \begin{cases} 0 & \text{if } |w - a| < s_1 \text{ or } |w - a| > s_2, \\ 1 & \text{if } s_1 < |w - a| < s_2. \end{cases}$$

In particular, the winding number equals 0 everywhere outside A. Therefore by Cauchy's Theorem,

$$\int_\Gamma f = 0; \quad \text{and} \quad f(z) = \frac{1}{2\pi i} \int_\Gamma \frac{f(w)}{w - z} dw \quad \text{for } z \text{ inside } \Gamma. \tag{3}$$

The first equation may be written as

$$\frac{1}{2\pi i} \int_{|w-a|=s_2} f(w) dw \; - \; \frac{1}{2\pi i} \int_{|w-a|=s_1} f(w) dw \; = \; 0. \tag{4}$$

The assertion that (4) holds whenever $r_1 < s_1 < s_2 < r_2$ is equivalent to the statement that the function H given by

$$H(s) := \int_{|w-a|=s} f(w) dw \quad \text{is constant for } r_1 < s < r_2. \tag{5}$$

So (5) holds whenever f is holomorphic on A. We will use this result twice in the rest of the proof.

The second equation in (3) may be written as

$$f(z) = \underbrace{\frac{1}{2\pi i} \int_{|w-a|=s_2} \frac{f(w)}{w - z} dw}_{=:f_2(z)} - \underbrace{\frac{1}{2\pi i} \int_{|w-a|=s_1} \frac{f(w)}{w - z} dw}_{=:f_1(z)}. \tag{6}$$

We will use the computations introduced in Proposition 2.6.3. In the first integral (z being fixed), replace $\dfrac{1}{w - z}$ with the series in powers of $z - a$ that converges to $\dfrac{1}{w - z}$ uniformly for $|w - a| = s_2$, obtaining

$$\frac{1}{2\pi i} \int_{|w-a|=s_2} \frac{f(w)}{w - z} dw = \frac{1}{2\pi i} \int_{|w-a|=s_2} f(w) \left(\sum_{k=0}^{\infty} \frac{(z-a)^k}{(w-a)^{k+1}} \right) dw.$$

Thus

$$f_2(z) = \sum_{k=0}^{\infty} c_k(z-a)^k \quad \text{for } |z-a| < s_2, \quad \text{where}$$

$$c_k := \frac{1}{2\pi i} \int_{|w-a|=s_2} \frac{f(w)}{(w-a)^{k+1}} dw. \tag{7}$$

For each k, since (5) applies with the function $w \mapsto \dfrac{f(w)}{(w-a)^{k+1}}$ in the role of f, the expression (7) is independent of s_2. Therefore the series is the same for every $s_2 < r_2$, and it must converge for $|z-a| < r_2$.

In the second integral of (6), z being fixed, replace $\dfrac{1}{w-z}$ with the series in negative powers of $z-a$ that converges to $\dfrac{1}{w-z}$ uniformly for $|w-a| = s_1$, obtaining

$$-\frac{1}{2\pi i} \int_{|w-a|=s_1} \frac{f(w)}{w-z} dw = -\frac{1}{2\pi i} \int_{|w-a|=s_1} f(w) \left(-\sum_{k=1}^{\infty} \frac{(w-a)^{k-1}}{(z-a)^k} \right) dw$$

Thus

$$f_1(z) = \sum_{k=1}^{\infty} c_{-k}(z-a)^{-k} \quad \text{for } |z-a| > s_1, \quad \text{where}$$

$$c_{-k} = \frac{1}{2\pi i} \int_{|w-a|=s_1} \frac{f(w)}{(w-a)^{-k+1}} dw. \tag{8}$$

For each k, the integral in (8) is independent of s_1. So the series is the same for every $s_1 > r_1$, and it must converge for $|z-a| > r_1$.

We have shown that when the coefficients are given by (2), the representation (1) holds. It remains to show that if the representation holds, the coefficients must be given by (2). To do so, fix an integer n and write (1) as a combined sum, with w in the role of z, thus:

$$f(w) = \sum_{k=-\infty}^{\infty} c_k(w-a)^k.$$

The convergence is uniform on the circle $|w-a| = s$ if $r_1 < s < r_2$. Multiply both sides by $(w-a)^{-n-1}$ and integrate over that circle, obtaining

$$\int_{|w-a|=s} \frac{f(w)}{(w-a)^{n+1}} dw = \sum_{k=-\infty}^{\infty} c_k \int_{|w-a|=s} (w-a)^{k-n-1} dw.$$

The integrals on the right-hand side are all zero except for the one with $k = n$, which equals $2\pi i$. Equation (2) follows. $\qquad \square$

3.3.2. Remark. We already knew (by 2.6.14) that the functions f_1 and f_2 of (6), being Cauchy integrals, were holomorphic. In the proof of that Proposition, to prove holomorphy, we generated a positive-power series centered at an arbitrary point z_0 not on the path. The Laurent series, in which both positive and negative powers appear, is centered at the one point a, the common center of the two circles. Another thing that's new here is our ability to identify $f_2 + f_1$ as f.

3.3.3. Remark. If you have the usual instincts of a calculus-course veteran, you are looking around for computational procedures to master (cranks to turn). Perhaps you are looking at the Laurent series, and wondering how to evaluate those integrals that give the coefficients. Consider instead the logic of Theorem 3.3.1 and think about how formula (2) might turn out to be a labor-saving device for evaluating the integrals. There is often an easy method for finding the Laurent series converging to f on the annulus. You then know all the coefficients. Therefore you know the values of the integrals. This section and Chapter 4 will offer plenty of examples. However, finding values is not the primary point. The more important point is simply that under certain conditions, holomorphic functions can be represented in a certain form, and we can use that fact to draw conclusions about how holomorphic functions behave. We will develop some powerful methods for evaluating integrals. You will be able to use them with confidence if you understand some theory well enough to apply it. Then it will be a matter more of finesse than of computational grinding. So be patient.

3.3.4. Remark. Given O, the domain of a holomorphic f, and a point $a \in O$, there may be many nonoverlapping annuli $\text{Ann}_a(r_1, r_2)$ which are subsets of O. For example, the cosecant, given by

$$\csc z := \frac{\cos z}{\sin z},$$

is holomorphic on $O := \mathbb{C} \setminus \{n\pi \mid n \in \mathbb{Z}\}$; it is undefined at each of the points $n\pi$. The set O contains $\text{Ann}(k\pi, (k+1)\pi)$ for $k = 0, 1, 2, \cdots$. On each of those annuli, there is a different Laurent series. The first annulus on the list is $\text{Ann}(0, \pi)$, which is the punctured disk $D'(0, \pi)$.

   In what follows, we will make use of Laurent series only on annuli with inner radius 0, that is, punctured disks.

B   Behavior Near an Isolated Singularity in the Plane

The function f has an **isolated singularity** at a if it is holomorphic on the punctured disk $D'(a, r)$ for some $r > 0$, but not holomorphic at a. There are then three ways in which f can behave near a.

   (i)   The finite limit $\lim_{z \to a} f(z)$ exists.

   (ii) $\lim_{z \to a} f(z) = \infty$.

   (iii)   Neither (i) nor (ii) holds.

If (i) holds, then f has a **removable singularity** at a. Define (or re-define) $f(a)$ to equal the limit. Thus modified, f is continuous at a, which implies that it is holomorphic at a; recall Corollary 3.1.9.

To discuss the three behaviors further, and present the necessary terminology, we need to refer to the Laurent series of f on $D'(a, r)$:

$$f(z) = \underbrace{\sum_{k=-\infty}^{-1} c_k(z-a)^k}_{=:f_1(z)} + \underbrace{\sum_{k=0}^{\infty} c_k(z-a)^k}_{=:f_2(z)} \quad \text{for} \quad z \in D'(a, r). \tag{9}$$

The **principal part** of f at a is the sum $f_1(z)$ of the negative-power series, which converges for all $z \neq a$.

We will be able to determine which behavior holds—(i), (ii), or (iii)—by looking at the coefficients c_k.

Evidently f has a removable singularity at a if and only if $f_1 \equiv 0$, in other words $c_k = 0$ for all $k < 0$. We then remove the singularity by defining $f(a) := c_0$, and then f_2 is the Taylor series for f.

The function f has a **pole** at a if c_k is nonzero for one or more, but only finitely many, negative values of k. If $-m$ is the least value of k such that c_k is nonzero, then the **order** of the pole is m. The principal part is then a polynomial of degree m in powers of $1/(z-a)$:

$$f_1(z) = c_{-m}(z-a)^{-m} + c_{-m+1}(z-a)^{-m+1} + \cdots + c_{-1}(z-a)^{-1}.$$

Observe that f_1 is holomorphic on $\mathbb{C} \setminus \{a\}$; and that $f - f_1 = f_2$ has a removable singularity at a. Observe also that the function given by $F(z) := (z-a)^m f(z)$ has a removable singularity at a, with $F(a) := c_{-m}$. So

$$f(z) = (z-a)^{-m} F(z), \quad \text{where } F \text{ is holomorphic and nonzero at } a. \tag{10}$$

Thus we can obtain a holomorphic function from f either by subtracting its principal part from it, or by multiplying $f(z)$ by $(z-a)^m$. When f has a pole at a, it follows from (10) that (ii) holds.

The function f has an **essential singularity** at a if c_k is nonzero for infinitely many negative values of k. Theorem 3.3.6 will establish that the behavior of f is then quite spectacular, and in particular, $\lim_{z \to a} f(a)$ fails to exist. So we have case (iii).

Let g be a non-constant holomorphic function on $D(a, r)$. If $g(a) = 0$, then g has a **zero** at a. We will point out the correspondence between zeros of g and poles of $1/g$. Consider the Taylor series:

$$g(z) = \sum_{k=0}^{\infty} c_k(z-a)^k \quad \left(\text{where of course } c_k = \frac{g^{(k)}(a)}{k!}\right).$$

Let m be the least value of k such that c_k is nonzero. Then g has a **zero of order** m at a, and we can write $g(z) = (z-a)^m(c_m + c_{m+1}(z-a) + \cdots)$. More briefly:

$$g(z) = (z-a)^m G(z), \quad \text{where } G \text{ is holomorphic and nonzero at } a. \tag{11}$$

Let $f := 1/g$, which has a removable singularity at a. Then (10) holds, with $F := 1/G$, which is holomorphic and nonzero at a. Evidently g has a zero of order m at a if and only if f has a pole of order m at a.

The next result is an improvement on Corollary 3.1.9.

3.3.5. Proposition. *If f has an isolated singularity at a, and if f is bounded on $D'(a, \delta)$ for some $\delta > 0$, then the singularity is removable.*

Proof. Let $B := \max\{|f(z)| \mid z \in D'(a, \delta)\}$. Then (2), the integral formula for the coefficients, allows this estimate (by 2.5.5):

$$\text{For each } s \text{ with } 0 < s < \delta, \quad |c_k| \le \frac{Bs}{s^{k+1}}.$$

For every $k < 0$, the right-hand side tends to zero as $s \to 0$, and therefore $c_k = 0$. $\square$

When f has an essential singularity at a, the behavior near the point is wild and crazy, in a sense made precise by the next Theorem. Exercise 40 offers some information about its history.

3.3.6. Theorem (Casorati-Weierstrass). *If f has an essential singularity at a, then $\overline{f[D'(a, \delta)]} = \mathbb{C}$ for every $\delta > 0$. In other words, for every $w \in \mathbb{C}$ there is a sequence $z_j \to a$ such that $f(z_j) \to w$.*

Proof. Suppose not. Then there would exist $w \in \mathbb{C}$, $\delta > 0$, and $\epsilon > 0$ such that $|f(z) - w| \ge \epsilon$ whenever $0 < |z - a| < \delta$. The function given by

$$g(z) := \frac{1}{f(z) - w}$$

would have an isolated singularity at a and would be bounded by $1/\epsilon$ on $D'(a, \delta)$. Therefore g would have a removable singularity at a; for some $n \ge 0$, we could write $g(z) = (z - a)^n G(z)$, where G would be holomorphic at a, with $G(a) \ne 0$. Then the function $h := 1/G$ would also be holomorphic near a and nonzero at a, and $f(z) = w + (z - a)^{-n} h(z)$. Thus f would have either a pole of order n at a, if $n > 0$; or a removable singularity, if $n = 0$; neither of which can be the case. $\square$

3.3.7. Remark. If for some $s > 0$ and every real θ_0, the function f is holomorphic on the set

$$\{a + re^{i\theta} \mid 0 < r < s, \theta_0 < \theta < \theta_0 + 2\pi\}$$

but cannot be extended to a holomorphic function on $D'(a, s)$, then a is a **branch point** of f. As you see, a branch point is not an isolated singularity in the sense we have been dealing with, but is a feature peculiar to the logarithm and certain other functions.

C   Examples: Classifying Singularities, Finding Residues

Let f be holomorphic on $D'(a, r)$, so that it has a Laurent series, as in (9). The **residue** of f at a is the coefficient c_{-1}:

$$\operatorname{Res}(f, a) := c_{-1} = \frac{1}{2\pi i} \int_{|z-a|=s} f(z)dz \quad (\text{if } 0 < s < r).$$

We will sometimes write $\operatorname{Res}(f(z), a)$, putting the expression $f(z)$ in place of the function's name f.

In each of the next few Examples, we will consider a function's behavior at an isolated singularity. We will generate the Laurent series, classify the singularity, identify the principal part, and find the residue.

3.3.8. Example. *Let* $g(z) := \dfrac{e^{z/2}}{z^3}$. *Classify the singularities, if any, and find the residues at each.*

Solution. The function is holomorphic on the punctured plane. The function $z \mapsto z^3$ has a zero at the origin of order 3, so that $g(0)$ is undefined. We know the Taylor series for the exponential function:

$$e^w = 1 + w + \frac{w^2}{2!} + \frac{w^3}{3!} + \cdots + \frac{w^n}{n!} + \cdots \quad (w \in \mathbb{C}). \tag{12}$$

Putting $z/2$ into the role of w, we obtain

$$e^{z/2} = 1 + \frac{z}{2} + \frac{z^2}{2^2 2!} + \frac{z^3}{2^3 3!} + \cdots + \frac{z^n}{2^n n!} + \cdots \quad (z \in \mathbb{C}).$$

For each nonzero z, dividing both sides by z^3 gives

$$g(z) = \underbrace{z^{-3} + \frac{1}{2}z^{-2} + \frac{1}{8}z^{-1}}_{g_1(z)} + \underbrace{\frac{1}{2^3 3!} + \cdots + \frac{z^{n-3}}{2^n n!} + \cdots}_{g_2(z)}, \tag{13}$$

and we can now see exactly the nature of the "trouble" at zero; it's a pole of order 3. As always for principal parts at poles, $g_1(z)$ tends to ∞ as $z \to 0$, but g_1 is holomorphic elsewhere, and $g_1(z) \to 0$ as $z \to \infty$. The function $g - g_1$ has a removable singularity at 0; when we remove it, we have $g - g_1 = g_2$, which is holomorphic on the whole plane. The residue of g at 0 is $1/8$.   □

3.3.9. Example. *Let* $h(z) := z^2 e^{1/z}$. *Classify the singularity at* 0 *and find* $\operatorname{Res}(h, 0)$.

Solution. Since for $z \neq 0$,

$$e^{1/z} = \sum_{n=0}^{\infty} \frac{z^{-n}}{n!} = \cdots + \frac{1}{n!}z^{-n} + \cdots + \frac{1}{3!}z^{-3} + \frac{1}{2!}z^{-2} + z^{-1} + 1,$$

we have

$$h(z) = \underbrace{\cdots + \frac{1}{n!}z^{-(n-2)} + \cdots + \frac{1}{4!}z^{-2} + \frac{1}{3!}z^{-1}}_{=:h_1(z)} + \underbrace{\frac{1}{2!} + z + z^2}_{=:h_2(z)}.$$

Thus h has an essential singularity at 0, and the residue is $1/6$. The principal part of h at 0 is the function h_1, which is wild and crazy near 0; h_2 is holomorphic on $\mathbb{C}$.   □

3.3.10. Example. *Evaluate* $\dfrac{1}{2\pi i}\displaystyle\int_{|z|=3} \dfrac{\sin z^2}{z^4}\,dz.$

Solution. By Theorem 3.1.6, the integral equals the coefficient on z^3 in the Taylor series at 0 for the function $z \mapsto \sin z^2$, which is entire. To obtain that Taylor series, begin with the Taylor series for the sine:

$$\sin z = z - \frac{z^3}{3!} + \frac{z^5}{5!} - \cdots + \frac{(-1)^k z^{2k+1}}{(2k+1)!} + \cdots; \tag{14}$$

$$\text{so} \qquad \sin z^2 = z^2 - \frac{z^6}{3!} + \frac{z^{10}}{5!} - \cdots + \frac{(-1)^k z^{4k+2}}{(2k+1)!} + \cdots.$$

The coefficient on z^3 is clearly 0.   □

3.3.11. Example. *The function given by*

$$F(z) := \frac{\sin z - z}{z^3},$$

has an isolated singularity at 0, *since* $F(0)$ *is left undefined by the expression on the right. Classify the singularity.*

Solution. By looking at the Taylor series (14), we can write the Taylor series for F:

$$F(z) = -\frac{1}{3!} + \frac{z^2}{5!} - \cdots + \frac{(-1)^k z^{2k-2}}{(2k+1)!} + \cdots.$$

Thus the singularity turns out to be removable. As soon as $F(0)$ is defined to be $-1/6$, the function F becomes holomorphic at 0 as well as everywhere else in the plane.   □

3.3.12. Example. *Consider the holomorphic function given by*

$$b(z) := \frac{1}{z^2 + 1}.$$

The domain of b is $\mathbb{C} \setminus \{i, -i\}$. Find the Laurent series on each annulus centered at 0 or at i. At i, classify the singularity and find the residue.

Solution. Given any annulus contained in the domain of b, the usual manipulations of geometric series (see subsection 2.6B) will yield the Laurent series that represents

b on that annulus. To study the singularity at i, the relevant Laurent series is the one on the annulus $\text{Ann}_i(0, 2)$.

Let's begin by looking at the two Laurent series on annuli centered at 0:

$$b(z) = 1 - z^2 + z^4 - z^6 + \cdots \qquad \text{for } |z| < 1, \tag{15}$$

$$b(z) = \frac{1}{z^2(1 + z^{-2})} = z^{-2} - z^{-4} + z^{-6} - z^{-8} + \cdots \quad \text{for } |z| > 1. \tag{16}$$

The first series tells us that b is holomorphic at 0, which we already knew. For emphasis: The second series, converging on the annulus $\text{Ann}(1, \infty)$ has no bearing on the discussion of behavior near 0.

One can generate the Laurent series on annuli centered at i, which means on the annuli $\text{Ann}_i(0, 2)$ and $\text{Ann}_i(2, \infty)$, starting with this decomposition of $b(z)$:

$$b(z) = \frac{1}{(z - i)(z + i)} = -\frac{i/2}{z - i} + \frac{i/2}{z + i}.$$

On the first annulus, $\left| \dfrac{z - i}{2i} \right| < 1$, so

$$
\begin{aligned}
b(z) &= -\frac{i/2}{z - i} + \frac{i/2}{2i\left(1 + \frac{z-i}{2i}\right)} \\
&= -\frac{i/2}{z - i} + \frac{1}{4} \sum_{k=0}^{\infty} \frac{(z - i)^k}{(2i)^k} \qquad \text{for } z \in \text{Ann}_i(0, 2). \tag{17}
\end{aligned}
$$

On the second annulus, $\left| \dfrac{2i}{z - i} \right| < 1$, so

$$
\begin{aligned}
b(z) &= -\frac{i/2}{z - i} + \frac{i/2}{(z - i)\left(1 + \frac{2i}{z-i}\right)} \\
&= -\frac{i/2}{z - i} + \frac{i}{2} \sum_{k=0}^{\infty} \frac{(-2i)^k}{(z - i)^{k+1}} \\
&= -\frac{1}{4} \sum_{n=2}^{\infty} \frac{(-2i)^n}{(z - i)^n} \qquad \text{for } z \in \text{Ann}_i(2, \infty). \tag{18}
\end{aligned}
$$

By looking at (17) (not (18)!), we can see that the singularity at i is a pole of order 1, and that the residue is $-i/2$. $\qquad\square$

Sometimes all that we need to do is to find the value of the residue of f at a pole a. This may be quite easy, since it is a matter of evaluating just one coefficient in the Laurent series. If the order of the pole is m, and we write $f(z) = (z - a)^{-m} F(z)$ as in (10), then we may substitute for F its Taylor series:

$$f(z) = (z-a)^{-m}F(z) = (z-a)^{-m}\sum_{k=0}^{\infty}\frac{F^{(k)}(a)}{k!}(z-a)^k$$

$$= \sum_{k=0}^{\infty}\frac{F^{(k)}(a)}{k!}(z-a)^{k-m},$$

from which we can pick out the coefficient on $(z-a)^{-1}$:

$$\text{Res}(f,a) = \frac{F^{(m-1)}(a)}{(m-1)!}. \tag{19}$$

Of course, if $m = 1$, then the right-hand side is simply $F(a)$.

3.3.13. Example. *The function given by* $f(z) := \dfrac{z^5}{(z^2-1)^2}$ *has two poles of order 2. They occur at* $+1$ *and* -1. *Find the residues.*

Solution. Applying the method of (19) with $m = 2$ to find the residue at $+1$, one obtains

$$f(z) = (z-1)^{-2}\left(\frac{z^5}{(z+1)^2}\right)$$

$$= (z-1)^{-2}\left(\frac{z^5}{(z+1)^2}\bigg|_{z=1} + \frac{d}{dz}\frac{z^5}{(z+1)^2}\bigg|_{z=1}(z-1) + \cdots\right).$$

The coefficient on $(z-1)^{-1}$ is thus

$$\frac{(z+1)^2 5z^4 - z^5 2(z+1)}{(z+1)^4}\bigg|_{z=1} = 1.$$

   An alternative method is to carry out algebraic decomposition, or let a computer algebra system do it (the *Mathematica* command is Apart):

$$\frac{z^5}{(z^2-1)^2} = \frac{1}{4(z-1)^2} + \frac{1}{z-1} + z - \frac{1}{4(z+1)^2} + \frac{1}{z+1}.$$

This is not in the form of a Laurent series, since not every summand is a multiple of a power of $z-a$ for the same a. However, the right-hand side may be recognized as the principal part at 1 plus a function holomorphic at 1; or as a function holomorphic at -1 plus the principal part at -1. Thus one can see that $\text{Res}(f,1) = 1$, and that $\text{Res}(f,-1) = 1$. □

3.3.14. Example. *Apply* (19) *to the function* b *of Example 3.3.12.*

Solution. The function given by

$$b(z) = \frac{1}{z^2+1} = \frac{1}{(z-i)(z+i)}$$

has two poles of order 1. By (19) with $m = 1$ applied at i and then at $-i$,

$$\text{Res}\left(\frac{1}{z^2+1}, i\right) = \frac{1}{z+i}\bigg|_{z=i} = \frac{1}{2i} = \frac{-i}{2}; \tag{20}$$

$$\text{Res}\left(\frac{1}{z^2+1}, -i\right) = \frac{1}{z-i}\bigg|_{z=-i} = \frac{1}{-2i} = \frac{i}{2}. \tag{21}$$

□

The following result gives the same idea in a slightly different form, for the special case when the order of the pole is 1.

3.3.15. Proposition. *Let g and h be holomorphic at a, and suppose that*

$$g(a) \neq 0, \ h(a) = 0, \ and \ h'(a) \neq 0.$$

Then the function $f := \dfrac{g}{h}$ has a pole of order 1 at a, and

$$\text{Res}(f, a) = \frac{g(a)}{h'(a)},$$

which may be written as

$$\text{Res}(f, a) = \lim_{z \to a} \frac{g(z)(z-a)}{h(z)}.$$

Proof. We may write

$$f(z) = \frac{g(z)}{(z-a)\left(h'(a) + \frac{h''(a)}{2!}(z-a) + \cdots\right)},$$

which has the form $\dfrac{F(z)}{z-a}$, where F is holomorphic and nonzero at a. Therefore the coefficient on $(z-a)^{-1}$ in the Laurent series for f at a is $F(a) = g(a)/h'(a)$.   □

3.3.16. Example. *Apply the Proposition to the function b of Example 3.3.12.*

Solution. By the Proposition,

$$\text{Res}\left(\frac{1}{z^2+1}, a\right) = \frac{1}{2z}\bigg|_{z=a},$$

and one obtains (20) again.   □

3.3.17. Example. *Let $f(z) := \dfrac{z^2}{z^6+1}$. Find the poles and the residues.*

Solution. The denominator has six zeros, each of order 1: $a_k = e^{\pi i/6} e^{\pi i k/3}$ for $k = 0, 1, 2, 3, 4, 5$. By the Proposition,

$$\operatorname{Res}(f, a_k) = \left.\frac{z^2}{6z^5}\right|_{z=a_k} = \frac{-i(-1)^k}{6}.$$

Thus the residues at the six points are, respectively, $\dfrac{-i}{6}$, $\dfrac{i}{6}$, and $\dfrac{-i}{6}$, in the upper half-plane, and then $\dfrac{i}{6}$, $\dfrac{-i}{6}$, and $\dfrac{i}{6}$ in the lower half-plane. □

3.3.18. Example. *Let* $f(z) := \dfrac{z}{(z^2 + 1)(z^2 + 2z + 2)}.$ *find the poles and the residues at* i *and at* $-1 + i$.

Solution. The poles of f occur at $\pm i$ and $-1 \pm i$. They are all of order 1, so once again we can use 3.3.15 to find residues. At i we can view f as g/h where $g(z) := \dfrac{z}{z^2 + 2z + 2}$ and $h(z) := z^2 + 1$. Thus

$$\operatorname{Res}\left(\frac{z}{(z^2 + 1)(z^2 + 2z + 2)}, i\right) = \left.\frac{z}{2z(z^2 + 2z + 2)}\right|_{z=i}$$

$$= \frac{i}{2i(i^2 + 2i + 2)} = \frac{1 - 2i}{10}.$$

At $-1 + i$ it is convenient to consider f as g/h where $g(z) := \dfrac{1}{z^2 + 1}$ and $h(z) := z^2 + 2z + 2$. Thus

$$\operatorname{Res}\left(\frac{z}{(z^2 + 1)(z^2 + 2z + 2)}, -1 + i\right) = \left.\frac{z}{(z^2 + 1)(2x + 2)}\right|_{z=-1+i}$$

$$= \frac{-1 + i}{(1 - 2i)2i} = \frac{-1 + 3i}{10}.$$

Exercise 18 asks you to find the residues at the other two poles. □

3.3.19. Example. *Let* $f(z) := \dfrac{z^2}{(z^2 + a^2)^3}$, *where* $a > 0$. *Find* $\operatorname{Res}(f, ai)$.

Solution. We may write $f(z)$ as $(z - ai)^{-3}$ times $F(z) := z^2(z + ai)^{-3}$. The function F is holomorphic at ai.

$$f(z) = \frac{z^2}{(z - ai)^3(z + ai)^3}$$

$$= (z - ai)^{-3}\left(F(ai) + F'(ai)(z - ai) + \frac{F''(ai)}{2!}(z - ai)^2 + \cdots\right).$$

Thus $\operatorname{Res}(f, ai) = \dfrac{F''(ai)}{2} = -\dfrac{i}{16a^3}.$ □

For a pole of order 1, the next result gives a useful relation between the residue and the integral over a circular arc, subtending a fixed angle, whose radius tends to zero.

3.3.20. Proposition. *If f has a pole of order 1 at a, and if $\theta_1 < \theta_2$, then*

$$\text{Res}(f,a) = \lim_{r \to 0^+} \frac{1}{(\theta_2 - \theta_1)i} \int_{\text{Arc}(r,\theta_1,\theta_2)} f(z)dz.$$

Proof. We know that $f(z) = \dfrac{\text{Res}(f,a)}{(z-a)} + g(z)$, where g is holomorphic at a. Therefore

$$\int_{\text{Arc}(r,\theta_1,\theta_2)} f(z)dz = \int_{\theta_1}^{\theta_2} \frac{\text{Res}(f,a)}{re^{it}} rie^{it}dt + \int_{\text{Arc}(r,\theta_1,\theta_2)} g(z)dz.$$

As $r \to 0^+$, the first integral on the right tends to $(\theta_2 - \theta_1)i\text{Res}(f,a)$. The second integral tends to 0, since g is bounded near a and the length of the arc is $(\theta_2 - \theta_1)r$. $\square$

3.3.21. Proposition. *If f has an isolated singularity at a, and m is a nonzero complex number, and $g(z) := f(z/m)$, then g has an isolated singularity at ma, and*

$$\text{Res}(g,ma) = m\text{Res}(f,a).$$

Proof. For z near a,

$$f(z) = \sum_{k=-\infty}^{\infty} c_k(z-a)^k; \quad \text{so} \quad g(z) = \sum_{k=-\infty}^{\infty} c_k m^{-k}(z-ma)^k$$

for z near ma. In the power series for g, the coefficient on $(z-ma)^{-1}$ is $c_{-1}m$. $\square$

D Behavior Near a Singularity at Infinity

Let g be a holomorphic function on $\text{Ann}(R,\infty)$ for some $R > 0$. Consider the Laurent series of g:

$$g(z) = \underbrace{\sum_{k=-\infty}^{0} c_k z^k}_{=:g_1(z)} + \underbrace{\sum_{k=1}^{\infty} c_k z^k}_{=:g_2(z)} \quad \text{for } z \in \text{Ann}(R,\infty), \tag{22}$$

$$\text{where} \quad c_k = \frac{1}{2\pi i} \int_{|w|=s} \frac{g(w)}{w^{n+1}}dw \quad \text{if } s > R. \tag{23}$$

The **residue** of g at ∞ is minus the coefficient on z^{-1}:

$$\text{Res}(g,\infty) := -c_{-1} = -\frac{1}{2\pi i} \int_{|z|=s} g(z)dz.$$

We may write $\mathrm{Res}(g(z), \infty)$, giving the expression $g(z)$ for the function value in place of the name g of the function.

For the sake of a useful comparison, we will associate with g the function f given by

$$f(z) := g\left(\frac{1}{z}\right) = \sum_{k=-\infty}^{\infty} c_{-k} z^k. \tag{24}$$

Then f is holomorphic on $D'(0, 1/R)$.

If $g(\infty)$ is defined to equal some complex number w, and if $\lim_{z \to \infty} g(z) = w$, then g is **holomorphic at infinity**. Such is the case if and only if the function f, with $f(0) := w$, is holomorphic at 0.

If g is not holomorphic at ∞, then g has an **isolated singularity** at ∞. The **principal part** of g at ∞ is the sum $g_2(z)$ of the positive-term series, which converges for all $z \in \mathbb{C}$. There are three mutually exclusive possibilities:

(i) g has a **removable singularity** at ∞ if $c_k = 0$ for all $k > 0$. Then $g = g_1$, and the finite limit $\lim_{z \to \infty} g(z)$ exists and equals c_0. Define $g(\infty)$ to equal that limit; g, thus modified, is holomorphic at ∞. Notice that g has a removable singularity at ∞ if and only if f, given by (24), has a removable singularity at 0.

(ii) g has a **pole of order** m at ∞ if $c_k \neq 0$ for one or more, but only finitely many, positive values of k, and if m is the greatest value of k such that c_k is nonzero. Then:

- $g(z) = z^m G(z)$, where G is holomorphic and nonzero at ∞.
- $g - g_2$ has a removable singularity at ∞.
- $\lim_{z \to \infty} g(z) = \infty$.

Notice that g has a pole of order m at ∞ if and only if f has a pole of order m at 0.

(iii) g has an **essential singularity** at ∞ if c_k is nonzero for infinitely many positive values of k. Theorem 3.3.24 below will establish that in this case, $\lim_{z \to \infty} g(z)$ does not exist. Notice that g has an essential singularity at ∞ if and only if f has an essential singularity at 0.

If g is holomorphic at ∞, and if $g(\infty) \equiv c_0 = 0$, then g has a **zero** at ∞. Suppose that g is non-constant. Let m be the least value of $-k$ such that $c_{-k} \neq 0$. Then g has a **zero of order** m at ∞. We may write:

$$g(z) = \sum_{k=-\infty}^{-m} c_{-k} z^{-k} = z^{-m}(c_{-m} + c_{-m-1} z^{-1} + \cdots).$$

More briefly: $g(z) = z^{-m} G(z)$ where G is holomorphic and nonzero at ∞.

3.3.22. Proposition. *Let g have an isolated singularity at ∞.*

(i) If g is bounded on $\mathrm{Ann}(S, \infty)$ for some S:

$$|g(z)| \leq B \quad for \ |z| > S,$$

then the singularity is removable.

(ii) If for some constants S, B, and $a > 1$,

$$|g(z)| \leq \frac{B}{|z|^a} \quad for \ |z| > S,$$

then the singularity is removable; g becomes holomorphic at ∞ when we define $g(\infty) := 0$; and $\mathrm{Res}(g, \infty) = 0$.

Proof. In case (i), the integral formula for the coefficients allows this estimate (by 2.5.5):

$$\text{For each } s \text{ with } S < s < \infty, \ |c_k| \leq \frac{Bs}{s^{k+1}}.$$

For every $k > 0$, the right-hand side tends to zero as $s \to \infty$, and therefore $c_k = 0$.

In case (ii), the hypothesis is stronger, forcing $g(z)$ to tend to zero with a certain speed as $z \to \infty$. The result is to force two more coefficients to be zero. The estimate gives:

$$\text{For each } s \text{ with } S < s < \infty, \ |c_k| \leq \frac{Bs}{s^a s^{k+1}} = Bs^{-a-k}.$$

For every $k \geq -1$, the right-hand side tends to zero as $s \to \infty$, and therefore $c_k = 0$. In particular, $c_{-1} = c_0 = 0$. $\qquad\square$

3.3.23. Example. *Let P and Q be polynomials, with no common zeros, with degrees m and n, respectively:*

$$P(z) := \sum_{k=0}^{m} a_k z^k \quad and \quad Q(z) := \sum_{k=0}^{n} b_k z^k, \quad where \ a_m \neq 0 \ and \ b_n \neq 0.$$

Let $g = P/Q$. Establish these results:

(i) If $n = m$, then g has a removeable singularity, and becomes holomorphic at ∞ when we define $g(\infty) := a_m/b_n$.

(ii) If $n \geq m + 1$, then g has a removeable singularity, and becomes holomorphic at ∞ when we define $g(\infty) := 0$.

(iii) If $n \geq m + 2$, then $\mathrm{Res}(g, \infty) = 0$.

Solution. The poles of g are precisely the zeros of Q. Let R_0 be sufficiently large so that they are all inside the circle $|z| = R_0$. Then

$$g(z) = \frac{P(z)}{Q(z)} = \frac{z^m \left(a_m + \frac{a_{m-1}}{z} + \cdots + \frac{a_0}{z^m} \right)}{z^n \left(b_n + \frac{b_{n-1}}{z} + \cdots + \frac{b_0}{z^n} \right)} \quad \text{for } |z| > R_0. \qquad (25)$$

Therefore $\lim_{|z| \to \infty} g(z)$ is a_m/b_n in case (i), 0 in case (ii). The result is clear in those two cases.

In case (iii), 2.5.5 applies. It follows from (25)

$$\max_{|z|=R} |g(z)| \leq R^{m-n} \frac{|a_m| + \left(\frac{|a_{m-1}|}{R} + \cdots + \frac{|a_0|}{R^m} \right)}{|b_n| - \left(\frac{|b_{n-1}|}{R} + \cdots + \frac{|b_0|}{R^n} \right)}.$$

For every ϵ with $0 < \epsilon < |b_n|$, the last quantity is bounded by $R^{m-n} \frac{|a_m| + \epsilon}{|b_n| - \epsilon}$ when R is sufficiently large. For such R, then,

$$\left| \int_{|z|=R} g(z) dz \right| \leq 2\pi R^{1+m-n} \frac{|a_m| + \epsilon}{|b_n| - \epsilon},$$

which tends to zero as $\mathbb{R} \to \infty$ provided $n \geq m + 2$. The proof is complete.

Notice that for $R > R_0$, the integral $\frac{1}{2\pi i} \int_{|z|=R} g(z) dz$ is constant and equal to $-\mathrm{Res}(g, \infty)$. It is also equal to the sum of the residues of g at its poles in $\mathbb{C}$. $\qquad \square$

The following Theorem is a repetition of 3.3.6 for the case of the point at ∞.

3.3.24. Theorem. *If g has an essential singularity at ∞, then $\overline{g[\mathrm{Ann}(S, \infty)]} = \mathbb{C}$ for every $S > R$. In other words, for every $w \in \mathbb{C}$, there is a sequence $z_j \to \infty$ such that $g(z_j) \to w$.*

Proof. Suppose not. Then there would exist $w \in \mathbb{C}, S > 0$, and $\epsilon > 0$ such that $|z| > S \implies |g(z) - w| \geq \epsilon$. Then the function given by

$$f(z) := \frac{1}{g(z) - w} \quad (|z| > S)$$

would have an isolated singularity at ∞ and would be bounded by $1/\epsilon$. The singularity would then be removable, and for some $n \geq 0$ we would be able to write $f(z) = z^{-n} h(z)$, where h is holomorphic and nonzero at ∞. Then $k := 1/h$ would also be holomorphic and nonzero at ∞, and $g(z) = w + z^n k(z)$. It would follow that g has a removable singularity if $n = 0$, a pole of order n if $n \geq 1$, neither of which can be the case. $\qquad \square$

E   A Digression: Picard's Great Theorem

We digress to tell you about an advanced result which may interest you. A non-constant entire function $F(z) = c_0 + c_1 z + \cdots$ falls into one of two classes: Either there is a largest integer n such that c_n is nonzero, in which case F is a polynomial and has a pole of order n at ∞; or else c_k is nonzero for infinitely many values of k, so that the singularity at infinity is essential. By Liouville's Theorem 3.1.19, the range of a non-constant polynomial is $\mathbb{C}$. Is that still true when F is a non-polynomial entire function? We know from Theorem 3.3.24 that for every $\zeta \in \mathbb{C}$ there exists a sequence $z_k \to \infty$ such that $F(z_k) \to \zeta$; but that doesn't imply ζ is in the range of F. Looking for evidence among familiar examples, we find that the answer is No, since the range of the exponential does not contain 0—though it contains every other complex number.

Picard's Great Theorem, which we will now state without proof, gives a rather satisfying answer; it turns out that the exponential represents the worst case. The Theorem comprises two statements, which you will find say the same thing about an essential singularity, whether it is at infinity or at a finite point.

3.3.25. Theorem. *If F is holomorphic on* $\mathrm{Ann}(R, \infty)$ *and has an essential singularity at* ∞, *then either*

$$F[\mathrm{Ann}(S, \infty)] = \mathbb{C} \quad \text{for every } S \geq R;$$

or there is one complex number ω such that

$$F[\mathrm{Ann}(S, \infty)] = \mathbb{C} \setminus \{\omega\} \quad \text{for every } S \geq R.$$

If f is holomorphic on $D'(a, r)$ and has an essential singularity at the point $a \in \mathbb{C}$, then either

$$f[D'(a, s)] = \mathbb{C} \quad \text{whenever } 0 < s \leq r;$$

or there is one complex number ω such that

$$f[D'(a, s)] = \mathbb{C} \setminus \{\omega\} \quad \text{whenever } 0 < s \leq r.$$

From the first statement it follows easily that at most one complex number is absent from the range of a non-constant entire function. This result is known as Picard's Little Theorem.

For proofs, see [22] or [68]. The original papers by Émile Picard (1856–1941) are [52] and [53]. Larry Zalcman [71] found an elementary proof for a variation on Picard's Theorem.

Exercises

1. Suppose that two Laurent series

$$\sum_{k=-\infty}^{\infty} b_k (z - a)^k \quad \text{and} \quad \sum_{k=-\infty}^{\infty} c_k (z - a)^k$$

converge to the same sum, a holomorphic function $f(z)$, on some annulus. Explain how we know that $b_k = c_k$ for all k.

2. Find the Laurent series for the function $z \mapsto \dfrac{1}{1-z}$, which is holomorphic on $\mathbb{C} \setminus \{1\}$, on each of the annuli $\mathrm{Ann}(0, 1)$ and $\mathrm{Ann}(1, \infty)$.

3. Let $f(z) := (z - 1)^{-3}$. Find the Laurent series for f on $\mathbb{C} \setminus \{1\}$, and find $\mathrm{Res}(f, 1)$.

4. Explain when a Laurent series is also a Taylor series.

5. Prove that the function H given by (5) is constant by showing that $H'(s) \equiv 0$.

6. Explain how to obtain the series representations (15) and (16).

7. Find the Laurent series for the function $z \mapsto \dfrac{1}{z^2 + 4}$ on each of the annuli $\mathrm{Ann}(0, 2)$ and $\mathrm{Ann}(2, \infty)$.

8. Explain how to obtain the series representations (17) and (18).

9. Evaluate the integral $\displaystyle\int_{|z-i|=1} \dfrac{dz}{(z^2 + 1)(z - i)^3}$ quickly and easily, using (17) and (2).

10. Explain: If f has a removable singularity at a, then $\mathrm{Res}(f, a) = 0$; but the converse is untrue.

11. Find the Laurent series for the function $z \mapsto \dfrac{1}{z^2 + 4}$ on the annulus $\mathrm{Ann}_{2i}(0, 4)$.

12. Find the Laurent series for the function $b(z) := \dfrac{1}{z^2 + 1}$ on each of the annuli $\mathrm{Ann}_{-i}(0, 2)$ and $\mathrm{Ann}_{-i}(2, \infty)$.

13. Find the Laurent series for the function $f(z) := \dfrac{1}{(z - 1)(z - 2)}$ on each of the annuli $\mathrm{Ann}(1, 2), \mathrm{Ann}(0, 1), \mathrm{Ann}(2, \infty)$, and $\mathrm{Ann}_1(0, 1)$.

14. Classify the singularity, and find the principal part and the residue for the function $z \mapsto \dfrac{\sin z - z}{z^3}$ at 0; and then do the same at ∞.

15. Let $f(z) = \dfrac{\sin z}{z^2 + 1}$.

   (a) Find the first five nonzero summands of the Taylor series for f at 0.

   (b) Find the summands involving the powers z^{-1} and z^1 in the Laurent series for f on the annulus $\mathrm{Ann}(1, \infty)$.

(c) Find the first three nonzero summands in the Laurent series for f on the annulus $\text{Ann}_i(0, 2)$.

16. Find $\text{Res}\left(\dfrac{e^{az} - e^{bz}}{z^2}, 0\right)$.

17. In each case, classify the isolated singularity of the given function, and find the residue. If there is a pole, determine the order.

(a) $z \mapsto \dfrac{z+1}{z-1}$ at 1.

(d) $z \mapsto \dfrac{1}{z^2+4}$ at $2i$.

(b) $z \mapsto \dfrac{z}{z^4+1}$ at i.

(e) $z \mapsto \dfrac{z^5}{(z^2+1)^2}$ at i.

(c) $z \mapsto \dfrac{1}{(z^2+1)^2}$ at i.

(f) $z \mapsto \dfrac{\sin z}{z^4}$ at 0.

18. Let f be as in Example 3.3.18. Find $\text{Res}(f, -i)$ and $\text{Res}(f, -1 - i)$. What is the sum of the residues at all four poles?

19. Let $f(z) := \dfrac{\cos z}{1 + z + z^2}$. Evaluate the residues at each of the two singularities.

20. Explain: If f has a removable singularity at ∞, then $\text{Res}(f, \infty) = 0$; but the converse is untrue.

21. In each case, classify the isolated singularity of the given function; find the principal part; and find the residue. If there is a pole, determine the order.

(a) $z \mapsto \sin\dfrac{1}{1-z}$ at 1.

(c) $z \mapsto \dfrac{1}{1 - \cos z}$ at 0.

(b) $z \mapsto \dfrac{1}{1 - e^z}$ at $2\pi i$.

(d) $z \mapsto \dfrac{1}{\cos z - \sin z}$ at $\frac{\pi}{4}$.

22. In each case, classify the isolated singularity of the given function; find the principal part; and find the residue. If there is a pole, determine the order.

(a) $z \mapsto \cos\dfrac{1}{z}$ at 0.

(c) $z \mapsto \cot z$ at 0.

(b) $z \mapsto \dfrac{\tan z}{z - \frac{\pi}{2}}$ at $\frac{\pi}{2}$.

(d) $z \mapsto \dfrac{\sin z}{1 - \cos z}$ at 0.

23. The function given by $f(z) := \dfrac{\log_{-1 \mapsto i\pi} z}{(z+5)^2(z+1)}$ has a pole of order 2 at -5. Find $\text{Res}(f, -5)$.

24. In each case, classify the isolated singularity of the given function; find the principal part; and find the residue. If there is a pole, determine the order.

(a) $z \mapsto \dfrac{1}{(\operatorname{Log} z)^2}$ at 1.

(b) $z \mapsto \dfrac{(z-1)^2}{(\operatorname{Log} z)^2}$ at 1.

25. Evaluate $\operatorname{Res}\left(\dfrac{\log z}{1+z^2}, i\right)$ and $\operatorname{Res}\left(\dfrac{\log z}{(1+z^2)^2}, i\right)$. Notice how the answer depends on which logarithm you use.

26. In each case, determine whether the given function has an isolated singularity at infinity; if so, classify it and find the residue. If it is a pole, determine the order.

(a) $z \mapsto \dfrac{z^4+4}{e^z}$.

(b) $z \mapsto e^{-1/z^2}$.

(c) $z \mapsto \dfrac{1-e^z}{1+e^z}$.

27. In each case, classify the singularity at ∞ of the given function, and find the residue. If it's a pole, determine the order.

(a) $z \mapsto \cos \dfrac{1}{z}$.

(c) $z \mapsto \cot z$.

(b) $z \mapsto \dfrac{6z^7}{1+z^3}$.

(d) $z \mapsto \dfrac{z^3+4z^2-5}{3z^4-z-14}$

28. Verify each evaluation of the residue at an essential singularity:

(a) $\operatorname{Res}\left(\sin \dfrac{1}{z}, 0\right) = 1$.

(c) $\operatorname{Res}\left(z^4 \sin \dfrac{1}{z}, 0\right) = \dfrac{1}{5!}$.

(b) $\operatorname{Res}\left(\sin \dfrac{1}{z^2}, 0\right) = 0$.

(d) $\operatorname{Res}\left((z^2 - z + 1)\sin \dfrac{1}{z}, 0\right) = \dfrac{5}{6}$.

29. Find $\operatorname{Res}(e^z \sin \frac{1}{z}, 0)$.

30. Let $f(z) := \dfrac{1}{(z-1)(z-2)z^2}$. Find the residue of f at each of its four isolated singularities. What is the sum of the residues?

31. Let $f(z) := \dfrac{z^3 e^{isz}}{(z+1)(z-1)(z^2+4)}$. Find the residue of f at each of its four isolated singularities.

32. Let $f(z) := \dfrac{e^{zt}z}{(z-1)(z+2)(z+3)^2}$. Find the residue of f at each of its three isolated singularities.

33. Let $f(z) := \dfrac{\sin z}{z(z-\pi)(z+\pi)}$. What is the radius of convergence of the series

$$\sum_{n=0}^{\infty} \dfrac{f^{(n)}(i)}{n!}(z-i)^n?$$ Explain.

34. Let g be holomorphic on $\mathrm{Ann}(S, \infty)$, and suppose that for some constant M and some number $\epsilon > 0$,

$$|g(z)| \le \frac{M}{|z|^{1+\epsilon}}.$$

Prove that $\mathrm{Res}(g, \infty) = 0$ and that the singularity at ∞ is removable.

35. Show that where f has a pole, e^f has an essential singularity.

36. The residue at ∞ was defined as an integral over a circle centered at 0. This definition shows a bias in favor of the point 0. Could we just as well use the circle $|z - a| = s$ for some point $a \ne 0$, provided s is sufficiently large? Explain.

37. Let $u : \mathbb{C} \to \mathbb{R}$. Suppose that u is harmonic and positive everywhere. Prove that u is constant.

38. Let f and g be entire functions such that $|f(z)| \le |g(z)|$ for all $z \in \mathbb{C}$. What conclusions can you draw?

39. Let g and h be holomorphic at a, $h(a) = h'(a) = 0$, $g(a) \ne 0$, and $h''(a) \ne 0$. Prove that g/h has a pole of order 2 at a and that

$$\mathrm{Res}\left(\frac{g}{h}, a\right) = \frac{2g'(a)}{h''(a)} - \frac{2g(a)h'''(a)}{3h''(a)^2}.$$

40. The result known as the Casorati-Weierstrass Theorem was discovered and published by Casorati in 1868, and by Weierstrass in 1876. One may make a controversy over the attribution, since closely related results were produced by Sokhotzky in 1873 in his dissertation, and by Briot and Bouquet in the first (1859) edition of their book [6, p. 39], though not in the second. Can the result, more or less, also be found in Cauchy's work? If you care to investigate the historical record, read the remarks and pursue the references given in [57, p. 309], especially [48].

Help on Selected Exercises

1. Study Theorem 3.3.1 and its proof.

3. The Laurent series: $f(z) = (z - 1)^{-3}$. The residue is the coefficient on $(z - 1)^{-1}$ in that Laurent series. The only nonzero coefficient is on the power -3. So $\mathrm{Res}(f, 1) = 0$.

5. Write $H(s)$ as a pullback and take the derivative:

$$H(s) = \int_0^{2\pi} f(a + se^{it})sie^{it}dt, \qquad \text{so}$$

$$H'(s) = \int_0^{2\pi} \frac{d}{ds}\left[f(a + se^{it})s\right] ie^{it}dt. \qquad \text{(Why is this valid?)}$$

$$= \cdots = \frac{1}{s}\int_{|w-a|=s}\left[f'(w)(w - a) + f(w)\right] dw,$$

which equals 0 because the integrand has an antiderivative, namely, $w \mapsto f(w)(w - a)$.

7. For $|z| < 2$,

$$\frac{1}{z^2 + 4} = \frac{1}{4\left(1 + \left(\frac{z}{2}\right)^2\right)} = \frac{1}{4}\sum_{k=0}^{\infty}(-1)^k\left(\frac{z}{2}\right)^{2k}.$$

For $|z| > 2$,

$$\frac{1}{z^2 + 4} = \frac{1}{z^2\left(1 + \frac{4}{z^2}\right)} = \frac{1}{z^2}\sum_{k=0}^{\infty}(-1)^k\left(\frac{2}{z}\right)^{2k}$$

$$= \sum_{k=0}^{\infty}(-1)^k 4^k z^{-2-2k}.$$

9. Apply (2) with $f(w) = \dfrac{1}{w^2 + 1}$ and $k = 2$. The integral to be evaluated equals $2\pi i c_2$. By looking at (17), we find that $2\pi i c_2 = -\dfrac{\pi i}{8}$.

13.   • $f(z) = -\displaystyle\sum_{k=-\infty}^{-1} z^k + \sum_{k=0}^{\infty}(-2^{-k-1})z^k$ for $1 < |z| < 2$.

• $f(z) = \displaystyle\sum_{k=0}^{\infty}(1 - 2^{-k-1})z^k$ for $|z| < 1$. It's a Taylor series. There's no singularity at 0.

• $f(z) = \displaystyle\sum_{k=-\infty}^{-1}(2^{1-k} - 1)z^k$ for $2 < |z| < \infty$.

• $f(z) = \dfrac{-1}{z - 1} - \displaystyle\sum_{k=0}^{\infty}(z - 1)^k$.

15. See Exercise 5 of Section 3.1, on page 199. For $|z| < 1$, $\dfrac{\sin z}{1 + z^2}$ equals the product of two convergent series:

$$\left(z - \frac{z^3}{3!} + \frac{z^5}{5!} - \frac{z^7}{7!} + \frac{z^9}{9!} - \cdots \right) (1 - z^2 + z^4 - z^6 + z^8 - \cdots)$$

$$= z - \left(-1 - \frac{1}{3!} \right) z^3 + \left(1 + \frac{1}{3!} + \frac{1}{5!} \right) z^5$$

$$+ \left(-1 - \frac{1}{3!} - \frac{1}{5!} - \frac{1}{7!} \right) z^7 + \left(1 + \frac{1}{3!} + \frac{1}{5!} + \frac{1}{7!} + \frac{1}{9!} \right) z^9 + \cdots$$

For $1 < |z| < \infty$, $\dfrac{\sin z}{z^2(1 + z^{-2})}$ equals the sum of these two convergent series:

$$\left(z - \frac{z^3}{3!} + \frac{z^5}{5!} - \frac{z^7}{7!} + \cdots \right) (z^{-2} - z^{-4} + z^{-6} - \cdots)$$

$$= \cdots \left(1 + \frac{1}{3!} + \frac{1}{5!} + \cdots \right) z^{-1} + \left(-\frac{1}{3!} - \frac{1}{5!} - \frac{1}{7!} - \cdots \right) z + \cdots.$$

17. (a) Order 1, residue 2. (b) Order 1, residue $-1/3$. (c) Order 2, residue $i/4$. (d) Order 1, residue $1/4i$. (e) Order 2, residue -1. (f) Order 3, residue $-1/6$.

19. $\operatorname{Res}\left(f, -\dfrac{1}{2} \pm i \dfrac{\sqrt{3}}{2} \right) = \dfrac{\cos\left(-\frac{1}{2} \pm i \frac{\sqrt{3}}{2} \right)}{\pm i \sqrt{3}}.$

23. $\dfrac{d}{dz} \dfrac{\log z}{z + 1} \Big|_{z=-5} = \cdots = \dfrac{\frac{4}{5} - \ln 5 - i\pi}{16}.$

29. Obtain the answer as a convergent series, using the result of Exercise 6 of Section 3.1 (page 199).

30. The singularity of f at ∞ is removable, and $\operatorname{Res}(f, \infty) = 0$. Proposition 3.3.15 applies to the poles of order 1 at 1 and 2 and gives

$$\operatorname{Res}(f, 1) = \frac{1}{(z - 2)z^2} \Big|_{z=1} = -1 \quad \text{and}$$

$$\operatorname{Res}(f, 2) = \frac{1}{(z - 1)z^2} \Big|_{z=2} = \frac{1}{4};$$

but it does not apply to poles of order 2. We know that $\operatorname{Res}(f, 0)$ equals the coefficient on z^{-1} in the Laurent series for f at 0. To generate the series, recalling the methods of 2.6B, we may begin by writing

$$\frac{1}{(z-1)(z-2)} = \frac{1}{1-z} - \frac{1}{2-z} = \sum_{k=0}^{\infty} z^k - \frac{1}{2}\sum_{k=0}^{\infty}\left(\frac{z}{2}\right)^2$$

$$= \frac{1}{2} + \frac{3z}{4} + \frac{7z^2}{8} + \cdots + \left(1 - \frac{1}{2^{k+1}}\right)z^k + \cdots; \quad \text{then}$$

$$f(z) = \frac{1}{2z^2} + \frac{3}{4z} + \frac{7}{8} + \cdots + \left(1 - \frac{1}{2^{k+1}}\right)z^{k-2} + \cdots.$$

Thus $\text{Res}(f,0) = 3/4$. (See also Exercise 39.)

31. See Example 4.4.6, page 288.

32. See Example 4.7.3, page 314.

35. An outline: Suppose that f has a pole at 0. If $D(0,r)$ is contained in the domain of f, then $f[D(0,r)] \subset D(\infty, S)$ for some S. The latter set contains the strip $\{x + iy \mid 2\pi k \le y < 2\pi(k+1)\}$ for all sufficiently large k, and the exponential maps the strip onto $\mathbb{C} \setminus \{0\}$. Therefore $e^f[D(0,r)] \supset \mathbb{C} \setminus \{0\}$.

39. For a proof, and more general formulas along this line, see [38, Section 4.1].

3.4 THE RESIDUE THEOREM AND THE ARGUMENT PRINCIPLE

A Meromorphic Functions and the Extended Plane

Consider these six holomorphic functions:

Function:	Domain:	Range:	
$G_1(z) := z^2$	$\mathbb{C}$	$\mathbb{C}$	
$G_2(z) := \dfrac{1}{z^3}$	$\mathbb{C} \setminus \{0\}$	$\mathbb{C} \setminus \{0\}$	
$G_3(z) := \dfrac{z+2}{3z+4}$	$\mathbb{C} \setminus \{-4/3\}$	$\mathbb{C}$	(1)
$G_4(z) := \dfrac{1}{z^2+1}$	$\mathbb{C} \setminus \{i, -i\}$	$\mathbb{C} \setminus \{0\}$	
$z \mapsto \sin z$	$\mathbb{C}$	$\mathbb{C}$	
$z \mapsto e^z$	$\mathbb{C}$	$\mathbb{C} \setminus \{0\}$	

So far in this book, the domain and range of a function have usually been understood to be subsets of $\mathbb{C}$. Accordingly, in each row of the table above, you can surmise the identity of the domain and range by looking at the expression in z which gives the value of the function at z. For example, in the case of G_3, you understand the domain to be the set of all $z \in \mathbb{C}$ for which $\dfrac{z + 2}{3z + 4}$ makes sense as an element of $\mathbb{C}$. We will now modify this understanding.

We have discussed the possible behaviors of a holomorphic function near a singularity, whether it occurs at a point of $\mathbb{C}$ or at infinity. There was a great similarity in the discussions of those two cases. We will adopt some new definitions that allow us to speak of the two cases as one. We will use the six examples above to show how the new terminology works.

Recall that the extended plane $\hat{\mathbb{C}}$ is the complex plane with the point at infinity adjoined:

$$\hat{\mathbb{C}} := \mathbb{C} \cup \{\infty\}.$$

The use of the symbol $D(\infty, r)$ for the set $\{z \in \hat{\mathbb{C}} \mid |z| > r\}$ suggests a notion of a "disk centered at ∞," analogous to a disk centered at a point of $\mathbb{C}$. We will use it to define an open set in $\hat{\mathbb{C}}$, and limits of sequences in $\hat{\mathbb{C}}$, without using separate phrases for the point at infinity. Thus:

- The set $\Omega \subseteq \hat{\mathbb{C}}$ is **open** if for every $p \in \Omega$, there exists $\delta > 0$ such that $D(p, \delta) \subseteq \Omega$.

- For $q \in \hat{\mathbb{C}}$, the statement $\lim_{k \to \infty} z_k = q$ means that for every $\epsilon > 0$, eventually $z_k \in D(q, \epsilon)$.

- The point $q \in \hat{\mathbb{C}}$ is a **boundary point** of the set $\Omega \subseteq \hat{\mathbb{C}}$ if for every $\delta > 0$, the set $D(q, \delta)$ contains at least one point of Ω and at least one point of $\hat{\mathbb{C}} \setminus \Omega$.

All that's lacking now is a way to visualize $\hat{\mathbb{C}}$ such that the point ∞ looks like all the other points. The answer to that need is the Riemann sphere, which we will discuss in Section 3.6.

3.4.1. Proposition. *$\hat{\mathbb{C}}$ is compact.*

Proof. Let $\{z_k\}$ be a sequence in $\hat{\mathbb{C}}$. If it is bounded, then it has a subsequence converging to a point in $\mathbb{C}$. If it is not bounded, then it has a subsequence converging to ∞. Thus in every case, it has a subsequence converging to a point of $\hat{\mathbb{C}}$.   □

From now on, whenever it makes sense, we will allow the point ∞ to belong to either the domain or range of a function. We will often consider functions

$$G : \Omega \to \hat{\mathbb{C}}, \text{ where } \Omega \text{ is an open subset of } \hat{\mathbb{C}}.$$

Thus we may have $G(z) = \infty$ for one or more $z \in \Omega$, or even $G \equiv \infty$.

Whenever the limit $q = \lim_{z \to \infty} G(z)$ is meaningful, with $q \in \hat{\mathbb{C}}$, we will understand ∞ to be in the domain of G, with $G(\infty) = q$. And if $\lim_{z \to p} G(z) = \infty$ for some $p \in \hat{\mathbb{C}}$, we will understand p to be in the domain of G, with $G(p) = \infty$. To state all this in a unified way: Whenever G has a removable singularity or a pole at $p \in \hat{\mathbb{C}}$, so that the limit $q := \lim_{z \to p} G(z)$ exists ($q \in \hat{\mathbb{C}}$), we will write $G(p) := q$ and say that G is continuous at p.

With this new understanding, how does the table (1) change? For the first four of the functions, $\hat{\mathbb{C}}$ is both the domain and range, thus:

$$
\begin{aligned}
G_1(\infty) &= \infty. \\
G_2(\infty) &= 0 \quad \text{and} \quad G_2(0) = \infty. \\
G_3(\infty) &= 1/3 \quad \text{and} \quad G_3(-4/3) = \infty. \\
G_4(\infty) &= 0 \quad \text{and} \quad G_4(i) = G_4(-i) = \infty.
\end{aligned}
\tag{2}
$$

On the other hand, the domain and range of the sine and exponential remain the same, since in each case the singularity at infinity is essential; $\sin \infty$ and e^∞ are not meaningful.

The holomorphic functions G_k, after we extend their domains and ranges, no longer fit the definition of "holomorphic;" we need a new word.

Let Ω be an open connected subset of $\hat{\mathbb{C}}$. The function $f : \Omega \to \hat{\mathbb{C}}$ is **meromorphic** on Ω if, at each point of that set, it either is holomorphic or has a pole; or, if $f \equiv \infty$.

3.4.2. Proposition. *Let $p \in \hat{\mathbb{C}}$.*

(i) *If f is holomorphic and has a zero of order $m > 0$ at p, then $1/f$ is meromorphic and has a pole of order m at p.*

(ii) *If f is meromorphic and has a pole of order $m > 0$ at p, then $1/f$ is holomorphic and has a zero of order m at p.*

(iii) *If f is meromorphic on the open connected set Ω, then $1/f$ is also meromorphic on Ω.*

(iv) *Let f be meromorphic at p. Then f has a pole of order m at p if and only if f' has a pole of order $m + 1$ at p.*

(v) *Let f be meromorphic at p. If f has either a pole or a zero at p, then f'/f has a pole of order 1 at p.*

(vi) *Let f be meromorphic on the open connected set $\Omega \subseteq \hat{\mathbb{C}}$, and let A be the set of its poles in Ω. Then:*

    (a) *A is a countable set.*

    (b) *The accumulation points of A are on the boundary of Ω.*

    (c) *The set $\Omega \setminus A$ is open.*

    (d) *If K is a compact subset of Ω, then $A \cap K$ is a finite set.*

Proof. We prove two parts of the Proposition, and leave the others as Exercises 4 and 5.

(i) Case 1, when $p \in \mathbb{C}$: For z near p, $f(z) = (z - p)^m h(z)$, where h is holomorphic and nonzero near p. Then $1/h$ is also holomorphic and nonzero near p, and we have $\dfrac{1}{f(z)} = (z - p)^{-m} \dfrac{1}{h(z)}$ for z near p. Case 2, when $p = \infty$: For z near ∞, $f(z) = z^{-m} h(z)$, where h holomorphic and nonzero near ∞. Then $1/h$ is also holomorphic and nonzero near ∞, and we have $\dfrac{1}{f(z)} = z^m \dfrac{1}{h(z)}$ for z near ∞.

(iii) If $f \equiv \infty$, then $1/f \equiv 0$; if $f \equiv 0$, then $1/f \equiv \infty$. The result follows from those implications and parts (i) and (ii).

□

B   The Residue Theorem

3.4.3. Theorem. *Let f be holomorphic on the open connected set $O \subseteq \mathbb{C}$ except for a finite number of isolated singularities $a_1, \cdots, a_J$. Let Γ be a contour contained in O that avoids the singularities of f and such that*

$$W_\Gamma(\zeta) = 0 \quad \text{for all } \zeta \in \mathbb{C} \setminus O.$$

Then

$$\frac{1}{2\pi i} \int_\Gamma f = \sum_{j=1}^{J} W_\Gamma(a_j) \operatorname{Res}(f, a_j). \tag{3}$$

Proof. We may suppose that $J \geq 1$ and that none of the singularities a_j is removable. Otherwise, f would be holomorphic on O, and Cauchy's Theorem would imply that the integral in (3) equals zero.

For each j, the principal part g_j of f at the singularity a_j has the form

$$g_j(z) = \sum_{m=1}^{m_j} \frac{c_{j,-m}}{(z - a_j)^m},$$

where, of course,

$$c_{j,-1} = \operatorname{Res}(f, a_j).$$

If the singularity is a pole, then m_j is its order; if the singularity is essential, then $m_j = \infty$. In either case, the function g_j is holomorphic on $\hat{\mathbb{C}}$ except at a_j, and

$$\frac{1}{2\pi i} \int_\Gamma g_j = c_{j,-1} W_\Gamma(a_j).$$

The function $f - g_j$ is holomorphic at a_j, in the sense that it has a removable singularity and we choose to remove it. The function $f - \sum_{j=1}^{J} g_j$ is holomorphic on O, in the sense that it has only removable singularities in O, and we remove them all. By Cauchy's Theorem,

$$\int_\Gamma \left(f - \sum_{j=1}^J g_j \right) = 0.$$

Equation (3) follows. $\square$

The next result establishes the converse of the obvious statement that every rational function is meromorphic on $\hat{\mathbb{C}}$.

3.4.4. Corollary. *A function that is meromorphic on the extended plane $\hat{\mathbb{C}}$ must be a rational function; and the sum of its residues is zero.*

Proof. Let f be meromorphic on $\hat{\mathbb{C}}$. Then f has at most a finite number of poles, because otherwise the poles would have an accumulation point in $\hat{\mathbb{C}}$, and f would not be meromorphic at such a point.

If f has no poles, then, being holomorphic on $\hat{\mathbb{C}}$, it is a bounded entire function on $\mathbb{C}$ and hence a constant.

If f has a pole at infinity, let g_∞ be its principal part, which of course is a polynomial. If f does not have a pole at infinity, let $g_\infty := 0$. In either case, $f - g_\infty$ is holomorphic at infinity. If f has no poles in $\mathbb{C}$, then $f - g_\infty$ is holomorphic on $\hat{\mathbb{C}}$, hence f equals g_∞ plus a constant.

Consider now the case when f has poles $a_1, \cdots, a_J$ in $\mathbb{C}$. Let g_j be the principal part of f at a_j. Then g_j is a polynomial in powers of $1/(z - a_j)$, and is thus holomorphic on $\hat{\mathbb{C}}$ except at a_j. The function $f - g_j$ is holomorphic at a_j.

The function

$$f - g_\infty - \sum_{j=1}^J g_j \tag{4}$$

is holomorphic on $\hat{\mathbb{C}}$, and is hence a constant c. Therefore

$$f \equiv c + g_\infty + \sum_{j=1}^J g_j,$$

which is a rational function.

To prove the last assertion in the Corollary, let Γ consist of a circle $|z| = r$ which has all the finite poles a_j, if any, inside it. Then

$$\frac{1}{2\pi i} \int_\Gamma f = \sum_{j=1}^J \mathrm{Res}(f, a_j)$$

by the Residue Theorem, and equals $-\mathrm{Res}(f, \infty)$ by definition. Therefore

$$\mathrm{Res}(f, \infty) + \sum_{j=1}^J \mathrm{Res}(f, a_j) = 0,$$

and the proof is complete. □

3.4.5. Remark. The process in the proof just given, by which we begin with a rational function and write it in the form (4) justifies the assertion in calculus texts that every rational function can be written as a sum of "partial fractions." The discussion is usually limited to real-valued coefficients. See [67, Section 7.4].

3.4.6. Example. *Evaluate* $\dfrac{1}{2\pi i} \displaystyle\int_{|z|=3/2} f(z)\,dz$, *where*

$$f(z) := \frac{1}{(z-1)(z-2)z^2}.$$

Solution. By the Residue Theorem, the integral equals $\mathrm{Res}(f,0) + \mathrm{Res}(f,1)$. By 3.4.4,

$$\mathrm{Res}(f,0) + \mathrm{Res}(f,1) = -\mathrm{Res}(f,\infty) - \mathrm{Res}(f,2). \tag{5}$$

The singularity of f at ∞ is removable, and $\mathrm{Res}(f,\infty) = 0$. By 3.3.15,

$$\mathrm{Res}(f,2) \;=\; \left.\frac{1}{(z-1)z^2}\right|_{z=2} \;=\; \frac{1}{4}.$$

Therefore the left-hand side of (5) equals $-1/4$. □

3.4.7. Proposition. *Let f be a one-to-one meromorphic function from $\hat{\mathbb{C}}$ into $\hat{\mathbb{C}}$. Then f must be of the form*

$$z \mapsto \frac{az+b}{cz+d} \qquad \text{where } ad - bc \neq 0.$$

Proof. The function f has at most one pole, since it is one-to-one. It must have one pole, for the following reason. If it had no pole, then it would be holomorphic and hence of course continuous on $\hat{\mathbb{C}}$, which is compact by 3.4.1. Therefore $f[\hat{\mathbb{C}}]$ would be compact, by 2.4.7. Thus f would be a bounded entire function on $\mathbb{C}$ and therefore constant by Liouville's Theorem 3.1.19, which cannot be the case.

So f has exactly one pole. If the pole is at ∞, and if g is the principal part of f at ∞, then $f - g$ is holomorphic on $\hat{\mathbb{C}}$ and hence equals a constant b; so $f = g + b$. The degree of the polynomial g is 1, because if it were greater than 1, then f would not be one-to-one. So we can write $f(z) = az + b$ for some $a \neq 0$, and let $c = 0, d = 1$.

If the pole of f is at some point $p \in \mathbb{C}$, and if g is the principal part of f at p, then as before $f = g + k$ for some constant k. Then g is a polynomial in powers of $\dfrac{1}{z-p}$. If the degree of that polynomial (the order of the pole) were greater than 1, then f would not be one-to-one. So for some constant $q \neq 0$,

$$f(z) \;=\; \frac{q}{z-p} + k \;=\; \frac{kz + (q - kp)}{z - p},$$

which is also of the form claimed. □

C   Multiplicity and Valence

Let f be meromorphic at $a \in \hat{\mathbb{C}}$. The **multiplicity** of f at a, denoted by $m(f, a)$, is defined as follows.

- If f is constant on a neighborhood of a, then $m(f, a) := \infty$.

- If f has a pole at a, then $m(f, a)$ is the order of the pole.

- If f is holomorphic and non-constant at a, then $m(f, a)$ is the order of the zero that the function $z \mapsto f(z) - f(a)$ has at a.

Let f be a meromorphic function on the open connected set $\Omega \subseteq \hat{\mathbb{C}}$. The **valence** of f is the counting function $\nu_f : \hat{\mathbb{C}} \to \mathbb{Z}^+ \cup \{\infty\}$ defined as follows. For each $q \in \hat{\mathbb{C}}$, $\nu_f(q)$ is the number of times f takes on the value q, counted by multiplicity:

$$\nu_f(q) := \sum \{m(f, p) \mid p \in \Omega \text{ and } f(p) = q\}.$$

If f is constant, $f \equiv q$, then $\nu_f(q) = \infty$ and $\nu_f(w) = 0$ if $w \neq q$. If f is non-constant and $q \in \mathbb{C}$, then $\nu_f(q)$ is the sum of the orders of the zeros of the function $z \mapsto f(z) - q$ in Ω; and $\nu_f(\infty)$ is the sum of the orders of the poles of f in Ω.

We will write $\nu_{f|S}$ for the valence of the restriction of f to a subset S of Ω. Thus

$$\nu_{f|S}(q) = \sum \{m(f, p) \mid p \in S \text{ and } f(p) = q\}.$$

We offer some examples using "ν_f" and "$\nu_{f|S}$" in sentences. From Section 2.2 we already understand the exponential and the sine. For each we may select a set S such that the restricted function $f|S$ is one-to-one and has the same range as f. Thus for the exponential, the natural domain is $\mathbb{C}$, and a good choice for S is the set

$$S := \{z \mid 0 \leq \operatorname{Im} z < 2\pi\}$$

Then:

$$\nu_{\exp}(w) = \begin{cases} \infty & \text{if } w \in \mathbb{C} \setminus \{0\}, \\ 0 & \text{if } w = 0 \text{ or } w = \infty. \end{cases}$$

$$\nu_{\exp|S}(w) = \begin{cases} 1 & \text{if } w \in \mathbb{C}, \\ 0 & \text{if } w = 0 \text{ or } w = \infty, \end{cases}$$

For the sine, let

$$S := \left\{ z = x + iy \ \middle| \ y \geq 0 \text{ and } -\tfrac{\pi}{2} \leq x \leq \tfrac{\pi}{2}; \text{ or, } y < 0 \text{ and } -\tfrac{\pi}{2} < x < \tfrac{\pi}{2} \right\}.$$

Then

$$\nu_{\sin}(w) = \begin{cases} \infty & \text{if } w \in \mathbb{C}, \\ 0 & \text{if } w = \infty. \end{cases}$$

$$\nu_{\sin|S}(w) = \begin{cases} 1 & \text{if } w \in \mathbb{C}, \\ 0 & \text{if } w = \infty. \end{cases}$$

D   Valence for a Rational Function

As the next two Propositions make clear, the valence of every rational function is constant.

3.4.8. Proposition. *Let $P : \hat{\mathbb{C}} \to \hat{\mathbb{C}}$, where P is a polynomial of degree $n > 0$, then $\nu_P \equiv n$.*

Proof. We may write $P(z) = \sum_{k=0}^{n} c_k z^k$, where the leading coefficient c_n is nonzero. By the Fundamental Theorem of Algebra, P may be factored as follows:

$$P(z) = c_n \prod_{j=1}^{J} (z - a_j)^{n_j},$$

where the complex numbers a_j are the J distinct zeros of P. Then $1 \leq J \leq n$; and the sum of the orders of the zeros, $n_j := m(P, a_j)$, is n. Thus $\nu_P(0) = n$. For every complex number w, the function given by $P_1(z) := P(z) - w$ is also a polynomial of degree n, so $\nu_{P_1}(0) = n$; and $\nu_P(w) = \nu_{P_1}(0)$. And P has a pole of order n at infinity: $\nu_P(\infty) = m(P, \infty) = n$.   □

3.4.9. Proposition. *Consider a non-constant rational function $f := P/Q$, and let n and m be the degrees of P and Q respectively. Then $\nu_f \equiv \max(n, m)$.*

Proof. If Q is constant ($m = 0$), then the previous Proposition gives the result. Suppose now that Q is non-constant, and that P and Q have leading coefficients c and 1, respectively. We must show that $\nu_f(w) = \max(m, n)$ for all $w \in \hat{\mathbb{C}}$.

Case 1, when $w = 0$: The zeros of f that lie in the plane $\mathbb{C}$ are precisely the zeros of P. The sum of their orders is n. Therefore $\nu_f(0)$ equals n plus the order of the zero at ∞, if there is one. If $m < n$, then $f(\infty) = \infty$; if $m = n$, then $f(\infty) = c$; therefore in either of those cases, f has no zero at ∞, and $\nu_f(0) = n$. If $m > n$, then f has a zero of order $m - n$ at ∞ so that $\nu_f(0) = n + (m - n) = m$.

Case 2, when $w \in \mathbb{C} \setminus \{0\}$: If g is the rational function given by

$$g(z) = \frac{P(z) - wQ(z)}{Q(z)},$$

then by Case 1 applied to g, $\max(m, n) = \nu_g(0)$, which equals $\nu_f(w)$.

Case 3, when $w = \infty$: By Case 1, $\max(m, n) = \nu_{1/f}(0)$, which is the same as $\nu_f(\infty)$.   □

E   The Argument Principle: Integrals That Count

Let γ be a smooth curve whose parametrization is of the rather special form

$$z(\theta) := r(\theta)e^{i\theta} \quad \text{for } \theta_0 \leq \theta \leq \theta_1,$$

where $r(\theta) > 0$. The parameter θ is an argument of $z(\theta)$; as it increases, $z(\theta)$ moves counterclockwise around the origin. Consider the integral of $z \mapsto 1/z$ over γ; using the pullback, we find that

$$\int_\gamma \frac{dz}{z} = \ln \frac{r(\theta_1)}{r(\theta_0)} + i(\theta_1 - \theta_0).$$

If γ is a closed curve, then $r(\theta_1) = r(\theta_0)$. So the integral equals $i(\theta_1 - \theta_0)$, which may not be 0. It equals $i2\pi n$, where n is the number of times $z(\theta)$ winds around the origin. This is an elementary instance of the idea which appears in the definition of the winding number; we use an integral to count for us the net number of times an arbitrary closed path γ winds around the origin:

$$\frac{1}{2\pi i} \int_\gamma \frac{dz}{z} = W_\gamma(0). \tag{6}$$

The next Theorem is called the "Argument Principle," and represents a further application of the same idea.

Let g be meromorphic on the open set $O \subseteq \mathbb{C}$. The **logarithmic derivative** of g is the function $z \mapsto g'(z)/g(z)$. It is also meromorphic on O. In fact, it is holomorphic on O except at the zeros and poles of g. Let's take a look at those points. If g has a pole of order I or a zero of order Z at the point $a \in O$, then for z near a,

$$g(z) = (z-a)^{-I}G(z) \qquad \text{or} \qquad g(z) = (z-a)^Z G(z),$$

respectively, where G is holomorphic and nonzero at a. We can deal with both cases at once, letting m denote $-I$ or Z, as the case may be. For z near a,

$$g(z) = (z-a)^m G(z), \quad \text{so} \quad g'(z) = m(z-a)^{m-1}G(z) + (z-a)^m G'(z).$$

Now compute the logarithmic derivative:

$$\frac{g'(z)}{g(z)} = \frac{m}{z-a} + \frac{G'(z)}{G(z)}.$$

Thus g'/g has a pole of order 1 at a, and

$$\mathrm{Res}\left(\frac{g'}{g}, a\right) = m, \quad \text{which, of course, equals} \quad \frac{1}{2\pi i} \int_{|z-a|=r} \frac{g'}{g}$$

for small enough r. Thus at a zero of g, the residue of g'/g is the order of the zero; and at a pole of g, the residue of g'/g is minus the order of the pole.

For example, if k is a nonzero integer and $g(z) := z^k$, then the logarithmic derivative of g is $z \mapsto kz^{-1}$, whose integral over the circle $|z| = r$ is k. This value is, you will notice, the order of the zero of g at 0 if $k > 0$, or minus the order of the pole of g at 0 if $k < 0$.

Let γ be a closed path contained in O, given by $z : [a,b] \to \mathbb{C}$. Suppose that g has no zeros or poles on γ. Let $g[\gamma]$ denote the curve given by

$$w := g \circ z : t \to g(z(t)) \text{ for } a \le t \le b.$$

Rewrite (6) with $g[\gamma]$ in the role of γ:

$$\frac{1}{2\pi i} \int_{g[\gamma]} \frac{dw}{w} = W_{g[\gamma]}(0).$$

This integral counts the number of times the curve $g[\gamma]$ winds around 0. The pullback of this integral is

$$\frac{1}{2\pi i} \int_a^b \frac{g'(z(t))z'(t)}{g(z(t))} dt,$$

which is also the pullback of the integral

$$\frac{1}{2\pi i} \int_\gamma \frac{g'}{g},$$

which is the integral of the logarithmic derivative of g over γ.

Thus the integral of g'/g is a winding number for a certain curve, and hence is related to the change in the argument as a point moves over that curve.

Consider the special case when γ is the circle $|z| = r$ and $g(z) = z^k$ for some integer $k \neq 0$. The circle is of course given by $z(t) := re^{it}$ $(0 \leq t \leq 2\pi)$. Then $g[\gamma]$ is the curve given by $t \mapsto r^k e^{itk}$, which is the circle $|z| = r^k$ traversed k times, and

$$\frac{1}{2\pi i} \int_\gamma \frac{g'}{g} = \frac{1}{2\pi i} \int_{g[\gamma]} \frac{dw}{w} = k.$$

There is only one pole for g'/g inside γ. It occurs where g has either a pole (if $k < 0$) or a zero (if $k > 0$). Thus the integral of the logarithmic derivative gives us the order of the zero or minus the order of the pole. As you will see, the next Theorem is a result of this kind, but addresses a considerably more general situation.

Let g be meromorphic on $\bar{D}(z_0, r)$, with no zeros or poles on the circle $|z - z_0| = r$. By the remarks above, and by the Residue Theorem, the integral of the logarithmic derivative around the circle,

$$\frac{1}{2\pi i} \int_{|z-z_0|=r} \frac{g'(z)}{g(z)} dz,$$

equals the number of zeros of g inside the circle, minus the number of poles—both counted by multiplicity.

Consider, for example, $z \mapsto \sin z$ in the role of g. We know where the zeros are (2.2C); there is a zero of order 1 at each integer multiple of π on the real axis, and no others. Therefore,

$$\frac{1}{2\pi i} \int_{|z|=r} \cot z\, dz = 2N + 1 \quad \text{if } N\pi < r < (N+1)\pi, \text{ for } N = 0, 1, 2, \cdots.$$

3.4.10. Theorem (The Argument Principle). *Let g be meromorphic on the open set $O \subseteq \mathbb{C}$. Let Γ be a contour in O which avoids all poles and zeros of g, and such that*

$$W_\Gamma(\zeta) = 0 \quad \text{for all } \zeta \in \mathbb{C} \setminus \Omega.$$

Inside Γ, let g have a pole of order I_j at the point a_j, for $1 \le j \le J$; and a zero of order Z_k at the point b_k, for $1 \le k \le K$; and no other poles or zeros. Then

$$\frac{1}{2\pi i} \int_\Gamma \frac{g'}{g} = \sum_{k=1}^{K} W_\Gamma(b_k) Z_k - \sum_{j=1}^{J} W_\Gamma(a_j) I_j. \tag{7}$$

If φ is holomorphic on O, then

$$\frac{1}{2\pi i} \int_\Gamma \frac{\varphi g'}{g} = \sum_{k=1}^{K} \varphi(b_k) W_\Gamma(b_k) Z_k - \sum_{j=1}^{J} \varphi(a_j) W_\Gamma(a_j) I_j. \tag{8}$$

Proof. The right-hand side of (7) includes exactly one summand for each pole of g'/g. The poles of g'/g are the points b_k and a_j, and we know the residues: $\text{Res}(g'/g, b_k) = Z_k$ and $\text{Res}(g'/g, a_j) = -I_j$. Apply the Residue Theorem 3.4.3 with g'/g in the role of f, and (7) follows.

For each pole c of g'/g,

$$\text{Res}\left(\frac{\varphi g'}{g}, c\right) = \varphi(c)\text{Res}\left(\frac{g'}{g}, c\right).$$

Apply the Residue Theorem with $\varphi g'/g$ in the role of f, and (8) follows. $\qquad\square$

3.4.11. Remark. Consider the case when W_Γ takes no values other than 0 or 1. Let V denote the inside of Γ. Then (7) equals the number of zeros in V minus the number of poles in V, each counted by multiplicity. Now consider the special case when g is holomorphic and thus has no poles. Then (7) equals the number of zeros of g, counted by multiplicity, that occur in V. We may write the result in terms of valence:

$$\frac{1}{2\pi i} \int_\Gamma \frac{g'(z)}{g(z)} dz = \nu_{g|V}(0).$$

For an arbitrary complex number w, we can just as easily use the Argument Principle to count the points at which g takes on the value w, since those points are the zeros of the function $z \mapsto g(z) - w$, and the logarithmic derivative of that function is $g'(z)/(g(z) - w)$. Thus

$$\frac{1}{2\pi i} \int_\Gamma \frac{g'(z)}{g(z) - w} dz = \nu_{g|V}(w).$$

Applying (8) with $\varphi(z) := z$, we find that

$$\frac{1}{2\pi i} \int_\Gamma \frac{z g'(z)}{g(z) - w} dz = \sum \{zm(g, z) \mid z \in V \text{ and } g(z) = w\}.$$

If g is one-to-one on V, so that there is an inverse function $(g|V)^{-1}$ defined on $g[V]$, then we have an integral representation for the inverse:

$$\frac{1}{2\pi i}\int_\Gamma \frac{zg'(z)}{g(z)-w}dz = (g|V)^{-1}(w) \quad \text{for } w \in g[V].$$

For example, let g be the sine, and let Γ be contained in the open vertical strip $\{x+iy \mid -\frac{\pi}{2} < x < \frac{\pi}{2}\}$, on which the sine is one-to-one. Then

$$\frac{1}{2\pi i}\int_\Gamma \frac{z\cos z}{\sin z - w}dz = \begin{cases} \arcsin w, & \text{if } w \in \sin[V]; \\ 0, & \text{if } w \notin \sin[V]. \end{cases}$$

Consider, as another example, $g(z) := z^2$. Let Γ be a contour contained in the upper half-plane, on which z is one-to-one. Then a holomorphic square root function with values in the upper half-plane is given by

$$w^{1/2} = \frac{1}{2\pi i}\int_\Gamma \frac{2z^2}{z^2 - w}dz \quad \text{for } w \in g[V].$$

3.4.12. Example. *Evaluate* $\dfrac{1}{2\pi i}\displaystyle\int_{|z-2|=3} \dfrac{dz}{z-4}$.

Solution. There are several ways to see that this integral equals 1.

- It is the winding number of the circle $|z-2|=3$ with respect to the point 4.

- The function $z \mapsto 1/(z-4)$ has one pole inside the circle $|z-2|=3$, and by the Residue Theorem the integral equals the residue of the function at that pole.

- The integrand has the form f'/f, where $f(z) = z - 4$, and by the Argument Principle the integral counts the number of zeros of f inside the circle.

$\square$

For a polynomial P of degree n, the integral of $P'/(P-w)$ over the circle $|z|=r$ counts the number of zeros of $z \mapsto P(z) - w$, counted by multiplicity, inside the circle; and for r sufficiently large, the integral equals n. We can get some help in locating the zeros of polynomials and other functions from the following result.

3.4.13. Theorem (Rouché's Theorem). *Let* f, g *be holomorphic on an open set* $O \subseteq \mathbb{C}$. *Let* Γ *be a contour in* O *that avoids the zeros of* f. *If* g/f *has a logarithm on* Γ, *which it does, for example, if*

$$|f(z) - g(z)| < |f(z)| \quad \text{for } z \in \Gamma, \tag{9}$$

then

$$\int_\Gamma \frac{f'}{f} = \int_\Gamma \frac{g'}{g}. \tag{10}$$

Proof. The inequality (9) holds on some open set U containing Γ:

$$\left| 1 - \frac{g(z)}{f(z)} \right| < 1 \quad \text{for } z \in U.$$

Thus the values of g/f on U lie in the disk $D(1,1)$, and on that disk there is a logarithm. On U, then, $\log(g/f)$ is holomorphic, and

$$\left(\log \frac{g}{f} \right)' = \frac{(g/f)'}{g/f} = \frac{fg' - gf'}{f^2} \frac{f}{g} = \frac{g'}{g} - \frac{f'}{f}.$$

Thus $\dfrac{g'}{g} - \dfrac{f'}{f}$ has an antiderivative on U, hence its integral over Γ must equal 0 and (10) follows. $\qquad\qquad\square$

When the zeros of one holomorphic function have been located, Rouché's Theorem is somewhat helpful in finding the zeros of another. Consider, for example, the function $f(z) := z^5 + 6z^3 + 8$. It is easy to find inequalities like (9) on three different circles:

$$|f(z) - 8| < 8 \qquad\qquad \text{for } |z| = 1;$$
$$|f(z) - 6z^3| < |6z^3| \equiv 48 \quad \text{for } |z| = 2;$$
$$|f(z) - z^5| < |z^5| \equiv 243 \quad \text{for } |z| = 3.$$

If follows that f has no zeros in $\bar{D}(0,1)$, three in $\text{Ann}(1,2)$, and two more in $\text{Ann}(2,3)$. Applied with a bit more care, the method would locate the zeros a bit more precisely.

3.4.14. Theorem (Hurwitz's Theorem). *Let $\{f_n\}$ be a sequence of one-to-one holomorphic functions on the open connected set $O \subseteq \mathbb{C}$, and let $f_n \to f$ subuniformly on O. Then f is either constant or one-to-one on O.*

Proof. By 3.1.22, f is holomorphic on O. Suppose that f is non-constant and not one-to-one. Then there exist distinct points $a, a_1 \in O$ such that $f(a) = f(a_1)$. There exists $r > 0$ such that the two disks $\bar{D}(a,r)$ and $\bar{D}(a_1,r)$ are disjoint and contained in O, and also small enough so that $f(z) \neq f(a)$ if $0 < |z - a| \leq r$ (see 3.1.16).

By 3.2.7, $f_n' \to f'$ subuniformly. It follows that

$$\lim_{n \to \infty} \frac{f_n'(z)}{f_n(z) - f(a)} = \frac{f'(z)}{f(z) - f(a)} \quad \text{uniformly for } |z - a| = r.$$

Therefore

$$\lim_{n \to \infty} \frac{1}{2\pi i} \int_{|z-a|=r} \frac{f_n'(z)}{f_n(z) - f(a)} dz = \frac{1}{2\pi i} \int_{|z-a|=r} \frac{f'(z)}{f(z) - f(a)} dz.$$

The integral on the right equals 1, so the integrals on the left, being integer-valued, eventually equal 1. So eventually f_n takes on the value $f(a)$ somewhere on $D(a,r)$. By similar reasoning, f_n also takes on the value $f(a)$ somewhere on $D(a_1,r)$. Thus eventually f_n is not one-to-one, contrary to hypothesis. $\qquad\qquad\square$

Exercises

1. Show that G_3, in (1), is one-to-one from $\hat{\mathbb{C}}$ onto $\hat{\mathbb{C}}$; do so by solving for z in terms of w in

$$w = \frac{z+2}{3z+4}.$$

2. Show that G_4, in (1), is two-to-one from $\mathbb{C}$ onto $\hat{\mathbb{C}} \backslash \{0\}$, and that $m(G_4, \infty) = 2$ (so that we may say that it has a double zero at ∞). Do so by solving for z in terms of w in $wz^2 + w - 1 = 0$.

3. Verify the statements in (2).

4. Write proofs of parts (ii), (iv), and (v) of Proposition 3.4.2.

5. Prove part (vi) of Proposition 3.4.2. .

6. Let f be holomorphic and g meromorphic on the open set $\Omega \subseteq \hat{\mathbb{C}}$. Prove that fg is meromorphic on Ω. You will need to formulate a definition of such a product.

7. Prove that if two functions are holomorphic on the open set $\Omega \subseteq \hat{\mathbb{C}}$, then their quotient is meromorphic on Ω. You will need to formulate a definition of such a quotient.

8. Explain in detail Case 2 in the proof of Proposition 3.4.9. Notice that the polynomial $z \mapsto P(z) - wQ(z)$ may not have the same degree as P.

9. Explain in detail Case 3 in the proof of Proposition 3.4.9.

10. Explain why a non-constant meromorphic function on $\hat{\mathbb{C}}$ must have at least one pole.

11. Explain the following statement: The composition of two meromorphic functions is meromorphic.

12. True or False? If f is meromorphic on $\hat{\mathbb{C}}$ and has a pole only at infinity, then $\mathrm{Res}(f, \infty) = 0$. Justify your answer.

13. Let f be meromorphic at $p \in \mathbb{C}$. Prove that if f has a pole of order 1 at p and φ is holomorphic at p, then $\mathrm{Res}(\varphi f, p) = \varphi(p)\mathrm{Res}(f, p)$. How is this result related to 3.3.15? Explain why the statement is false if $p = \infty$.

14. True or False? If f is meromorphic on $\hat{\mathbb{C}}$ and $\mathrm{Res}(f, \infty) = 0$, then f has a pole only at infinity. Justify your answer.

15. Classify the three singularities in $\hat{\mathbb{C}}$ of the function $z \mapsto \dfrac{2}{z^2 + 4iz - 1}$.

16. Evaluate $\displaystyle\int_{|z|=3/2} \frac{dz}{(z-1)(z-2)^3 z^2}$. Explain how Corollary 3.4.4 applies.

17. Evaluate $\displaystyle\int_{|z|=r} \frac{\sin z}{z^2+1}\,dz$ in terms of r.

18. Let $\displaystyle I(r) := \int_{|z|=r} \frac{z^2+1}{\sin z}\,dz$. Then $I(r)$ is undefined when r is a multiple of π. For each integer n, $I(r)$ has a constant value for $\pi n < r < \pi(n+1)$. Find the value as a function of n.

19. For $r > 0$ not an odd multiple of $\pi/2$, explain how to evaluate $\displaystyle\int_{|z|=r} \tan z\,dz$ by using the Residue Theorem, and also by using the Argument Principle.

20. Evaluate the integral $\displaystyle\int_{|z|=r} \sec z\,dz$ when $r > 0$ is not an odd multiple of $\pi/2$.

21. In the Examples of this Section, we've used contours for which the winding number is everywhere 0 or 1. This Exercise will remind you of what happens in other cases. Let γ be the contour of Figure 2.5-2, page 163 and Exercise 11 in Section 2.5. Evaluate the following integrals.

(a) $\displaystyle\int_\gamma z^3\,dz$　　　　(d) $\displaystyle\int_\gamma \frac{z^3}{z-6}\,dz$　　　　(g) $\displaystyle\int_\gamma \cot z\,dz$

(b) $\displaystyle\int_\gamma \frac{dz}{z^3}$　　　　(e) $\displaystyle\int_\gamma \frac{e^{2z}}{(z-6)^2}\,dz$　　　　(h) $\displaystyle\int_\gamma \frac{dz}{z^2+36}$

(c) $\displaystyle\int_\gamma \frac{dz}{z-2}$　　　　(f) $\displaystyle\int_\gamma \frac{z^3}{(z-4i)^3}\,dz$　　　　(i) $\displaystyle\int_\gamma \frac{z^3}{z^2+36}\,dz$

22. In subsection 2.2C we located all the solutions of $\sin z = 2$. Let Γ be a closed path such that W_Γ takes on no values other than 0 and 1. Then the number of those solutions inside Γ is the value of

$$\frac{1}{2\pi i}\int_\Gamma \frac{\cos z}{\sin z - 2}\,dz.$$

If Γ is the circle $|z| = r$, for what values of r is the integral equal to 8?

23. Let Q_R be the rectangle with vertices $-R - \pi i, R - \pi i, R + \pi i$, and $-R + \pi i$. Let ∂Q_R be its counterclockwise boundary. Prove that

$$\operatorname{Log} w = \lim_{R\to\infty} \frac{1}{2\pi i}\int_{\partial Q_R} \frac{ze^z}{e^z - w}\,dz,$$

except when w is a non-positive real number.

24. If $f(z) = z$, then z is a **fixed point** of f. Let f be holomorphic on $\bar{D}(0,1)$ such that $|f(z)| < 1$ for $|z| = 1$. How many fixed points must f have in the unit disk?

25. Show that the polynomial $p(z) := z^4 + 4z^3 + z^2 + 1$ has exactly one zero in the annulus $1 < |z| < 5$. Can you say something more precise?

26. Let $\{f_n\}$ be a sequence of holomorphic functions on the open connected set $O \subseteq \mathbb{C}$, and let $w \in \mathbb{C}$, such that for all n, $\nu_{f_n}(w) = 0$. Prove that if f_n converges subuniformly on O to f, then either $f \equiv w$ or $\nu_f(w) = 0$.

27. Let $R > 0$. Let $p_n(z) = \sum_{k=0}^{n} \dfrac{z^n}{n!}$. Prove that for all n sufficiently large, p_n has no zeros in $\bar{D}(0, R)$.

28. In Rouché's Theorem, a condition that would serve in place of (9) is

$$|f(z) + g(z)| < |f(z)| + |g(z)| \quad \text{for } z \in \Gamma,$$

because then g/f takes on no nonnegative real values. Explain.

29. Use Rouché's Theorem to prove the Fundamental Theorem of Algebra.

Help on Selected Exercises

17. The integral equals 0 for $0 < r < 1$ and $\dfrac{\pi i(e^2 - 1)}{2e}$ for $r > 1$.

18. The integral equals 1 for $0 < r < 1$, and $1 - 2(\pi^2 + 1)$ for $\pi < r < 2\pi$.

19. $\displaystyle\int_{|z|=r} \tan z\, dz = -i4\pi k$ for $(2k - 1)\frac{\pi}{2} < r < (2k + 1)\frac{\pi}{2}$.

21. (a) 0; (b) 0; (c) 5; (d) 6^3; (e) $2e^1 2$; (f) $36i$; (g) 1; (h) 0; (i) $-72\pi i$.

27. The smallest value of $|e^z|$ for $|z| \leq R$ is e^{-R}. For n sufficiently large, $|p_n(z) - e^z| < e^{-R}$ for $|z| \leq R$. Apply Rouché's Theorem.

3.5   MAPPING PROPERTIES

Let f be a non-constant holomorphic function on an open connected set. The next Theorem states that f is an open mapping, which leads immediately to the fact that f cannot attain its maximum modulus at any point of its domain; and to other related results. Our proof of the open mapping property uses counting integrals taken over circles and their images. This approach yields also the fact that when a holomorphic function has an inverse, the inverse is holomorphic too; as well as some very clear information about how holomorphic functions in general behave locally. We will give you now a preview of these mapping properties, with some examples to hold in mind when we address the general case.

The power mapping $z \mapsto z^m$ is easy to describe (recall subsection 2.2A). It has multiplicity m at the origin and maps the punctured plane $\mathbb{C} \setminus \{0\}$ m-to-one onto itself. For arbitrary constants a, b, and $c \neq 0$, the mapping $z \mapsto b + c(z - a)^m$ may be described in the same way. It has multiplicity m at a and maps $\mathbb{C} \setminus \{a\}$ m-to-one onto $\mathbb{C} \setminus \{b\}$.

We shall prove that if f is an arbitrary non-constant holomorphic function, and the point a is in its domain, and $m := m(f, a)$, then the mapping

$$z \mapsto f(z) \equiv f(a) + c_m(z - a)^m + \cdots ,$$

behaves locally like the mapping

$$z \mapsto f(a) + c_m(z - a)^m$$

in the sense that for small $r > 0$, it maps the punctured disk $D'(a, r)$ m-to-one onto some punctured open neighborhood of $f(a)$. Thus f is always locally easy to describe, even though its mapping behavior on the whole of its domain may be much more complicated than that of $z \mapsto z^m$.

Consider, as an example, the mapping given by $f(z) := z + z^2 + z^3$. In this case $a = 0$, $f(0) = 0$, and $m = 1$. There are three zeros, all of order 1, located at 0, $e^{2\pi i/3}$, and $e^{-2\pi i/3}$. In the left-hand column of Figure 3.5-1 are pictured three circles with center 0 and radii $r = 1.2, 0.8$, and 0.4. In each picture, the zeros of f are indicated by the three grey dots. Next to each circle $|z| = r$ is shown its image γ_r, in the right-hand column of the Figure. To indicate which part of the circle is mapped to which part of its image, five dots of different sizes are shown on the circle, and the image of each dot is shown as a dot of the same size.

Notice that the three pictures in the right-hand column have different scales; try to visualize how γ_r changes as r changes continuously from 1.2 down to 0.4; it gets smaller and the loops go away. Notice these facts: All three zeros of f lie inside the circle $|z| = 1.2$; the image $\gamma_{1.2}$ of that circle winds around 0 three times. But f has only one zero inside each of the circles $|z| = 0.8$ and $|z| = 0.4$, and their images $\gamma_{0.8}$ and $\gamma_{0.4}$ wind around 0 only once. Those are not coincidences, but indicate the general pattern.

Using the same scheme, Figure 3.5-2 treats the mapping given by $f(z) := z^3 + z^4 + z^5$. In this case, $a = 0$, $f(0) = 0$, and $m = 3$. There are five zeros, counting by multiplicity; three of them are at 0, and the other two are at $e^{2\pi i/3}$ and $e^{-2\pi i/3}$. Notice these facts: Five zeros lie inside the circle $|z| = 1.2$, and its image $\gamma_{1.2}$ winds around 0 five times. There are only three zeros inside each of the two smaller circles, and the image of each winds around 0 only three times.

In each of the six cases you may ask, What is the image of $D(0, r)$? You might speculate that it is the inside of γ_r, and you would be correct. Consider the bottom row in Figure 3.5-1. Is the function one-to-one from $D(0, 0.4)$ onto the inside of $\gamma_{0.4}$? You might conjecture that is is, and you would be correct. We now present some Theorems which provide a way to deal with such questions.

3.5.1. Theorem. *Let f be a non-constant holomorphic function defined on the open connected set $O \subseteq \mathbb{C}$. Then f is an open mapping. That is, if $U \subseteq O$ and U is open, then $f[U]$ is open.*

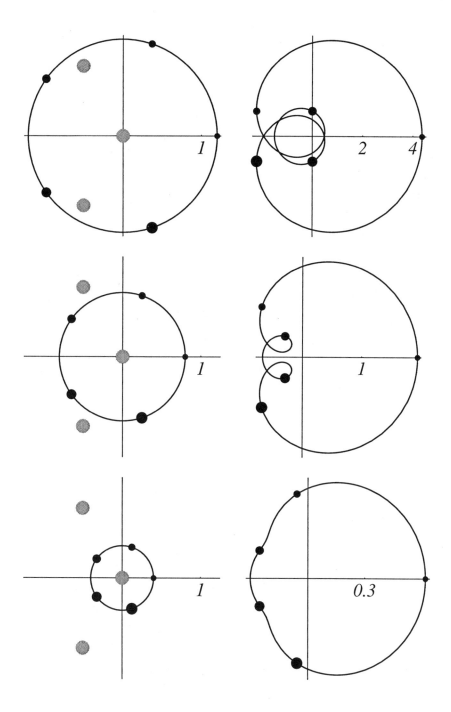

Figure 3.5-1   On the left, three circles $|z| = r$ ($r = 1.2, 0.8, 0.4$). To the right of each circle is its image γ_r under the mapping $z \mapsto z + z^2 + z^3$. The grey dots indicate the three zeros of the function.

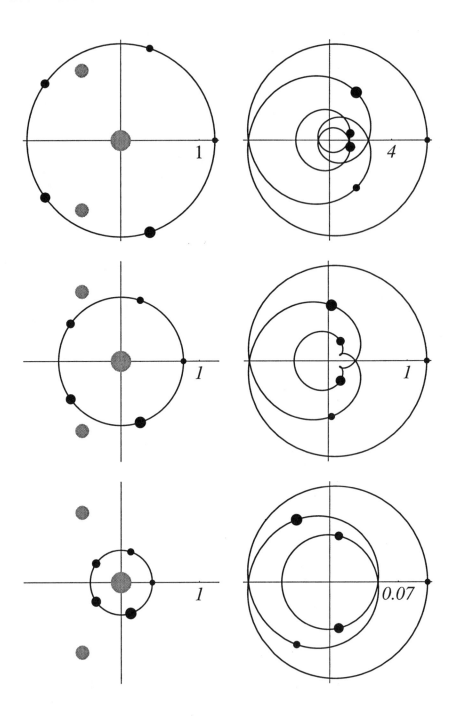

Figure 3.5-2 On the left, three circles $|z| = r$ ($r = 1.2, 0.8, 0.4$). To the right of each is its image γ_r under the mapping $z \mapsto z^3 + z^4 + z^5$. The grey dots indicate the zeros of the function. The zero at the origin is of order 3.

Proof. To prove that $f[U]$ is open, it suffices to find, for every point $a \in U$, an open set S such that $f(a) \in S \subseteq f[U]$. Since U is open,

$$\bar{D}(a, r) \subset U$$

for sufficiently small positive r. By 3.1.16, in selecting r we may require also that

$$f(z) \neq f(a) \quad \text{for } 0 < |z - a| \leq r. \tag{1}$$

(In each of the examples of Figures 3.5-1 and 3.5-2, the requirement (1) would rule out $r = 1.2$, because 0 is not the only point inside the circle at which the function equals 0. Note that in those examples, $a = f(a) = f'(a) = 0$.)

Let C or C_r denote the circle $|z - a| = r$, given by

$$z(t) := a + re^{it} \quad (0 \leq t \leq 2\pi),$$

and let γ or γ_r denote the image $f[C]$, which is the closed path given by

$$\zeta(t) := f(z(t)) \quad (0 \leq t \leq 2\pi).$$

By (1), the value $f(a)$ is taken on at a and at no other point of $\bar{D}(a, r)$. Therefore

$$\nu_{f|D(a,r)}(f(a)) = m(f, a).$$

For $w \notin \gamma$, consider the integral that counts the number of times γ winds around w, and its pullback:

$$W_\gamma(w) = \frac{1}{2\pi i} \int_\gamma \frac{d\zeta}{\zeta - w} = \frac{1}{2\pi i} \int_0^{2\pi} \frac{f'(z(t))z'(t)dt}{f(z(t)) - w} \tag{2}$$

Next, consider the integral over C which (by 3.4.11) equals the number of points in $D(a, r)$ at which $f = w$, counted by multiplicity:

$$\frac{1}{2\pi i} \int_C \frac{f'(z)dz}{f(z) - w} = \nu_{f|D(a,r)}(w). \tag{3}$$

The integral on the right-hand side of (2) is the pullback also of the integral in (3). Therefore the two integrals, and the two counts, are the same:

$$W_\gamma(w) = \nu_{f|D(a,r)}(w) \quad \text{for all } w \notin \gamma. \tag{4}$$

This elegant equality explains our observations, preceding the statement of the Theorem, about the examples in Figures 3.5-1 and 3.5-2.

Let S be the component of $\mathbb{C} \setminus \gamma$ that contains $f(a)$. Then S is necessarily an open set. By 2.5.17, page 154, W_γ is constant on S:

$$W_\gamma(w) = \nu_{f|D(a,r)}(w) = m(f, a) \quad \text{for } w \in S.$$

To say that $\nu_{f|D(a,r)}(w)$ is nonzero for $w \in S$ is to say that $S \subseteq f[D(a,r)] \subseteq f[U]$, and the Theorem is proved.

It may or may not be true that $S = f[D(a, r)]$. In the two Figures, the bottom row of 3.5-1 represents the only case in which equality holds.   □

3.5.2. Theorem. *Let f be a non-constant holomorphic function defined on the open connected set $O \subseteq \mathbb{C}$. Let $a \in O$, $m = m(f, a)$. Then there exist open sets, Q containing a and S containing $f(a)$, such that f is m-to-1 from $Q \setminus \{a\}$ onto $S \setminus \{f(a)\}$.*

Proof. We repeat the proof of Theorem 3.5.1, except that in the choice of $r > 0$ we impose the additional requirement that

$$f'(z) \neq 0 \quad \text{for } 0 < |z - a| \leq r, \tag{5}$$

which we can do by 3.1.16, applied to the function f'. Then we define C, γ, and S as before. (In each of the examples of Figures 3.5-1 and 3.5-2, the requirement (5) would rule out $r = 0.8$.)

The set $f^{-1}[S]$ is open because f is continuous and S is open. Therefore the set $Q := f^{-1}[S] \cap D(a, r)$ is open. Evidently

$$\nu_{f|Q}(w) = \nu_{f|D(a,r)}(w) = m \quad \text{for } w \in S. \tag{6}$$

By (5), $m(f, b) = 1$ for all $b \in Q \setminus \{a\}$. Therefore (6) implies that f maps $Q \setminus \{a\}$ m-to-one onto $S \setminus \{f(a)\}$.   □

3.5.3. Theorem. *Let f be a non-constant holomorphic function defined on the open connected set $O \subseteq \mathbb{C}$. Let $a \in O$ and $f'(a) \neq 0$, so that $m(f, a) = 1$. Then there exist a disk $D(a, r)$ and an open set S containing $f(a)$, such that f maps $D(a, r)$ one-to-one onto S. The inverse g of $f|D(a, r)$ is holomorphic on S; and*

$$g'(w) = \frac{1}{f'(z)} \quad \text{whenever } z \in D(a, r) \text{ and } w = f(z).$$

Proof. Since $f'(a)$ is nonzero, by Theorem 3.5.2 there is an open set Q, containing a, on which f is one-to-one. We repeat the proof of Theorem 3.5.1, except that in the choice of $r > 0$ we impose the stronger requirement that $\bar{D}(a, r) \subset Q$, which assures that

$$f \text{ is one-to-one on } \bar{D}(a, r), \tag{7}$$

and thus that $f|\bar{D}(a, r)$ has an inverse, which we will denote by g. Then we define C, γ, and S as before.

Condition (7) implies (1), so that $f(a)$ is not on γ; but it implies something more, namely that for every $b \in D(a, r)$, $f(b)$ is not on γ. Since f is continuous and $D(a, r)$ is connected, $f[D(a, r)]$ is connected. It is a connected subset of $\mathbb{C} \setminus \gamma$ and hence must be contained in S. Therefore $f[D(a, r)] = S$. It remains to prove the last statement of the Theorem.

Since f is an open mapping, g is continuous on $\bar{D}(a, r)$. The function given by

$$g_1(w) := \frac{1}{2\pi i} \int_\gamma \frac{g(\zeta)}{\zeta - w} d\zeta \quad (w \in S)$$

is holomorphic because it is a Cauchy integral (2.6.14). Its pullback to $[0, 2\pi]$ is also the pullback of a certain integral over C:

$$g_1(w) = \frac{1}{2\pi i} \int_\gamma \frac{g(\zeta)}{\zeta - w} d\zeta = \frac{1}{2\pi i} \int_0^{2\pi} \frac{g(\zeta(t))\zeta'(t)}{\zeta(t) - w} dt$$

$$= \frac{1}{2\pi i} \int_0^{2\pi} \frac{z(t)f'(z(t))z'(t)}{f(z(t)) - w} dt$$

$$= \frac{1}{2\pi i} \int_C \frac{zf'(z)}{f(z) - w} dz,$$

which by 3.4.11 equals the value of z in $D(a, r)$ such that $f(z) = w$, namely $g(w)$. So $g_1 = g$. Therefore g is holomorphic on S. Since $g(f(z)) = 1$ it follows that $g'(f(z)) \cdot f'(z) = 1$. $\qquad\square$

The next Theorem is specific about how the function described in the last two Theorems is constructed, at least locally.

3.5.4. Theorem. *Let f be a non-constant holomorphic function defined on the open connected set $O \subseteq \mathbb{C}$. If $a \in O$ and $m := m(f, a)$, then there is an open set U containing a, and a function F which is one-to-one and holomorphic on U, such that*

$$f(z) = f(a) + F(z)^m \quad \text{for } z \in U,$$

so that f is m-to-1 on $U \setminus \{a\}$.

Proof. For z near a, we may write $f(z) = f(a) + (z - a)^m h(z)$, where h is holomorphic at a and $h(a) \neq 0$. There exists $\delta > 0$ such that $|h(z) - h(a)| < |h(a)|$ for $|z - a| < \delta$. There is a logarithm on the disk $D(h(a), |h(a)|)$. Let

$$F(z) := (z - a)e^{\frac{1}{m} \log h(z)} \quad \text{for } |z - a| < \delta.$$

Then F is holomorphic, $F(a) = 0$,

$$f(z) = f(a) + F(z)^m \quad \text{for } |z - a| < \delta,$$

and

$$F'(a) = e^{\frac{1}{m} \log h(a)} \neq 0.$$

Therefore by Theorem 3.5.3, applied to F, there is a disk $D(a, r)$, with $r \leq \delta$, such that F is one-to-one from $D(a, r)$ onto an open set containing 0. There is a disk $D(0, \alpha)$ contained in $F[D(a, r)]$; let $U := F^{-1}[D(0, \alpha)]$. The map $w \mapsto w^m$ is m-to-one on $D'(0, \alpha)$. Therefore the composition $z \mapsto F(z) \mapsto F(z)^m$ is m-to-one on $U \setminus \{a\}$, and

$$f(z) = f(a) + F(z)^m \quad \text{for } z \in U.$$

$\qquad\square$

3.5.5. Theorem. *Let f be a one-to-one holomorphic function on the open connected set $O \subseteq \mathbb{C}$. Then f' has no zeros in O.*

Proof. If there were a point $a \in O$ with $f'(a) = 0$, then f would be m-to-one on some punctured neighborhood of a where $m := m(f, a) > 1$, by 3.5.2. In particular f would not be one-to-one on O. $\square$

3.5.6. Theorem (The Maximum Modulus Principle.). *Let f be a holomorphic function on the open connected set $O \subseteq \mathbb{C}$. Then $|f(z)|$ cannot attain a local maximum at any point of O. If f has no zeros in O, then $|f(z)|$ cannot attain a local minimum at any point of O.*

Proof. See Proposition 1.4.6, page 56. $\square$

3.5.7. Theorem (Schwarz's Lemma). *Let f be holomorphic on the unit disk D, such that $f(0) = 0$ and $|f(z)| \le 1$ for all $z \in D$. Then*

$$|f(z)| \le |z| \quad \text{for all } z \in D.$$

Moreover, there are just two possibilities: Either

$$f(z) = e^{i\theta}z \quad \text{for all } z \in D \tag{8}$$

for some real number θ; or

$$|f'(0)| < 1, \quad \text{and} \quad |f(z)| < |z| \quad \text{for all } z \in D'. \tag{9}$$

Proof. The function q given by

$$q(z) := \begin{cases} \dfrac{f(z)}{z} & \text{for } z \ne 0, \\[2ex] f'(0) & \text{for } z = 0 \end{cases}$$

is holomorphic on D (recall 3.1.10). Let $0 < s < 1$. A continuous function on a compact set must attain its maximum at some point of the set. Therefore the supremum of $|q|$ on $\bar{D}(0, s)$ is attained at some point z_0 on the boundary, $|z_0| = s$. (If it were attained at some point on the interior, then q would be constant, and the supremum would still be attained on the boundary.) Therefore

$$|q(z)| \le \frac{1}{s} \quad \text{for } |z| \le s.$$

That being true for each $s < 1$, it follows that

$$\sup_{z \in D} |q(z)| \le 1.$$

Either $|q(z)| < 1$ for all $z \in D$, so that (9) holds; or else $|q|$ attains the value 1 somewhere on D, which implies that q is constant, so that (8) holds. $\square$

3.5.8. Corollary. *Let f be a one-to-one holomorphic function from D onto D such that $f(0) = 0$. Then $f(z) \equiv e^{i\theta} z$ for some real θ.*

Proof. Schwarz's Lemma applies to f, and we may conclude that $|f(z)| \leq |z|$ for all $z \in D$. Schwarz's Lemma applies also to f^{-1}, and we may conclude that $|z| \leq |f(z)|$ for all $z \in D$. It follows that (8) must hold. $\qquad\square$

Exercises

1. In Figures 3.5-1 and 3.5-2, in each picture of an image curve γ_r, identify the set S. In which cases does $S = f[D(0, r)]$?

2. In each of the Figures 3.5-1 and 3.5-2, consider how γ_r changes as r varies from 0.4 to 1.2. There is a value of r for which γ_r is not smooth; it has cusps. What is that value of r? How are the cusps related to the zeros of f?

3. Let f be a holomorphic function on $\bar{D}(0, 1)$, such that $|f(z)| > 3$ for $|z| = 1$ and $f(0) = 1 - 2i$. Must f have a zero in the unit disk?

4. Let f be a one-to-one holomorphic mapping from the unit disk onto itself, $f(0) = 0, f'(0) > 0$. Prove that $f(z) \equiv z$.

5. In each case, state whether the assertion is true or false, and justify your answer with a proof or counterexample.

   (a) Let f be holomorphic on an open connected set $O \subseteq \mathbb{C}$. Let $a \in O$. Let $\{z_k\}$ and $\{\zeta_k\}$ be two sequences contained in $O \setminus \{a\}$ and converging to a, such that for every $k = 1, 2, \cdots$, $f(z_k) = f(\zeta_k)$ and $z_k \neq \zeta_k$. Then $f'(a) = 0$.

   (b) Let g be an entire function such that $\dfrac{1}{2\pi i} \displaystyle\int_{|z|=R} \dfrac{g'}{g} = 0$ for all $R > 1000$. Then g is constant.

   (c) Let u be a real-valued harmonic function on $D(0, 1)$, and let γ be a closed curve in that disk. Then $\int_\gamma u = 0$.

6. Prove the following version of Schwarz's Lemma: Let f be holomorphic on $D(a, r)$, and let

$$B := \sup_{|z-a|<r} |f(z) - f(a)|.$$

Then there are just two possibilities: Either

$$f(z) - f(a) = \frac{Be^{i\theta}}{r}(z - a) \quad \text{for all } z \in D(a, r)$$

for some real number θ; or

$$|f'(a)| < \frac{B}{r} \quad \text{and} \quad |f(z) - f(a)| < \frac{B}{r}|z - a| \quad \text{for all } z \in D'(a, r).$$

Help on Selected Exercises

3. Yes. Otherwise, $1/f$ would be holomorphic on $\bar{D}(0,1)$ and would attain its maximum modulus on the unit circle. That maximum would be less than $1/3$, and yet $|1/f(0)| = 1/\sqrt{5} > 1/3$.

5(b). Consider the exponential.

3.6 THE RIEMANN SPHERE

For the kind of function we are studying, the point ∞ may occur in the domain as well as any point of $\mathbb{C}$. There is a substantial correspondence between a function's possible behaviors near ∞, and its possible behaviors near a point of $\mathbb{C}$.

Likewise, for the functions we are studying, the point ∞ may occur in the range as well as any point of $\mathbb{C}$. There is a substantial correspondence between the properties of a function at a pole, and its properties at a point where it takes on a finite value.

We have ways of describing and picturing a situation which are different, depending on whether ∞ or a point of $\mathbb{C}$ is involved. That can be inefficient. Sometimes we can cover both cases with fewer words, for example by using the term "meromorphic." To the same end, this optional section presents the Riemann sphere S^2, which offers a new mental picture of the extended plane $\hat{\mathbb{C}} \equiv \mathbb{C} \cup \{\infty\}$ which you may find interesting and helpful.

Let S^2 be the sphere in $\mathbb{R}^3$ of center $(0,0,0)$ and radius 1. It consists of all the points (x, y, ζ) such that

$$x^2 + y^2 + \zeta^2 = 1. \tag{1}$$

Consider also the complex plane imbedded in $\mathbb{R}^3$; let each point $z \equiv x + iy$ in $\mathbb{C}$ be identified with the point $(x, y, 0)$. Thus the plane $\mathbb{C}$ cuts through the middle of S^2.

We will talk about S^2 as though it were the surface of the Earth. We are going to identify the points of S^2 with those of $\hat{\mathbb{C}}$. The North Pole $N = (0,0,1)$ will correspond to ∞, the South Pole to the origin, the equator to the unit circle. The northern and southern hemispheres will be identified with the outside and the inside of the unit circle, respectively.

We will achieve this identification by means of a one-to-one mapping τ from $\hat{\mathbb{C}}$ onto S^2. First of all, we define $\tau(\infty)$ to be the North Pole $N = (0,0,1)$. Then, for $z = x + iy \in \mathbb{C}$, which is identified with $(x, y, 0)$, we proceed as follows to identify $\tau(z)$. Consider the line L_z that contains the two points $N = (0,0,1)$ and $(x, y, 0)$. There are two points in the intersection of L_z with S^2. One of them is N. The other one is defined to be $\tau(z)$. Make a sketch!

From the geometric description of the map, several properties are evident:

- If $|z| > 1$, then $\tau(z)$ is in the northern hemisphere; if $|z| < 1$, then $\tau(z)$ is in the southern hemisphere.

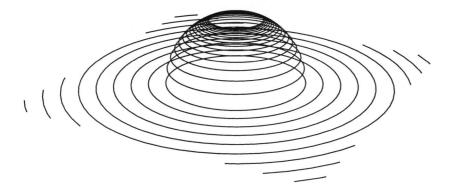

Figure 3.6-1   Seventeen circles on the Riemann sphere which are the images under τ of the circles in the plane with center at the origin and radii $1, 1.25, 1.5, \cdots, 5$. The circles in the plane are only partially shown.

- If C is the circle in the plane with center at the origin and radius $r > 0$, then $\tau[C]$ is a circle on S^2 lying in a plane parallel to $\mathbb{C}$; that plane is above $\mathbb{C}$ if $r > 1$, below it if $r < 1$. See Figures 3.6-1 and 3.6-2.

- Let M be a line that lies in $\mathbb{C}$. Then $\tau[M]$ is the circle on S^2 which is the intersection of S^2 with the plane that contains the line M and the point N.

- Conversely, every circle on S^2 that contains N is the image under τ of some line in the plane. See Figure 3.6-4.

To obtain further results, we need to analyze the mapping a bit further. The line L_z may be parametrized by

$$t \mapsto (tx, ty, 1 - t) \quad (t \in \mathbb{R}).$$

For every point $(\xi, \eta, \zeta) \equiv (tx, ty, 1 - t)$ on the line, evidently $t = 1 - \zeta$ and thus

$$\xi = (1 - \zeta)x \quad \text{and} \quad \eta = (1 - \zeta)y. \tag{2}$$

It follows that for $\zeta \neq 1$,

$$z = \frac{\xi + i\eta}{1 - \zeta} \quad \text{and} \quad z\bar{z} = \frac{\xi^2 + \eta^2}{(1 - \zeta)^2}. \tag{3}$$

Now let $(\xi, \eta, \zeta) = \tau(z)$, that is, the point $(\xi, \eta, \zeta) \neq (0, 0, 1)$ satisfying both (1) and (2). From (3),

$$z\bar{z} = \frac{1 - \zeta^2}{(1 - \zeta)^2} \equiv \frac{1 + \zeta}{1 - \zeta}, \quad \text{so} \quad \zeta = \frac{z\bar{z} - 1}{z\bar{z} + 1} \quad \text{and} \quad 1 - \zeta = \frac{2}{z\bar{z} + 1}.$$

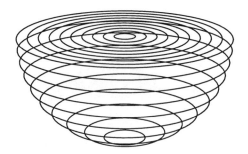

Figure 3.6-2 In the complex plane, the ten circles centered at the origin with radii running from .1 to 1; on the Riemann sphere, their images under τ.

So $\tau(z) \equiv (\xi, \eta, \zeta)$ is defined by

$$
\begin{aligned}
\xi &= \frac{z + \bar{z}}{z\bar{z} + 1} = \frac{2x}{x^2 + y^2 + 1}, \\
\eta &= \frac{i(\bar{z} - z)}{z\bar{z} + 1} = \frac{2y}{x^2 + y^2 + 1}, \\
\zeta &= \frac{z\bar{z} - 1}{z\bar{z} + 1} = \frac{x^2 + y^2 - 1}{x^2 + y^2 + 1}.
\end{aligned}
\tag{4}
$$

The inverse of τ is the **stereographic projection**:

$$
\tau^{-1}(\xi, \eta, \zeta) = \left(\frac{\xi}{1 - \zeta}, \frac{\xi}{1 - \zeta}, 0 \right).
$$

This projection from a sphere onto a plane was used for celestial maps as far back as Ptolemy. It is isogonal; that is, circles on the sphere meet at the same angles as their images on the plane. See Exercise 3.

3.6.1. Proposition. *Let K be a circle which lies on the Riemann sphere S^2. Then $K = \tau[Q]$, where Q is either a line or a circle in the complex plane, depending on whether K contains the North Pole.*

Proof. Let K be a circle contained in S^2. Then K is the intersection of some plane with S^2. Then the plane is the locus in 3-space of an equation

$$
A\xi + B\eta + C\zeta + D = 0,
\tag{5}
$$

where the three coefficients A, B, C are real and do not all vanish. Since the distance from the plane to the origin must be less than 1, we know that

$$
\frac{|D|}{\sqrt{A^2 + B^2 + C^2}} < 1, \quad \text{or} \quad D^2 < A^2 + B^2 + C^2.
$$

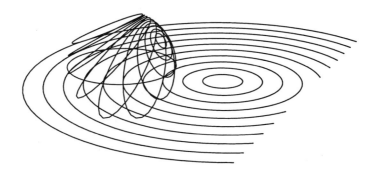

Figure 3.6-3 In the complex plane, shown partially, ten circles centered at 2 with radii .35, .7, 1.05, $\cdots$, 3.5; on the Riemann sphere, their images under τ. The equator of the Riemann sphere is also shown.

To make use of the relation $(\xi, \eta, \zeta) = \tau(x, y, 0)$, we make the substitutions indicated by (4) in equation (5), which then becomes

$$A2x + B2y + (x^2 + y^2)(C + D) = C - D.$$

If $C + D = 0$, this is the equation of a line Q in the plane, and $\tau[Q] = K$. Notice that $C + D = 0$ is precisely the condition for K to contain the North Pole. If $C + D \neq 0$, then (5) is equivalent to the equation

$$x^2 + y^2 + \frac{2A}{C + D}x + \frac{2B}{C + D}y = \frac{C - D}{C + D},$$

which, after completing the squares, becomes

$$\left(x + \frac{A}{C + D}\right)^2 + \left(y + \frac{B}{C + D}\right)^2 = \frac{A^2 + B^2 + C^2 - D^2}{(C + D)^2} =: R^2$$

or

$$\left| z - \left(\frac{-A - iB}{C + D}\right) \right| = R,$$

which defines a circle Q in the plane such that $\tau[Q] = K$. $\qquad \square$

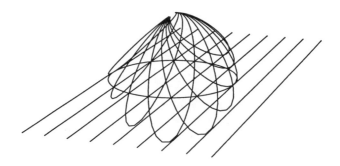

Figure 3.6-4   In the complex plane, the nine parallel lines (shown partially) with equations $x = k/3$ for $k = -3, -2, \cdots, 5$; on the Riemann sphere, their images under τ. The equator of the Riemann sphere is also shown.

Exercises

1. Prove that if Q is a line or a circle in the z-plane, then there is a circle K on the Riemann sphere such that $\tau[Q] = K$.

2. Consider two antipodal points on the Riemann sphere: (ξ, η, ζ) and $(-\xi, -\eta, -\zeta)$. Prove that their stereographic projections z and w are related by $z\bar{w} = -1$. Give a geometric interpretation.

3. Prove that the stereographic projection is isogonal.

4. The **chordal metric** in the extended plane is defined by

$$\chi(z, w) = |\tau(z) - \tau(w)|.$$

In other words, we are defining the chordal distance between z and w to be the ordinary distance in $\mathbb{R}^3$ between the corresponding two points on the Riemann sphere. Verify that

$$\chi(z, w) = \frac{2|z - w|}{\sqrt{|z|^2 + 1}\sqrt{|w|^2 + 1}} \quad \text{for } z, w \in \mathbb{C}$$

and that

$$\chi(z, \infty) = \frac{2}{\sqrt{|z|^2 + 1}}.$$

5. Let C be the circle in the plane with center at the origin and radius $r > 0$. What is the radius of the circle $\tau[C]$? In what plane does it lie?

6. If C is a circle in the plane, and if the circle $\tau[C]$ winds around the North Pole on the Riemann sphere, what can you conclude about C?

7. For each line M in the plane shown in Figure 3.6-4, what is the equation of the plane that contains $\tau[M]$?

Help on Selected Exercises

1. If Q is a line in the z-plane, then its equation may be written in one of the forms

$$ax + by = 0 \quad \text{or} \quad ax + by = 1,$$

where a and b are real constants, not both equal to 0. This line is a subset of the plane in $\mathbb{R}^3$ given by the equation

$$ax + by = 0 \quad \text{or} \quad ax + by + z = 1,$$

respectively, which is a plane containing N. The intersection of that plane with S^2 is, of course, a circle which is $\tau^{-1}[Q]$.

If Q is a circle in the z-plane, then by the discussion in subsection 2.1G, its equation may be written as

$$|z - \beta| = r,$$

with $r > 0$, or equivalently as

$$z\bar{z} - \beta\bar{z} - \bar{\beta}z + k = 0,$$

where $k = |\beta|^2 - r^2$. If we choose A, B, C, D such that

$$C + D = 1, \quad \beta = -A + iB, \quad \text{and} \quad D - C = |\beta|^2 - r^2,$$

then the plane whose equation is (5) will intersect S^2 in a circle which is $\tau^{-1}[Q]$.

3. If M is a line in the plane, then $\tau[M]$ is a circle on the sphere that contains the North Pole N. The tangent line to the circle $\tau[M]$ at the point N is parallel to M. It may help to point out that $\tau[M]$ lies in the plane that contains the line M and the point N. Now let M_1 and M_2 be two lines in the plane that meet at some point z, at angle θ. Their images under τ are circles on S^2 which meet both at N and at $\tau(z)$. The tangent lines to $\tau[M_1]$ and $\tau[M_2]$ at N are respectively parallel to M_1 and M_2, so the angle between them must be θ. Hence likewise for the tangent lines at $\tau[z]$.

4

The Residue Calculus

The Residue Calculus is a collection of methods for finding the exact values of real integrals—by which we mean, integrals over intervals of the real line. Here are some examples for which the methods work:

(a) $\displaystyle\int_0^{\pi/2} \frac{dt}{2 + \sin^2 t}$

(b) $\displaystyle\int_0^{\infty} \frac{dx}{x^2 + 6x + 8}$

(c) $\displaystyle\int_0^{\infty} \frac{dx}{\sqrt{x}(x^2 + 1)}$

(d) $\displaystyle\int_0^{\infty} \frac{dx}{(1 + x^{3/4})x^{3/4}}$

(e) $\displaystyle\int_0^{\infty} \frac{\sin x}{x}\,dx$

(f) $\displaystyle\int_{-\infty}^{+\infty} \frac{x^3 \cos x}{(x + 1)(x - 1)(x^2 + 4)}\,dx$

The methods are nothing more than applications of the Residue Theorem 3.4.3. No new theorems appear in this chapter, just techniques for relating real integrals to complex integrals whose values we know by the Residue Theorem. Some of the results bear the name "Proposition." Each of them treats a class of applications in a unified manner. For the reader, the point is not to remember such a Proposition as if it were a new and separate idea, but to see how the result flows from the Residue Theorem.

To understand this chapter, it is essential that you know how to write down a pullback for a given complex integral, as discussed in Section 2.5; and that you be able to find singularities and evaluate residues, as explained in Section 3.3. Those are fundamental skills. If you wish to become proficient in the uses of the Residue Calculus, you must work problems on your own in a thoughtful and unhurried fashion. At the end of each section, we give you some to do.

Two references which give more extensive accounts of Residue Calculus techniques, both for evaluating integrals and for summing series, are [2], [40], and [41].

It is not proposed that the methods presented here should supersede and replace other available techniques for evaluating integrals. Some of the integrals, like (b) above, will yield to techniques from first-year calculus. Many will yield to a knowledge of Fourier transforms; see, for example, [31, Chapter 3]. Computer algebra systems like *Mathematica* will produce answers for many of the problems, including those for which the result depends on one or more parameters. And of course, there are applications in which numerical evaluations are preferable or at least adequate.

If the problem is to evaluate a real integral over a compact interval, sometimes the integral can be recognized as the pullback of a complex integral over a contour and then evaluated using the Residue Theorem. Section 4.1 deals with some cases of that kind. But most of the problems of this chapter require further ideas and techniques. To prepare the way for those, Section 4.2 will give useful examples of how one can estimate and take limits of complex integrals. Section 4.3 evaluates integrals of rational functions over the real line. Section 4.4 treats several kinds of integrals that involve the exponential, including some that compute Fourier transforms. Section 4.5 discusses integrals involving a logarithm. In Section 4.6, we make one example the occasion to introduce Σ, the Riemann surface for the logarithm. Integration on Σ achieves with elegance the evaluations of other types of integrals, including some that compute Mellin transforms. Section 4.7 presents the complex inversion formula for the Laplace transform.

It is never a bad idea to begin work on an integral by making whatever *easy* observations one can think of. For example, we may be able to sketch a region of which the integral gives the area; or somehow estimate the answer; or check convergence questions. Sometimes a relaxed, unpressured preliminary look at the problem will save us work, or provide a plausibility criterion by which to check the answer we get. Accordingly, in approaching a problem, we will sometimes begin with remarks which you may find elementary, or obvious, or inessential. For example, the complex methods themselves will often establish the convergence of an improper integral, and yet we may choose to provide a comment as to how one could answer the convergence question at the outset.

4.1 INTEGRALS OF TRIGONOMETRIC FUNCTIONS OVER A COMPACT INTERVAL

Whenever F is a rational function of two variables and the function $t \mapsto F(\cos t, \sin t)$ is well defined on $[0, 2\pi]$, the integral $\int_0^{2\pi} F(\cos t, \sin t)dt$ is the pullback of a complex integral over the unit circle which can be evaluated using the Residue Theorem.

4.1.1. Example. *Use the Residue Calculus to evaluate*

$$\int_0^{2\pi} \frac{dt}{a - \cos t} \qquad (a > 1).$$

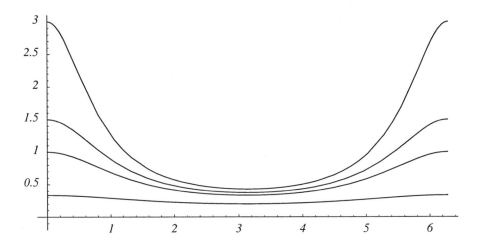

Figure 4.1-1   Graphs on the interval $[0, 2\pi]$ of the functions $t \mapsto 1/(a - \cos t)$ for $a = 4, 2, 5/3$, and $4/3$.

Solution. Before we set out to do what's asked, let's make a few observations. The condition $a > 1$ assures that $a - \cos t$ has no zeros, so the integrand is well defined on the interval of integration $[0, 2\pi]$. Figure 4.1-1 shows the graph of the integrand for several values of a. Since the value of the integrand lies between $1/(a + 1)$ and $1/(a - 1)$, the answer will be at least $2\pi/(a + 1)$ and at most $2\pi/(a - 1)$.

The trick is to recognize the integral as a pullback of a complex integral over the unit circle. The circle is parametrized by $z(t) := e^{it}$. Thus $dz = z'(t)dt = ie^{it}dt$; and

$$\cos t = \frac{e^{it} + e^{-it}}{2} = \frac{1}{2}\left(z(t) + \frac{1}{z(t)}\right).$$

To convert the pullback to the complex integral, we follow this procedure:

$$\text{Substitute } z \text{ for } e^{it} \text{ and } \frac{dz}{iz} \text{ for } dt. \tag{1}$$

Thus

$$\int_0^{2\pi} \frac{dt}{a - \cos t} = \int_{|z|=1} \frac{dz}{iz\left(a - \frac{1}{2}\left(z + \frac{1}{z}\right)\right)} = \frac{-2}{i}\int_{|z|=1} \frac{dz}{z^2 - 2az + 1}.$$

The function $z \mapsto 1/(z^2 - 2az + 1)$ has two poles, $a \pm \sqrt{a^2 - 1}$. Both are of order 1 and lie on the real axis; only one of them is inside the unit circle, and the residue at that one is (by 3.3.15)

$$\text{Res}\left(\frac{1}{z^2 - 2az + 1}, a - \sqrt{a^2 - 1}\right) = \left.\frac{1}{2z - 2a}\right|_{z=a-\sqrt{a^2-1}} = \frac{-1}{2\sqrt{a^2 - 1}}.$$

By the Residue Theorem, the integral equals $-2/i$ times $2\pi i$ times the residue:

$$\int_0^{2\pi} \frac{dt}{a - \cos t} = \frac{2\pi}{\sqrt{a^2 - 1}}.$$

$\square$

4.1.2. Example. *Let $a > 0, b > 0$. Evaluate* $\displaystyle\int_0^{2\pi} \frac{dt}{a^2 \sin^2 t + b^2 \cos^2 t}.$

Solution. Some easy observations to make before diving in: First, if $a = b$, the value is $2\pi/a^2$; so we may suppose $a \neq b$. Second, if the constants a and b are both multiplied by a nonzero number k, the effect on the integral's value is to divide it by k^2. When we arrive at an answer, we should look to see if it has that property.

We do (1) and find that the integral equals

$$\frac{4}{i} \int_{|z|=1} \frac{z\,dz}{(b^2 - a^2)z^4 + 2(b^2 + a^2)z^2 + (b^2 - a^2)},$$

which indeed equals $2\pi/a^2$ if $a = b$. For the sake of momentary relief from clutter, let

$$c := \frac{b^2 + a^2}{b^2 - a^2}.$$

Then the integral becomes

$$\frac{4}{i(b^2 - a^2)} \int_{|z|=1} \frac{z\,dz}{z^4 + 2cz^2 + 1}.$$

The integrand has four poles: $\pm\sqrt{-c \pm \sqrt{c^2 - 1}}$. Of those, two are inside the circle $|z| = 1$, and the residue is the same at each:

$$\text{Res}\left(\frac{z}{z^4 + 2cz^2 + 1}, \pm\sqrt{-c + \sqrt{c^2 - 1}}\right) = \left.\frac{1}{4(z^2 + c)}\right|_{z = \pm\sqrt{-c + \sqrt{c^2 - 1}}}$$

$$= \frac{1}{4\sqrt{c^2 - 1}}.$$

With a bit of computation, we find the answer:

$$\frac{4}{i(b^2 - a^2)} 2\pi i \frac{2}{4\sqrt{c^2 - 1}} = \frac{2\pi}{ab}.$$

$\square$

Exercises

1. Let $a > 0$. Show that these three integrals and the one in Example 4.1.1 are all equal, using properties of the trigonometric functions and appropriate changes

of variable. Then, pretending you do not already know the answer, evaluate (c) by the method of Example 4.1.1.

(a) $\displaystyle\int_0^{2\pi} \frac{dt}{a + \cos t}$ (b) $\displaystyle\int_0^{2\pi} \frac{dt}{a - \sin t}$ (c) $\displaystyle\int_0^{2\pi} \frac{dt}{a + \sin t}$

2. Let $a > 1$. Show that $\displaystyle\int_0^{2\pi} \frac{\cos t}{a + \cos t}\, dt = 2\pi\left(1 - \frac{a}{\sqrt{a^2 - 1}}\right)$.

3. Let $a > 0$. Evaluate $\displaystyle\int_0^{\pi/2} \frac{dt}{a + \sin^2 t}$.

4. In each case, write the given real integral as a complex integral over the circle $|z| = 1$.

(a) $\displaystyle\int_0^{\pi/2} \cos^2 t\, dt$.   (c) $\displaystyle\int_0^{2\pi} \sin^4 t\, dt$.

(b) $\displaystyle\int_0^{2\pi} \frac{dt}{(a + \cos t)^2}\quad (a > 1)$.   (d) $\displaystyle\int_0^{2\pi} \frac{\sin^2 3t}{1 + \sin 2t}\, dt$.

5. Use the Residue Theorem to evaluate exactly the complex integral obtained in each part of Exercise 4.

6. Derive the formula

$$\frac{1}{2\pi}\int_0^{2\pi} \cos^{2n} t\, dt = \frac{1 \cdot 3 \cdot 5 \cdots (2n - 1)}{2 \cdot 4 \cdot 6 \cdots (2n)} \equiv \frac{(2n)!}{2^{2n}(n!)^2}.$$

7. Assuming $0 < a < 1$, evaluate $\displaystyle\int_0^{2\pi} \frac{dt}{1 - 2a\cos t + a^2}$.

8. Evaluate the integral in Exercise 7 in the case when $a > 1$.

9. Assuming $0 < a < 1$, evaluate $\displaystyle\int_0^{2\pi} \frac{\cos^2 2t}{1 - 2a\cos t + a^2}\, dt$.

10. Let $a > 0, b > 0$. Show that $\displaystyle\int_0^{2\pi} \frac{dt}{(a^2 \sin^2 t + b^2 \cos^2 t)^2} = \frac{\pi(a^2 + b^2)}{4a^3 b^3}$.

11. Let a be a nonzero real number. Show that $\displaystyle\int_{-\pi/2}^{\pi/2} \tan(t + ia)\, dt = \frac{i\pi a}{|a|}$.

12. Find out whether your computer algebra system can exactly solve problems like those above. For example, you might try *Mathematica's* command Integrate.

Help on Selected Exercises

1. (c) The integral is the pullback of

$$\int_{|z|=1} \frac{1}{a + \frac{1}{2i}\left(z - \frac{1}{z}\right)} \frac{dz}{iz} = 2 \int_{|z|=1} \frac{dz}{z^2 + 2aiz - 1}.$$

The integrand has poles at $-ai \pm i\sqrt{a^2 - 1}$. The only pole inside the unit circle $|z| = 1$ is $i(-a + \sqrt{a^2 - 1})$. So the answer is

$$2 \cdot 2\pi i \cdot \operatorname{Res}\left(\frac{1}{z^2 + 2aiz - 1}, i\left(-a + \sqrt{a^2 - 1}\right)\right)$$

$$= 4\pi i \cdot \left. \frac{1}{2z + 2ai}\right|_{z=i(-a+\sqrt{a^2-1})} = \frac{2\pi}{\sqrt{a^2 - 1}}.$$

2. Notice that $\dfrac{\cos t}{a + \cos t} = 1 - \dfrac{a}{a + \cos t}.$

3. By observing the symmetries of the integrand, we find that

$$\int_0^{\pi/2} \frac{dt}{a + \sin^2 t} = \frac{1}{4} \int_0^{2\pi} \frac{dt}{a + \sin^2 t},$$

which, when we do (1) and simplify, becomes

$$\int_{|z|=1} f(z)dz, \quad \text{where } f(z) := \frac{iz}{z^4 - (4a + 2)z^2 + 1}.$$

Using the quadratic formula, we find that z is a root of the denominator of f if and only if $z^2 = 2a+1\pm 2\sqrt{a(a+1)}$. With a bit of care, we can establish that z is a root lying within the circle $|z| = 1$ if and only if $z^2 = 2a+1-2\sqrt{a(a+1)}$. So there are two poles of the integrand within that circle, and the residue is the same at both:

$$\operatorname{Res}\left(f, \pm\sqrt{2a + 1 - 2\sqrt{a(a+1)}}\right)$$

$$= \left. \frac{i}{4z^2 - 2(4a + 2)}\right|_{z^2 = 2a+1-2\sqrt{a(a+1)}} = \frac{-i}{8\sqrt{a(a+1)}}.$$

So the integral equals $2\pi i$ times the sum of those two residues:

$$\int_0^{\pi/2} \frac{dt}{a + \sin^2 t} = \frac{\pi}{2\sqrt{a(a+1)}}.$$

6. This problem is easy enough without a conversion to a complex integral. A key portion of the calculation, using the binomial formula, is as follows:

$$\int_0^{2\pi} (e^{it} + e^{-it})^{2n}dt = \sum_{k=0}^{2n} \binom{2n}{k} \int_0^{2\pi} e^{i(2n-k)t}e^{-ikt}dt,$$

and the only nonzero summand is the one in which $k = n$.

7. This evaluation may be done by the method of Example 4.1.1. Or, a change of variable will reduce it to that Example. Later, you may recognize its connection with the Poisson integral formula, treated in Section 5.5.

4.2   ESTIMATING COMPLEX INTEGRALS

Nearly every Residue Calculus application in the rest of this chapter uses a result or a technique from one of the Examples in this section. Proposition 2.5.5 provides one straightforward way to estimate an integral. Other results require something more subtle.

In Section 2.2, we presented in detail the exponential function $z \mapsto e^z$. One of its most prominent features is that it maps the left half-plane into the unit disk, and its modulus tends to 0 as z goes to the left. More precisely, $|e^{x+iy}| = e^x \to 0$ as $x \to -\infty$. That fact is the basis for the results in the next two Examples, which concern variants of the exponential function, namely $z \mapsto e^{iz}$ and $z \mapsto e^{z^2}$. These Examples prepare the way for Jordan's Lemma 4.2.5 and for the evaluation of the Fresnel integrals in Example 4.4.7, respectively. Exercises 1–5 are intended to help you understand them.

4.2.1. Example. *Show that*

$$\lim_{\epsilon \to 0^+} \int_{\text{Arc}(\epsilon,0,\pi)} \frac{e^{iz}}{z}\,dz = \pi i, \quad \text{and that} \tag{1}$$

$$\lim_{R \to \infty} \int_{\text{Arc}(R,0,\pi)} \frac{e^{iz}}{z}\,dz = 0. \tag{2}$$

Solution. The function $z \mapsto e^{iz}/z$ has a pole of order 1 at 0, and $\text{Res}(e^{iz}/z, 0) = 1$. Therefore (1) follows from 3.3.20. The limit would still be πi even if we used other half-circles: $\text{Arc}(\epsilon, t_0, t_0 + \pi)$ for arbitrary t_0.

On the other hand, what makes (2) plausible is that the semicircular arcs lie in the upper half-plane, where e^{iz} is small: $|e^{i(x+iy)}| = e^{-y} \to 0$ as $y \to \infty$. The proof requires a somewhat delicate estimate because of the points on $\text{Arc}(R, 0, \pi)$ that are near the x-axis. Consider the pullback:

$$\int_{\text{Arc}(R,0,\pi)} \frac{e^{iz}}{z}\,dz = \int_0^\pi \frac{e^{iRe^{it}}}{Re^{it}} Rie^{it}\,dt = i\int_0^\pi e^{iR\cos t} e^{-R\sin t}\,dt.$$

The modulus of the integrand is $e^{-R\sin t}$, and the sine has a certain symmetry on the interval $[0, \pi]$.

$$\left| \int_{\text{Arc}(R,0,\pi)} \frac{e^{iz}}{z}\,dz \right| \le \int_0^\pi e^{-R\sin t}\,dt = 2\int_0^{\pi/2} e^{-R\sin t}\,dt.$$

We need to be careful here because there is no bound on the integrand, uniform throughout the t-interval, that tends to 0 as $R \to \infty$. Although the exponent $-R \sin t$ is negative on most of the interval, it is near 0 (and thus the integrand is near 1) for t near 0.

Sketch the graphs of $t \mapsto \sin t$ and $t \mapsto 2t/\pi$, and you will see that

$$\sin t \geq \frac{2t}{\pi} \quad \text{and hence} \quad e^{-R \sin t} \leq e^{-2Rt/\pi} \quad \text{for } 0 \leq t \leq \frac{\pi}{2}.$$

Therefore

$$\int_0^{\pi/2} e^{-R \sin t} dt \; \leq \; \int_0^{\pi/2} e^{-2Rt/\pi} dt \; = \; \frac{\pi}{R}(1 - e^{-R}) \to 0,$$

and (2) follows.   □

4.2.2. Example. *Show that* $\displaystyle \lim_{R \to \infty} \int_{\text{Arc}(R,0,\frac{\pi}{4})} e^{-z^2} dz = 0.$

Solution. Sketch the graphs of $t \mapsto \cos t$ and $t \mapsto 1 - \frac{2t}{\pi}$, and you will see that

$$\cos t \geq 1 - \frac{2t}{\pi} \quad \text{for } 0 \leq t \leq \frac{\pi}{2}. \tag{3}$$

We may estimate the integral as follows:

$$\int_{\text{Arc}(R,0,\frac{\pi}{4})} e^{-z^2} dz = \left| \int_0^{\pi/4} e^{-R^2(\cos 2\theta + i \sin 2\theta)} i R e^{i\theta} d\theta \right|$$

$$= R \int_0^{\pi/4} e^{-R^2 \cos 2\theta} d\theta$$

$$= \frac{R}{2} \int_0^{\pi/2} e^{-R^2 \cos t} dt \quad (\text{letting } 2\theta = t)$$

$$\leq \frac{R}{2} \int_0^{\pi/2} e^{-R^2(1 - \frac{2}{\pi}t)} dt \quad (\text{by (3)})$$

$$= \frac{Re^{-R^2}}{2} \int_0^{\pi/2} e^{\frac{2R^2 t}{\pi}} dt$$

$$= \frac{\pi}{4R}(1 - e^{-R^2}) \to 0 \quad \text{as } R \to \infty.$$

□

4.2.3. Example. *Let g be holomorphic at 0, and let* $t_0 < t_1$. *Show that*

$$\lim_{\epsilon \to 0+} \int_{\text{Arc}(\epsilon, t_0, t_1)} g(z) \log z \, dz = 0.$$

Solution. For z near 0, there is a bound on $g(z)$, say, $|g(z)| < B$. The branch of the logarithm is left unspecified; let's say we are using the one such that $\log \epsilon e^{it} = \ln \epsilon + i(2\pi n + t)$. Then the pullback of the integral may be estimated as follows:

$$\left| \int_{t_0}^{t_1} g(\epsilon e^{it}) \log(\epsilon e^{it}) i\epsilon e^{it} dt \right| \leq B\epsilon \int_{t_0}^{t_1} |\ln \epsilon + i2\pi n + it| \, dt$$

$$\leq B\epsilon((|\ln \epsilon| + 2\pi n)(t_1 - t_0) + \frac{1}{2}(t_1^2 - t_0^2)) \to 0 \quad \text{as } \epsilon \to 0^+.$$

$\square$

4.2.4. Example. *Let g be holomorphic at ∞. Suppose also that for some constants $B > 0$ and $a > 1$,*

$$|g(z)| < \frac{B}{R^a} \quad \text{when } |z| = R$$

for all sufficiently large R. (For example, g might be a rational function P/Q, where the degree of Q exceeds the degree of P by at least 2.) Let $t_0 < t_1$. Show that

$$\lim_{R \to \infty} \int_{\text{Arc}(R,t_0,t_1)} g(z) dz = 0, \quad \text{and that}$$

$$\lim_{R \to \infty} \int_{\text{Arc}(R,t_0,t_1)} g(z) \log z \, dz = 0.$$

Solution. By 2.5.5,

$$\left| \int_{\text{Arc}(R,t_0,t_1)} g(z) dz \right| \leq R(t_1 - t_0) \frac{B}{R^a},$$

which tends to 0 as $R \to \infty$. To obtain the second limit statement, we estimate the pullback as in the preceding example:

$$\left| \int_{t_0}^{t_1} g(Re^{it}) \log(Re^{it}) i Re^{it} dt \right| \leq \frac{B}{R^{a-1}} \int_{t_0}^{t_1} |(\ln R + i2\pi n + it)| \, dt$$

$$\leq \frac{B}{R^{a-1}}((\ln R + 2\pi n)(t_1 - t_0) + \frac{1}{2}(t_1^2 - t_0^2)) \to 0 \quad \text{as } R \to \infty.$$

$\square$

4.2.5. Proposition (Jordan's Lemma). *For some $r > 0$, let f be holomorphic on the set $A := \{z = x + iy \mid y \geq 0 \text{ and } |z| > r\}$, such that*

$$\lim_{S \to \infty} \sup_{z \in A, |z| \geq S} |f(z)| = 0.$$

Let $s > 0$. Then

$$\lim_{K \to \infty, L \to \infty} \int_{[L, L+iL, -K+iL, -K]} f(z) e^{isz} dz = 0. \tag{4}$$

Also,

$$\lim_{R\to\infty} \int_{\text{Arc}(R,0,\pi)} f(z)e^{isz}dz = 0. \tag{5}$$

Proof. The path in (4) consists of three of the four segments bounding a rectangle, which need not be centered on the y-axis; we are not limited to the case when $K = L$. Make a sketch.[1] It is easy to prove (4), even though there are three parts to deal with:

$$\left| \int_{[L,L+iL]} f(z)e^{isz}dz \right| = \left| i\int_0^L f(L+iy)e^{i(sL+siy)}dy \right|$$

$$< \max_{0\le y\le L} |f(L+iy)| \int_0^L e^{-sy}dy$$

$$< \max_{0\le y<\infty} |f(L+iy)|s^{-1} \to 0 \text{ as } L\to\infty;$$

$$\left| \int_{[L+iL,-K+iL]} f(z)e^{isz}dz \right| = \left| \int_{-K}^L f(x+iL)e^{isx-sL}dx \right|$$

$$< \max_{-K\le x\le L} |f(x+iL)|e^{-sK}2K \to 0 \text{ as } K\to\infty;$$

$$\left| \int_{[-K+iL,-K]} f(z)e^{isz}dz \right| = \left| \int_0^L f(-K+iy)e^{i(-sK+siy)}dy \right|$$

$$< \max_{0\le y\le L} |f(-K+iy)| \int_0^L e^{-sy}dy$$

$$< \max_{0\le y<\infty} |f(-K+iy)|s^{-1} \to 0 \text{ as } K\to\infty.$$

In (5), the path is a symmetrical semicircle. Since integrals of f are independent of path in the set A,

$$\int_{\text{Arc}(R,0,\pi)} f(z)e^{isz}dz = \int_{[R,R+iR,-R+iR,-R]} f(z)e^{isz}dz,$$

which tends to 0 as $R\to\infty$ (it's a special case of (4)).   □

Exercises

1. In each case, prove the given limit statement.

(a) $\displaystyle\lim_{R\to\infty} \int_{\text{Arc}(R,-\pi,\pi)} \frac{e^{iz}}{z} dz = 2\pi i.$

---

[1] We have made an artificial choice in making the height of the rectangle equal to L; it could as well be K, or $K + L$, and the only point is that in taking the limit the height must not grow much more slowly than K or L.

(b) $\displaystyle \lim_{R \to \infty} \int_{\text{Arc}(R, \frac{\pi}{2}, \frac{3\pi}{2})} \frac{e^z}{z}\, dz = 0.$

2. Discuss why 2.5.5 was not used in Examples 4.2.1 and 4.2.2.

3. How does the integral $\displaystyle \int_{\text{Arc}(R,0,\pi)} e^{isz} dz$ behave as $R \to \infty$?

4. Prove that $\displaystyle \lim_{R \to \infty} \int_{[R, Re^{i\pi/4}]} e^{-z^2} dz = 0.$

5. Example 4.2.2 established the limit statement

$$\lim_{R \to \infty} \int_{\text{Arc}(R,\theta_0,\theta_1)} e^{-z^2} dz = 0$$

in the case when $\theta_0 = 0$ and $\theta_1 = \pi/4$. For what other intervals $[\theta_0, \theta_1]$ does the statement hold?

6. Write a version of 4.2.5 for $s < 0$.

4.3   INTEGRALS OF RATIONAL FUNCTIONS OVER THE LINE

First-year calculus texts (for example, [67, Section 7.4]) discuss rational functions $f := P/Q$ on the real line, for arbitrary polynomials P and Q, and show you how to find, for every interval $I \subseteq \mathbb{R}$ on which Q has no zeros, a function F such that $F' = f$ on I. If $[a, b] \subset I$, then of course

$$\int_a^b f(x)dx \;=\; F(x)\big|_a^b.$$

In this section we consider the case when $I = \mathbb{R}$. If the degree of Q is at least 2 more than the degree of P, then the improper integral converges, and

$$\int_{-\infty}^{+\infty} f(x)dx \;=\; \lim_{R \to \infty} F(x)\Big|_{-R}^{R}.$$

For example,

$$\int_{-\infty}^{+\infty} \frac{dx}{x^2 + 1} \;=\; \lim_{R \to \infty} \arctan x \big|_{-R}^{R} \;=\; \pi.$$

In the general case, one can write f as a sum of such easy cases. The method is called decomposition into partial fractions. For example, consider the integral

$$\int_{-\infty}^{+\infty} f(x)dx, \quad \text{where } f(x) := \frac{x}{(x^2 + 1)(x^2 + 2x + 2)}. \tag{1}$$

The decomposition breaks the integrand $f(x)$ into four parts:

$$f(x) = \frac{-x-4}{5(x^2+2x+2)} + \frac{x+2}{5(x^2+1)}$$

$$= -\frac{1}{10}\frac{2x+2}{x^2+2x+2} - \frac{3}{5}\frac{1}{x^2+2x+2} + \frac{1}{10}\frac{2x}{x^2+1} + \frac{2}{5}\frac{1}{x^2+1}.$$

Each of those four summands has an antiderivative involving the natural logarithm or the arctangent.

The Residue Calculus provides an alternative method for evaluating such integrals.

4.3.1. Proposition. *Let P and Q be polynomials with no common zeros, such that the degree of Q is at least 2 more than the degree of P, and such that Q has no zeros on the real axis. Let f be the meromorphic function on $\hat{\mathbb{C}}$ given by*

$$f(z) := \frac{P(z)}{Q(z)}.$$

Let σ equal $2\pi i$ times the sum of the residues of f at its poles in the upper half-plane. Let τ equal $2\pi i$ times the sum of the residues of f at its poles in the lower half-plane. Then

$$\int_{-\infty}^{+\infty} f(x)dx = \sigma = -\tau. \tag{2}$$

Proof. The conditions on P and Q guarantee that the integral converges. We know that

$$\int_{-\infty}^{+\infty} f(x)dx = \lim_{R\to\infty}\int_{-R}^{+R} f(x)dx.$$

Let R be sufficiently large so that all of the poles of f that lie in the upper half-plane are inside the contour $[-R, R] + \mathrm{Arc}(R, 0, \pi)$. By the Residue Theorem, then,

$$\int_{-R}^{R} f(x)dx + \int_{\mathrm{Arc}(R,0,\pi)} f(z)dz = \sigma.$$

To prove (2), it suffices now to prove that

$$\lim_{R\to\infty}\int_{\mathrm{Arc}(R,0,\pi)} f(z)dz = 0,$$

which was done in 4.2.4.

A proof that the integral equals $-\tau$ may be given similarly, using the contour $[R, -R] - \mathrm{Arc}(R, \pi, 2\pi)$, which goes around a half disk in the lower half-plane. Or just note that $\sigma = -\tau$ by 3.4.4. $\qquad\square$

4.3.2. Example. *Evaluate* $\displaystyle\int_{-\infty}^{+\infty} \frac{dx}{x^2+1}.$

Solution. By the Proposition, the integral equals

$$2\pi i \operatorname{Res}\left(\frac{1}{z^2+1}, i\right) = 2\pi i \frac{1}{2i} = \pi.$$

$\square$

4.3.3. Example. *Evaluate* $\displaystyle\int_{-\infty}^{+\infty} \frac{x^2}{x^6+1} dx.$

Solution. You might notice that the change of variables $t = x^3$ reduces this problem to the one just solved. But suppose you don't. Proposition 4.3.1 applies. The function given by

$$f(z) := \frac{z^2}{z^6+1}$$

has three poles of order 1 in the upper half-plane, and the sum of the three residues is $-i/6$ (3.3.17). Thus

$$\int_{-\infty}^{+\infty} \frac{x^2}{x^6+1} dx = 2\pi i\left(\frac{-i}{6}\right) = \frac{\pi}{3}.$$

$\square$

4.3.4. Example. *Evaluate the integral* (1).

Solution. First, some preliminary observations: The integral is not necessarily positive, since the integrand is not everywhere so. Your calculus skills (or your computer) allow you to sketch the integrand, and then you can tell that, in fact, the integral is negative.

The function f has two poles of order 1 in the upper half-plane, and the two residues add up to $i/10$ (3.3.18). Therefore by 4.3.1, the integral equals $-\pi/5$. $\square$

4.3.5. Example. *Evaluate* $\displaystyle\int_0^\infty f(x)dx,$ *where* $f(x) = \dfrac{1}{x^3+1}.$

Solution. The function f has a pole of order 1 at each of the points $e^{i\pi/3}, -1$, and $e^{-i\pi/3}$. Since one of those is on the negative real axis, 4.3.1 does not apply. Consider the contour γ_R consisting of the line segment $[0, R]$, followed by the circular arc $\operatorname{Arc}(R, 0, \alpha)$, and finally the line segment $[Re^{i\alpha}, 0]$. Make a sketch, and notice that provided $R > 1$ and $\frac{\pi}{3} < \alpha < \pi$, the contour γ_R goes once around the pole of f at $e^{i\pi/3}$, but does not go around either of the other two. (So that we can motivate for you the selection of α, we leave it otherwise unspecified for the moment.) By the Residue Theorem, then,

$$\int_{\gamma_R} f = 2\pi i \operatorname{Res}\left(\frac{1}{z^3+1}, e^{i\pi/3}\right) = \frac{2\pi i}{3e^{i2\pi/3}}. \tag{3}$$

Also, the integral is the sum of its three parts:

$$\int_{\gamma_R} f = \int_{[0,R]} f + \int_{\text{Arc}(R,0,\alpha)} f + \int_{[Re^{i\alpha},0]} f.$$

The limit as $R \to \infty$ of the first integral is the answer to our problem. The limit of the second integral is 0, by 4.2.4, regardless of α. Consider now the pullback of the integral over part (3) of the contour, parametrized by $x \mapsto xe^{i\alpha}$ ($R \geq x \geq 0$):

$$\int_{\text{Arc}(R,0,\alpha)} f = \int_R^0 \frac{e^{i\alpha}}{x^3 e^{3i\alpha} + 1} dx.$$

The idea is to choose α so that this integral is a constant times the integral over $[0, R]$. A good choice is $\alpha := 2\pi/3$:

$$\int_{\text{Arc}(R,0,\frac{2\pi}{3})} f = -e^{i2\pi/3} \int_0^R \frac{dx}{x^3 + 1}.$$

Then

$$\int_{\gamma_R} f = (1 - e^{i2\pi/3}) \int_0^R \frac{dx}{x^3 + 1} + \int_{\text{Arc}(R,0,\frac{2\pi}{3})} f$$

$$\to (1 - e^{i2\pi/3}) \int_0^\infty \frac{dx}{x^3 + 1} \quad \text{as } R \to \infty.$$

(4)

Putting (3) together with (4), we find that

$$\int_0^\infty \frac{dx}{x^3 + 1} = \frac{2\pi i}{3(e^{i2\pi/3} - e^{-i2\pi/3})} = \frac{\pi}{3 \sin \frac{2\pi}{3}} = \frac{2\pi}{3\sqrt{3}}.$$

Exercises 11–14 are variations on this Example.   □

Exercises

1. Evaluate $\displaystyle\int_{-\infty}^{+\infty} \frac{dx}{x^2 + x + 2}$.

2. Evaluate $\displaystyle\int_{-\infty}^{+\infty} \frac{dx}{x^4 + 10x^2 + 9} dx$.

3. Evaluate $\displaystyle\int_{-\infty}^{+\infty} \frac{x^2 - x + 2}{x^4 + 10x^2 + 9} dx$.

4. Evaluate $\displaystyle\int_{-\infty}^{+\infty} \frac{x^4}{x^6 + 1} dx$.

5. Show that $\displaystyle\int_{-\infty}^{+\infty} \frac{dx}{x^2 - 2x + 5} = \frac{\pi}{2}$.

6. Evaluate $\displaystyle\int_0^\infty \frac{dx}{(1+x^2)^2}$.

7. Let $a > 0$. Show that $\displaystyle\int_{-\infty}^{+\infty} \frac{dx}{(x^2+a^2)^3} = \frac{3\pi}{8a^5}$.

8. Let $a > 0$. Evaluate $\displaystyle\int_{-\infty}^{+\infty} \frac{x^2}{x^4+a^4}\,dx$.

9. Let $a > 0$. Show that $\displaystyle\int_0^\infty \frac{x^2}{(x^2+a^2)^3}\,dx = \frac{\pi}{8a^3}$.

10. Evaluate $\displaystyle\int_{-\infty}^{+\infty} \frac{x^2}{(x^2+4)(x^2+9)}\,dx$.

11. By generalizing the method of 4.3.5, show that

$$\int_0^\infty \frac{dx}{x^n+1} = \frac{\pi}{n\sin\frac{\pi}{n}} \quad \text{for every integer } n \geq 2.$$

12. Discuss whether the integral $\displaystyle\int_0^\infty \frac{dx}{1+x^b}$ can be evaluated using the method of 4.3.5 for an arbitrary real number $b > 1$.

13. Carry out the first-year calculus method for evaluating the integral (1).

14. Discuss the difficulty that arises if one tries to use the method of 4.3.5 to evaluate the integral $\displaystyle\int_0^\infty \frac{dx}{x^2+6x+8}$. Find a first-year calculus method for doing it.

Help on Selected Exercises

9. See Example 3.3.19.

10. The integral equals $2\pi i$ times the sum of the residues in the upper half-plane of the function given by

$$f(z) := \frac{z^2}{z^4+13z^2+36} \equiv \frac{z^2}{(z-2i)(z+2i)(z-3i)(z+3i)}.$$

There are four poles of order one. The residues can be obtained by this equation: $\mathrm{Res}(f, a) = \left.\dfrac{z^2}{4z^3+26z}\right|_{z=a}$. The residues at $2i$ and $3i$ are $i/5$ and $-3i/10$. The appropriate computation shows that the integral equals $\pi/5$.

12. See 4.5.2.

14. It yields to decomposition into partial fractions; briefly,

$$\int_0^\infty \frac{dx}{x^2 + 6x + 8} = \lim_{X \to \infty} \frac{1}{2} \int_0^X \left(\frac{1}{x+2} - \frac{1}{x+4} \right) dx = \ldots = \frac{\ln 2}{2}.$$

Proposition 4.6.2 will provide another method, rather different from that of 4.3.5.

4.4 INTEGRALS INVOLVING THE EXPONENTIAL

4.4.1. Example. *Use the Residue Calculus to show that*

$$\int_{-\infty}^\infty \frac{e^{ax}}{1 + e^x} \, dx = \frac{\pi}{\sin \pi a} \quad \text{if } 0 < a < 1; \text{ and that} \tag{1}$$

$$\int_{-\infty}^\infty \frac{e^{ax}}{1 + e^{bx}} \, dx = \frac{\pi}{b \sin \frac{\pi a}{b}} \quad \text{if } 0 < a < b. \tag{2}$$

Solution. You may wish first to assure yourself that the two improper integrals are convergent. They are so, because each of the integrands is well defined, positive, and continuous on $\mathbb{R}$; and tends to 0 exponentially as $x \to \infty$ and as $x \to -\infty$, thanks to the conditions on a and b.

You may suspect that the two integrals are related, and you should look for a change of variable to find the connection. Indeed, if we try $x = t/b$ we find that

$$\int_{-\infty}^\infty \frac{e^{ax}}{1 + e^{bx}} \, dx = \frac{1}{b} \int_{-\infty}^\infty \frac{e^{at/b}}{1 + e^t} \, dt.$$

Therefore (2) follows from (1).

To prove (1), we will use the fact that

$$\int_{-\infty}^{+\infty} \frac{e^{ax}}{1 + e^x} dx = \lim_{X \to \infty} \int_{-X}^X \frac{e^{ax}}{1 + e^x} dx. \tag{3}$$

The real integral on the right is a pullback of the complex integral $\int_{[-X,X]} f$, where

$$f(z) := \frac{e^{az}}{1 + e^z}.$$

There are several steps in the method, and more ways than one to carry it out. The technique, stated briefly, is as follows. For each X, we will devise a contour γ_X of which the segment $[-X, X]$ is one part, such that $\int_{\gamma_X} f$ can be evaluated by the Residue Theorem, and such that the limit in (3) is revealed when we take the limit as $X \to \infty$.

The poles of f are all of order 1 and occur at the odd multiples of πi. By 3.3.15,

$$\text{Res}(f, \pi i k) = \left. \frac{e^{az}}{e^z} \right|_{z = \pi i k} = -e^{\pi i a k}.$$

Let Y be a fixed positive number which is not an odd multiple of π. For $X > 0$, consider the contour, the boundary of a rectangle, comprising four line segments:

$$\gamma_X = [-X, X, X + iY, -X + iY, -X].$$

Make a sketch. By the Residue Theorem, the integral of f over γ_X equals $2\pi i$ times the sum of the residues inside γ_X. As you can see, the sum is independent of X but depends on the choice of Y, because Y determines how many poles of f lie inside γ_X. The integral is the sum of four integrals over line segments—the one appearing in (3) and the following three:

$$\int_{[X,X+iY]} f = \int_0^Y f(X + iy)i\,dy = i\int_0^Y \frac{e^{a(X+iy)}}{1 + e^{X+iy}}\,dy; \tag{4}$$

$$\int_{[X+iY,-X+iY]} f = -\int_{-X}^X f(x + iY)\,dx = -\int_{-X}^X \frac{e^{ax}e^{iaY}}{1 + e^x e^{iY}}\,dx; \tag{5}$$

$$\int_{[-X+iY,-X]} f = -\int_0^Y f(-X + iy)i\,dy = -i\int_0^Y \frac{e^{a(-X+iy)}}{1 + e^{-X+iy}}\,dy. \tag{6}$$

Regardless of Y, the integrals (4) and (6) both tend to 0 as $X \to \infty$ because

$$\left| \int_0^Y \frac{e^{a(X+iy)}}{1 + e^{X+iy}}\,dy \right| < \max_{0 \leq y \leq Y} \left| \frac{e^{a(X+iy)}}{1 + e^{X+iy}} \right| Y \leq \frac{e^{aX}Y}{e^X - 1}, \quad \text{and}$$

$$\left| \int_0^Y \frac{e^{a(-X+iy)}}{1 + e^{-X+iy}}\,dy \right| < \max_{0 \leq y \leq Y} \left| \frac{e^{a(-X+iy)}}{1 + e^{-X+iy}} \right| Y \leq \frac{e^{-aX}Y}{1 - e^{-X}}.$$

We have postponed until now the choice of Y. If Y is chosen to be 2π times an integer n, then $f(x + iY)$ is a constant multiple of $f(x)$:

$$f(x + i2\pi n) = \frac{e^{ax}e^{ia2\pi n}}{1 + e^x e^{i2\pi n}} = e^{ia2\pi n}f(x) \quad \text{for all } x.$$

Then (5) equals a constant times the integral over $[-X, X]$. Take $Y = 2\pi$. Then

$$\int_{[X+i2\pi,-X+i2\pi]} f = -e^{i2\pi a}\int_{-X}^X f(x)\,dx.$$

With this choice of Y, the only singularity inside γ_X is at $i\pi$, and $\mathrm{Res}(f, i\pi) = -e^{i\pi a}$. Therefore

$$\int_{\gamma_X} f = -2\pi i e^{i\pi a}$$

$$= (1 - e^{i2\pi a})\int_{-X}^X f(x)\,dx + \int_{[X,X+i2\pi]} f + \int_{[-X+i2\pi,-X]} f$$

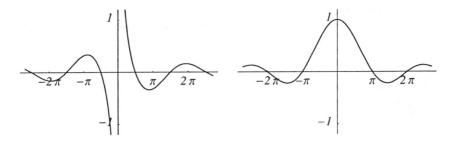

Figure 4.4-1   Graphs of the functions $x \mapsto x^{-1} \cos x$ and $x \mapsto x^{-1} \sin x$

for every $X > 0$. When we take the limit as $X \to \infty$, we find that

$$-2\pi i e^{\pi i a} = (1 - e^{i2\pi a}) \int_{-\infty}^{+\infty} f(x)\,dx.$$

Therefore

$$\int_{-\infty}^{+\infty} f(x)\,dx = \frac{-2\pi i e^{i\pi a}}{1 - e^{i2\pi a}} = \frac{2i\pi}{e^{i\pi a} - e^{-i\pi a}} = \frac{\pi}{\sin \pi a}.$$

$\square$

4.4.2. Example. *Evaluate if possible the real integrals*

$$\int_{-\infty}^{+\infty} \frac{\cos x}{x}\,dx \quad and \quad \int_{0}^{\infty} \frac{\sin x}{x}\,dx. \tag{7}$$

Proof. The two integrands' graphs are represented in Figure 4.4-1. The integrand $x \mapsto x^{-1} \sin x$ is continuous at 0 if we declare it equal to 1 there. So the integral is improper only because the interval is infinite. Since the integrand is an even function, it suffices to study its integral over $[0, \infty)$. The integral does not converge absolutely, but it does converge, essentially because the series

$$\int_{0}^{\infty} \frac{\sin x}{x}\,dx = \sum_{n=0}^{\infty} \int_{n\pi}^{(n+1)\pi} \frac{\sin x}{x}\,dx \tag{8}$$

converges by the Alternating Series Test. Therefore we expect to find a finite value for the integral by evaluating a limit:

$$\int_{-\infty}^{+\infty} \frac{\sin x}{x}\,dx = 2 \lim_{R \to \infty} \int_{0}^{R} \frac{\sin x}{x}\,dx;$$

or, what turns out to be the workable choice,

$$= 2 \lim_{\epsilon \to 0^+, R \to \infty} \int_{\epsilon}^{R} \frac{\sin x}{x}\,dx.$$

The integrand $x \mapsto x^{-1} \cos x$ is an odd function which is continuous except at 0:

$$\lim_{x \to 0+} \frac{\cos x}{x} = +\infty \quad \text{and} \quad \lim_{x \to 0-} \frac{\cos x}{x} = -\infty.$$

Its integral over $[0, \infty)$ is thus improper for two reasons—the misbehavior of the integrand at 0, and the infinite interval of integration. We would call the integral convergent if and only if both of these limits existed (and were finite):

$$\lim_{R \to \infty} \int_{\pi/2}^{R} \frac{\cos x}{x} dx \quad \text{and} \quad \lim_{\epsilon \to 0+} \int_{\epsilon}^{\pi/2} \frac{\cos x}{x} dx.$$

The first is like (8) and exists, but the second limit diverges to $+\infty$. Therefore the integrals

$$\int_{0}^{\infty} \frac{\cos x}{x} dx \quad \text{and} \quad \int_{-\infty}^{+\infty} \frac{\cos x}{x} dx$$

are both divergent. Nevertheless, just because the integrand is an odd function, the second integral does converge in a weak and delicate sense; that is, if we take limits in a precisely symmetric fashion:

$$\lim_{\epsilon \to 0+, R \to \infty} \left(\int_{-R}^{-\epsilon} \frac{\cos x}{x} dx + \int_{\epsilon}^{R} \frac{\cos x}{x} dx \right) = 0;$$

The principal value of the integral, as discussed in subsection 1.5B, exists, and

$$\text{P.V.} \int_{-\infty}^{+\infty} \frac{\cos x}{x} dx = 0. \tag{9}$$

We will tie the two integrals (7) together and evaluate both by applying the Residue Theorem to the function $f(z) = e^{iz}/z$, which is holomorphic on $\mathbb{C}'$. We write it in terms of x and y.

$$f(x + iy) = \frac{e^{-y} \cos x}{x + iy} + i \frac{e^{-y} \sin x}{x + iy}.$$

On the real axis, its real and imaginary parts are the functions we are interested in:

$$f(x) = \frac{\cos x}{x} + i \frac{\sin x}{x} \qquad (x \neq 0).$$

Consider the contour consisting of two line segments and two semicircular arcs: $[-R, -\epsilon], -\text{Arc}(\epsilon, 0, \pi), [\epsilon, R]$, and $\text{Arc}(R, 0, \pi)$. Make a sketch. We have designed the contour so that it does not go around 0, where f has its only pole. Therefore the integral of f over the contour is 0:

$$0 = \int_{[-R, -\epsilon]} f - \int_{\text{Arc}(\epsilon, 0, \pi)} f + \int_{[\epsilon, R]} f + \int_{\text{Arc}(R, 0, \pi)} f.$$

Taking the limit as $\epsilon \to 0^+$ and $R \to \infty$, and using 4.2.1, we find that

$$
\begin{aligned}
0 &= \lim_{\epsilon \to 0^+, R \to \infty} \left(\int_{-R}^{-\epsilon} f(x)dx + i \int_{\epsilon}^{R} f(x)dx \right) - \pi i \\
&= \text{P.V.} \int_{-\infty}^{+\infty} \frac{\cos x}{x} dx + i \int_{-\infty}^{+\infty} \frac{\sin x}{x} dx - \pi i,
\end{aligned}
$$

which confirms (9) and also gives $\displaystyle \int_{-\infty}^{+\infty} \frac{\sin x}{x} dx = \pi$. $\qquad\square$

A Integrals Giving Fourier Transforms

The next Proposition provides a rule by which one can find values for a large class of integrals. You may notice similarities with 4.3.1. The integrals evaluated will look familiar if you have studied Fourier transforms, so we remind you of the basic definitions. If f is an integrable function on $\mathbb{R}$, the **Fourier transform** of f is the function $\hat{f}$ given by

$$
\hat{f}(s) := \int_{-\infty}^{+\infty} f(x)e^{-isx}dx.
$$

Variants on that definition are often used; see Exercise 24 for an example. Under certain conditions, f is recoverable from $\hat{f}$ by the **inverse Fourier transform**, given by

$$
f(x) := \frac{1}{2\pi} \int_{-\infty}^{+\infty} \hat{f}(s)e^{ixs}ds.
$$

For a treatment of Fourier transform theory, see [31] or, at a more advanced level, [32].

4.4.3. Proposition. *Let f be holomorphic on $\mathbb{C}$ except for a finite number of singularities, none of which is on the real axis; and such that $\lim_{|z|\to\infty} f(z) = 0$. Let $a_1, \cdots, a_M$ be the singularities of f in the upper half-plane, and let $b_1, \cdots, b_N$ be the ones in the lower half-plane. For each nonzero real number s, let*

$$
\sigma_s = 2\pi i \sum_{m=1}^{M} \text{Res}(f(z)e^{isz}, a_m) \quad and \quad \tau_s = 2\pi i \sum_{n=1}^{N} \text{Res}(f(z)e^{isz}, b_n).
$$

Then

$$
\int_{-\infty}^{+\infty} f(x)e^{isx}dx = \begin{cases} \sigma_s & \text{if } s > 0, \\ -\tau_s & \text{if } s < 0. \end{cases} \tag{10}
$$

If $f(x)$ is real whenever x is real, then $\sigma_s = -\bar{\tau}_s$; and

$$\int_{-\infty}^{+\infty} f(x)\cos sx\,dx = \operatorname{Re}\sigma_s = -\operatorname{Re}\tau_s; \quad \text{and}$$

$$\int_{-\infty}^{+\infty} f(x)\sin sx\,dx = \operatorname{Im}\sigma_s = \operatorname{Im}\tau_s. \tag{11}$$

Proof. Notice that the singularities of $z \mapsto f(z)e^{isz}$ are the same as those of f. Of course, the value of the residue varies with s.

Case 1, when $s > 0$: Let K and L be positive numbers and consider the contour $[-K, L, L + iL, -K + iL, -K]$. Make a sketch. For K and L sufficiently large, all of the singularities a_m are inside the contour. So by the Residue Theorem,

$$\int_{[-K,L,L+iL,-K+iL,-K]} f(z)e^{isz}\,dz = \sigma_s.$$

By Jordan's Lemma 4.2.5, the integral over the segments other than $[-K, L]$ tend to 0. It follows that

$$\lim_{K\to\infty,L\to\infty} \int_{-K}^{L} f(z)e^{isz}\,dz = \sigma_s,$$

which establishes the convergence (which is not necessarily absolute) of the integral in (10) to σ_s.

Case 2, when $s < 0$: The change of variable $x = -u$ gives

$$\int_{-\infty}^{+\infty} f(x)e^{isx}\,dx = \int_{-\infty}^{+\infty} f(-u)e^{-isu}\,du.$$

The singularities of $z \mapsto f(-z)$ in the upper half-plane are precisely the points $-b$ such that b is a singularity of f in the lower half-plane; and by 3.3.21,

$$\operatorname{Res}(f(-z)e^{-isz}, -b) = -\operatorname{Res}(f(z)e^{isz}, b).$$

Therefore by an application of Case 1, with $-s$ in the role of s, and $z \mapsto f(-z)$ in the role of f, the second half of (10) is established.

We leave the last sentence of the Proposition as Exercise 18.   □

The Proposition applies, for example, when f is the rational function P/Q, where P and Q are two polynomials with no common zeros, such that the degree of Q exceeds that of P, and such that Q has no zeros on the real axis. Notice that if the degree of Q is only 1 more than the degree of P, then the integrals do not converge absolutely.

4.4.4. Example. *Let $a > 0$. Evaluate* $\displaystyle\int_{-\infty}^{+\infty} \frac{\cos x}{x^2 + a^2}\,dx$ *and* $\displaystyle\int_{-\infty}^{+\infty} \frac{\sin x}{x^2 + a^2}\,dx.$

Solution. Easy observations: Both integrals converge absolutely. The second integral is 0 because the integrand is odd. Proposition 4.4.3 applies, and

$$\int_{-\infty}^{+\infty} \frac{e^{ix}}{x^2 + a^2} dx = 2\pi i \mathrm{Res}\left(\frac{e^{iz}}{z^2 + a^2}, ia\right) = 2\pi i \frac{e^{iz}}{2z}\bigg|_{z=ia} = \frac{\pi e^{-a}}{a}.$$

So the first integral equals π/ae^a, and the second one equals 0.   □

4.4.5. Example. *Evaluate* $I := \int_{-\infty}^{+\infty} \frac{\cos x}{x - i} dx$ *and* $J := \int_{-\infty}^{+\infty} \frac{\sin x}{x - i} dx.$

Solution. We make use of Proposition 4.4.3. Note that (11) does not apply. Using (10), we will evaluate

$$A := \int_{-\infty}^{+\infty} \frac{e^{ix}}{x - i} dx \quad \text{and} \quad B := \int_{-\infty}^{+\infty} \frac{e^{-ix}}{x - i} dx.$$

Then $I = (A + B)/2$ and $J = (A - B)/2i$. We obtain

$$A = 2\pi i \mathrm{Res}\left(\frac{e^{iz}}{z - i}, i\right) = 2\pi i\, e^{iz}\big|_{z=i} = \frac{2\pi i}{e}.$$

And $B = 0$, since $z \mapsto z - i$ has no zeros in the lower half-plane. Finally, then,

$$I = \frac{\pi i}{e} \quad \text{and} \quad J = \frac{\pi}{e}.$$

□

4.4.6. Example. *Evaluate in terms of* s, *for* $s \neq 0$,

$$I(s) := \int_{-\infty}^{+\infty} \frac{x^3 \cos sx}{(x + 1)(x - 1)(x^2 + 4)} dx \quad \text{and}$$

$$J(s) := \int_{-\infty}^{+\infty} \frac{x^3 \sin sx}{(x + 1)(x - 1)(x^2 + 4)} dx.$$

Solution. Because of the infinite limits at $x = \pm 1$, the integrals do not converge. However, principal values exist. One might observe at the outset that the integrand in $I(s)$ is an odd function of x, so that $I(s) \equiv 0$. But we need to work a bit to evaluate $J(s)$. Let

$$f(z) = \frac{z^3 e^{isz}}{(z + 1)(z - 1)(z^2 + 4)}.$$

Notice that for real x, $\mathrm{Re}\, f(x)$ and $\mathrm{Im}\, f(x)$ are the integrands of the integrals $I(s)$ and $J(s)$; Figure 4.4-2 shows their graphs in the case $s = 1$. The residues are as follows:

$$\mathrm{Res}(f, \pm 1) = \frac{e^{\pm is}}{10}; \quad \mathrm{Res}(f, \pm 2i) = \frac{2e^{\mp 2s}}{5}.$$

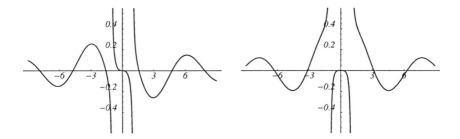

Figure 4.4-2 Graphs of functions considered in Example 4.4.6.

We take limits in a carefully chosen manner so as to obtain the principal value:

$$\text{P.V.} \int_{-\infty}^{+\infty} f = \lim_{\epsilon \to 0, R \to \infty} \int_{[-R,-1-\epsilon] + [-1+\epsilon,1-\epsilon] + [1+\epsilon,R]} f.$$

We put those three segments on the x-axis together with three semicircles to make up a contour, as shown in Figure 4.4-3. For all sufficiently large R and small ϵ, the contour goes around the only pole of f in the upper half-plane, which occurs at $2i$. Then by the Residue Theorem, the integral of f over that contour equals $2\pi i$ times the residue at $2i$:

$$\int_{[-R,-1-\epsilon] + [-1+\epsilon,1-\epsilon] + [1+\epsilon,R]} f - \int_{\text{Arc}_{-1}(\epsilon,0,\pi)} f$$

$$- \int_{\text{Arc}_1(\epsilon,0,\pi)} f + \int_{\text{Arc}(R,0,\pi)} f = \frac{4\pi i e^{-2s}}{5}.$$

Now take the limits, using 3.3.20 for the integrals over the two small arcs:

$$\text{P.V.} \int_{-\infty}^{+\infty} f - \frac{\pi e^{-is}}{10} - \frac{\pi e^{is}}{10} + 0 = \frac{4\pi i e^{-2s}}{5}.$$

Thus

$$\text{P.V.} \int_{-\infty}^{+\infty} f = \frac{\pi i}{5}(\cos s + 4e^{-2s}),$$

hence

$$I(s) = 0 \text{ and } J(s) = \frac{\pi}{5}(\cos s + 4e^{-2s}).$$

□

4.4.7. Example. *Establish the values of the Fresnel integrals:*

$$\int_0^\infty \cos x^2 dx = \int_0^\infty \sin x^2 dx = \sqrt{\frac{\pi}{8}}.$$

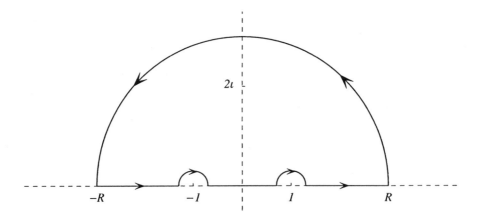

Figure 4.4-3   A contour for use in Example 4.4.6.

Solution. We will make use of the fact (see Exercise 12) that

$$\int_0^\infty e^{-x^2} dx = \frac{\sqrt{\pi}}{2}. \tag{12}$$

Consider the contour consisting of two line segments and a circular arc: $[0, R]$, $\text{Arc}(R, 0, \pi/4)$, and $[Re^{i\pi/4}, 0]$. Since $z \mapsto e^{-z^2}$ is an entire function, its integral around that contour equals 0 for every R; and the integral is the sum of three parts, whose behaviors as $R \to \infty$ are as follows:

$$\int_{\beta_R} e^{-z^2} dz \to 0 \quad \text{(Example 4.2.2)};$$

$$\int_{[Re^{i\pi/4}, 0]} e^{-z^2} dz = -\int_0^R e^{-it^2} e^{i\pi/4} dt$$

$$= -e^{i\pi/4} \int_0^R (\cos t^2 - i \sin t^2) dt \to \frac{-1-i}{\sqrt{2}} \int_0^\infty (\cos t^2 - i \sin t^2) dt;$$

$$\text{and} \quad \int_{[0, R]} e^{-z^2} dz \to \frac{\sqrt{\pi}}{2} \quad \text{by (12)}.$$

The sum of the three limits is 0, so

$$\int_0^\infty \cos t^2 dt - i \int_0^\infty \sin t^2 dt = \frac{\sqrt{\pi}}{2} \cdot \frac{\sqrt{2}}{1+i} = \sqrt{\frac{\pi}{8}}(1 - i),$$

and the result follows.   □

Exercises

1. Do Example 4.4.1 with $Y = 4\pi$ instead of 2π.

2. Let $b > 0$. Show that $\displaystyle\int_0^\infty e^{-x^2} \cos 2bx\, dx = \frac{\sqrt{\pi}}{2} e^{-b^2}$.

3. Let $a > 0$. Show that $\displaystyle\int_{-\infty}^{+\infty} \frac{\cos 3x}{x^2 + a^2}\, dx = \frac{\pi e^{-3a}}{a}$.

4. Let $a > 0$. Evaluate $\displaystyle\int_0^\infty \frac{\cos x}{(x^2 + a^2)^2}\, dx$.

5. Let $s \in \mathbb{R}$. Show that $\displaystyle\int_{-\infty}^{+\infty} \frac{e^{-isx}}{x^2 + 1}\, dx = \pi e^{-|s|}$.

6. Show that $\displaystyle\int_0^\infty \frac{x \sin x}{x^2 + 1}\, dx = \frac{\pi e^{-2}}{2}$.

7. Let $a > 0$. Show that $\displaystyle\int_0^\infty \frac{x^3 \sin ax}{x^4 + 1}\, dx = \frac{\pi}{2} e^{-a/\sqrt{2}} \cos \frac{a}{\sqrt{2}}$.

8. Let $a \geq 0, b \geq 0$. Evaluate $\displaystyle\int_0^\infty \frac{\cos ax - \cos bx}{x^2}\, dx$.

9. Let $a > 0, b > 0$. Evaluate $\displaystyle\int_\infty^{+\infty} \frac{\cos x}{(x^2 + a^2)(x^2 + b^2)}\, dx$.

10. Let $a > 0, b > 0$. Evaluate $\displaystyle\int_\infty^{+\infty} \frac{\sin x}{(x + a)^2 + b^2}\, dx$.

11. Let $b > 0$. Show that $\displaystyle\int_{-\infty}^{+\infty} \frac{\cos bx}{x^2 + x + 1}\, dx = \frac{2\pi \cos \frac{b}{2}}{\sqrt{3} e^{b\sqrt{3}/2}}$.

12. Prove (12).

13. There is a way to use the Residue Calculus to establish (12). Here is a brief sketch; provide the details. First show by a change of variable that (12) is equivalent to the equation $\displaystyle\int_{-\infty}^{+\infty} e^{-\pi x^2}\, dx = 1$. Apply the Residue Theorem to the integral of $f(z) := \dfrac{e^{i\pi z}}{\sin \pi z}$ around the parallelogram with vertices $\pm\frac{1}{2} \pm Re^{i\pi/4}$.

14. Let $s > 0$. In each case, prove the given limit statement. Explain the relevance of Jordan's Lemma.

(a) $\displaystyle\lim_{R \to \infty} \int_{\text{Arc}(R, -\pi, \pi)} \frac{z e^{isz}}{z^2 + 1}\, dz = 2\pi i \cosh s$.

(b) $\displaystyle\lim_{R \to \infty} \int_{\text{Arc}(R, \frac{\pi}{2}, \frac{3\pi}{2})} \frac{z e^{sz}}{z^2 + 1}\, dz = 0$.

15. Evaluate the integral $\displaystyle\int_0^\infty \frac{\sin^2 x}{x^2}\,dx$.

16. Evaluate if possible the real integrals $\displaystyle\int_{-\infty}^{+\infty} \frac{\cos 2x}{x}\,dx$ and $\displaystyle\int_0^\infty \frac{\sin 2x}{x}\,dx$.

17. Show that $\displaystyle\int_{-\infty}^{+\infty} \frac{e^{ax} - e^{bx}}{1 - e^x}\,dx = \pi(\cot a\pi - \cot b\pi)$ if $0 < a, b < 1$.

18. Justify the last sentence in Proposition 4.4.3.

19. Evaluate $\displaystyle\int_{-\infty}^{+\infty} \frac{\sin \pi x}{x^2 - 1}\,dx$.

20. Show that $\displaystyle\int_{-\infty}^{+\infty} \frac{\cos \frac{\pi x}{2}}{x^2 - 1}\,dx = \pi$.

21. Show that $\displaystyle\int_0^\infty \frac{\sin x}{x(x^2 + 1)}\,dx = \frac{\pi(1 + \cos 1)}{2}$.

22. Show that $\displaystyle\int_0^\infty e^{-x^2} \cos x^2\,dx = \frac{\sqrt{\pi(1 + \sqrt{2})}}{4}$.

23. Let f be integrable on $\mathbb{R}$ and suppose that for some $N > 0$, $f(x) = 0$ whenever $|x| > N$. Then the Fourier transform may be regarded as defined on $\mathbb{C}$:

$$\hat{f}(z) = \int_{-\infty}^{+\infty} f(t)e^{-izt}dt \quad (z \in \mathbb{C}).$$

Show by checking the Cauchy-Riemann equations that $\hat{f}$ is an entire function, and find its Taylor series at 0.

24. Let $g(x) := \operatorname{sech} \pi x$. Using this definition of the Fourier transform:

$$\hat{f}(s) := \int_{-\infty}^{+\infty} f(x)e^{-i2\pi sx}dx,$$

show that $\hat{g}(s) = \operatorname{sech} \pi s$. This result is sometimes called Ramanujan's Identity, since Ramanujan published the result, though with a quite different proof, in 1915.

Help on Selected Exercises

2. Use (12). Consider the integral of $z \mapsto e^{-z^2}$ around the rectangle whose vertices are $-R, R, R + ib$, and $R - ib$; and let $R \to \infty$.

8. The answer is $\frac{\pi}{2}(b - a)$. Use the function $f(z) := \dfrac{e^{iaz} - e^{ibz}}{z^2}$ and the same

contours as in Example 4.4.2. See Exercise 16 in Section 3.3.

12. A proof that is frequently seen in real-analysis and calculus texts (like [8, p. 225] and [67, Exercise 15.4-34]) may be outlined as follows. Show that $\int_0^\infty e^{-x^2} dx$ equals the square root of the double integral

$$\iint_{Q_I} e^{-(x^2+y^2)} dx dy,$$

which is taken over the positive quadrant of the plane. Change variables; in terms of polar coordinates, the integral is easy to evaluate. A proof using only differentiation under the integral sign is found in [70].

13. Note that the integral along one of the four segments is

$$\int_{[\frac{1}{2}-Re^{i\pi/4}, \frac{1}{2}+Re^{i\pi/4}]} f = i \int_{-R}^{R} \frac{e^{i\pi t e^{i\pi/4}} e^{-\pi t^2}}{\cos(\pi t e^{i\pi/4})} dt.$$

The procedure is presented fully in [14], in [56, pp. 126–128], and in [29, pp. 124–125]. Its first appearance seems to have been in the 1933 paper [42].

15. One may use the same contour as in Example 4.4.2, and integrate over it the function $z \mapsto \dfrac{1 - e^{i2z}}{z^2}$. The integral equals $\pi/2$.

17. Use the contour $[-R, R, R + i\pi, -R + i\pi, -R]$.

19. Simply because the integrand is odd, the value is 0.

22. Proceed as in Example 4.4.7, using $\pi/8$ instead of $\pi/4$.

23. $\hat{f}(z) = \displaystyle\sum_{k=0}^{\infty} \left(\frac{(-i)^k}{k!} \int_{-\infty}^{+\infty} t^k f(t) dt \right) z^k.$

24. The procedure is somewhat like that of Example 4.4.1. Integrate the function $z \mapsto e^{-i2\pi sz} \operatorname{sech} \pi z$ over the rectangle $[-R, R, R + i, -R + i, -R]$ and let $R \to \infty$. The only pole inside the rectangle is at $i/2$.

4.5 INTEGRALS INVOLVING A LOGARITHM

4.5.1. Example. *Consider the function given by* $f(x) := \dfrac{\ln x}{1 + x^2}$. *Evaluate the integral* $\displaystyle\int_0^\infty f(x) dx$.

Solution. The integral of f is improper for two reasons: The interval is infinite; and $\lim_{x \to 0+} f(x) = -\infty$. Before reading further, you may wish to ask a computer to draw the graph on, say, the interval $[0.5, 2]$, and speculate as to whether the integral converges.

We can verify that both parts of the integral converge by comparison tests, as follows.

- For $0 < x \le 1$, we have $0 > f(x) > \ln x$. Since $\int_0^1 \ln x\, dx$ converges and equals -1, the integral $\int_0^1 f(x)dx$ must also converge, and its value is between 0 and -1.

- For $1 \le x < \infty$, we have $0 \le f(x) < \dfrac{x^{1/2}}{1+x^2} < x^{-3/2}$. Since $\int_1^\infty x^{-3/2} dx$ converges, $\int_1^\infty f(x)dx$ must also converge.

Thus our integral equals

$$\int_0^\infty f(x)dx = \lim_{\epsilon \to 0^+, R \to \infty} \int_\epsilon^R f(x)dx, \tag{1}$$

which we know to be finite. A calculus student might produce the discussion so far. The Residue Calculus allows us essentially to avoid it. (But see Exercise 2.)

What we need to do is to evaluate the limit (1). If the value is finite, then we know that the integral converges. The technique is to devise a contour $\gamma(\epsilon, R)$ of which the interval $[\epsilon, R]$ is a part, and to extend f to a function that is meromorphic inside and on γ, such that we can apply the Residue Theorem to good effect.

To extend f, we must choose a logarithm that is real on the positive real axis. We choose the one on $V_{-\pi/2}$:

$$\log(re^{i\theta}) = \ln r + i\theta \quad \text{for } r > 0, \quad -\frac{\pi}{2} < \theta < \frac{3\pi}{2}.$$

Then let $f(z) := (\log z)/(1 + z^2)$ for $z \in V_{-\pi/2}$. We will integrate f over this contour, made up of two intervals and two semicircles in the upper half-plane:

$$\gamma(\epsilon, R) = [-R, -\epsilon] - \text{Arc}(\epsilon, 0, \pi) + [\epsilon, R] + \text{Arc}(R, 0, \pi).$$

Make your own sketch. We have selected the contour so that the origin is outside it, for good reason. In Example 4.4.2 we were using the function $z \mapsto e^z/z$, which has a pole of order 1 at 0; in this example, our function f has worse than a singularity at the origin—it has a branch point.

We require that $0 < \epsilon < 1 < R < \infty$, so that $\gamma(\epsilon, R)$ goes around i, which is the one pole of f in the upper half-plane. Applying the Residue Theorem, we can now state the value of the integral of f over the contour; and we also write that integral as the sum of its four parts:

$$\int_{\gamma(\epsilon,R)} f \;=\; 2\pi i \mathrm{Res}(f,i) \;=\; \frac{\pi^2 i}{2} \quad \text{(by 3.3.15)}$$

$$= \int_{[-R,-\epsilon]} f \;+\; \int_{-\mathrm{Arc}(\epsilon,0,\pi)} f \;+\; \int_{[\epsilon,R]} f \;+\; \int_{\mathrm{Arc}(R,0,\pi)} f.$$

The integrals of f over the two intervals are related as follows:

$$\int_{[-R,-\epsilon]} f \;=\; \int_\epsilon^R \frac{\ln x + i\pi}{1+x^2}\, dx \;=\; \int_{[\epsilon,R]} f \;+\; i\pi \int_\epsilon^R \frac{dx}{1+x^2}.$$

Therefore

$$\frac{\pi^2 i}{2} \;=\; 2\int_{[\epsilon,R]} f \;+\; i\pi \int_\epsilon^R \frac{dx}{1+x^2} \;+\; \int_{-\mathrm{Arc}(\epsilon,0,\pi)} f \;+\; \int_{\mathrm{Arc}(R,0,\pi)} f.$$

Now take the limit as $\epsilon \to 0$ and $R \to \infty$. We know (4.3.2) that

$$\lim_{\epsilon\to 0^+, R\to\infty} \int_\epsilon^R \frac{dx}{1+x^2} \;=\; \pi/2.$$

The integrals of f over the two arcs tend to 0, by Examples 4.2.3 and 4.2.4. So the result is that

$$\frac{\pi^2 i}{2} \;=\; 2\int_0^\infty f(x)dx \;+\; \frac{\pi^2 i}{2}, \quad \text{so} \quad \int_0^\infty f(x)dx \;=\; 0.$$

$\square$

4.5.2. Example. *Let $b > 1$. Show that* $\displaystyle\int_0^\infty \frac{dx}{1+x^b} = \frac{\pi}{b\sin\frac{\pi}{b}}.$

Solution. By means of the change of variable $x = e^t$, we find that

$$\int_0^\infty \frac{dx}{1+x^b} \;=\; \int_{-\infty}^\infty \frac{e^t}{1+e^{bt}}\, dt.$$

So this problem has already been solved, in Example 4.4.1. We will now solve it again, by a somewhat different method. The condition $b > 1$ assures convergence, and we know that

$$\int_0^\infty \frac{dx}{1+x^b} \;=\; \lim_{\epsilon\to 0, R\to\infty} \int_\epsilon^R \frac{dx}{1+x^b}.$$

The technique is to select a suitable contour $\gamma(\epsilon, R)$ of which the interval $[\epsilon, R]$ is a part; and a function f which is meromorphic inside and on that contour, and which agrees with $x \to 1/(1+x^b)$ on the interval $[\epsilon, R]$. We will now propose our selections, asking you to be patient until we show you how they work.

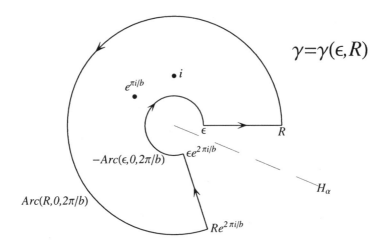

Figure 4.5-1   The contour $\gamma(\epsilon, R)$ used in Example 4.5.2. In this picture, $b = \frac{5}{4}$.

The contour will consist of two line segments and two circular arcs:

$$\gamma(\epsilon, R) = [\epsilon, R] + \mathrm{Arc}(R, 0, 2\pi/b) + [Re^{i2\pi/b}, \epsilon e^{i2\pi/b}] - \mathrm{Arc}(\epsilon, 0, 2\pi/b).$$

Now to specify our choice of f. Choose α so that

$$\frac{2\pi}{b} < \alpha < 2\pi. \tag{2}$$

We will use the logarithm on $V_\alpha \equiv \mathbb{C} \setminus H_\alpha$ given by

$$\log re^{i\theta} = \ln r + i\theta \quad \text{if } r > 0 \text{ and } \alpha < \theta < 2\pi + \alpha. \tag{3}$$

Those choices determine a branch of $z \mapsto z^b$ and hence of f:

$$f(z) = \frac{1}{1 + e^{b\log z}} \quad (z \in V_\alpha).$$

Then f is meromorphic on V_α, and its only pole is at $e^{i\pi/b}$, which is inside the contour $\gamma(\epsilon, R)$ provided $0 < \epsilon < 1 < R$. We write the integral as the sum of its four parts, and then use the Residue Theorem to obtain its value:

$$\int_{\gamma(\epsilon, R)} f = \int_{[\epsilon, R]} f + \int_{\mathrm{Arc}(R, 0, 2\pi/b)} f$$

$$+ \int_{[Re^{i2\pi/b}, \epsilon e^{i2\pi/b}]} f + \int_{-\mathrm{Arc}(\epsilon, 0, 2\pi/b)} f \tag{4}$$

$$= 2\pi i \mathrm{Res}(f, e^{i\pi/b}) = -\frac{2\pi i e^{i\pi/b}}{b}.$$

The choice of the segment $[Re^{i2\pi/b}, \epsilon e^{i2\pi/b}]$ is a good one because the integral of f over it equals a constant times the integral of f over $[\epsilon, R]$, thus:

$$\int_{[Re^{i2\pi/b}, \epsilon e^{i2\pi/b}]} f = -\int_\epsilon^R \frac{e^{i2\pi/b}}{1 + e^{b(\ln x + i2\pi/b)}} dx = -e^{i2\pi/b} \int_{[\epsilon, R]} f. \quad (5)$$

The integrals over the two circular arcs both tend to 0:

$$\left| \int_{\text{Arc}(R,0,2\pi/b)} f \right| < \frac{2\pi R}{b} \frac{1}{R^b - 1} \to 0 \quad \text{as } R \to \infty;$$

$$\left| \int_{\text{Arc}(\epsilon,0,2\pi/b)} f \right| < \frac{2\pi \epsilon}{b} \frac{1}{1 - \epsilon^b} \to 0 \quad \text{as } \epsilon \to 0.$$

When we take the limit of (4) as $\epsilon \to 0$ and $R \to \infty$, we find that

$$(1 - e^{i2\pi/b}) \int_0^\infty \frac{dx}{1 + x^b} = -\frac{2\pi i e^{i\pi/b}}{b},$$

and hence

$$\int_0^\infty \frac{dx}{1 + x^b} = \frac{2\pi i e^{i\pi/b}}{b(e^{i2\pi/b} - 1)} = \frac{2\pi i}{b(e^{i\pi/b} - e^{-i\pi/b})} = \frac{\pi}{b \sin \frac{\pi}{b}}.$$

□

4.5.3. Example. *Show that*

$$\int_0^\infty \frac{dx}{(1 + x^b) x^c} = \frac{\pi}{b \sin \frac{\pi(1-c)}{b}} \quad \text{if } c < 1 < b + c.$$

Solution. By the change of variable $x = e^t$, the result follows from Example 4.4.1, but we will do the problem by the method of Example 4.5.2. The conditions on b and c imply convergence, and

$$\int_0^\infty \frac{dx}{(1 + x^b) x^c} = \lim_{\epsilon \to 0^+, R \to \infty} \int_\epsilon^R \frac{dx}{(1 + x^b) x^c}.$$

Consider first the case when $b > 1$. We can then use the same contour $\gamma = \gamma(\epsilon, R)$ as in Example 4.5.2 and Figure 4.5-1. We let $f(z) = (1 + z^b)^{-1} z^{-c}$ for $z \in V_\alpha$, using the same logarithm as before, given by (3), so that

$$f(re^{i\theta}) = \frac{1}{(1 + e^{b(\ln r + i\theta)}) e^{c(\ln r + i\theta)}} \quad \text{where } -\alpha < \theta < 2\pi - \alpha.$$

The function f is meromorphic on and inside the contour, and the only singularity inside is the pole of order 1 at $e^{i\pi/b}$. We write the integral as the sum of its four parts,

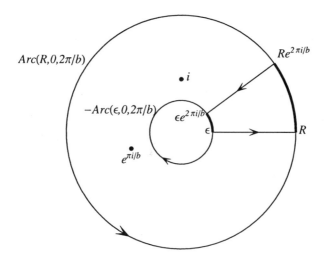

Figure 4.5-2   The contour $\gamma(\epsilon, R)$ for Example 4.5.3, in a case when $b < 1$. It appears to be unsatisfactory.

and then use the Residue Theorem to obtain its value:

$$
\int_{\gamma(\epsilon,R)} f = \int_{[\epsilon,R]} f + \int_{\text{Arc}(R,0,2\pi/b)} f
$$
$$
+ \int_{[Re^{i2\pi/b},\epsilon e^{i2\pi/b}]} f + \int_{-\text{Arc}(\epsilon,0,2\pi/b)} f \tag{6}
$$
$$
= 2\pi i \operatorname{Res}(f, e^{i\pi/b}) = -\frac{2\pi i e^{\pi i(1-c)/b}}{b}.
$$

The integrals of f on the two line segments are related as follows:

$$
\int_{[Re^{i2\pi/b},\epsilon e^{i2\pi/b}]} f = -\int_{\epsilon}^{R} \frac{e^{i2\pi/b}dx}{\left(1+e^{b(\ln x + i2\pi/b)}\right)e^{c(\ln x + i2\pi/b)}}
$$
$$
= -e^{i2\pi(1-c)/b} \int_{[\epsilon,R]} f. \tag{7}
$$

Compare (7) with (5). Take the limit of (6) as $\epsilon \to 0$ and $R \to \infty$. The integral over each of the circular arcs tends to 0, and the desired result follows.

If $b \le 1$, there seem to be difficulties, starting with the impossibility of (2). The trouble shows up in various ways. In Figure 4.5-2, which shows γ in the case when $b = 10/11$, γ overlaps itself. What's worse, if $b < 1/2$, then γ goes around the pole more than once. The difficulty seems to be that the Residue Theorem does not apply, because there is no well defined f meromorphic inside and on γ to apply it to.

One may escape from the difficulty by choosing $n > 1/b$ and making the change of variable $x = t^n$, thereby reducing the problem to the case $b \ge 1$. But let's pretend we don't notice that possibility.

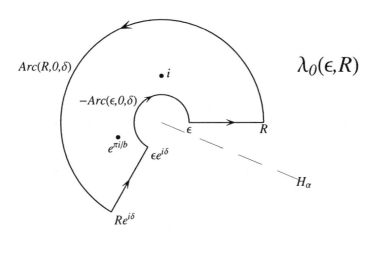

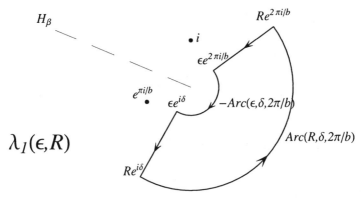

Figure 4.5-3   Example 4.5.3 uses the contours $\lambda_0(\epsilon, R)$ and $\lambda_1(\epsilon, R)$, and two functions f_0 and f_1 which are the same except that they involve two different logarithms. The two agree on the segment $[\epsilon e^{i\delta}, Re^{i\delta}]$. Section 4.6 will describe a more elegant way to proceed in such problems.

Here is a way to overcome the difficulty for the case when $1/2 < b \leq 1$. Choose δ and α such that $\pi/b < \delta < 2\pi$ and $2\pi - \delta < \alpha < 0$. See the top picture in Figure 4.5-3. Let f_0 be the branch of $z \mapsto 1/((1 + z^b)z^c)$ defined on V_α using this logarithm:

$$\log re^{i\theta} = \ln r + i\theta \quad \text{for } \alpha < \theta < 2\pi + \alpha.$$

Then f_0 is meromorphic on and inside the contour $\lambda_0 = \lambda_0(\epsilon, R)$, and the Residue Theorem gives

$$\int_{\lambda_0} f_0 = 2\pi i \frac{e^{i(1-c)\pi/b}}{b}.$$

Choose β such that $0 < \beta < \pi/b$. See the bottom picture in Figure 4.5-3. Let f_1 be the branch of $z \mapsto 1/((1 + z^b)z^c)$ defined on V_β using this logarithm:

$$\log re^{i\theta} = \ln r + i\theta \quad \text{for } \beta < \theta < 2\pi + \beta.$$

Then f_1 is holomorphic inside and on the contour $\lambda_1 = \lambda_1(\epsilon, R)$, so

$$\int_{\lambda_1} f_1 = 0.$$

Notice that

$$\int_{[Re^{i2\pi/b}, \epsilon e^{i2\pi/b}]} f_1 = -e^{i2\pi(1-c)/b} \int_\epsilon^R \frac{dx}{(1 + x^b)x^c}.$$

Since $f_1 = f_0$ on $[\epsilon e^{i\delta}, Re^{i\delta}]$,

$$\int_{[Re^{i\delta}, \epsilon e^{i\delta}]} f_0 + \int_{[\epsilon e^{i\delta}, Re^{i\delta}]} f_1 = 0.$$

Therefore

$$\int_{\lambda_0(\epsilon, R)} f_0 + \int_{\lambda_1(\epsilon, R)} f_1 = 72\pi i \frac{e^{i(1-c)\pi/b}}{b}$$

$$= (1 - e^{i2\pi(1-c)/b}) \int_\epsilon^R \frac{dx}{(1 + x^b)x^c}$$

$$+ \int_{-\text{Arc}(\epsilon, 0, \delta)} f_0 + \int_{-\text{Arc}(\epsilon, \delta, 2\pi/b)} f_1$$

$$+ \int_{\text{Arc}(R, 0, \delta)} f_0 + \int_{\text{Arc}(R, \delta, 2\pi/b)} f_1.$$

Take the limit as $\epsilon \to 0$ and $R \to \infty$, and the result follows. $\qquad\square$

Every case with $0 < b < 1$ can be dealt with similarly, using some finite number of different contours and logarithms. But that would be tedious. The inelegance of it all is due to the fact that we are using the logarithm in an unnatural setting. In the next Section we will present a simpler technique using the idea of a Riemann surface.

Exercises

1. Generalize Example 4.5.1 by showing that for every $a > 0$,

$$\int_0^\infty \frac{\ln x}{a^2 + x^2} dx = \frac{\pi \ln a}{2a}.$$

2. For Example 4.5.1, the clever calculus student does not need the Residue Calculus. Find the value of $\int_0^\infty \frac{\ln x}{1 + x^2} dx$ by making the substitution $x = 1/u$.

3. Show that $\int_0^\infty \frac{\ln x}{(1 + x^2)^2} dx = -\frac{\pi}{4}$.

4. Show that $\int_0^\infty \frac{dx}{\sqrt{x}(x^2 + 1)} = \frac{\pi}{\sqrt{2}}$ and that $\int_0^\infty \frac{dx}{\sqrt{x}(x^2 + 4)} = \frac{\pi}{4}$.

5. Evaluate $\int_0^\infty \frac{\ln^2 x}{1 + x^2} dx$.

6. Evaluate $\int_1^\infty \frac{dx}{x\sqrt{x^2 - 1}}$.

7. The **beta function** is the function of two variables given by

$$B(p, q) := \int_0^1 t^{p-1}(1 - t)^{q-1} dt \quad (p > 0, q > 0).$$

Carry out the substitution $t = \dfrac{1}{x + 1}$ and show that $B(p, 1 - p) = \dfrac{\pi}{\sin p\pi}$.

8. Let $a > 0, b > 0$. Show that $\displaystyle\int_0^\infty \frac{\ln(ax)}{x^2 + b^2} dx = \frac{\pi \ln(ab)}{2b}$.

Help on Selected Exercises

3. You may wish to ask a computer to plot the function on the interval $[0.5, 2]$, say, and compare the graph with that of the function in Example 4.5.1. This convergent improper integral equals $\displaystyle\lim_{\epsilon \to 0^+, R \to \infty} \int_\epsilon^R g(x) dx$. If you use the method of Example 4.5.1, you will obtain the intermediate result

$$2 \int_0^\infty g(x) dx + \pi i \int_0^\infty \frac{dx}{(1 + x^2)^2} = 2\pi i \text{Res}(g, i).$$

5. The value is $\pi^3/8$. One may use the contour $\gamma(\epsilon, R)$ as in 4.5.1.

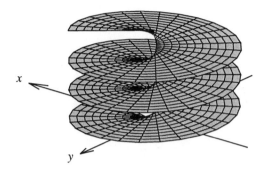

Figure 4.6-1   A partial picture of the Riemann surface of the logarithm

6. Sometimes freshman calculus methods are best. Make the change of variable $x = \sec\theta$, and the integral becomes $\int_0^{\pi/2} d\theta$.

8. One may use the contour $\gamma(\epsilon, R)$ as in 4.5.1 for $0 < \epsilon < b < R$. A byproduct of the procedure is the fact that $\int_0^\infty \dfrac{dx}{x^2 + b^2} = \dfrac{\pi}{2b}$.

4.6   INTEGRATION ON A RIEMANN SURFACE

We indicated how to deal with the integral of Example 4.5.3,

$$\int_0^\infty \frac{dx}{(1 + x^b)x^c} \qquad (c < 1 < b + c),$$

by piecing together a number of contours, each contained within the domain of a selected branch of the logarithm. The smaller b gets, the more branches must be used. As we pointed out, the integral converts into that of Example 4.4.1. We hope you noticed that the two procedures correspond closely; and that the same work is being done in each. But it is easier to write it down for the integrand involving the exponential than for the one involving the logarithm, at least by the method we used.Riemann surface, integration on a

In 1851, Riemann set forth in his Inaugural Dissertation [58] a new way of looking at the logarithm, and at other holomorphic functions with more than one branch. Confined to the plane, Example 4.5.3 engages in a good bit of tedium to consider one branch at a time of the logarithm, and to use various selected sets $V_\alpha \subset \mathbb{C}'$ as domains. Riemann saw this procedure as awkward and unnatural. Riemann proposed instead, for every such function, to construct a unique and natural geometric object to serve as its domain, on which it is one-to-one. Such an object is now called a Riemann surface. In this book we will offer just one example of a Riemann surfaces and thereby give a hint of the general theory. An excellent elementary text on the subject is [30]. Detlef Laugwitz's article [37] provides a historical account of Riemann's dissertation and its influence. For an advanced discussion, see [23, Chapter XVIII].

The following is a concrete description of the Riemann surface Σ of the logarithm. Consider an infinite set of copies of $\mathbb{C}'$ ("sheets") stacked one above another, one sheet assigned to each integer n. Cut every sheet along the positive real axis, so that quadrant IV of sheet n is no longer attached to quadrant I of sheet n. Next, join the top edge of quadrant IV in sheet n to the bottom edge of quadrant I in sheet $n + 1$. The resulting object is Σ. A partial picture appears in Figure 4.6-1.

Let $r > 0$. The mapping $t \to re^{it}$ maps $\mathbb{R}$ one-to-one into Σ. When t starts at 0, the point re^{it} is on sheet 0; as t increases, the point moves around a circle in sheet 0 until t reaches 2π; then the point is on sheet 1; and so forth.

In other words, Σ is a two-dimensional surface imbedded in 3-space and may be parametrized by the one-to-one mapping

$$(r, t) \mapsto (r\cos t, r\sin t, t) \qquad (r > 0, t \in \mathbb{R}).$$

The first two coordinates $(r\cos t, r\sin t)$ may be regarded as a nonzero complex number z, and the third coordinate t as merely an index to tell us which sheet z lives on. We will drop the third coordinate and write merely $(r\cos t, r\sin t)$ or re^{it} to denote a point on Σ, allowing the value of t to tell us which sheet the point is on. It is on sheet n when $2\pi n \leq t < 2\pi(n + 1)$.

Riemann's appealing idea was to regard Σ as the natural domain of the logarithm. It is a one-to-one holomorphic mapping from Σ onto $\mathbb{C}$:

$$\log re^{it} = \ln r + it.$$

Its inverse, which maps $\mathbb{C}$ one-to-one onto Σ, is the exponential function.

Riemann's revelation was that complex analysis can be done on a Riemann surface just as well as on $\mathbb{C}$ and its open subsets. For example, Cauchy's Theorem and the Residue Theorem hold on Σ.

4.6.1. Example. *Show that*

$$\int_0^\infty \frac{dx}{(1 + x^b)x^c} = \frac{\pi}{b \sin \frac{\pi(1-c)}{b}} \qquad \text{if } c < 1 < b + c.$$

Solution. In what follows, we repeat the work of 4.5.3, but it becomes simpler because we work in the setting of Σ. The function $z \mapsto \log z$, and hence also the functions $z \mapsto z^b$ and $z \mapsto z^c$, are well defined functions on Σ. Consider the function given by

$$f(z) := \frac{1}{(1 + z^b)z^c} = \frac{1}{(1 + e^{b \log z})e^{c \log z}} \quad \text{for } z \in \Sigma \qquad (c < 1 < b + c)$$

$$\text{or:} \quad f(re^{it}) := \frac{1}{(1 + r^b e^{itb})r^c e^{itc}}.$$

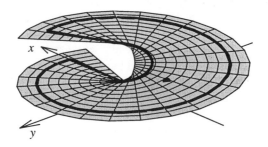

Figure 4.6-2 A contour $\lambda(\epsilon, R)$ on the Riemann surface of the logarithm, going around one pole of a function f

For $R > 1 > \epsilon > 0$, let $\lambda(\epsilon, R)$ be the contour on Σ that consists of two line segments and two arcs:

$$\lambda(\epsilon, R) = [0, R] + \text{Arc}(R, 0, 2\pi/b) + [Re^{2\pi i/b}, \epsilon e^{2\pi i/b}] - \text{Arc}(\epsilon, 0, 2\pi/b).$$

The contour is pictured in Figure 4.6-2 for the case when $b = 10/11$, so that sheet 0 and sheet 1 are involved. The smaller b is, the more sheets of Σ the contour travels on. No matter how small b is, there is only one contour to consider, and only one log to deal with instead of many branches.

The singularities of f are poles of order 1 which occur when $z^b = -1$, and there are infinitely many on Σ. There is precisely one pole inside the contour $\lambda(\epsilon, R)$, namely at $e^{i\pi/b}$. So we may use the Residue Theorem to write down the value of the integral; and we may also write it as the sum of its four parts (two of them as pullbacks):

$$\int_{\lambda(\epsilon, R)} f = 2\pi i \text{Res}(f, e^{i\pi/b}) = -\frac{2\pi i e^{\pi i(1-c)/b}}{b}$$

$$= \int_{\text{Arc}(R, 0, 2\pi/b)} f + \int_\epsilon^R \frac{dx}{(1 + x^b)x^c}$$

$$- \int_{\text{Arc}(\epsilon, 0, 2\pi/b)} f - e^{i2\pi(1-c)/b} \int_\epsilon^R \frac{dx}{(1 + x^b)x^c}.$$

The first and third parts tend to zero as $R \to \infty$ and $\epsilon \to 0^+$, respectively. Thus we arrive again at the conclusion

$$\lim_{\epsilon \to 0^+, R \to \infty} \int_{\lambda(\epsilon, R)} f = -\frac{2\pi i e^{\pi i(1-c)/b}}{b} = \left(1 - e^{i2\pi(1-c)/b}\right) \int_0^\infty \frac{dx}{(1 + x^b)x^c},$$

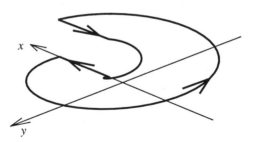

Figure 4.6-3 $\alpha(\epsilon, R)$, another contour on the Riemann surface of the logarithm, for use in Examples 4.6.2 and 4.6.4. The surface itself is shown in Figure 4.6-2, but is not shown here.

which gives the result.   □

The next Proposition provides a way to evaluate integrals on $[0, \infty)$ of rational functions which have poles on the negative reals. Notice that the hypothesis of Proposition 4.3.1 ruled out such problems; but that we dealt with two of them in Example 4.3.5 and in Exercise 14 of Section 4.3, by different means.

4.6.2. Proposition. *Let P and Q be polynomials with no common zeros, such that Q has no zeros on the nonnegative real axis, and such that the degree of Q exceeds that of P by at least 2. Let*

$$f(z) := \frac{P(z)\log z}{Q(z)}.$$

Let the zeros of Q be denoted by $a_k := r_k e^{it_k}$ for $k = 1, 2, \cdots, n$, where $r_k > 0$ and $0 < t_k < 2\pi$. Then

$$\int_0^\infty \frac{P(x)}{Q(x)} dx = -\sum_{k=0}^{n} \mathrm{Res}(f, a_k). \tag{1}$$

In particular, if the zeros of Q are all of order 1, then

$$\int_0^\infty \frac{P(x)}{Q(x)} dx = -\sum_{k=0}^{n} (\ln r_k + it_k) \frac{P(a_k)}{Q'(a_k)}. \tag{2}$$

Proof. Notice that the integral (1) being evaluated does not involve a logarithm; the log is introduced as part of the device that provides the evaluation. Consider the contour on Σ, shown in Figure 4.6-3, given by

$$\alpha(\epsilon, R) = [\epsilon, R] + \mathrm{Arc}(R, 0, 2\pi) + [Re^{i2\pi}, \epsilon e^{i2\pi}] - \mathrm{Arc}(\epsilon, 0, 2\pi).$$

Except for the third part, it lies in sheet 0. The third part, $[Re^{i2\pi}, \epsilon e^{i2\pi}]$, is the copy of the segment $[\epsilon, R]$, traversed backwards, that lies in sheet 1. The function f is meromorphic on Σ and has n poles on every sheet. Require ϵ to be sufficiently small and R sufficiently large so that contour winds once around each of the n poles on sheet 0. We may write the integral's value according to the Residue Theorem, and then as the sum of its four parts, two of them as pullbacks, as follows:

$$\int_{\alpha(\epsilon,R)} f = 2\pi i \sum_{k=0}^{n} \text{Res}(f, a_k) \tag{3}$$

$$= \int_{\epsilon}^{R} \frac{\ln x P(x)}{Q(x)} dx + \int_{\text{Arc}(R,0,2\pi)} f$$

$$- \int_{\epsilon}^{R} \frac{(\ln x + i2\pi)P(x)}{Q(x)} dx + \int_{\text{Arc}(\epsilon,0,2\pi)} f.$$

As may be deduced from Examples 4.2.3 and 4.2.4, the second and fourth integrals tend to zero as $\epsilon \to 0^+$ and $R \to \infty$, respectively. It follows that

$$\lim_{\epsilon \to 0^+, R \to \infty} \int_{\alpha(\epsilon,R)} f = -i2\pi \int_{0}^{\infty} \frac{P(x)}{Q(x)} dx. \tag{4}$$

Compare (3) and (4), and (1) follows. If the pole at $a_k = re^{it_k}$ is of order 1, then by 3.3.15,

$$\text{Res}(f, a_k) = \frac{\log(r_k e^{it_k})P(a_k)}{Q'(a_k)} = (\ln r_k + it_k)\frac{P(a_k)}{Q'(a_k)},$$

and (2) follows. □

4.6.3. Example. *Using 4.6.2, show that* $\displaystyle\int_{0}^{\infty} \frac{dx}{x^2 + 6x + 8} = \frac{\ln 2}{2}.$

Solution. The poles of $z \mapsto (z^2 + 6z + 8)^{-1}$ occur at -4 and -2. Both are of order 1. The residues are $-1/2$ and $1/2$, respectively. By (2), then, the integral equals

$$- \left((\ln 4 + \pi i)\left(-\frac{1}{2}\right) + (\ln 2 + \pi i)\left(\frac{1}{2}\right) \right) = \frac{\ln 2}{2}.$$

See [20, Example 4.6.5] for a different presentation of this method. The integral appeared also in Exercise 14 of Section 4.3, on page 281. □

A Mellin Transforms

The function g given by

$$g(a) = \int_{0}^{\infty} x^{a-1} f(x) dx \tag{5}$$

is the **Mellin transform** of f. Often, one practical way to find a value of g is to integrate over a contour on Σ, very much as in Proposition 4.6.2. The next result deals with cases when f is a rational function.

4.6.4. Proposition. *Let $a > 0$, not an integer. Let P and Q be polynomials with no common zeros, such that $Q(x) \neq 0$ when $x > 0$. Let*

$$h(z) = \frac{z^{a-1}P(z)}{Q(z)}.$$

Let p and q be the degrees of P and Q, respectively. If $Q(0) = 0$, let ℓ be the order of the zero; if not, let $\ell = 0$. If $P(0) = 0$, let k be the order of the zero; if not, let $k = 0$. Suppose that

$$a + p - q < 0 \quad and \quad a + k - \ell > 0. \tag{6}$$

Let $a_1, \cdots, a_n$ denote the zeros of Q, excluding 0 if it is one. Then

$$\int_0^\infty h(x)dx = \frac{-\pi}{\sin \pi a} \sum_{k=0}^n \operatorname{Res}(h, a_k). \tag{7}$$

Proof. We again use the Riemann surface Σ, and we integrate h over the contour $\alpha(\epsilon, R)$ shown in Figure 4.6-3. On Σ, the power of $z = re^{it}$ is well defined: $z^{a-1} = e^{(a-1)(\ln r + it)}$. The Residue Theorem gives

$$\int_{\alpha(\epsilon, R)} h = 2\pi i \sum_{k=0}^n \operatorname{Res}(h, a_k)$$

$$= \int_\epsilon^R h(x)dx + \int_{|z|=R} h(z)dz$$

$$- \int_\epsilon^R e^{i2\pi(a-1)}h(x)dx - \int_{|z|=\epsilon} h(z)dz,$$

provided $0 < \epsilon < \min_k |a_k|$ and $\max_k |a_k| < R$. Take the limits as $R \to \infty$ and $\epsilon \to 0$. The conditions (6) suffice to imply that the second and fourth integrals tend to 0, and we find that

$$(1 - e^{i2\pi a}) \int_0^\infty h(x)dx = 2\pi i \sum_{k=0}^n \operatorname{Res}(h, a_k),$$

and (7) follows. □

Exercises

1. Evaluate $\displaystyle\int_0^\infty \frac{dx}{x^2 + 1}$ using 4.6.2.

2. Evaluate $\displaystyle\int_0^\infty \frac{dx}{x^3 + 1}$ using 4.6.2.

3. Evaluate $\displaystyle\int_0^\infty \frac{dx}{(x + 5)^2(x + 1)}$, both by decomposition into partial fractions, and by 4.6.2.

4. Show that the equality $\displaystyle\int_0^\infty \frac{x^{1/3}}{1 + x^2}\,dx = \frac{\pi}{\sqrt{3}}$ can be obtained either from 4.6.4 or from 4.5.3.

5. Evaluate $\displaystyle\int_0^\infty \frac{dt}{t^2 - 2t}$.

6. Let $0 < a < 1$. Show that

$$\int_0^\infty \frac{x^{a-1}}{1 + x}\,dx = \int_0^\infty \frac{x^{-a}}{1 + x}\,dx = \int_0^\infty \frac{x^a}{x(x + 1)}\,dx = \pi \csc \pi a.$$

7. Show that $\displaystyle\int_0^\infty \frac{x^{1/2}}{x^2 + 2x + 1}\,dx = \frac{\pi}{2}$.

8. Show that $\displaystyle\int_0^\infty \frac{x^{a-1}}{x^3 + 1}\,dx = \frac{\pi}{3\sin(\pi a/3)}$ for $0 < a < 3$.

9. Show how the Mellin transform of f, given by (5), is related to the Fourier transform of $t \mapsto f(e^t)$.

Hints for Selected Exercises

2. The poles of $z \mapsto \dfrac{1}{z^3 + 1}$ occur at $e^{i\pi/3}, -1$, and $e^{i5\pi/3}$, and are all of order 1. The residues are $e^{i4\pi/3}/3, 1/3$, and $e^{i2\pi/3}/3$, respectively. By (2), then, the integral equals

$$-\left(\frac{i\pi}{3}\frac{1}{3}e^{i4\pi/3} + i\pi\frac{1}{3} + i\frac{5\pi}{3}\frac{1}{3}e^{i2\pi}3\right) = \frac{2\pi}{3}.$$

3. $\dfrac{5\ln 5 - 4}{80}$.

5. Try the change of variable $t = x + 4$.

6. These equalities result from Example 4.5.3.

9. Change of variable, $x = e^t$.

4.7   THE COMPLEX INVERSION FORMULA FOR THE LAPLACE TRANSFORM

Students usually gain an appreciation for this transform first from introductory courses on differential equations. Two widely used textbooks that cover Laplace transform methods for solving initial value problems in mechanical and electrical vibrations are [3, Chapter 6] and [44, Chapter 7]. This Section is intended primarily for those who have previously learned the basics of Laplace transforms from such courses or such books. Our objective is to introduce the complex inversion formula, which you may especially appreciate if you have used tables to find the inverse Laplace transform. In case the Laplace transform is new to you, we will provide a summary of what you need to know.

For a self-contained treatment of the Laplace transform at a sophisticated mathematical level, see [38, Chapter 8]. For an advanced treatment that is very much concerned with applications, see [19, Chapter 8].

Let $f : \mathbb{R}^+ \to \mathbb{C}$, such that

- f is piecewise continuous on $[0, T]$ for every $T > 0$; and

- f is of **exponential order**, that is, there exist constants $a \in \mathbb{R}$, $K > 0$, and $M > 0$, such that

$$|f(t)| \le K e^{at} \quad \text{for } t \ge M.$$

Then the **Laplace transform** of f is the function $\widetilde{f}$ given by

$$\widetilde{f}(s) := \int_0^\infty e^{-st} f(t) dt, \tag{1}$$

which is defined at least for all real $s > a$. For example, if $f(t) = e^{iat}$, then

$$\widetilde{f}(s) = \int_0^\infty e^{-st} e^{iat} dt \ = \ \frac{1}{s - ia}.$$

It is easy to verify that the Laplace transform is linear; the transform of $c_1 f_1 + c_2 f_2$ is $c_1 \widetilde{f}_1 + c_2 \widetilde{f}_2$.

If $f(t) := \cos at \equiv \frac{1}{2}(e^{iat} + e^{-iat})$, then

$$\widetilde{f}(s) = \frac{1}{2} \left(\frac{1}{s - ia} + \frac{1}{s + ia} \right) = \frac{s}{s^2 + a^2}.$$

If $f(t) := \sin at \equiv \frac{1}{2i}(e^{iat} - e^{-iat})$, then

$$\widetilde{f}(s) = \frac{1}{2i} \left(\frac{1}{s - ia} - \frac{1}{s + ia} \right) = \frac{a}{s^2 + a^2}.$$

The operator $f \mapsto \tilde{f}$ enjoys the important property, which one may prove from (1) using integration by parts, that the transforms of f and of its derivative f' are related as follows:

$$\tilde{f'}(s) = s\tilde{f}(s) - f(0),$$

and thus also

$$\widetilde{f''}(s) = s^2\tilde{f}(s) - sf(0) - f'(0),$$

and so forth for higher order derivatives.

Let f_1 and f_2 be two piecewise continuous functions of exponential order. If $f_1(t) = f_2(t)$ for all but a finite number of values of t, then we regard them as the same; and $\tilde{f}_1 = \tilde{f}_2$. If for some number a, $\tilde{f}_1(s) = \tilde{f}_2(s)$ for all $s > a$, then f_1 and f_2 are the same. In other words, the Laplace transform operator is one-to-one; $\tilde{f}$ determines f.

The usefulness of the Laplace transform is as follows. One can often change a differential equation in an unknown function y into an equivalent algebraic equation in $\tilde{y}$. One can then solve for $\tilde{y}$. At that point, the problem is to find y when we know $\tilde{y}$, which means to find the inverse Laplace transform of $\tilde{y}$.

Here is an example, taken from [3, p. 301]. Consider the initial value problem

$$y'' + y = \sin 2t, \quad y(0) = 2, \quad y'(0) = 1, \tag{2}$$

which represents an undamped vibrating system with a sinusoidal driving force. Take the Laplace transform of that equation, obtaining

$$s^2\tilde{y}(s) - 2s - 1 + \tilde{y}(s) = \frac{2}{s^2 + 4}.$$

Solve for $\tilde{y}$:

$$\tilde{y}(s) = \frac{2s^3 + s^2 + 8s + 6}{(s^2 + 1)(s^2 + 4)} = \frac{2s}{s^2 + 1} + \frac{5/3}{s^2 + 1} - \frac{2/3}{s^2 + 4}, \tag{3}$$

where the last expression is the decomposition into partial fractions. We have identified, not the solution, but the Laplace transform of the solution. In this case, by studying the information above about the transforms of $t \mapsto \cos at$ and $t \mapsto \sin at$, we can recognize (3) as the transform of this function:

$$y(t) = 2\cos t + \frac{5}{3}\sin t - \frac{1}{3}\sin 2t.$$

Thus (2) is solved.

We will now explain how the Residue Calculus can be used, in many cases, to find the inverse Laplace transform—to find f when $\tilde{f}$ is known. We first point out that the integral (1) is convergent and $\tilde{f}(s)$ is defined, not merely for all real $s > a$, but for all complex $s = u + iv$ with $u > a$. Furthermore, $\widetilde{f}$ is holomorphic on the half plane $\{u + iv \mid u > a\}$; its derivative is given by

$$\frac{d\widetilde{f}}{ds}(s) = \int_0^\infty e^{-st}(-tf(t))dt,$$

which, by the way, is the Laplace transform of $t \mapsto -tf(t)$.

So far in this book, we have not used "s" to stand for a complex variable. We do so here because it's often the symbol for a point in the domain of the Laplace transform.

Let's say that F is the Laplace transform of f; we know F; and we wish to find f. Then F will be a holomorphic function at least on some right-half plane $u > c$. In many cases, F extends to a holomorphic function on $\mathbb{C}$ except for a finite number of singularities. The following Theorem applies in some of those cases, including the case when F is a rational function P/Q and the degree of Q exceeeds that of P. For example, the function (3) is meromorphic with poles only at $\pm i, \pm 2i$.

4.7.1. Theorem. *Let F be holomorphic on $\mathbb{C} \setminus \{a_1, \cdots, a_n\}$, such that*

$$\sup_{|s| \geq S} |F(s)| \to 0 \quad as \ \ S \to \infty. \tag{4}$$

Let $a = \max_k \operatorname{Re} a_k$.

(i) *Then $F(s) = \widetilde{f}(s)$ for $\operatorname{Re} s > a$, where f is given by the complex inversion formula:*

$$f(t) := \sum_{k=0}^n \operatorname{Res}(e^{st} F(s), a_k) \quad for \ \ t \geq 0. \tag{5}$$

(ii) *For every real number $\alpha > a$, the inverse transform can be recovered from the values of F on the vertical line $u = \alpha$ by the Bromwich integral:*

$$f(t) = \frac{1}{2\pi} \int_{-\infty}^{+\infty} e^{(\alpha + iv)t} F(\alpha + iv)dv, \tag{6}$$

often written in this manner:

$$f(t) = \frac{1}{2\pi i} \int_{\alpha - i\infty}^{\alpha + i\infty} e^{st} F(s)ds. \tag{7}$$

Proof. Fix s, with $\operatorname{Re} s > a$. Select a real number α such that $a < \alpha < \operatorname{Re} s$. Let R be sufficiently large so that all the singularities a_k of F are inside the disk $D(\alpha, R)$ and thus of course inside its left half; and so that s is also inside that disk and thus of course inside its right half.

See Figure 4.7-1. The circle $|\zeta - \alpha| = R$ is the sum of the two contours

$$\gamma_R := [\alpha - iR, \alpha + iR] + \operatorname{Arc}_\alpha\left(R, \frac{\pi}{2}, \frac{3\pi}{2}\right) \quad \text{and}$$

$$\rho_R := \operatorname{Arc}_\alpha\left(R, -\frac{\pi}{2}, \frac{\pi}{2}\right) + [\alpha + iR, \alpha - iR].$$

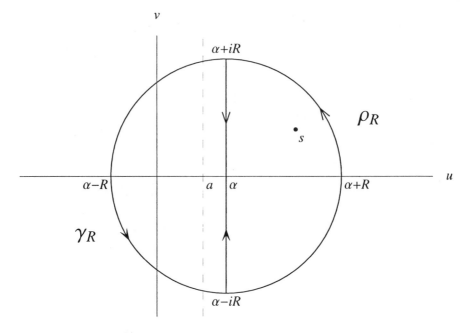

Figure 4.7-1   The circle $|z - \alpha| = R$, used in the proof of 4.7.1, is the sum of two paths γ_R and ρ_R, each of which is the boundary of a half-disk.

Thus

$$\int_{|\zeta - \alpha| = R} \frac{F(\zeta)}{\zeta - s} d\zeta = \int_{\rho_R} \frac{F(\zeta)}{\zeta - s} d\zeta + \int_{\gamma_R} \frac{F(\zeta)}{\zeta - s} d\zeta. \tag{8}$$

To prove part (i) of the Theorem, it suffices to show that the integral on the left tends to 0 as $R \to \infty$, and that the integrals on the right equal $2\pi i F(s)$ and $-2\pi i \tilde{f}(s)$, respectively (independently of R).

The integral on the left may be estimated using 2.5.5:

$$\left| \int_{|\zeta - \alpha| = R} \frac{F(\zeta)}{\zeta - s} d\zeta \right| \leq 2\pi R \max_{|\zeta - \alpha| = R} \left| \frac{F(\zeta)}{\zeta - s} \right|$$

$$\leq \frac{2\pi R \max_{|\zeta - \alpha| = R} |F(\zeta)|}{(R - |s|)},$$

which by (4) tends to 0 as $R \to \infty$.

Since F is holomorphic on and inside ρ_R, Cauchy's Theorem gives the value of the first integral on the right of (8):

$$2\pi i F(s) = \int_{\rho_R} \frac{F(\zeta)}{\zeta - s} d\zeta.$$

To evaluate the last integral on the right of (8), several steps are required. First, we apply the Residue Theorem to the function $\zeta \mapsto e^{\zeta t}F(\zeta)$, which for every t is, like F, holomorphic on $\mathbb{C} \setminus \{a_1, \cdots, a_n\}$. By the Residue Theorem and the definition of f, we obtain that

$$f(t) = \frac{1}{2\pi i} \int_{\gamma_R} e^{\zeta t}F(\zeta)d\zeta. \tag{9}$$

Next we use the definition of the Laplace transform of f:

$$\tilde{f}(s) = \lim_{T \to \infty} \int_0^T e^{-st}f(t)dt. \tag{10}$$

For each T,

$$\int_0^T e^{-st}f(t)dt = \int_0^T e^{-st}\left(\frac{1}{2\pi i}\int_{\gamma_R} e^{\zeta t}F(\zeta)d\zeta\right)dt$$

$$= \frac{1}{2\pi i}\int_{\gamma_R} F(\zeta)\left(\int_0^T e^{(\zeta-s)t}dt\right)d\zeta$$

$$= \frac{1}{2\pi i}\int_{\gamma_R} F(\zeta)\frac{e^{(\zeta-s)T}-1}{\zeta-s}d\zeta.$$

As $T \to \infty$, the quotient that appears in the last integrand converges to $\dfrac{-1}{\zeta-s}$ uniformly for ζ on γ_R, because $\mathrm{Re}\,(\zeta-s) \leq \alpha - \mathrm{Re}\,s$ for ζ on γ_R. By way of (10), therefore, we have found that

$$2\pi i\tilde{f}(s) = -\int_{\gamma_R} \frac{F(\zeta)}{\zeta-s}d\zeta.$$

Part (i) is proved. To prove part (ii), consider (9) written out in full:

$$f(t) = \frac{1}{2\pi i}\int_{[\alpha-iR,\alpha+iR]} e^{\zeta t}F(\zeta)d\zeta + \underbrace{\frac{1}{2\pi i}\int_{\mathrm{Arc}_\alpha(R,\pi/2,3\pi/2)} e^{\zeta t}F(\zeta)d\zeta}_{=:I(R)}.$$

Take the limit as $R \to \infty$, and (6) follows as soon as we show that $I(R) \to 0$. We take its pullback and make an estimate as follows:

$$|I(R)| = \left|\frac{1}{2\pi}\int_{\pi/2}^{3\pi/2} e^{\alpha+Re^{it}}F(\alpha+Re^{it})Rie^{it}dt\right|$$

$$\leq \frac{Re^\alpha}{2\pi}\max_t|F(\alpha+Re^{it})|\int_{\pi/2}^{3\pi/2} e^{R\cos t}dt.$$

Since (recalling the technique for 4.2.1)

$$\int_{\pi/2}^{3\pi/2} e^{R\cos t}\,dt \;=\; 2\int_{0}^{\pi/2} e^{-R\sin t}\,dt \;\leq\; \frac{2\pi}{R}(1 - e^{-R}),$$

and in view of (4), the integral $I(R)$ indeed tends to 0 as $R \to \infty$. □

4.7.2. Example. *Find the inverse Laplace transform of $F(s) = \dfrac{s}{s^2 + 1}$.*

Solution. Whenever the transform F is the ratio P/Q of two polynomials and the degree of Q is at least 1 plus the degree of P, Theorem 4.7.1 applies. In this case, the poles of F are at $\pm i$. The needed residues are easy to find:

$$\operatorname{Res}\left(\frac{se^{ts}}{s^2 + 1}, \pm i\right) \;=\; \frac{se^{ts}}{2s}\bigg|_{s=\pm i} \;=\; \frac{e^{\pm it}}{2}.$$

By (5), then, $f(t) = \frac{1}{2}(e^{it} + e^{-it}) = \cos t$. □

4.7.3. Example. *Find the inverse Laplace transform of*

$$F(s) := \frac{s}{(s^2 + s - 2)(s + 3)^2}.$$

Solution. Theorem 4.7.1 applies. The poles occur at $1, -2,$ and -3, and the residues are as follows:

$$\operatorname{Res}\left(\frac{e^{st}s}{(s-1)(s+2)(s+3)^2}, 1\right) = \frac{e^{t}}{48};$$

$$\operatorname{Res}\left(\frac{e^{st}s}{(s-1)(s+2)(s+3)^2}, -2\right) = \frac{-2e^{-2t}}{-3}; \quad \text{and}$$

$$\operatorname{Res}\left(\frac{e^{st}s}{(s-1)(s+2)(s+3)^2}, -3\right) = \frac{d}{ds}\frac{e^{st}t}{s^2 + s - 2}\bigg|_{s=-3}$$

$$= -\frac{11}{16}e^{-3t} - \frac{3}{4}te^{-3t}.$$

Therefore $f(t) = \dfrac{1}{48}e^{t} + \dfrac{2}{3}e^{-2t} - \dfrac{11}{16}e^{-3t} - \dfrac{3}{4}te^{-3t}.$ □

Inverse transforms like those in the two Examples above can also be found by using a basic table of Laplace transforms. They can also be done quickly using *Mathematica*. Theorem 4.7.1 allows one to deal with other, more difficult cases. A stronger version can be proved to deal with functions F which have infinitely many singularities. There are still other situations that will yield to applications of the Residue Theorem.

For example, Gonzàlez-Valasco [19, Example 8.17, pp. 289–290] solves in detail the following boundary value problem. One seeks a time-dependent temperature distribution $u(x,t)(x \geq 0, t \geq 0)$ for a semi-infinite bar with initial temperature T.

The end of the bar at $x = 0$ loses heat by convection to the surrounding medium, assumed to be at zero temperature. He shows that for each x, the Laplace transform of the solution $t \mapsto u(x, t)$ is given, for certain constants k and h, by

$$U(x, s) = \frac{T}{s} - \frac{T}{s}e^{-x\sqrt{s/k}} + \frac{T}{\sqrt{sk}(\sqrt{s/k} + h)}e^{-x\sqrt{s/k}}.$$

The difficulty is in finding the inverse Laplace transform for the last summand. The task is done using the Residue Theorem and a contour made up of suitable arcs and segments.

Exercises

1. In each case, show in detail how to use Theorem 4.7.1 to find the inverse Laplace transform of F, where

   (a) $F(s) := \dfrac{2s^3 + s^2 + 8s + 6}{(s^2 + 1)(s^2 + 4)}$.

   (b) $F(s) := \dfrac{s^2}{s^3 - 1}$.

   (c) $F(s) := \dfrac{s + 1}{(s + 1)^2 + \pi^2}$.

2. Let $c > 0$. Evaluate $\mathrm{Res}\left(\dfrac{e^{st}e^{-ct}}{s}, 0\right)$. Does the function $F(s) := \dfrac{e^{-cs}}{s}$ satisfy the hypotheses of Theorem 4.7.1?

3. Find the inverse Laplace transform of $F(s) := \dfrac{1}{s(e^s + 1)}$.

Hints for Selected Exercises

1. (a) See (3) and Example 4.7.2.   (b) $e^{-t}\cos \pi t$.

2. No. Note that F is the Laplace transform of a Heaviside step function:

$$u_c(t) := \begin{cases} 1 & \text{if } t \geq c, \\ 0 & \text{if } t < c. \end{cases}$$

3. A modified version of Theorem 4.7.1 is needed. Instead of using the contours ρ_R, γ_R for all sufficiently large R, use a sequence of values of R such that the contours avoid the singularities of F.

5

Boundary Value Problems

The title of this chapter could be understood to promise a vast array of subject matter. In fact, though, the chapter offers only an introduction to the basics of two-dimensional problems, of the kind that yield to the conformal mapping and Poisson integral methods. There will be some students whose already-well-formed curiosity will take them beyond what we do here, and who need to be advised about further topics and other sources. We offer the following comments especially for them.

- Many conformal mapping problems can be solved by hand, given a knowledge of the basics and a bit of practice. Developing such a skill is somewhat like learning to use an integral table, while constructing a partial one in your mind and in your notes. Accordingly, as you read the chapter, you may wish to build yourself a partial dictionary of conformal mappings.

- Available reference works on conformal mappings include the inexpensive [34], a dictionary compiled for the British Admiralty during the years 1944–1948. A more modern work, a "manual on the visualization of a conformal mapping on a display," is [28], which includes historical remarks, a summary of the theory, a large number of excellent computer-generated graphics, and a computer program called CONFORM on a floppy disk.

- There is an extensive literature on computational methods for generating conformal mappings. One starting point is the book [36], written for a two-semester graduate course for students in mathematics and engineering. It calls for a background in complex analysis and numerical analysis, as well as a working knowledge of *Mathematica*.

- If you want merely to see additional and different examples on the same level as we present here, then you might consult other elementary complex analysis textbooks. Three that are currently popular are [7], [39], and [59].

- As we said at the outset, the chapter title connotes a much broader field of study than just two-dimensional problems susceptible to the methods of this chapter, and could be understood to point in several other directions. We mention a small sample, [19] and [24].

5.1 EXAMPLES

A Easy Problems

We consider problems of this kind: Given an open connected set O, find a harmonic function $u : O \to \mathbb{R}$ that behaves in some specified manner on the boundary. In some cases, if you search the list of harmonic functions that you are familiar with, you will easily find one that does what is required.

In each of the Examples, the physics of the situation entails assumptions that guarantee the uniqueness of the solution. A mathematical result will be described in Section 5.6, to the effect that there is exactly one bounded harmonic function that satisfies the given boundary conditions.

The Examples in this section are stated as temperature distribution problems, although the mathematical models will serve for other applications as well. Some of the problems were previewed in Section 1.1.

5.1.1. Example. *Consider the horizontal strip*

$$O := \{z = x + iy \mid 0 < y < 1\}$$

shown in Figure 5.1-1 as a mathematical model for a long flat strip of heat-conducting material. Its bottom edge is maintained at $0°$, and its top edge at $100°$. Assume that the system has reached a steady state; that is, the temperature distribution is no longer changing with time. Find the function on O that gives the temperature at each point.

Solution. In mathematical terms: We seek a bounded function $u : \overline{O} \to \mathbb{R}$ such that u is continuous on $\overline{O}$ and harmonic on O, and such that

$$u(x,0) = 0 \quad \text{and} \quad u(x,1) = 100 \quad \text{for all } x. \tag{1}$$

It is easy to think of an answer: $u(x,y) := 100y$. A harmonic conjugate of u is given by $v(x,y) := -100x$.

Let's interpret our answer in terms of the physical problem. The level curves of u, the horizontal lines, are the isotherms; for example, the temperature is $50°$ on the line given by $y = 1/2$. The level curves of v are the vertical line segments; they are the paths of heat flow. □

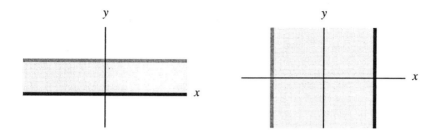

Figure 5.1-1 A horizontal strip O and a vertical strip U, referred to in Examples 5.1.1 and 5.1.3, respectively.

5.1.2. Remark. Every Example in this section is stated as a problem in two real variables, but represents a physics problem in three real variables such that the functions of interest are dependent on only two. For example, we could consider the set O of Example 5.1.1 as a mathematical model for a two-dimensional cross-section of a solid slab, one unit thick, of which one side is kept at 0° and the other at 100°.

5.1.3. Example. *Consider the vertical strip shown in Figure 5.1-1,*

$$U := \{z = x + iy \mid -\tfrac{\pi}{2} < x < \tfrac{\pi}{2}\}.$$

Find a function u which is bounded and continuous on $\overline{U}$ and harmonic on U such that for all y, $u(-\tfrac{\pi}{2},y) = 100$ and $u(\tfrac{\pi}{2},y) = 0$.

Solution. This is a mathematical statement of the same physical problem as in Example 5.1.1, but with a different choice of coordinates. Every function of the form $x \mapsto ax + b$ is harmonic, and we easily find one that satisfies the boundary conditions: $u(x,y) := 50 - \tfrac{100}{\pi}x$. $\qquad\square$

5.1.4. Example. *On the annulus $1 < |z| < 5$, find the steady-state temperature distribution with these boundary values: The temperature is maintained at 100° on the inner circle $|z| = 1$, and at 500° on the outer circle $|z| = 5$.*

Solution. It is convenient to describe the set and the solution in terms of polar coordinates; the set is the annulus $A := \{re^{i\theta} \mid 1 \le r \le 5\}$, and we want

$$u(e^{i\theta}) = 100 \quad\text{and}\quad u(5e^{i\theta}) = 500 \quad\text{for all } \theta.$$

It is reasonable to look for a function that depends only on r, and not on θ. As you may recall, the only harmonic functions u with $u_\theta = 0$ are those of the form $re^{i\theta} \mapsto a \ln r + b$. It is easy to find one that works:

$$u(re^{i\theta}) = 100 + \frac{400}{\ln 5} \ln r.$$

There is no harmonic conjugate on the whole annulus, but for each α, the function given by

$$v(re^{i\theta}) = \frac{400}{\ln 5}\theta \quad (\alpha < \theta < \alpha + 2\pi)$$

will serve on $A \cap V_\alpha$. The isotherms are of course circles with center 0, and the paths of heat flow are radial lines. It is no surprise that the temperature at the middle of the annulus, which is $100 + \frac{400 \ln 3}{\ln 5} \approx 373°$, is greater than the average of the boundary temperatures, since the outer boundary is larger and therefore has more influence. It would be an absent-minded mistake to propose the solution $u(re^{i\theta}) = 100r$, since it is not harmonic!   □

5.1.5. Example. *Consider the upper half-plane $\mathbb{Y}$ as a mathematical model for a large, flat heat-conducting plate, with the real axis representing its boundary. In each case, a function a is given. For each $x \in \mathbb{R}$ at which a is continuous, the temperature at x is maintained at $a(x)°$. Find the steady-state temperature distribution $z \mapsto u(z)$ on $\mathbb{Y}$.*

(i) $a(x) = \begin{cases} 80° & for\ x < 0, \\ 20° & for\ x > 0. \end{cases}$

(ii) $a(x) = \begin{cases} 50° & for\ x < 1, \\ 100° & for\ 1 < x < 2, \\ 10° & for\ 2 < x. \end{cases}$

(iii) $a(x) = \begin{cases} a_0 & for\ x < m_1, \\ a_1 & for\ m_1 < x < m_2, \\ a_2 & for\ m_2 < x. \end{cases}$

(iv) $a(x) = \begin{cases} a_0 & for\ x < m_1; \\ a_k & for\ m_k < x < m_{k+1},\ for\ k = 1, 2, \cdots, K - 1; \\ a_K & for\ m_K < x. \end{cases}$

Solution. These problems are all quite easy, if we make use of the argument $\arg_{1 \mapsto 0}$ on $V_{-\pi/2}$. This function is harmonic on $V_{-\pi/2}$ and in particular on $\mathbb{Y} \setminus \{0\}$. It equals π on the negative reals, 0 on the positive reals. We may use it to devise the solution for part (i):

$$u(z) = \frac{60}{\pi} \arg z + 20.$$

A harmonic conjugate is $v(z) = -\frac{60}{\pi} \ln |z|$. The level curves of u and v are, respectively, rays and circular arcs, and they are the isotherms and paths of heat flow. Make a sketch.

Part (ii) specifies a different constant value on each of the three intervals $I_0 = (-\infty, 1)$, $I_1 = (1, 2)$, and $I_2 = (2, \infty)$. For a solution, consider a function of this form:

$$u(z) = A \frac{\arg(z - 1)}{\pi} + B \frac{\arg(z - 2)}{\pi} + C.$$

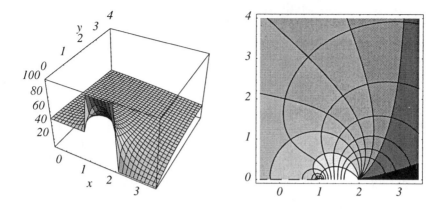

Figure 5.1-2   On the left, the graph over $[-.5, 3.5] \times [0, 4]$ of the temperature distribution $u(x, y)$ found in Example 5.1.5, part (ii). On the right, some isotherms and paths of heat flow. The paths of heat flow begin at points of the hot interval $(1, 2)$ and end at other points of the boundary. The isotherms have one end at the point $x = 2$ on the boundary. Starting just above the cold interval $(2, \infty)$ on that edge, the isotherms in the picture correspond to the temperatures $15°, 30°, 35°, 40°, 45°, 50°, 60°,$ and $75°$.

Its values for real z are easy to compute; it equals $A + B + C$ on I_0, $B + C$ on I_1, and C on I_2. Solving the problem is a matter of finding A, B, and C so that

$$A + B + C = 50, \quad B + C = 100, \quad \text{and} \quad C = 10,$$

which is easy to do. We find the solution to be

$$u(z) = -50 \frac{\arg(z - 1)}{\pi} + 90 \frac{\arg(z - 2)}{\pi} + 10.$$

Notice that u is harmonic on the set

$$P := \{x + iy \mid y > 0; \text{or } y \le 0 \text{ and } x \in I_0 \cup I_1 \cup I_2\},$$

which is to say on the whole plane with two vertical half lines removed. A harmonic conjugate of u on P is

$$v(z) = -\frac{50}{\pi} \ln|z - 1| + \frac{90}{\pi} \ln|z - 2|.$$

Figure 5.1-2 shows some of the isotherms and paths of heat flow.

Part (iii) generalizes part (ii). If $A + B + C = a_0$, $B + C = a_1$, and $C = a_2$, then $A = a_0 - a_1$, $B = a_1 - a_2$, and $C = a_2$. So the solution is

$$u(z) = (a_0 - a_1) \frac{\arg(z - m_1)}{\pi} + (a_1 - a_2) \frac{\arg(z - m_2)}{\pi} + a_2.$$

Part (iv) generalizes part (iii) to the case of $K + 1$ intervals. The solution is

$$u(z) = \frac{1}{\pi} \sum_{k=1}^{K} (a_{k-1} - a_k) \arg(z - m_k) + a_K.$$

$\square$

Boundary value problems do not always specify values of u itself on ∂O as in the examples above. They sometimes specify the rate at which $u(q)$ is changing as q moves across the boundary, at some or all points. To develop notation in which to state precisely this type of condition, let γ be a smooth curve which forms part of ∂O, and let $q \in \gamma$. The notation $\frac{\partial u}{\partial \vec{\mathbf{n}}}(q)$ stands for the directional derivative of u at q in the direction normal to γ and toward the outside of O. If the gradient of u is definable at q, then $\frac{\partial u}{\partial \vec{\mathbf{n}}}(q)$ equals the normal component of the gradient of u, given by the dot product **grad** $u \cdot \vec{\mathbf{n}}$.

A **Dirichlet problem** is a boundary value problem in which the values of u on ∂O are specified. A **Neumann problem** is one in which the values of $\frac{\partial u}{\partial \vec{\mathbf{n}}}$ are specified on ∂O. A **Dirichlet-Neumann problem** is one in which the values of u are specified on part of the boundary, and the values of $\frac{\partial u}{\partial \vec{\mathbf{n}}}$ are specified on another part.

In temperature distribution problems, it is often specified that some part γ of the boundary is insulated, which means physically that there is zero heat flow across γ at each of its points: $\frac{\partial u}{\partial \vec{\mathbf{n}}}(q) = 0$ for $q \in \gamma$. Such a requirement is a **homogeneous Neumann condition**.

5.1.6. Example. *Consider the vertical half-strip*

$$O_2 := \{ q = s + it \mid -\tfrac{\pi}{2} < s < \tfrac{\pi}{2} \text{ and } t > 0 \}$$

shown on the left in Figure 5.1-3 as a mathematical model for a strip of heat-conducting material. The left-hand edge of the strip is maintained at $100°$, the right-hand edge at $0°$. The bottom edge is insulated. Find the function on O_2 that gives the steady-state temperature at each point.

Solution. In mathematical terms: We seek a bounded continuous function u_2 on $\overline{O}_2$, harmonic on O_2, satisfying these conditions:

$$u_2\left(-\frac{\pi}{2}, t\right) = 100 \quad \text{and} \quad u_2\left(\frac{\pi}{2}, t\right) = 0 \quad \text{for all } t > 0; \text{ and}$$

$$\frac{\partial u_2}{\partial t}(s, 0) = 0 \quad \text{for} \quad -\frac{\pi}{2} < s < \frac{\pi}{2}.$$

The solution that we pointed out for Example 5.1.3 works also for this one. The solution u_2 and a harmonic conjugate v_2 are as follows:

$$u_2(s, t) := 50 - \frac{100}{\pi} s, \qquad v_2(s, t) := -\frac{100}{\pi} t.$$

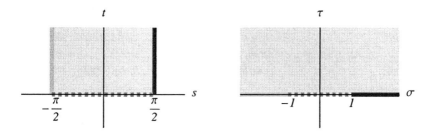

Figure 5.1-3 A vertical half-strip O_2 and the upper half plane Υ, referred to in Examples 5.1.6 and 5.1.7, respectively.

We could write them as functions of one complex variable $q := s + it$ instead of two real variables, as follows:

$$u_2(q) := 50 - \frac{100}{\pi} \operatorname{Re} q, \qquad v_2(q) := -\frac{100}{\pi} \operatorname{Im} q. \tag{2}$$

Make a sketch! The isotherms and paths of heat flow are vertical half-lines and horizontal line segments, respectively. □

B The Conformal Mapping Method

Two open subsets of the extended plane, O_1 and O_2, are **conformally equivalent**, and we write $O_1 \cong O_2$, if there exists a holomorphic function f on O_1 which maps O_1 one-to-one onto O_2. The function f is then a **conformal equivalence** of the two sets. The mapping $z \mapsto z$ is a conformal equivalence; and the inverse of a conformal equivalence is also one; and the composition of two conformal equivalences is also one. Therefore the relation $\cong$ is indeed an equivalence relation among open subsets of $\mathbb{C}$.

Suppose that we seek a harmonic function u_1 to solve a boundary value problem on an open set O_1. In the Examples above, we found solutions with little difficulty. It is not always so easy. Sometimes it is helpful to change variables in the domain. Suppose that we can find a conformal equivalence f from O_1 onto another open set O_2. Suppose furthermore that f extends to a one-to-one continuous map from $\overline{O}_1$ onto $\overline{O}_2$. Then the boundary value problem on O_1 is equivalent to a new boundary value problem on O_2.

The process is useful if the new problem is easier to solve. If u_2 solves the new problem, then $u_1 := u_2 \circ f$ solves the old problem. This assertion rests on the fact, proved in subsection 1.6D, that if u_2 is harmonic on O_2, then u_1 is harmonic on O_1. An easier proof is as follows. If u_2 is harmonic, then u_2 is locally the real part of a holomorphic function g. Since g and f are holomorphic, $g \circ f$ is also holomorphic and hence its real part u_1 is harmonic.

5.1.7. Example. *Let the upper half plane* $\mathbb{Y} := \{\zeta = \sigma + i\tau \mid \tau \geq 0\}$, *with the real axis as its boundary, be a mathematical model for a large flat sheet of heat-conducting material (right-hand picture, Figure 5.1-3). Suppose that the interval* $(-1, 1)$ *is insulated, and the infinite intervals* $(-\infty, -1)$ *and* $(1, \infty)$ *are maintained at* $100°$ *and* $0°$, *respectively. Find the function on* $\mathbb{Y}°$ *that gives the steady-state temperature at each point.*

Solution. In preparation for this Example, we discussed the sine and its inverse in Section 2.2 and Exercises 25 and 26 of that section. A suitable branch of $\sin^{-1}$ provides a change of variable, $\zeta \mapsto q := \sin^{-1} \zeta$, which allows us to see that this problem is equivalent to Example 5.1.6. The mapping $\sin^{-1}$ (the branch that takes $i \sinh 1$ to i) is a one-to-one continuous mapping from $\mathbb{Y}$ onto $\overline{O}_2$, holomorphic on $\mathbb{Y}°$. Furthermore, it preserves the boundary conditions, in the following sense. It maps the hot interval $(-\infty, -1)$ to the hot part of ∂O_2; the insulated interval $(-1, 1)$ to the insulated part of ∂O_2; and the cold interval $(1, \infty)$ to the cold part of ∂O_2. Since the function u_2 given by (2) solves the problem in Example 5.1.6, it follows that $u_2 \circ \sin^{-1}$ solves this one. Thus the solution and a harmonic conjugate are as follows:

$$u_1(\zeta) := u_2(\sin^{-1}(\zeta)), \qquad v_1(\zeta) := v_2(\sin^{-1}(\zeta)). \tag{3}$$

The level curves of v_1, which are the upper halves of certain ellipses, are the paths of heat flow. The level curves of u_1, which are the upper halves of certain hyperbolas, are the isotherms. Such ellipses and hyperbolas have appeared in Figure 1.1-11, page 14, and also in Figure 2.2-3, page 108.

The form of the answer given by (3) is neat and elegant, and you may find it entirely satisfactory for your purposes. But if you wish, you can write u_1 and v_1 in terms of real-valued functions of the real variables σ and τ, using the relations (17) and (18) on page 113 in Section 2.2. The resulting expressions were given as equations (4) and (5) on page 15 in subsection 1.1F (where the variables were called x and y instead of σ and τ). $\qquad\square$

5.1.8. Example. *Let* $Q := \{x + iy \mid x > 0, \ y > 0\}$, *the open first quadrant. The interval* $(1, \infty)$ *on the* x*-axis is maintained at* $0°$. *The interval* $(1, \infty)$ *on the* y*-axis is maintained at* $100°$. *The rest of the boundary is insulated. Find the harmonic function that gives the steady-state temperature distribution in Q.*

Solution. This is the same as Example 1.1.1. See Figure 1.1-4, page 6. The mapping $z \mapsto \zeta := z^2$, which we may write as

$$x + iy \mapsto \sigma + i\tau = (x^2 - y^2) + i2xy,$$

makes this problem equivalent to Example 5.1.7. That is, it is a one-to-one continuous map from $\overline{Q}$ onto $\mathbb{Y}$, holomorphic on Q; and it maps the hot, cold, and insulated parts of ∂Q respectively onto the hot, cold, and insulated parts of the real axis as specified in the previous Example.

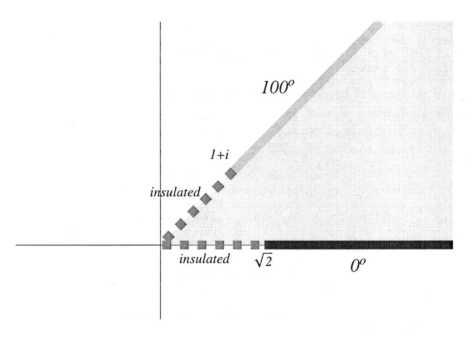

100°

1+i

insulated

insulated $\sqrt{2}$

0°

Figure 5.1-4   A sector of the plane, with mixed boundary conditions; see Exercise 11.

Since u_1, given by equation (3), solves Example 5.1.7, it follows that the temperature distribution for the present Example, and a harmonic conjugate, are given respectively by

$$u(z) := u_1(z^2), \qquad v(z) := v_1(z^2). \tag{4}$$

In the method of conformal mapping, the general idea is to find a change of variable that transforms the problem into one that is easy to solve. Perhaps we should not call Example 5.1.7 "easy to solve," but we took advantage of the fact that we had already solved it.

What if we set out to solve Example 5.1.8, and did not have the solution to 5.1.7 at hand? We would need to be sufficiently familiar with the mappings $z \mapsto \zeta := z^2$ and $\zeta \mapsto q := \sin^{-1} \zeta$ to realize that if we put them together, we obtain a two-step change of variable $z \mapsto \zeta \mapsto q$ which transforms the problem into Example 5.1.6, which is indeed easy to solve. Thus we have the solution and its harmonic conjugate

$$u(z) := u_2(\sin^{-1} z^2), \qquad v(z) := v_2(\sin^{-1} z^2), \tag{5}$$

which are the same functions as in (4). If we wish, we can write u and v in terms of real-valued functions of x and y; see the formulas (2) in Section 1.1, which were used to generate the graphs in Figures 1.1-4 (page 6) and 1.1-5 (page 8).   □

Exercises

1. In Example 5.1.3, find a harmonic conjugate of u, and sketch and label a sample of isotherms and paths of heat flow.

2. In Example 5.1.4, explain fully the statement about "harmonic functions u with $u_\theta = 0$."

3. Find the solution for Example 5.1.4 with these boundary conditions: $u(e^{i\theta}) = 400$ and $u(5e^{i\theta}) = 0$ for all θ.

4. In Example 5.1.5, part (i), sketch the isotherms for 35°, 50°, and 85°. Show that $\lim_{y\to\infty} u(x+iy) = 50$ for every x.

5. In Example 5.1.5, part (ii), find $\lim_{y\to\infty} u(x+iy)$, which is independent of x. What is the significance of this limit? Find its value also in parts (iii) and (iv), in terms of the values a_k.

6. In Figure 5.1-2, the picture on the right, which of the isotherms has a vertical asymptote? Explain the significance.

7. Solve the steady-state temperature distribution problem for the upper half plane with boundary values given by

$$a(x) = \begin{cases} 0^o & \text{if } |x| > 3 \text{ or } -2 < x < 2, \\ 100^o & \text{if } 2 < |x| < 3. \end{cases}$$

Sketch a few isotherms and paths of heat flow. If you lack a computer to help you do so, make an educated guess.

8. Let O be the horizontal strip of Example 5.1.1. Find an unbounded harmonic function on O satisfying (1). Show, then, that the solution of the boundary value problem is not unique when we do not insist on a *bounded* harmonic function.

9. Find a function which is harmonic on the annulus $1 < |z| < 2$ and continuous on its closure, which equals 100 on the inner circle and 0 on the outer circle.

10. Find a function $u(re^{i\theta})$ which is harmonic on the annulus $1 < r < 2$ and continuous on its closure, such that $u(e^{i\theta}) = 100 \cos 2\theta$ and $u(2e^{i\theta}) = 400 \cos 2\theta$ for all θ. Interpreting the question as a temperature-distribution problem, dentify the isotherms and paths of heat flow.

11. Solve the Dirichlet-Neumann problem indicated in Figure 5.1-4.

12. Fix α, $0 < \alpha < 2\pi$. Let

$$\mathcal{U}_\alpha = \{re^{i\theta} \mid r > 0 \text{ and } 0 < \theta < \alpha\}.$$

Exercise 11 dealt with a boundary value problem on $\mathcal{U}_\alpha$ in the special case when $\alpha = \pi/4$. Discuss how to solve the same boundary value problem in the general case.

Help on Selected Exercises

2. Review Section 1.6.

4. Write $u(x + iy)$ as $20 + \frac{60}{\pi} \operatorname{arccot} \frac{x}{y}$ and take the limit as $y \to \infty$.

6. The 30° isotherm has a vertical asymptote. The reason is suggested by Exercise 5.

8. Consider $\operatorname{Im} \cosh \pi z \equiv \sinh \pi x \sin \pi y$.

10. $u(z) = \operatorname{Re} 100z^2$.

5.2 THE MÖBIUS MAPS

In using the conformal mapping method, we become interested in the following general problem: Given connected open sets O_1 and O_2, what one-to-one holomorphic functions f from O_1 onto O_2, if any, can we find? Section 6.3 proves the existence of such functions under certain conditions. But we are concerned here with the ability to produce a satisfactory f "by hand" for various important special cases. We wish to increase your supply of understandable and usable conformal mappings. You already have at your disposal the power functions, the sine, the cosine, and the exponential. This section will present a further class of mappings which are well worth the effort to learn about. A beautiful and much more extensive treatment of them appears in [10, Part I].

If T is a function from $\hat{\mathbb{C}}$ to $\hat{\mathbb{C}}$ given by

$$ T(z) := \frac{az + b}{cz + d} \qquad \text{for } z \in \hat{\mathbb{C}} \tag{1} $$

for some $a, b, c, d \in \mathbb{C}$ such that $ad - bc \neq 0$, then T is a **Möbius map**.[1] If $c \neq 0$, then we interpret (1) to say that $T(\infty) = a/c$ and that $T(-d/c) = \infty$. If $c = 0$, then we understand $T(\infty)$ to be ∞. By requiring $ad - bc$ to be nonzero, we assure that (1) makes sense and gives a non-constant function.

The following are the **basic** Möbius maps:

- $z \mapsto z$, the identity mapping;

- $z \mapsto z + k$, translation by k (for each $k \in \mathbb{C}$);

- $z \mapsto Kz$, magnification by K (for each $K \in \mathbb{C}'$); and

- $z \mapsto \dfrac{1}{z}$, inversion.

---

[1]Other names for a Möbius map: Möbius transformation, linear fractional transformation, fractional linear transformation, homography, homographic transformation, bilinear mapping.

There is some overlap in the listing, since the identity mapping is the same as translation by 0 or magnification by 1.

We will now show that the Möbius map (1), if it is not basic, must be the composition of two or more basic Möbius maps. If $c = 0$, then T is the composition of

$$z \mapsto \frac{a}{d}z \quad \text{and} \quad z \mapsto z + \frac{b}{d}.$$

That is, T is the same as magnification by a/d, followed by translation by b/d. If $c \neq 0$, then

$$T(z) \;=\; \frac{\frac{a}{c}(cz + d) + b - \frac{ad}{c}}{cz + d} \;=\; \frac{a}{c} + \frac{b - \frac{ad}{c}}{cz + d}.$$

Thus T is the composition of

$$z \mapsto cz, \quad z \mapsto z + d, \quad z \mapsto \frac{1}{z},$$

$$z \mapsto \left(b - \frac{ad}{c}\right)z, \quad \text{and} \quad z \mapsto z + \frac{a}{c}.$$

Next we will present some facts about Möbius maps. Then we will develop techniques for using them. The types of problems which we need to know how to solve are as follows:

- Given something that you want a mapping to do, determine whether there is a Möbius map that does it, and if so, find it.

- Given a Möbius map, describe systematically "what it does," that is, which objects are mapped to which.

5.2.1. Proposition. *Every Möbius map T is a one-to-one meromorphic function from $\hat{\mathbb{C}}$ onto $\hat{\mathbb{C}}$. Conversely, every one-to-one meromorphic function on $\hat{\mathbb{C}}$ is a Möbius map.*

Proof. The first statement follows from the fact that T is the composition of basic Möbius maps, each of which is clearly a one-to-one meromorphic functions from $\hat{\mathbb{C}}$ onto $\hat{\mathbb{C}}$. The second statement repeats Propositon 3.4.7. ☐

When we write "**Circle**" with a capital "C," we will mean "a line or a circle." A line in this context is always understood to contain the point ∞.

This section is written assuming you have not studied Section 3.6. If you have done so, you will profit from interpreting the Möbius maps as maps from the Riemann sphere to itself. Note first of all that the Circles are precisely the objects in $\hat{\mathbb{C}}$ that correspond to circles on the Riemann sphere.

5.2.2. Proposition. *Möbius maps preserve Circles. That is, if C is a Circle and T is a Möbius map, then $T[C]$ is a Circle.*

Proof. The result follows from the fact that T is the composition of basic Möbius maps. Translations and magnifications clearly map lines and circles to lines or circles. We established in subsection 2.1H that inversion does so also. □

5.2.3. Proposition. *The composition of two Möbius maps is a Möbius map.*

Proof. This result follows from Proposition 5.2.1, since the composition of two one-to-one meromorphic functions from $\hat{\mathbb{C}}$ onto $\hat{\mathbb{C}}$ must also be one-to-one and meromorphic. Here we will give another proof, one which will help make a connection between Möbius maps and matrices (see Exercises 4-7). Let T be given by (1) and S by

$$S(z) := \frac{a_1 z + b_1}{c_1 z + d_1} \quad \text{for } z \in \hat{\mathbb{C}}.$$

Then $S \circ T$ may be computed to show that it has the form of a Möbius map:

$$S \circ T(z) := S(T(z)) = \frac{a_1 \frac{az+b}{cz+d} + b_1}{c_1 \frac{az+b}{cz+d} + d_1} = \frac{(a_1 a + b_1 c)z + (a_1 b + b_1 d)}{(c_1 a + d_1 c)z + (c_1 b + d_1 d)}.$$

□

5.2.4. Proposition. *The inverse of a Möbius map is a Möbius map.*

Proof. We claim that the inverse of the Möbius map T given by (1) is given by

$$T^{-1}(w) = \frac{-dw + b}{cw - a}. \tag{2}$$

By computation, $T^{-1}(T(z)) = z$ for all z and $T(T^{-1}(w)) = w$ for all w. □

If $T(z) = z$, then z is called a fixed point of T. If T has exactly one fixed point, it is **parabolic**. If it has exactly two fixed points, it **loxodromic**.

5.2.5. Proposition. *If T is a Möbius map other than the identity, then T has at most two fixed points.*

Proof. If T is given by (1), then z is a fixed point of T if and only if

$$\frac{az + b}{cz + d} = z. \tag{3}$$

Case 1, when $c = 0$: Then ∞ is a fixed point; and for $z \in \mathbb{C}$, (3) is equivalent to

$$(a - d)z = b.$$

If $a - d = b = 0$, then T would be the identity. If $a - d \neq 0$, then there is one fixed point in addition to ∞, namely, $b/(a - d)$. If $a - d = 0$ and $b \neq 0$, then ∞ is the only fixed point.

Case 2, when $c \neq 0$: Then (3) is equivalent to

$$z \in \mathbb{C} \quad \text{and} \quad cz^2 + (d - a)z - b = 0.$$

Thus there are one or two fixed points, obtainable by the quadratic formula.   □

5.2.6. Proposition. *If T and U are Möbius maps which agree at three distinct points in $\hat{\mathbb{C}}$: $T(z_k) = U(z_k)$ for $k = 1, 2, 3$, then $T = U$.*

Proof. The map $U^{-1}T$ has three distinct fixed points: $U^{-1}(T(z_k)) = z_k$ for $k = 1, 2, 3$. By 5.2.5 it must be the identity: $U^{-1} \circ T = I$. Therefore $U \circ U^{-1} \circ T = U \circ I$, that is, $T = U$.   □

We will write

$$T \; : \; z_1, z_2, z_3 \;\rightarrow\; w_1, w_2, w_3 \tag{4}$$

to mean that $T(z_k) = w_k$ for $k = 1, 2, 3$, where we assume always that $\{z_1, z_2, z_3\}$ and $\{w_1, w_2, w_3\}$ are two sets, each containing three distinct points of $\hat{\mathbb{C}}$. By 5.2.6, there exists at most one Möbius map T satisfying (4). Does there always exist one such T? The answer is Yes, as the next Lemma and Proposition will show.

5.2.7. Remark. The specification (4) conveys much more information about the Möbius map T than merely what it does at three points. Three points determine a Circle. If C is the Circle determined by the points z_k, then its image $T[C]$ is the Circle determined by the points w_k. Furthermore, the order of the points z_k establishes a direction along C, and likewise for the points w_k, along $T[C]$. By the conformality of a Möbius map, the region on the left of C is mapped by T to the region on the left of $T[C]$.

5.2.8. Lemma. *Let $\{z_1, z_2, z_3\}$ be a set of three distinct points of $\hat{\mathbb{C}}$. Then there exists a Möbius map $U : z_1, z_2, z_3 \rightarrow 0, \infty, 1$.*

Proof. In each of the four possible cases, the map U is easy to devise.

- Case 1, when $z_3 = \infty$: Let $U(z) := \dfrac{z - z_1}{z - z_2}$.

- Case 2, when $z_2 = \infty$: Let $U(z) := \dfrac{z - z_1}{z_3 - z_1}$.

- Case 3, when $z_1 = \infty$: Let $U(z) := \dfrac{z_3 - z_2}{z - z_2}$.

- Case 4, when z_1, z_2, and z_3 are all finite: Let $U(z) := \dfrac{z_3 - z_2}{z_3 - z_1} \dfrac{z - z_1}{z - z_2}$.

The expression for $U(z)$ in Case 4 can be interpreted so that it covers all four cases. For example, if $z_1 = \infty$, then cancel the two factors that include z_1.   □

5.2.9. Proposition. *Let* $\{z_1, z_2, z_3\}$ *and* $\{w_1, w_2, w_3\}$ *be two sets, each containing three distinct points of* $\hat{\mathbb{C}}$. *Then there exists a unique Möbius map*

$$T : z_1, z_2, z_3 \rightarrow w_1, w_2, w_3. \tag{5}$$

Proof. We already know by 5.2.6 that there is at most one such T. The following argument will show that there is one. By the Lemma, there is a Möbius map

$$U : z_1, z_2, z_3 \rightarrow 0, \infty, 1.$$

Again by the Lemma, there is a Möbius map

$$V : w_1, w_2, w_3 \rightarrow 0, \infty, 1.$$

The desired map is

$$T := V^{-1} \circ U. \tag{6}$$

$\square$

5.2.10. Remark. The proofs above suggest a practical procedure for finding the T satisfying (5) whenever the points z_k and w_k are given. The maps U and V used in the proof of Proposition 5.2.9 are always easy to devise, as in the proof of Lemma 5.2.8. From (6) it follows that $V \circ T = U$. In other words,

$$V(w) \equiv U(z), \tag{7}$$

where w stands for $T(z)$. Solve (7) for w in terms of z, and you have identified T. In many cases, T can be found with even less trouble, as in the next Example.

5.2.11. Example. *Identify the Möbius map*

$$U : -1, 1, \infty \rightarrow 0, \infty, -1.$$

What is the image of the upper half-plane? Describe the images of the circles that contain the points -1 *and* $+1$. *What is the image of the unit disk? Find the fixed points.*

Solution. The proof of Lemma 5.2.8 shows the way. To assure that -1 is mapped to 0, we put $z + 1$ in the numerator. To make $+1$ go to ∞, we put $z - 1$ in the denominator. Under the resulting Möbius map $z \mapsto \dfrac{z + 1}{z - 1}$, the point ∞ goes to $+1$; to make it go to -1 instead, we multiply by -1. Thus

$$U(z) = \frac{z + 1}{1 - z}.$$

By 5.2.2, U must map the real axis onto a Circle. It is the Circle containing $0, \infty$, and -1, which means that it is also the real axis. As x moves from -1 to 1 to ∞ along the real axis $\mathbb{R}$, the upper half-plane $\mathbb{Y}$ is to the left; as $U(x)$ moves from 0 to

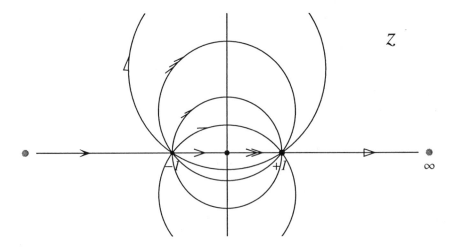

Figure 5.2-1 The real axis and four circles in the z-plane. Four points on the real axis are indicated by dots. The images of those objects, with corresponding marks, are shown in Figure 5.2-2. The mapping, given by $w := \dfrac{z+1}{1-z}$, is the subject of Example 5.2.11.

∞ to -1, the upper half-plane is again to the left, by the conformality of U. Hence $U[\mathbb{Y}] = \mathbb{Y}$.

The Circles through -1 and $+1$ are mapped to the Circles through 0 and ∞, which means that they are mapped to the lines through 0.

Figures 5.2-1 and 5.2-2 show a z-plane and a w-plane. With these figures, we establish some conventions for using a diagram to describe a Möbius map. A number of objects appear in the z-plane, and their respective images appear in the w-plane with the same markings. Thus four points are shown as dots with various sizes and shadings in Figure 5.2-1, and the image of each is shown in Figure 5.2-2 with the same size and shading. For example, 0 in the first figure and its image 1 in the second appear as medium-sized, very dark dots. The point ∞ is shown twice in each figure. Four types of arrowheads appear on the intervals $(-\infty, -1)$, $(-1, 0)$, $(0, 1)$, and $(1, \infty)$ in the first figure, and the same types appear on their respective images in the second figure. Also, four Circles appear in Figure 5.2-1. They are the Circles with centers $-.75i$, 0, $.75i$, and $1.5i$ that pass through -1 and $+1$. The image of each is a line, and it appears in Figure 5.2-2 with the same respective markings.

Observe that the image of the upper half of the unit disk D is the first quadrant of the w-plane; the image of the lower half is the fourth quadrant. The image of the upper half of the outside of the unit disk, where $|z| > 1$ and $y > 0$, is the second quadrant.

The fixed points are the roots of the equation $\dfrac{z+1}{1-z} = z$, namely, $\pm i$. $\qquad\square$

5.2.12. Example. *Identify the Möbius map*

$$G : -1, 0, 1 \;\rightarrow\; 1-i,\; 0,\; 1+i. \qquad\qquad (8)$$

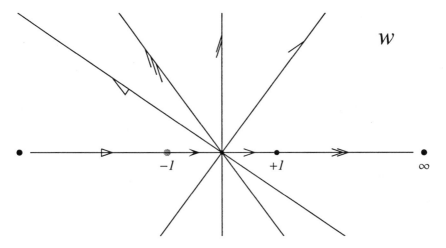

Figure 5.2-2   The images in the w-plane of the objects shown in the z-plane of Figure 5.2-2.

Find the images of the two axes.

Solution. We can deduce a few facts from the information given, even before finding the equation for $G(z)$. The triple of points $-1, 0, 1$ determines the real axis, and the upper half-plane is on the left as we travel through the triple. The triple $1 - i, 0, 1 + i$ determines the circle with center 1 and radius 1, and as we travel through the triple, the outside of the circle is on the left. So it is clear that G maps the real axis onto that circle, and takes the upper half-plane to the outside of that circle. Make a sketch.

Remark 5.2.10 points out a method for identifying G. Let

$$U : -1, 0, 1 \to 0, \infty, 1 \quad \text{and} \quad V : 1 - i, 0, 1 + i \to 0, \infty, 1.$$

These two maps are easily found, using the ideas in the proof of Lemma 5.2.8:

$$U(z) = \frac{z + 1}{2z} \quad \text{and} \quad V(w) = \left(\frac{1 + i}{2i}\right) \frac{w - (1 - i)}{w}.$$

Equation (6) now becomes

$$\left(\frac{1 + i}{2i}\right) \frac{w - (1 - i)}{w} = \frac{z + 1}{2z}. \tag{9}$$

Solving (9) for $w \equiv G(z)$ in terms of z, we obtain

$$G(z) = \frac{2z}{z - i}.$$

To check our work, we can easily verify that G satisfies (8).

The real and imaginary axes are mapped by G to two Circles that cross at right angles at $G(0) = 0$. The first Circle is the circle $|z - 1| = 1$, and since $G(i) = \infty$, the second Circle must be the real axis. $\square$

5.2.13. Example. *Find a conformal equivalence of the open upper half-plane* $\mathbb{Y}^o$ *and the unit disk D.*

Solution. We can easily produce a Möbius map $T : \mathbb{Y}^o \to D$. In fact, there are many such. If we select three points on the real axis, $z_1 < z_2 < z_3$, and three points on the unit circle, w_1, w_2, w_3, listed in a counterclockwise direction, then the Möbius map (4) will do what is required. We are free to choose the six points, and to choose them with a view to making the rest of our computations as simple as possible. Take, for example,

$$T : 0, 1, \infty \to 1, i, -1. \tag{10}$$

As Exercise 16 asks you to verify,

$$T(z) = \frac{i - z}{z + i}; \quad T^{-1}(z) = \frac{i(1 - z)}{z + 1}. \tag{11}$$

□

5.2.14. Example. *Let C be the crescent-shaped open set whose upper boundary is the upper half of the unit circle, and whose lower boundary is an arc of another circle through the points -1 and 1 whose center is $-ib$, where $b > 0$. Such a set C is shown in Figure 5.2-3, on the left. Find a conformal equivalence f of C and the unit disk D.*

Solution. We will proceed by steps. The first two steps will achieve a mapping of C onto $\mathbb{Y}^o$.

First of all, select a Möbius map that takes the upper boundary of C to the positive real axis. It is reasonable to specify

$$S : 1, i, -1 \to 0, 1, \infty. \tag{12}$$

From (12) we see immediately that S is the map T^{-1}, given in (11). Then S maps D onto $\mathbb{Y}^o$. It maps every circle through 1 to a line through 0. If θ is the angle between the two bounding circles, then $S[C]$ is the region shown on the right in Figure 5.2-3, the sector of the plane between the positive real axis and the half-line H_θ.

The second step is to map $S[C]$ onto $\mathbb{Y}^o$, using $z \mapsto z^{\pi/\theta}$, where this is understood to mean

$$re^{it} \mapsto r^{\pi/\theta} e^{it\pi/\theta} \quad \text{for } 0 < t < \theta.$$

The Möbius map T of Example 5.2.13, given in (11), provides the last step; it maps $\mathbb{Y}^o$ onto D. So

$$f(z) = T(S(z)^{\pi/\theta}) = \frac{i - \left(\frac{i(1-z)}{z+1}\right)^{\pi/\theta}}{\left(\frac{i(1-z)}{z+1}\right)^{\pi/\theta} + i}.$$

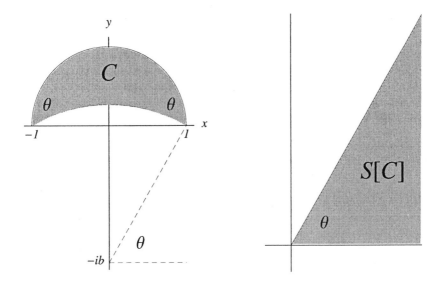

Figure 5.2-3   On the left, a crescent-shaped open set C. Its lower boundary is an arc of the circle with center ib; its upper boundary, the upper half of the unit circle. The two circles meet at angle θ. On the right, its image under the mapping S of Example 5.2.14, which consists of all points re^{it} with $0 < t < \theta$.

Our answer is not fully responsive to the question until we express θ in terms of b. An inspection of the picture reveals that $\theta = \arctan b$.   □

5.2.15. Example. *Find all Möbius maps T that take the upper half of the unit disk, $E = \{z = x + iy \mid |z| < 1, y > 0\}$, onto the first quadrant Q_I.*

Solution. Of course, the map U of Example 5.2.11 is one such map.

We begin by adopting notation for the various parts of Circles which form the boundaries of E and Q_I. Let E_1 be the upper half of the unit circle, and let E_2 be the interval $[-1, 1]$. Let L_1 be the positive half of the imaginary axis, and let L_2 be the positive half of the real axis. Make a sketch.

Let T be as described in the problem. Since T is a Möbius map, it is a one-to-one holomorphic mapping from $\hat{\mathbb{C}}$ onto itself. The sets E_1 and E_2 meet at right angles, at -1 and at 1. The images of E_1 and E_2 must be parts of Circles that meet at right angles both at $T(-1)$ and at $T(1)$. The image of $E_1 \cup E_2$ must be $L_1 \cup L_2$. The sets L_1 and L_2 meet at right angles only at 0 and at ∞. Therefore there are two possibilities.

Case 1, when $T(-1) = 0$ and $T(1) = \infty$. If $T(0)$ were on the imaginary axis, then $T[E]$ would lie to the left of that axis, which cannot be. So $T(0)$ is on the positive real axis; call it α. Now

$$T : -1, 0, 1 \;\rightarrow\; 0, \alpha, \infty.$$

Case 2, when $T(-1) = \infty$ and $T(1) = 0$. If $T(0)$ were on the real axis, then $T[E]$ would lie below the real axis, which cannot be. So $T(0)$ is of the form $i\beta$, with $\beta > 0$, and

$$T : -1, 0, 1 \rightarrow \infty, i\beta, 0.$$

It follows that either $T(z) = \alpha \dfrac{z+1}{1-z}$ for some $\alpha > 0$; or $T(z) = i\beta \dfrac{1-z}{z+1}$ for some $\beta > 0$.   □

5.2.16. Example. *Find a conformal equivalence of the first quadrant of the unit disk, the set $D_I := D \cap Q_I$, and the open upper half-plane $\mathbb{Y}^o$.*

Solution. A Möbius map cannot do this work by itself, because the boundary of D_I has three right angles, which a Möbius map would preserve; and the boundary of $\mathbb{Y}^o$ has none. So we will seek a solution which is a composition of two or more mappings that we know. Make sketches to keep track.

The mapping $z \mapsto z^2$ takes D_I to E, the upper half of the unit disk. From Example 5.2.11, we know that the Möbius map $z \mapsto \dfrac{z+1}{1-z}$ takes E to the positive quadrant. Finally, $z \mapsto z^2$ takes the positive quadrant to $\mathbb{Y}^o$. So the composition g of those three maps does what is required:

$$g(z) = \left(\frac{z^2+1}{1-z^2}\right)^2. \tag{13}$$

□

5.2.17. Example. *Find a conformal equivalence of D_1, the first quadrant of the unit disk, and the unit disk D.*

Solution. If T is the Möbius map of Example 5.2.13 and g is the map found in Example 5.2.16, then the composition $T \circ g$ does what is required.   □

5.2.18. Example. *Let $|a| < 1$. Discuss the Möbius map $N_a(z) := \dfrac{a-z}{1-\bar{a}z}$.*

Solution. Notice that

$$N_a : 0, a, \frac{1}{\bar{a}} \rightarrow a, 0, \infty.$$

One can show by computation that N_a is its own inverse: $N_a(N_a(z)) \equiv z$. This result also becomes clear (by 5.2.5) as soon as we notice that $N_a \circ N_a$ has three fixed points: 0, a, and ∞. One can show by computation that N_a maps the circle $|z| = 1$ to itself; if $z = e^{it}$, then

$$|N_a(z)| = \left|\frac{a-e^{it}}{1-\bar{a}e^{it}}\right| = \left|\frac{ae^{-it}-1}{1-\bar{a}e^{it}}\right| = 1.$$

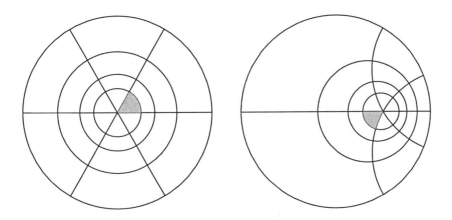

Figure 5.2-4   On the left, four concentric circles and three diameters. On the right, their images under the mapping $N_{1/2}$ of Example 5.2.18. Also, one region and its image are shaded, indicating the rotational effect of this mapping.

Since $N_a(0) = a$ and $N_a(\infty) = 1/\bar{a}$, each line through the origin is mapped by N_a to a circle through a and $1/\bar{a}$. Each circle centered at the origin is mapped to a circle that crosses at right angles every circle that contains the two points a and $1/\bar{a}$. Figure 5.2-4 indicates how $N_{1/2}$ acts on D.   □

The mapping N_a will be used for various technical purposes. Exercises 26 and 27 ask you to explore and compare the properties of the closely related mapping given by

$$M_a(z) := \frac{z + a}{\bar{a}z + 1}. \tag{14}$$

5.2.19. Proposition.   *If f is a one-to-one holomorphic mapping from D onto D, then $f(z) \equiv e^{i\theta} N_a(z)$ for some a with $|a| < 1$ and some real θ.*

Proof. If $f(0) = 0$, then $f(z) \equiv e^{i\theta} z$ for some real θ, by 3.5.8, so that the desired result holds with $a = 0$. Otherwise, there is a nonzero $a \in D$ such that $f(a) = 0$. Then 3.5.8 applies with $f \circ N_a$ in the role of f. Therefore

$$f(N_a(z)) \equiv e^{i\theta} z, \quad \text{hence} \quad f(z) \equiv e^{i\theta} N_a^{-1}(z) \equiv e^{i\theta} N_a(z).$$

□

5.2.20. Corollary.   *Let O be an open set such that there exists a Möbius map T from O onto D. Then every one-to-one holomorphic mapping g from O onto O is a Möbius map.*

Proof. Let $f := T \circ g \circ T^{-1}$. By 5.2.19, f is a Möbius map. Therefore $g \equiv T^{-1} \circ f \circ T$ is a Möbius map.   □

The following result is known as the symmetry principle for Möbius maps. It will be useful in the proof of the reflection principle, Theorem 5.7.3. Recall the discussion of reflection mappings J_A in 2.1H.

5.2.21. Proposition. *Let T be a Möbius map and let A be a Circle. Then*

$$T \circ J_A = J_{T[A]} \circ T. \tag{15}$$

Proof. Problems 35–37 in Section 2.1 indicate how this result can be proved using elementary plane geometry, together with the fact that Möbius maps are conformal and preserve Circles. We will give a different proof here. Consider the function

$$f := J_{T[A]} \circ T \circ J_A. \tag{16}$$

Since reflections are conjugate-meromorphic, f is meromorphic by 2.3.6. Since each function on the right hand side of (16) is one-to-one from $\hat{\mathbb{C}}$ onto $\hat{\mathbb{C}}$, f is a one-to-one meromorphic mapping from $\hat{\mathbb{C}}$ onto $\hat{\mathbb{C}}$. Therefore by 5.2.1, f is a Möbius map. For each point z on the circle A, let $w = T(z)$. We know that $J_A(z) = z$ and $J_{T[A]}(w) = w$. Therefore $f(z) = w = T(z)$ for every z on A. By 5.2.6, since f and T agree at many points, $f = T$. Therefore $T = J_{T[A]} \circ T \circ J_A$, so $T \circ J_A^{-1} = J_{T[A]} \circ T$. Since J_A is its own inverse, (15) follows. $\qquad\square$

Exercises

1. What does (1) say in the case when $ad - bc = 0$?

2. From (1), compute $T'(z)$, assuming $z \in \mathbb{C}$ and $T(z) \in \mathbb{C}$. Can T' have a zero?

3. Show that two Möbius maps are the same,

$$\frac{az + b}{cz + d} = \frac{a'z + b'}{c'z + d'} \qquad \text{for all } z,$$

   if and only if there exists a constant $m \neq 0$ such that $a' = ma, b' = mb, c' = mc$, and $d' = md$.

4. By means of equation (1), every nonsingular two-by-two matrix $\begin{pmatrix} a & b \\ c & d \end{pmatrix}$ determines a Möbius map. Given two matrices $M_1 := \begin{pmatrix} a_1 & b_1 \\ c_1 & d_1 \end{pmatrix}$ and $M_2 := \begin{pmatrix} a_2 & b_2 \\ c_2 & d_2 \end{pmatrix}$, determining Möbius maps T_1 and T_2 respectively, prove that the matrix product $M_2 M_1$ determines the Möbius map which is the composition $T_2 \circ T_1$.

5. If T_1 and T_2 are two Möbius maps, is it always true that $T_2 \circ T_1 = T_1 \circ T_2$?

6. Show that there is a one-to-one correspondence between the Möbius maps and the two-by-two matrices $\begin{pmatrix} a & b \\ c & d \end{pmatrix}$ with $ad - bc = 1$.

7. Use the matrix approach to verify that T^{-1}, given by (2), gives the inverse of T, defined by (1).

8. Let T be a Möbius map. Let z_1, z_2, z_3 and z be four distinct complex numbers, and let w_1, w_2, w_3, and w be their respective images under T. Prove that

$$\frac{(w_3 - w_2)(w - w_1)}{(w_3 - w_1)(w - w_2)} = \frac{(z_3 - z_2)(z - z_1)}{(z_3 - z_1)(z - z_2)}.$$

The expression on the right is the **cross-ratio** of the four points z_1, z_2, z_3 and z. We may state the result by saying that the cross-ratio of four points is invariant under a Möbius map.

9. Show that the Möbius maps form a group under composition.

10. For the Möbius map U of Example 5.2.11, what is the image of the imaginary axis? What are the images of the four quadrants?

11. In Figure 5.2-1, five Circles are shown that meet at -1. Each pair of Circles meets at a certain angle (that is, the angle between the tangents to the Circles at that point). Since U is conformal, the corresponding angle in Figure 5.2-2 is the same. From the information given, find each angle in radians, approximately.

12. Sketch diagrams like Figures 5.2-1 and 5.2-2 for the Möbius map $z \mapsto \dfrac{z-1}{z+1}$. What are the images of the four quadrants?

13. Sketch diagrams like Figures 5.2-1 and 5.2-2 for the Möbius map G of Example 5.2.12. What are the images under G of the four quadrants?

14. For each of the maps T described in Example 5.2.15, find the image of the segment $[0, i]$.

15. Devise a Möbius map that takes the upper half-plane onto the disk $D(1, 1)$.

16. Using the method described in Remark 5.2.10 and demonstrated in Example 5.2.12, show that the Möbius map determined by (10) is given by (11).

17. Under the mapping (11), what are the images of lines through the origin?

18. Verify that the Möbius map given by $U(z) := \dfrac{z+1}{1-z}$ maps D onto the open right half-plane. Find U^{-1}.

19. Using the result of Exercise 18, show that the Möbius map given by $V(z) := U^{-1}(-iz) \equiv \dfrac{z-i}{z+i}$ takes the open upper half-plane onto D. Discuss the difference between V and the map T of Example 5.2.13.

20. Let g be the mapping given by (13). Identify the image under g of each quadrant of the unit disk.

21. Find a conformal equivalence of each quadrant of the unit disk and $\mathbb{Y}^{\circ}$.

22. Show that each of the Möbius maps N_a $(a \neq 0)$ is related to a Möbius map N_s where $0 < s < 1$.

23. Let N_a be as in Example 5.2.18. Compute $N_a'(z)$ and verify that $N_a'(0) < 0$.

24. Let N_a be as in Example 5.2.18, with $a \neq 0$. Find the fixed points of N_a. Show the fixed points in a diagram in the cases $a = \frac{1}{2}$ and $a = \frac{1}{2}e^{i\pi/4}$.

25. In Figure 5.2-4, identify in the right-hand picture the image under $N_{1/2}$ of each of the three diameters shown in the left-hand picture.

26. Let M_a be the Möbius map given by (14). Re-write Example 5.2.18, dealing with M_a instead of N_a.

27. Make a sketch by hand of pictures like the ones in Figure 5.2-4, but for $M_{1/2}$. Discuss the differences.

28. Let f be a one-to-one holomorphic mapping from the disk $D(a, r)$ onto the disk $D(b, s)$. Prove that f is a Möbius map.

29. Explain and justify the statement in the proof of 15: Reflections are conjugate-meromorphic.

Help on Selected Exercises

10. Three points on the imaginary axis are $0, i$, and ∞; their images are $1, i$, and -1. Therefore the image of the imaginary axis is the unit circle in the w-plane, traversed counterclockwise. The images of the four quadrants are as follows:

$$U[Q_I] = \{w = u + iv \mid |w| > 1, v > 0\};$$
$$U[Q_{II}] = \{w = u + iv \mid |w| < 1, v > 0\};$$
$$U[Q_{III}] = \{w = u + iv \mid |w| < 1, v < 0\};$$
$$U[Q_{IV}] = \{w = u + iv \mid |w| > 1, v < 0\}.$$

11. The centers of the circles are identified in the text. Since each circle contains the points ± 1, its radius can be determined, and hence the point where it crosses the positive imaginary axis. The four crossings take place at $i/2, i, 2i$, and $(3 + \sqrt{13})i/2$. The images of those points under U can be computed. They are of the form $e^{i\theta_k}$, with $0 < \theta_1 < \theta_2 = \pi/2 < \theta_3 < \theta_4 < \pi$.

13. The images of the four quadrants are as follows:

$$G[Q_I] = \{w = u + iv \mid |w - 1| > 1, v > 0\};$$
$$G[Q_{II}] = \{w = u + iv \mid |w - 1| > 1, v < 0\};$$
$$G[Q_{III}] = \{w = u + iv \mid |w - 1| < 1, v < 0\};$$
$$G[Q_{IV}] = \{w = u + iv \mid |w - 1| < 1, v > 0\}.$$

16. We make use of the Möbius maps

$$U \; : \; 0, 1, \infty \; \to \; 0, \infty, 1 \quad \text{and} \quad V \; : \; 1, i, -1 \; \to \; 0, \infty, 1,$$

which can be found using the ideas in the proof of Lemma 5.2.8:

$$U(z) \;=\; \frac{z}{z-1}; \quad V(w) = \left(\frac{1+i}{2} \right) \frac{w-1}{w-i}.$$

Solve $V(w) = U(z)$ for w in terms of z, obtaining (11).

18. $U^{-1}(z) = \dfrac{z-1}{z+1}.$

20. $g[D_{II}] = g[D_{IV}] = Y_-^o$, and $g[D_{III}] = Y^o.$

22. If $a = se^{i\alpha}$, then $N_a(z) = e^{i\alpha} N_s(e^{-i\alpha} z).$

5.3 ELECTRIC FIELDS

A   A Point Charge in 3-Space

The problems of physics inherently involve three spatial dimensions. Their solutions are frequently harmonic functions $u(x, y, z)$ of three variables. The methods for finding such functions are different from those of this book, which deal with mathematical models in real x, y-space and produce harmonic functions $u(x, y)$ of two variables. Nevertheless, the models suffice to solve certain problems in physics.

We begin by discussing a problem for which they do not suffice. It is a basic boundary value problem from electrostatics. The problem is to describe an electric field. There are various systems of units which might be used. We will specify a choice of units, because it may make the discussion clearer to readers already conversant with the physics.

What we will call an electric field is also called an electric intensity field, or an electrostatic field.

Consider an electric charge of q coulombs located at the origin $(0, 0, 0)$ of $\mathbb{R}^3$. A coulomb is 6×10^6 electrons. Suppose that all the rest of 3-space is a vacuum, containing no electric charges. The charge q causes an **electric field** $\vec{\mathbf{E}}$, a vector field defined on $\mathbb{R}^3 \setminus \{(0, 0, 0)\}$. The boundary of this domain consists of two points, the origin and the point at infinity. At each point (x, y, z) in the domain, the magnitude of the vector $\vec{\mathbf{E}}(x, y, z)$ is that of the force per coulomb which would be exerted on a charge (of, say, q_1 coulombs) if it were located at (x, y, z). That magnitude is proportional to q and inversely proportional to the square of the distance $r = (x^2 + y^2 + z^2)^{1/2}$. The direction of the vector $\vec{\mathbf{E}}(x, y, z)$ is the direction in which the force acts. The field is radial, that is, the force acts along a line through $(0, 0, 0)$ and (x, y, z). It acts away from the origin if $qq_1 > 0$, and toward the origin if $qq_1 < 0$. Thus

$$\vec{\mathbf{E}}(x, y, z) \;=\; \frac{kq}{(x^2 + y^2 + z^2)^{3/2}} (x\vec{\mathbf{i}} + y\vec{\mathbf{j}} + z\vec{\mathbf{k}}) \;=\; \frac{kq}{r^2} \vec{\mathbf{r}},$$

where $\vec{r}$ denotes the unit vector pointing away from the origin, and where k is a certain positive constant, approximately 9×10^9 newton meter^2 per coulomb^2.

Thus if a charge of q_1 coulombs were located at the point (x, y, z), then the magnitude of the force acting upon it would be $\dfrac{kqq_1}{r^2}$ newtons, and the force would act radially, away from or toward the origin depending on the sign of qq_1.

Another way to describe the field adequately is to identify the potential. Recall 1.5D. Because the field is radial and its domain is charge-free, $\mathbf{curl} \ \vec{E} = \vec{0}$ and $\mathrm{div} \ \vec{E} = 0$. Therefore there is a harmonic function $u(x, y, z)$, the **potential**, such that

$$\vec{E} = -\mathbf{grad} \ u. \tag{1}$$

This being the only requirement on u, it is determined only up to an additive constant. We may take

$$u(x, y, z) \ = \ -\frac{kq}{r} \ = \ -\frac{kq}{(x^2 + y^2 + z^2)^{1/2}}. \tag{2}$$

The units of potential are volts (newton-meters per coulomb). To say that u is harmonic is to say that the 3-dimensional Laplacian of u vanishes:

$$u_{xx} + u_{yy} + u_{zz} = 0.$$

The **equipotential surfaces**, or **equipotentials**, are the sets defined by conditions $u(x, y, z) = c$. They are spheres. On each of them, the magnitude of the force exerted per unit charge is constant.

It is a matter of convention that the factor -1 appears in both equations (1) and (2), instead of in neither. The rule is that, the charges being positive, the force should act in the direction of decreasing potential.

B   Uniform Charge on One or More Long Wires

What might seem to be the straightforward 2-dimensional analogue of the mathematical-electrostatics model above is not. That is to say, one does not just place a point charge in the midst of a planar vacuum and invent a 2-dimensional physics. The function given by

$$u(x, y) := (x^2 + y^2)^{-1/2}$$

is not harmonic; the 2-dimensional Laplacian $u_{xx} + u_{yy}$ does not vanish, as Exercise 16 in Section 1.6 (page 81) asked you to show. However, a 2-dimensional model does fit the electrostatics problem presented in the next Example.

5.3.1. Example.   *Consider a line L in $\mathbb{R}^3$ as a mathematical model of a long straight wire carrying a uniform static charge of q coulombs per meter. Identify the resulting electric field $\vec{E}$ defined on the complement of L, assumed to be a charge-free vacuum. Find the potential, and describe the equipotentials and the paths of force.*

Solution. We will derive the answer from the discussion in subsection A. By symmetry considerations, the direction of the vector field will always be perpendicular to L, and its magnitude at a point P, at distance r from L, will depend only on r. We may as well take L to be the z-axis; note that "z" here stands for the third real variable. Then $r = (x^2 + y^2)^{1/2}$. Make a sketch. The magnitude of the force per coulomb at P due to the charge located at coordinate z on L is $\dfrac{kq}{z^2 + r^2} dz$; its component in the direction perpendicular to L is then

$$\frac{r}{(z^2 + r^2)^{1/2}} \cdot \frac{kq}{z^2 + r^2} dz,$$

of which the integral over $\mathbb{R}$ is $2kq/r$. Thus the field $\vec{\mathbf{E}}$ has no z-component, and its other two components depend only on x and y:

$$\vec{\mathbf{E}}(x,y) \;=\; \frac{2kq}{x^2 + y^2}(x\vec{\mathbf{i}} + y\vec{\mathbf{j}}) \;\equiv\; \frac{2kq\vec{\mathbf{r}}}{r}, \quad \text{where} \;\; \vec{\mathbf{r}} := \frac{x\vec{\mathbf{i}} + y\vec{\mathbf{j}}}{(x^2 + y^2)^{1/2}}.$$

Thus every planar cross-section of $\mathbb{R}^3$ that is perpendicular to the wire is the same for this problem; a 2-dimensional mathematical model is quite sufficient. The electrostatic potential is given by the harmonic function

$$u(x,y) := -kq\ln(x^2 + y^2) = -2kq\ln r = -2kq\ln|z|, \tag{3}$$

where in the last expression, the letter "z" has returned to its role as the complex variable $x + iy$. A harmonic conjugate of u, definable locally, is given by $v(x,y) := -kq\arg(x,y)$, and $u + iv$ is called the complex potential. The equipotentials are circles centered at the origin. The paths of force are lines to the origin. $\qquad\square$

If there are more wires than one passing through our x, y plane, the potential u is the sum of potentials like the one found above. The requirement on u is that for each of the wires, it should give the appropriate electric potential in the limit very near each of the wires; that is, as a point tends to that wire, the ratio of **grad** u to the electric potential due to that wire should tend to 1. See Example 5.3.6. In the next Example, there is still only one wire, but the domain is bounded by an surface of specified potential.

5.3.2. Example. *In real x, y, z-space, consider the z-axis as a mathematical model for a long wire which carries a stationary electric charge, consisting of q coulombs per meter. Let the condition $x^2 + y^2 = 1$ define a cylindrical surface which encloses the wire. The surface is charge-free; it is grounded; that is, its potential is 0. Inside the cylinder, except for the wire, there is a vacuum. By finding the potential, describe the electric field inside the cylinder that is caused by the electric charge on the wire.*

Solution. We seek a harmonic function u on the punctured unit disk D' such that $u(x,y) = 0$ when $x^2 + y^2 = 1$, and such that the gradient of u should give the appropriate electric field near $(0,0)$. Since $\ln 1 = 0$, equation (3) provides the

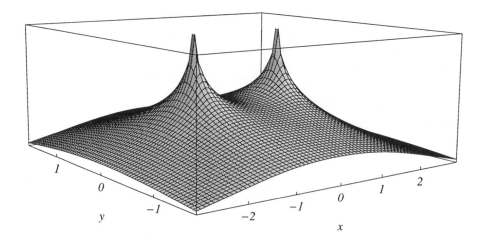

Figure 5.3-1   The graph of the potential for the electric field caused by two parallel positively charged wires, with equal charge densities $q_1 = q_2$. See Example 5.3.6.

solution. The picture on the left in Figure 5.2-4, page 337, shows a few of the equipotentials, which are circles centered at the origin. Equipotentials in our pictures are chosen so that between every two adjacent ones, the potential difference is the same. Also shown are a few of the paths of force, which are line segments running to the origin.   □

5.3.3. Example. *Consider the same problem as in Example 5.3.2, except that the wire is the line given by $x = \frac{1}{2}, y = 0$.*

Solution. We seek a harmonic function u_1 on the open set

$$O_1 = \{(x,y) \mid x^2 + y^2 < 1 \text{ and } (x,y) \neq (\tfrac{1}{2},0)\}$$

such that $u_1(x,y) = 0$ when $x^2 + y^2 = 1$, and such that the gradient of u should give the appropriate electric field near $(\frac{1}{2},0)$. A conformal equivalence of this problem and that of Example 5.3.2 can be found among the Möbius maps N_a of Example 5.2.18. The desired map is $N_{1/2}$, which takes O_1 to D'. The solution for D' is given by (3), so the solution for O_1 is $u \circ N_{1/2}$:

$$u_1(z) = -2kq \ln |N_{1/2}(z)| = -kq \ln \left(\frac{1 - 4x + 4x^2 + 4y^2}{4 - 4x + x^2 + y^2} \right).$$

Unless there is a need to have it in terms of x and y, we do not especially favor the last form of the answer. A harmonic conjugate, defined locally, is given by

$$v_1(z) = -2kq \arg N_{1/2}(z).$$

The picture on the right in Figure 5.2-4 shows several equipotentials and paths of force.   □

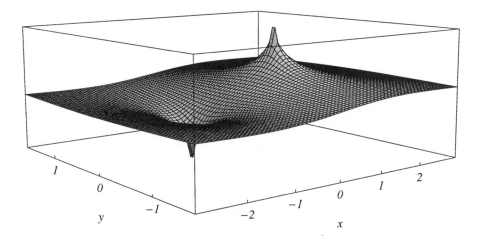

Figure 5.3-2 The graph of the potential for the electric field caused by two parallel charged wires, with charge densities of opposite sign, $q_2 = -q_1$. See Example 5.3.6.

5.3.4. Remark. The Examples in this section all involve unbounded objects in real (x, y, z)-space, defined by certain conditions stated in terms of x and y only, with z unrestricted. Therefore the objects are perpendicular to the x, y-plane, in which we deal with their cross-sections. Being unbounded in the z direction, such objects serve as good mathematical models for actual objects which are merely very long, provided the cross-section is far from the ends. The letter "z" plays its customary role as the third real variable, which is not likely to get confused with its other role as $x + iy$.

5.3.5. Example. *In real x, y, z-space, a static charge of q coulombs per meter resides on the line given by $x = 1, y = 1$ inside a vacuum in the region given by $x > 0, y > 0$. Its boundary is grounded, that is, maintained at 0 volts. Find the potential, and sketch a few equipotentials and paths of force.*

Solution. This problem will also reduce to that of 5.3.2, as soon as we find a conformal equivalence from

$$O_2 := \{(x, y) \mid x > 0, y > 0, \text{ and } (x, y) \neq (1, 1)\}$$

onto D'. A mapping that takes the positive quadrant onto the disk D and maps $(1, 1)$ to $(0, 0)$ is given by

$$g(z) := \frac{z^2 - 2i}{z^2 + 2i}. \tag{4}$$

Accordingly, a solution is the potential $z \mapsto u(g(z))$, or

$$u_2(z) := -2kq \ln \left| \frac{z^2 - 2i}{z^2 + 2i} \right|,$$

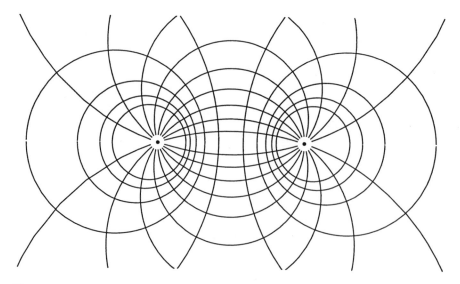

Figure 5.3-3   Circles of Apollonius, circles of Steiner: Equipotentials and paths of force for the electric field caused by two parallel wires with charges of opposite signs, as discussed in Example 5.3.6. Compare Figure 5.2-1 and Figure 1.1-7 on page 10.

and a harmonic conjugate of u_2 is

$$v_2(z) := -2kq \arg \left(\frac{z^2 - 2i}{z^2 + 2i} \right).$$

Figure 1.1-8, page 11, shows some equipotentials and paths of force.   □

5.3.6. Example. *Each of two parallel wires carries a uniform charge: q_j coulombs per meter for $j = 1, 2$, with $q_1 > 0$. The rest of 3-space is charge-free. Describe the resulting field in the two cases when $q_2 = -q_1$, and when $q_2 = q_1$.*

Solution. It suffices to describe the field in a cross-section of 3-space to which the two wires are perpendicular. We are free to choose the coordinate system. Let the wires be represented by the lines that meet the x, y-plane at right angles, at the points ± 1. Then the electric field is the sum of the fields caused by the two wires:

$$\vec{E}(x, y) = \frac{2kq_1}{\sqrt{(x+1)^2 + y^2}} \vec{r} + \frac{2kq_2}{\sqrt{(x-1)^2 + y^2}} \vec{s},$$

where $\vec{r}$ and $\vec{s}$ are the unit vectors pointing away from the points -1 and 1, respectively. One charge repels, the other attracts. The potential of the sum is the sum of the potentials,

$$u(z) = -2k(q_1 \ln|z + 1| + q_2 \ln|z - 1|),$$

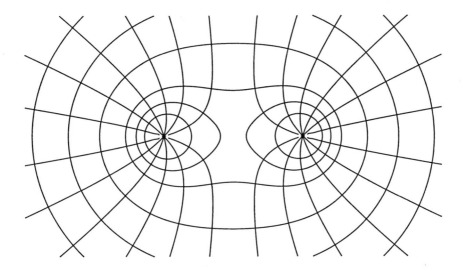

Figure 5.3-4   Equipotentials and paths of force for the electric field caused by two parallel wires with charges of the same sign, as discussed in Example 5.3.6. The equipotentials are the cassinian curves.

and the harmonic conjugate is the sum of the corresponding arguments, defined locally.

If $q_2 = -q_1$, then

$$u(z) = -2kq_1 \ln \left| \frac{z+1}{z-1} \right|.$$

The equipotentials are level curves defined by equations

$$\left| \frac{z+1}{z-1} \right| = c.$$

They are circles; and the paths of force are the circles that contain the two points -1 and 1. See Figure 5.3-3.

If $q_2 = q_1$, each of the two charges repels another positive charge. The equipotentials are given by equations $|z+1||z-1| = c$. These curves, called **cassinians**, are shown in Figure 5.3-4. For $0 < c < 1$, the curve is a **two-part Cassini's oval**; for $c = 1$, it may be described as a figure 8, the **lemniscate of Bernoulli**. For $c > 1$, it is a **one-part Cassini's oval**, a single closed curve which becomes more and more like a circle in shape as $c \to \infty$, reflecting the fact that at a great distance the field becomes more like what would be induced by a single charged wire. $\square$

C   Examples with Bounded Potentials

The next few Examples share mathematical models with temperature distribution problems like those considered in Section 5.1.

5.3.7. Example. *The open upper half-plane $\mathbb{Y}^o$ is a charge-free vacuum. The part of its boundary given by $y = 0, x > 0$ is grounded, maintained at 0 volts; the part given by $y = 0, x < 0$ is attached to a battery and maintained at 100 volts. Find the electric potential. Identify the equipotentials and paths of flux.*

Solution. We seek a harmonic function u on $\mathbb{Y}^o$ whose boundary values are 0 on the positive real axis, 100 on the negative. We find one easily, and write both it and a harmonic conjugate in terms of polar coordinates:

$$u(re^{i\theta}) = \frac{100}{\pi}\theta, \qquad v(re^{i\theta}) = \frac{100}{\pi}\ln r.$$

Equivalently,

$$u(z) = \frac{100}{\pi}\operatorname{Arg} z, \qquad v(z) = \frac{100}{\pi}\ln|z|. \tag{5}$$

The paths of flux are semicircular arcs centered at the origin, and the equipotentials are half-lines. For example, the equipotentials with values 90 volts and 80 volts are $H_{.9\pi}$ and $H_{.8\pi}$, respectively. Make a sketch. $\qquad\square$

5.3.8. Example. *The cylinder in 3-space defined by $x^2 + y^2 = 1$ is insulated along the lines $x = 1, y = 1$ and $x = 0, y = 1$. The part of the cylinder with $x > 0, y > 0$ is attached to a battery and maintained at 100 volts. The rest is grounded, maintained at 0 volts. The inside of the cylinder is a charge-free vacuum. Find the electric potential u_1. Identify the equipotentials and paths of flux.*

Solution. We seek a harmonic function u on the unit disk D with boundary values as shown in the picture on the right in Figure 1.1-1, page 3 (but with volts instead of degrees). This problem can be converted into that of Example 5.3.7, as follows. We will devise a conformal equivalence from D onto $\mathbb{Y}^o$, so that the 100-volt boundary goes to the negative real axis. The Möbius map

$$T : 1, i, -1 \;\rightarrow\; \infty, 0, 1, \quad \text{given by } T(z) = (1 - i)\frac{z - i}{z - 1}$$

will serve. Then the potential is $u_1 := u \circ T$. In terms of x and y:

$$u_1(x, y) = \frac{100}{\pi}\operatorname{arccot}\left(\frac{(x - 1)^2 + (y - 1)^2 - 1}{1 - x^2 - y^2}\right),$$

$$v_1(x, y) = -\frac{50}{\pi}\ln\left|\frac{2(x^2 + (y - 1)^2)}{(x - 1)^2 + y^2}\right|.$$

Some equipotentials and paths of force are shown in Figure 1.1-7, page 10. In a temperature-distribution problem described there, they were the isotherms and paths of heat flow. $\qquad\square$

5.3.9. Example. *Let $0 < r_1 < r_2$. Two cylinders in 3-space, given by $x^2 + y^2 = r_1^2$ and $x^2 + y^2 = r_2^2$, are maintained at -10 and $+10$ volts, respectively. The space* *between them is a charge-free vacuum. Find the potential for the electric field; sketch several equipotentials and paths of flux.*

Solution. The cross-section with the x, y-plane of the space between the cylinders is the annulus $\text{Ann}(r_1, r_2)$. The harmonic function $u(z) = C_1 + C_2 \ln |z|$ is the potential, provided we select the constants so that $C_1 + C_2 \ln r_2 = 10$ and $C_1 + C_2 \ln r_1 = -10$. The equipotentials are circles centered at 0, and the paths of force are portions of radial lines. □

5.3.10. Example. *The two cylinders in 3-space given by $(x - \frac{1}{2})^2 + y^2 = \frac{1}{16}$ and $x^2 + y^2 = 1$ are maintained at -10 and $+10$ volts, respectively. The space between them is a charge-free vacuum. Find the potential for the electric field.*

Solution. If we can find a Möbius map N_a, as defined in Example 5.2.18, that maps the disk $D(\frac{1}{2}, \frac{1}{4})$ to a disk $D(0, r)$ for some $r > 0$, then we will have converted the problem on $O := \{z \mid |z| < 1, |z - \frac{1}{2}| > \frac{1}{4}\}$ to an equivalent problem on the annulus $\text{Ann}(r, 1)$, a special case of Example 5.3.9. The potential on the annulus is given by $u(z) = C_1 + C_2 \ln |z|$, provided $C_1 + C_2 \ln 1 = 10$ and $C_1 + C_2 \ln r = -10$. Solving that pair of equations gives us

$$u(z) = 10 - \frac{20}{\ln r} \ln |z|.$$

It would be a careless mistake to think that $N_{1/2}$ is the desired mapping. It is true that $N_{1/2}(\frac{1}{2}) = 0$, but

$$N_{1/2}[D(\tfrac{1}{2}, \tfrac{1}{4})] = D(-\tfrac{2}{35}, \tfrac{12}{35}), \tag{6}$$

whose center is not 0.

Recall that N_a is its own inverse; and that if s is real, then $N_s(\bar{z}) \equiv \overline{N_s(z)}$. It follows that N_s maps every circle with center on the real axis to another such circle. We want to find s with $0 < s < 1$ such that $N_s[D(\frac{1}{2}, \frac{1}{4})] = D(0, r)$ for some $r > 0$. It suffices to satisfy the condition $N_s(\frac{1}{4}) = -N_s(\frac{3}{4})$, which becomes a quadratic in s with root

$$s_o = \frac{19 - \sqrt{105}}{16} \approx 0.5470656,$$

and then the desired value of r is

$$r_o = \frac{s - \frac{1}{4}}{1 - \frac{1}{4}s} \approx 0.3441312.$$

The desired potential is

$$u_1(z) = u(N_{s_o}(z)) = 10 - \frac{20}{\ln r_o} \ln \left| \frac{s_o - z}{1 - s_o z} \right|. \tag{7}$$

□

Exercises

1. Verify that (4) does what is claimed.

2. Solve Example 5.3.2, but where the wire is the line given by $x = .5, y = .5$.

3. The open upper half-plane $\mathbb{Y}^o$ is a charge-free vacuum. The part of its boundary given by $y = 0, |x| > 1$ is grounded, maintained at 0 volts; the part given by $y = 0, -1 < x < 1$ is attached to a battery and maintained at 100 volts. Find the electric potential. Identify the equipotentials and paths of flux.

4. Explain this remark: The potential graphs in Figure 5.3-1 carry much the same information about the electric fields as do Figures 5.3-3 and 5.3-4.

5. Figure 5.3-4 does not show the equipotential which is a figure 8. Where is it? What is the significance of the point where it intersects itself? There is a path of force which is the line segment $[-1, 1]$. What force would be exerted on an electric charge located at its midpoint?

6. Verify (6).

7. If $0 < s < 1$ and $0 < r < 1$, show that the image under N_s of the circle $|z| = r$ is the circle with center $\dfrac{s(1 - r^2)}{1 - s^2 r^2}$ and radius $\dfrac{r(1 - s^2)}{1 - s^2 r^2}$.

8. In Example 5.3.10, describe the equipotentials. The paths of force are parts of the circles that pass through what two points?

9. Interpret Example 5.3.10 as a temperature distribution problem. Identify the $0°$ isotherm.

Help on Selected Exercises

3. Consider the procedure developed in Example 5.1.5.

8. The paths of force are arcs of the circles that pass through s_o and $1/s_o$.

9. The $0°$ isotherm is the circle, with center on the real axis, that crosses the axis at approximately -0.0582556 and 0.8582576.

5.4   STEADY FLOW OF A PERFECT FLUID

We now resume the discussion begun in subsection 1.5D and continued in subsection 2.5I. Given an unbounded open connected set O in the plane, we seek a vector field $\vec{\mathbf{F}}$ whose value at each point $(x, y) \in O$ tells the speed and direction of the fluid flow at that point. For the problems we wish to consider, the boundary conditions are as follows. First, we require that at each point of the boundary, called the impenetrable

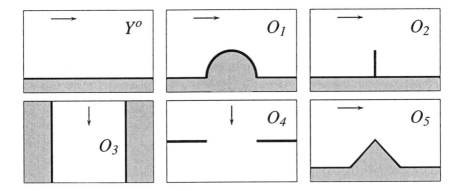

Figure 5.4-1   Each picture represents a fluid flow problem in the unshaded region.

wall, $\vec{F}$ shall have no component perpendicular to the boundary: $\vec{F} \cdot \vec{n} = 0$. Second, we impose some requirement as to how $\vec{F}$ behaves near ∞.

Figure 5.4-1 indicates a number of such problems. It bears repeating one more time that in all these cases, we speak in terms of a 2-dimensional mathematical model which is a cross-section of a 3-dimensional physical reality. The third component of the vector field is $\vec{0}$, so we treat $\vec{F}$ as having only two components; and each of those is independent of the third real variable, so we treat it as a function of two real variables (one complex variable). Also, the boundary of each region represented in the Figure has infinite extent, being unrestricted in the third real variable. For example, the boundary of O_2 is a cross-section of the 3-dimensional stream bed, the set

$$\{(x, y, z) \mid y > 0, \text{ or } x = 0 \text{ and } y > 1\}.$$

Thus it is a matter, not of a mere "stick" at the bottom, but of a vertical dam extending across the stream bed.

The vector field $\vec{F}$ will be -1 times the gradient of a function u which is harmonic on O:

$$\vec{F}(x, y) = -u_x(x, y)\vec{i} - u_y(x, y)\vec{j}.$$

Let v be a harmonic conjugate of u, and let $f := u + iv$. Then u is called the velocity potential, v is called a stream function, and f a complex potential. Notice that the function $F := -u_x - iu_y \equiv -u_x + iv_x$ is conjugate-holomorphic.

The level curves of u, called the equipotential curves, are orthogonal to the direction of flow, and they meet the boundary at right angles at every point where the boundary does not have a cusp. The level curves of v, called the streamlines, indicate the direction of flow, and the vector $\vec{F}$ is tangent to a streamline at each point. The speed, or magnitude of the velocity, of the flow is

$$\sigma = |\vec{F}| = |u_x + iv_x| = |f'|.$$

We may think of the problem as one of finding u, since every aspect of the solution can be generated from it. In terms of u, the boundary conditions consist of the homogeneous Neumann condition $\dfrac{\partial u}{\partial \vec{n}} = 0$, required to hold at most points on the boundary, and the stipulation that σ is bounded.

We can use our model to discuss the behavior of a fluid flow which is not quite perfect. However, when we interpret our findings physically, we need to remember that the model is then an approximation, which may or may not be good enough to justify confidence in what the model predicts. For example, if the solution to a problem tells us that $\sigma(z)$ becomes very large as z approaches a certain boundary point, we may suspect that the flow will not really remain irrotational near that point as the model predicts. You may wish to look at Figure 5.4-1 and speculate as to which of the boundaries might cause turbulence.

5.4.1. Example. *Consider the problem represented by the picture at the upper left in Figure 5.4-1. Here the set O is the open upper half plane $\mathbb{Y}^o$, the real axis is the impenetrable wall, and the flow at ∞ is c meters per second toward the right. Find the velocity potential.*

Solution. In this case, it is easy to think of a quite plausible solution, the one given by $u(x,y) = -cx$. Then a stream function is given by $v(x,y) = -cy$; the function $f(z) = -cz$ is a complex potential; and the velocity vector field is given by $\vec{F}(x,y) \equiv c\vec{i}$. The picture on the left in Figure 1.1-9, page 12, shows some equipotentials and streamlines of this flow. $\qquad\square$

5.4.2. Example. *Find the velocity potential that solves the problem for the region O_1 in Figure 5.4-1, where the flow at ∞ is c meters per second toward the right. Determine the speed of the flow near the top of the bump, and also in the corners at the base of the bump.*

Solution. We have a choice of coordinate system, and for convenience we take the boundary lines to be the intervals $(-\infty, -1]$ and $[1, \infty)$ of the real axis, and the bump to be the upper half of the unit disk. Thus we are being asked for the velocity near the points i, -1, and $+1$. We will solve the problem by finding a conformal equivalence with the problem of Example 5.4.1, for which a multiple of the function $z \mapsto -z$ is a complex potential. If g is a one-to-one holomorphic mapping from O_1 onto $\mathbb{Y}^o$, then a complex potential for the problem on O_2 will be $z \mapsto kg(z)$ for the correct choice of k, the one that makes the speed of the flow at ∞ equal to c.

We will show that the desired g is the function whose value at z is the average of z and its reciprocal:

$$g(z) := \frac{1}{2}\left(z + \frac{1}{z}\right).$$

You should find it easy to verify the following observations, which give an indication of how g behaves.

(i) The fixed points of $g : \hat{\mathbb{C}} \to \hat{\mathbb{C}}$ are ± 1.

(ii) $g'(z)$ is nonzero, and thus g is conformal, everywhere except at ± 1.

(iii) g maps each of the intervals $(-\infty, -1)$ and $(-1, 0)$ one-to-one onto $(-\infty, -1)$.

(iv) g maps each of the intervals $(1, \infty)$ and $(0, 1)$ one-to-one onto $(1, \infty)$.

(v) g maps each of two arcs of the unit circle, $\text{Arc}(1, 0, \pi)$ and $\text{Arc}(1, \pi, 2\pi)$, one-to-one onto the inverval $[-1, 1]$.

To verify that g maps O_1 onto $\mathbb{Y}^o$, write z in the form re^{it}:

$$g(re^{it}) = \frac{1}{2}\left(re^{it} + \frac{1}{r}e^{it}\right).$$

For a fixed value of $r > 0$, restricting g to the circle of radius r gives a parametrization:

$$t \mapsto \frac{1}{2}\left(r + \frac{1}{r}\right)\cos t + \frac{1}{2}\left(r - \frac{1}{r}\right)\sin t. \tag{1}$$

Thus g maps the upper half of the circle with center 0 and radius r onto the upper half E_r of an ellipse, provided $r > 1$. The union of the E_r, with r running from 1 to ∞, is $\mathbb{Y}^o$.

A complex potential is, for some constant k,

$$f_1(z) := kg(z) = \frac{k}{2}\left(z + \frac{1}{z}\right).$$

From it we may compute the speed of the flow:

$$\sigma(z) = |f_1'(z)| = \frac{|k|}{2}\left|1 - \frac{1}{z^2}\right|.$$

The limit as $z \to \infty$ is $|k|/2$, so we should take $k = -2c$. Thus

$$f_1(z) = -c\left(z + \frac{1}{z}\right) \quad \text{and} \quad \sigma(z) = c\left|1 - \frac{1}{z^2}\right|.$$

The speed σ is continuous at the boundary, and we easily find that it equals 0 at the corners and equals $2c$ at the top of the bump: $\sigma(\pm 1) = 0, \sigma(i) = 2c$.

The equipotential curves and the streamlines are the level curves of the velocity potential $u_1 = \text{Re } f_1$ and the stream function $v_1 = \text{Im } f_1$, respectively. The resulting orthogonal grid is represented in the right hand picture of Figure 1.1-9, page 12, drawn using the ContourPlot command of *Mathematica*. Depending on the software one uses, it may or may not be necessary to look at the two functions in terms of x and y. In any event, here are the two expressions:

$$u_1(x, y) = -\frac{c(x^3 + xy^2 + x)}{2(x^2 + y^2)}, \quad v_1(x, y) = -\frac{c(x^2y + y^3 - y)}{2(x^2 + y^2)}. \tag{2}$$

$\square$

5.4.3. Example. *Given that the speed of flow at ∞ is c meters per second in the region O_2 of Figure 5.4-1, find the velocity potential and the stream function, and evaluate the speed of flow at the top of the vertical barrier and in the corners.*

Solution. We will find a one-to-one holomorphic mapping h from O_2 onto $\mathbb{Y}^o$. A complex potential that solves the problem on $\mathbb{Y}^o$ is a multiple of $z \mapsto -z$, so a complex potential that solves the problem on O_2 will be $z \mapsto kh(z)$, where the constant k is chosen so that the speed at ∞ equals c.

We have a choice of coordinate system. Let's declare the boundary to be the real axis together with the segment $[0, i]$. We will devise the mapping h in three steps. Make sketches to keep track. First, $z \mapsto z^2$ maps O_2 onto

$$U := \{z = x + iy \mid x \geq -1 \text{ or } y \neq 0\} \equiv \mathbb{C} \setminus [-1, \infty).$$

Next, $z \mapsto z + 1$ moves everything to the right by 1 and maps U onto $V_0 \equiv \mathbb{C} \setminus H_0$. Finally, the square root function $re^{it} \mapsto \sqrt{r}e^{it/2}$ (for $0 < t < 2\pi$) maps V_0 onto $\mathbb{Y}^o$. The desired mapping h is the composition of those three, times the constant k:

$$h(z) = k\sqrt{z^2 + 1}; \qquad \text{so} \quad \sigma(z) = |h'(z)| = k\left|\frac{z}{\sqrt{z^2 + 1}}\right|.$$

Since $\sigma(z) \to k$ as $z \to \infty$, we may take $k = -c$. The limit of $\sigma(z)$ as $z \to i$ is ∞, so the speeds near the top of the boundary get arbitrarily large. The limit in either of the corners, as $z \to 0$, is 0.

Figure 1.1-10, page 13, shows a sampling of the equipotentials and streamlines, which are level curves of the velocity potential $u_2(z) = -c\,\mathrm{Re}\,h(z)$ and of the stream function $v_2(z) = -c\,\mathrm{Im}\,h(z)$, respectively. The expressions for those functions in terms of x and y are as follows:

$$u_2(x, y) = -\frac{c\,\mathrm{sign}(x)}{\sqrt{2}}\sqrt{x^2 - y^2 + 1 + \sqrt{(x^2 - y^2 + 1)^2 + (2xy)^2}},$$

$$v_2(x, y) = -\frac{1}{\sqrt{2}}\sqrt{-x^2 + y^2 - 1 + \sqrt{(x^2 - y^2 + 1)^2 + (2xy)^2}}. \tag{3}$$

The presence of the sign of x in the expression for u_2 reflects the need to specify the branch of the square root so that the second quadrant maps to itself. $\qquad\square$

Exercises

1. Verify the assertions (i)–(v) in Example 5.4.2.

2. Consider the mapping g in Example 5.4.2, and let $0 \leq \alpha < 2\pi$. Show that $g[H_\alpha]$ is half of a hyperbola with foci ± 1.

3. Show that for each $r > 0$, the ellipse parametrized by (1) has foci ± 1.

4. Describe the images under the mapping g (Example 5.4.2) of the upper and lower halves of the unit disk are the lower half-plane and upper half-plane, respectively.

5. In Example 5.4.2, the speed of flow at the top of the bump is $2c$. If $b > 0$ and the radius of the bump is b instead of 1, what then is the speed of flow at the top of the bump? Discuss the significance of your answer.

6. Explain the following remark. In Example 5.4.2, there is an alternative method for drawing the level curves of u_1 and v_1. Those curves are the images under g^{-1} of horizontal and vertical lines, and therefore can be easily parametrized.

7. Verify the expressions (2).

8. Verify the expressions (3).

5.5   USING THE POISSON INTEGRAL TO OBTAIN SOLUTIONS

A   The Poisson Integral on a Disk

To prepare a foundation for this method of solving Dirichlet problems, we will recount and put together what we know about certain integrals over the boundary of a disk. We will work with $D(0, 1)$ instead of an arbitrary disk $D(a, R)$ to avoid the distraction of extra clutter in the equations. In what follows, when we write down a complex integral over the circle $|w| = 1$, we will often also write down a pullback thereof to $[-\pi, \pi]$. We will often take advantage of the fact that $\bar{w}$ and $1/w$ are the same when $|w| = 1$.

Consider a holomorphic function $f = u + iv$ on $\bar{D}$. We know that by using the Cauchy integral, we can recover the values of f on the interior of the disk from the values of f on the boundary:

$$f(z) \equiv u(z) + iv(z) = \frac{1}{2\pi i} \int_{|w|=1} \frac{f(w)}{w - z} dw$$
$$= \frac{1}{2\pi} \int_{-\pi}^{\pi} \frac{f(e^{i\theta})}{1 - ze^{-i\theta}} d\theta \quad \text{for } |z| < 1. \tag{1}$$

We may reasonably ask: Can we, by means of integral formulas, recover (at least, up to an additive constant) the values of u, v, and f on the interior from the values of u on the boundary? After all, if u is known, then by 1.6.1, v and hence f are determined up to an additive constant. We will show that the answer is Yes.

If f is replaced in the integrand of (1) by an arbitrary piecewise continuous function h defined on the boundary, then (by 2.6.14) we still obtain a holomorphic function g on the complement of the circle:

$$g(z) = \frac{1}{2\pi i} \int_{|w|=1} \frac{h(w)}{w - z} dw = \frac{1}{2\pi} \int_{-\pi}^{\pi} \frac{h(e^{i\theta})}{1 - ze^{-i\theta}} d\theta \quad \text{for } |z| \neq 1; \tag{2}$$

and we may ask, What is g? How is it related to h? The factor in the integrand on the right is writeable as a geometric series, and then the order of integration and summation can be reversed, thus:

$$g(z) = \frac{1}{2\pi} \int_{-\pi}^{\pi} h(e^{i\theta}) \left(\sum_{k=0}^{\infty} z^k e^{-ik\theta} \right) d\theta$$

$$= \sum_{k=0}^{\infty} \left(\frac{1}{2\pi} \int_{-\pi}^{\pi} \overline{h(e^{i\theta})} e^{ik\theta} \, d\theta \right) z^k. \tag{3}$$

We point out some special cases. If $h(e^{i\theta}) = e^{in\theta}$, so that $\overline{h(e^{i\theta})} = e^{-in\theta}$, then $g(z) = 0$ if $n < 0$ and $g(z) = z^n$ if $n \geq 0$. Those results follow from the fact that

$$\frac{1}{2\pi} \int_{-\pi}^{\pi} e^{i(k-n)\theta} \, d\theta = \begin{cases} 0 & \text{if } n \neq k, \\ 1 & \text{if } n = k. \end{cases}$$

Evidently, then, when h is a combination of powers of w, the operator $h \mapsto g$ has the effect of replacing the coefficients on the negative powers by 0. For example, if $h(e^{i\theta}) = e^{-in\theta} + 1 + e^{in\theta} \equiv 1 + 2\cos n\theta$, then $g(z) = 1 + z^n$, that is, $g(re^{i\theta}) = 1 + r^n e^{in\theta}$.

We consider now another case in which, by means of (2), a not-necessarily-holomorphic h defined on the circle gives rise to a holomorphic g on the disk. Let $f = u + iv$ be a holomorphic function on $\bar{D}$, such as appeared in (1). Let $h = \bar{f} = u - iv$ (which is holomorphic only if f is constant). Then (3) becomes

$$g(z) = \frac{1}{2\pi i} \int_{|w|=1} \frac{\overline{f(w)}}{w - z} dw = \sum_{k=0}^{\infty} \left(\frac{1}{2\pi} \int_{-\pi}^{\pi} f(e^{i\theta}) e^{ik\theta} \, d\theta \right) z^k. \tag{4}$$

The function $w \mapsto f(w)w^{k-1}$ is holomorphic on the domain of f whenever k is a positive integer, so

$$\frac{1}{2\pi i} \int_{|w|=1} f(w)w^{k-1} dw = \frac{1}{2\pi} \int_{-\pi}^{\pi} f(e^{i\theta}) e^{ik\theta} d\theta = 0 \quad \text{for } k = 1, 2, \cdots.$$

Therefore the integrals in the sum (4) all equal 0, except for the one with $k = 0$; and we have established that the Cauchy integral of $\bar{f}$ is a constant function:

$$\overline{f(0)} = u(0) - iv(0) = \frac{1}{2\pi} \int_{-\pi}^{\pi} \frac{\overline{f(e^{i\theta})}}{1 - ze^{-i\theta}} d\theta \quad \text{for } |z| < 1. \tag{5}$$

Add (1) and (5). Since $f + \bar{f} = 2u$,

$$f(z) + u(0) - iv(0) = \frac{1}{2\pi} \int_{-\pi}^{\pi} u(e^{i\theta}) \frac{2}{1 - ze^{-i\theta}} d\theta. \tag{6}$$

We know from 1.6.2 or 3.1.4 that $u(0)$ equals the average value of u on the boundary:

$$u(0) = \frac{1}{2\pi} \int_{-\pi}^{\pi} u(e^{i\theta}) d\theta. \tag{7}$$

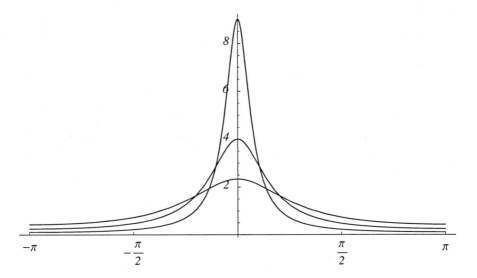

Figure 5.5-1   Poisson kernel graphs.

Subtract (7) from (6), using the fact that

$$\frac{2}{1 - ze^{-i\theta}} - 1 = \frac{1 + ze^{-i\theta}}{1 - ze^{-i\theta}} = \frac{e^{i\theta} + z}{e^{i\theta} - z},$$

and obtain

$$\text{Herglotz's Formula:} \quad f(z) - iv(0) = \frac{1}{2\pi} \int_{-\pi}^{\pi} u(e^{i\theta}) \frac{e^{i\theta} + z}{e^{i\theta} - z} d\theta. \tag{8}$$

Each of the integrals in (6) and (8) serves to generate, from the values of u on the boundary, a holomorphic function on the interior which differs from f by a constant. When we take the real part on both sides of (8), we get an integral that generates, from the values of u on the boundary, the function u itself on the interior:

$$\text{Poisson's Formula:} \quad u(z) = \frac{1}{2\pi} \int_{-\pi}^{\pi} u(e^{i\theta}) \frac{1 - |z|^2}{|e^{i\theta} - z|^2} d\theta. \tag{9}$$

Let $z = re^{it}$ for $0 \le r < 1$ and $t \in \mathbb{R}$, and (9) becomes

$$u(re^{it}) = \frac{1}{2\pi} \int_{-\pi}^{\pi} u(e^{i\theta}) P_r(t - \theta) d\theta, \tag{10}$$

$$\text{where} \quad P_r(x) := \frac{1 - r^2}{1 - 2r \cos x + r^2}. \tag{11}$$

The *Poisson kernel* is the set of functions $P_r (0 \le r < 1)$. Figure 5.5-1 shows the graphs of $P_{.4}, P_{.6},$ and $P_{.8}$ on $[-\pi, \pi]$, and may give you an idea about what happens as $r \to 1$.

5.5.1. Proposition. *The Poisson kernel has these properties:*

(i) $P_r(x) > 0$ *for* $x \in \mathbb{R}$ *and for* $0 \le r < 1$.

(ii) $\dfrac{1}{2\pi} \displaystyle\int_{-\pi}^{\pi} P_r(x)dx = 1$ *for* $0 \le r < 1$.

(iii) *For each* $\delta > 0$, $\sup\{P_r(x) \mid \delta \le x \le 2\pi - \delta\}$ *tends to* 0 *as* $r \to 1^-$.

Proof. The Poisson kernel is clearly positive everywhere; look at the form it takes in equation (9). To see (ii), apply (10) when $u \equiv 1$. As for (iii): For every $\delta > 0$ there exists $k > 0$ such that

$$\delta \le x \le 2\pi - \delta \implies \cos x < 1 - k.$$

Therefore for all $x \in [\delta, 2\pi - \delta]$,

$$P_r(x) < \frac{1 - r^2}{1 - 2r(1 - k) + r^2} = \frac{1 - r^2}{(1 - r)^2 + 2rk},$$

which tends to 0 as $r \to 1^-$. $\qquad\square$

Taking the imaginary part on both sides of (8) leads to

$$v(re^{it}) - iv(0) = \frac{1}{2\pi} \int_{-\pi}^{\pi} u(e^{i\theta}) \frac{2r\sin(t - \theta)}{1 - 2r\cos(t - \theta) + r^2} d\theta. \tag{12}$$

Therefore we can also recover the values v on the interior from the values of u on the boundary; (8), (9) or (10), and (12) answer the question that we asked at the outset.

The discussion could be carried out as well for f holomorphic on an arbitrary closed disk $\overline{D}(a, R)$ instead of $\overline{D}(0, 1)$. The result corresponding to (10), with $z = a + re^{it}$ for $0 \le r < R$, is

$$u(a + re^{it}) = \frac{1}{2\pi} \int_{-\pi}^{\pi} u(a + Re^{i\theta}) P_{r,R}(t - \theta)d\theta,$$

$$\text{where} \quad P_{r,R}(x) := \frac{R^2 - r^2}{R^2 - 2Rr\cos x + r^2} \qquad (0 \le r < R).$$

B Solutions on the Disk by the Poisson Integral

The Dirichlet problem for the unit disk may be stated as follows: Let h be defined on the boundary, continuous except at a finite number of points, and bounded. Find a bounded harmonic function u on the interior whose boundary values are given by h, in the sense that

$$\lim_{|z|<1, z \to w} u(z) = h(w) \tag{13}$$

for every point w on the boundary at which h is continuous.

One may certainly consider a broader class of boundary value functions h, but the treatment here will be adequate for our purposes.

The discussion above dealt with a seemingly much easier problem. In effect we *began* with a harmonic function u on $D(0, 1 + \epsilon)$. Letting h be the restriction of u to the unit circle, we found that we could recover u on the interior from h. But now we seem to begin with much less information. We are given a function h defined only on the boundary, perhaps with discontinuities; we must find u on D, show that u is harmonic, and prove (13).

Equation (2) is a starting point. It gives us a function g which is holomorphic, and hence harmonic, in the disk, though of course it is easy to find examples in which (13) fails if we were to try g for u. The process that led to equation (10) will yield the answer.

5.5.2. Theorem. *Let h be a bounded, piecewise continuous function defined on the boundary of the unit disk. Let*

$$u(re^{it}) = \frac{1}{2\pi} \int_{-\pi}^{\pi} P_r(t - \theta)h(e^{i\theta})d\theta \qquad (0 \le r < 1). \tag{14}$$

Then u is a bounded harmonic function on D. Its boundary values are given by h, in the sense that for every point $e^{i\theta_0}$ at which h is continuous,

$$\lim_{t \to \theta_0, r \to 1^-} u(re^{it}) = h(e^{i\theta_0}).$$

Proof. It suffices to prove the Theorem in the case when h is real-valued, because then we can apply it to the real and imaginary parts of an arbitrary h to obtain the general case.

The function g given by (2) is holomorphic on D. But consider instead the following function, which is $2g$ minus a constant:

$$g_1(z) := \frac{1}{2\pi} \int_{-\pi}^{\pi} h(e^{i\theta}) \left(\frac{2}{1 - ze^{-i\theta}} - 1 \right) d\theta. \tag{15}$$

It is also holomorphic. So if we take the real part, we obtain a harmonic function u. In fact, we obtain (14), noting again that

$$\text{Re} \left(\frac{2}{1 - ze^{-i\theta}} - 1 \right) = \frac{1 - |z|^2}{|e^{i\theta} - z|^2} = P_r(t - \theta) \quad \text{(where } z = re^{it}\text{)},$$

a computation that we used to get from (8) to (10).

The function u is bounded because h is bounded and

$$|u(re^{it})| \le \left| \frac{1}{2\pi} \int_{-\pi}^{\pi} P_r(t - \theta)h(e^{i\theta})d\theta \right|$$

$$\le \frac{1}{2\pi} \int_{-\pi}^{\pi} |P_r(t - \theta)|d\theta \cdot \max_{\theta} |h(e^{i\theta})|$$

$$= \max_{\theta} |h(e^{i\theta})|.$$

If $h(e^{i\theta})$ is replaced by its rotation $h(e^{i(\theta - \theta_0)})$ on the right hand side of equation (14), then the integral equals $u(re^{i(t - \theta_0)})$. Thus the Poisson integral of a rotation is the rotation of the Poisson integral. Therefore it suffices to prove the last statement of the Theorem in the case when $\theta_0 = 0$:

$$\lim_{t \to 0, r \to 1^-} u(re^{it}) = h(1).$$

Because of condition (ii) in Theorem 5.5.1,

$$|u(re^{it}) - h(1)| = \left| \frac{1}{2\pi} \int_{-\pi}^{\pi} P_r(t - \theta)[h(e^{i\theta}) - h(1)]d\theta \right|.$$

To estimate this difference, we break the interval of integration into two parts; for every choice of δ with $0 < \delta < \pi$,

$$|u(re^{it}) - h(1)| \leq \left| \frac{1}{2\pi} \int_{-\delta}^{\delta} P_r(t - \theta)[h(e^{i\theta}) - h(1)]d\theta \right|$$

$$+ \left| \frac{1}{2\pi} \int_{\delta}^{2\pi - \delta} P_r(t - \theta)[h(e^{i\theta}) - h(1)]d\theta \right|$$

$$\leq \left(\frac{1}{2\pi} \int_{\delta}^{2\pi - \delta} |h(e^{i\theta}) - h(1)|d\theta \right) \cdot \underbrace{\sup\{P_r(t - \theta) \mid \delta \leq \theta \leq 2\pi - \delta\}}_{=:A}$$

$$+ \left(\int_{-\delta}^{\delta} |P_r(t - \theta)|d\theta \right) \cdot \underbrace{\sup\{|h(e^{i\theta}) - h(1)| \mid -\delta \leq \theta \leq \delta\}}_{=:B}$$

$$\leq \underbrace{\left(\frac{1}{2\pi} \int_{0}^{2\pi} |h(e^{i\theta})|d\theta + |h(1)| \right)}_{=:K} \cdot A \quad + \quad B.$$

Let $\epsilon > 0$. We need to show that for all r sufficiently close to 1 and all t sufficiently small, $|u(re^{it}) - h(1)| < \epsilon$. Since h is continuous at 1, we know that $B < \epsilon/2$ for all sufficiently small δ. Fix such a δ. Require $|t| \leq \frac{\delta}{2}$, so that

$$A \leq \sup\{P_r(x) \mid \tfrac{\delta}{2} \leq x \leq 2\pi - \tfrac{\delta}{2}\}.$$

Therefore, by condition (iii) of Proposition 5.5.1, $A < \epsilon/(2K)$ for all r sufficiently close to 1.   □

5.5.3. Example. *Find the bounded harmonic function u on the unit disk whose boundary values are given by*

$$h(e^{i\theta}) = \begin{cases} 100 & for \ 0 < \theta < \frac{\pi}{2}, \\ 0 & for \ \frac{\pi}{2} < \theta < 2\pi. \end{cases}$$

Draw its graph.

Solution. By Theorem 5.5.2,

$$u(re^{it}) = \frac{1}{2\pi} \int_0^{\pi/2} \frac{1 - r^2}{1 - 2r\cos t + r^2} dt.$$

Commands for asking *Mathematica* to draw the graph:

$$u[r\_, t\_] := \frac{1}{2\pi} \text{NIntegrate} \left[\frac{1 - r^2}{1 - 2r\cos t + r^2}, \{t, 0, \tfrac{\pi}{2}\} \right];$$

ParametricPlot3D[$\{r\cos[t], r\sin[t], u[r, t]\}, \{r, 0, 0.98\}, \{t, 0, 2\pi\}$].

The result is shown in Figure 1.1-6, page 9.   □

Theorem 5.5.2 can be proved in the same manner for an arbitrary disk $\overline{D}(a, R)$ instead of for $D(0, 1)$. For the sake of convenient reference, we provide the general version:

5.5.4. Theorem. *Let h be a bounded, piecewise continuous function defined on the boundary of $D(a, R)$. Let*

$$u(a + e^{it}) = \frac{1}{2\pi} \int_{-\pi}^{\pi} P_{r,R}(t - \theta)h(a + Re^{i\theta})dt \qquad (0 \le r < R). \tag{16}$$

Then u is a bounded harmonic function on $D(a, R)$. Its boundary values are given by h, in the sense that for every $\theta_0 \in \mathbb{R}$ at which h is continuous,

$$\lim_{t \to \theta_0, r \to R^-} u(a + re^{it}) = h(a + Re^{i\theta_0}).$$

C   Geometry of the Poisson Integral

In Section 1.1, we described a geometric procedure for obtaining u on the interior from h on the boundary of the unit disk. Now we shall restate that procedure and show how it follows from the integral formula (9). This result is due to H. A. Schwarz ([64, pp. 144–171]); a nice account is given in Tristan Needham's book ([45, pp. 556–560]).

5.5.5. Proposition. *Let h be a function defined on the unit circle, continuous except at a finite number of points, and bounded. Let (14) give its extension to the unit disk. Fix the point $z := re^{it}$, with $0 \le r < 1$. For each point $w := e^{i\theta}$ on the unit circle, let w^* be the other point that lies both on the unit circle and on the line through w and z. Let h^* be the function on the boundary given by $h^*(w) := h(w^*)$. Then*

$$u(z) = \frac{1}{2\pi} \int_{-\pi}^{\pi} h^*(e^{i\theta})d\theta.$$

Proof. Observe that $w \mapsto w^*$ is its own inverse, so that $w^{**} \equiv w$ and $h^*(w^*) \equiv h(w)$. Let $\rho := |w - z|$ and $\sigma := |w^* - z|$. Figure 5.5-2 shows a typical case; there is an

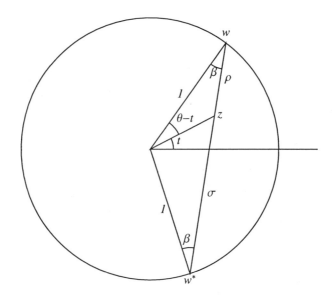

Figure 5.5-2 An illustration for the proof of Schwarz's geometric interpretation of the Poisson Integral Formula.

isosceles triangle whose sides have lengths 1, 1, and $\rho + \sigma$. Let β be its base angle. The line segment from 0 to z divides it into two triangles; applying to them the Law of Cosines gives us

$$r^2 = 1 + \rho^2 - 2\rho \cos \beta \quad \text{and} \quad r^2 = 1 + \sigma^2 - 2\sigma \cos \beta.$$

Therefore ρ and σ are the two roots of the quadratic equation

$$X^2 - 2X \cos \beta + (1 - r^2) = 0.$$

It follows that $\rho\sigma = 1 - r^2$. Therefore

$$\frac{\sigma}{\rho} = \frac{1 - r^2}{\rho^2}, \quad \text{which equals} \quad \frac{1 - |z|^2}{|e^{i\theta} - z|^2},$$

the Poisson kernel as it appears in (9). That integral equals the limit as $n \to \infty$ of Riemann sums

$$\frac{1}{2\pi} \sum_{k=0}^{n-1} h(w_k) \frac{1 - |z|^2}{|w_k - z|^2} (\theta_{k+1} - \theta_k), \tag{17}$$

where $\theta_k = \frac{2\pi}{n} k$ and $w_k = e^{i\theta_k}$; which remains true, as we ask you to prove in Exercise 4, if we somewhat modify the sums as follows:

$$u(z) = \lim_{n \to \infty} \frac{1}{2\pi} \sum_{k=0}^{n-1} h(w_k) \frac{1 - |z|^2}{|w_k - z|^2} |w_{k+1} - w_k|. \tag{18}$$

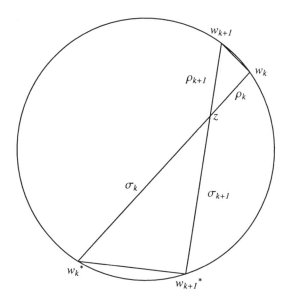

Figure 5.5-3   Similar triangles.

Because the two triangles in Figure 5.5-3 are similar,

$$\frac{|w_{k+1}^* - w_k^*|}{|w_{k+1} - w_k|} = \frac{\sigma_k}{\rho_k} = \frac{1 - |z|^2}{|w_k - z|^2}.$$

Therefore

$$u(z) = \lim_{n \to \infty} \frac{1}{2\pi} \sum_{k=0}^{n-1} h(w_k)|w_{k+1}^* - w_k^*|$$

$$= \lim_{n \to \infty} \frac{1}{2\pi} \sum_{k=0}^{n-1} h^*(w_k^*)|w_{k+1}^* - w_k^*| = \frac{1}{2\pi} \int_{-\pi}^{\pi} h^*(e^{i\theta})d\theta.$$

$\square$

D   Harmonic Functions and the Mean Value Property

Theorems 1.6.2 and 3.1.4 state that harmonic functions have the mean value property. We are now in a position to prove the converse.

5.5.6. Theorem. *Let f be a real-valued continuous function on the open set O. Suppose that f enjoys the mean value property on O. That is, for each $a \in O$, it is true for all sufficiently small $R > 0$ that the value of $f(a)$ equals the average value* *of f on the circle* $|z - a| = R$:

$$f(a) = \frac{1}{2\pi} \int_{-\pi}^{\pi} f(a + Re^{i\theta})d\theta.$$

Then f is harmonic on O.

Proof. Let $a \in O$ and let $R > 0$ be sufficiently small. It suffices to prove f harmonic on $D(a, R)$. By Theorem 5.5.4, the function u given by

$$u(a + re^{it}) = \frac{1}{2\pi} \int_{-\pi}^{\pi} f(a + Re^{i\theta})P_{r,R}(t - \theta)dt \qquad (0 \le r < R)$$

is harmonic on $D(a, R)$. If we define u to equal f on the boundary, then u is continuous on $\overline{D}(a, R)$. The proof will be complete as soon as we show that $u = f$ on $D(a, R)$.

We know that each of the two functions u and f has the mean value property on $D(a, R)$. Therefore $u - f$ also has it. By Theorem 1.6.5, then, $u - f$ cannot attain a local extremum at any point of $D(a, R)$ unless it is constant. But being continuous, it must attain both its extrema on the boundary, and we know that $u - f = 0$ on the boundary. Therefore $u - f \equiv 0$ on $\overline{D}(a, R)$. □

E   The Neumann Problem on a Disk

We will derive from Theorem 5.5.2 an integral formula for recovering an unknown harmonic function u on D from known boundary values of its normal derivative $\dfrac{\partial u}{\partial \vec{n}}$.

5.5.7. Theorem. *Let k be a bounded, piecewise continuous function defined on the boundary of the unit disk such that*

$$\frac{1}{2\pi} \int_{-\pi}^{\pi} k(e^{i\theta})d\theta = 0. \tag{19}$$

Let

$$u(re^{it}) = \frac{1}{2\pi} \int_{-\pi}^{\pi} \ln(1 - 2r \cos(t - \theta) + r^2)^{-1} k(e^{i\theta})d\theta \quad (0 \le r < 1). \tag{20}$$

Then u is a bounded harmonic function on D, and the boundary values of its radial derivative are given by k in the following sense: At every point $e^{i\theta_0}$ at which k is continuous,

$$\lim_{t \to \theta_0, r \to 1^-} \frac{\partial}{\partial r} u(re^{it}) = k(e^{i\theta_0}). \tag{21}$$

Proof. Let $E(r, x) = \ln|1 - re^{ix}|^{-2} = \ln(1 - 2r \cos x + r^2)^{-1}$. The function $z \mapsto \ln|1 - z|^{-2}$ is harmonic on D, since it is the real part of $z \mapsto \log(1 - z)^{-2}$, which is defined and holomorphic for Re $z < 1$. A computation shows that

$$\frac{\partial E}{\partial r}(r,x) = \begin{cases} \dfrac{P_r(x) - 1}{r} & \text{for } 0 < r < 1, \\[2mm] 1 & \text{if } r = 0. \end{cases} \qquad (22)$$

Equation (20) may be written

$$u(re^{it}) = \frac{1}{2\pi}\int_{-\pi}^{\pi} E(r, t - \theta)k(e^{i\theta})d\theta \quad (0 \le r < 1).$$

Then u is harmonic on D, since it is the real part of the holomorphic function

$$z \mapsto \frac{1}{2\pi}\int_{|w|=1} \log(1 - z\bar{w})^{-2}k(w)dw \quad (|z| < 1).$$

Then for $0 < r < 1$,

$$\frac{\partial u}{\partial r}(re^{it}) = \frac{1}{2\pi}\int_{-\pi}^{\pi} \frac{P_r(t - \theta) - 1}{r}k(e^{i\theta})d\theta \quad \text{(by (22))}$$

$$= \frac{1}{2\pi r}\int_{-\pi}^{\pi} P_r(t - \theta)k(e^{i\theta})d\theta \quad \text{(by (19))}.$$

Then (21) follows from Theorem 5.5.2. $\qquad\qquad\qquad\qquad\qquad\qquad\square$

F   The Poisson Integral on a Half-Plane, and on Other Domains

There is ample reason for the emphasis on the disk in this section, since it is a kind of universal setting for boundary value problems. Let O be an open set such that there is a conformal equivalence f of O and D, and such that f extends to a one-to-one continuous mapping from $\bar{O}$ onto $\bar{D}$. Then a problem on O reduces to a problem on D; if we can solve the latter, we can solve the former.

We may take a somewhat different view of that procedure, which is essentially a change of variables. Let us say that there is a particular set O on which we wish to solve Dirichlet problems. Then we can use the mapping f to change variables in the Poisson formula (14) for the disk and obtain a formula specially for O. In effect, we prove a version of Theorem 5.5.2 specially for O. In what follows, we do so for the upper half-plane.

5.5.8. Theorem. *Let a be a bounded, piecewise continuous function defined on the real axis. For $z = x + iy$ in the open upper half-plane $\mathbb{Y}^{\circ}$, let*

$$u(x,y) = \frac{1}{\pi}\int_{-\infty}^{+\infty} \frac{ya(s)}{(s - x)^2 + y^2}ds. \qquad (23)$$

Then u is a bounded harmonic function on $\mathbb{Y}^{\circ}$, and its boundary values are given by a in the sense that for every point $s_0 \in \mathbb{R}$ at which a is continuous,

$$\lim_{x \to s_0, y \to 0^+} u(x,y) = a(s_0).$$

Proof. We may suppose that a is real-valued. We will use a change of variables. Any Möbius map S from $\mathbb{Y}^\circ$ onto D will serve; here is a good choice:

$$S(z) = -\frac{z-i}{z+i}; \qquad S^{-1}(w) = -i\frac{w-1}{w+1}.$$

Let $h(e^{i\theta}) := a(S^{-1}(e^{i\theta}))$. Proceed as in the proof of 5.5.2. We obtain a holomorphic function g_1 as in (15):

$$
\begin{aligned}
g_1(w) &= \frac{1}{2\pi} \int_{-\pi}^{\pi} h(e^{i\theta}) \left(\frac{2}{1 - we^{-i\theta}} - 1 \right) d\theta \\
&\equiv \frac{1}{2\pi} \int_{-\pi}^{\pi} h(e^{i\theta}) \left(\frac{e^{i\theta} + w}{e^{i\theta} - w} - 1 \right) d\theta.
\end{aligned}
\tag{24}
$$

Change variables: Let z and s be defined by $w = S(z)$ and $e^{i\theta} = S(s)$, so that $ie^{i\theta}d\theta = S'(s)ds$. We obtain the holomorphic function

$$
\begin{aligned}
z \mapsto g_1(S(z)) &= \frac{1}{2\pi} \int_{-\infty}^{+\infty} h(S(s)) \left(\frac{S(s) + S(z)}{S(s) - S(z)} \right) \frac{S'(s)}{iS(s)} ds \\
&= \frac{1}{\pi} \int_{-\infty}^{+\infty} \left(\frac{sz + 1}{i(s - z)} \right) \frac{ds}{s^2 + 1}.
\end{aligned}
\tag{25}
$$

Taking the real part on both sides leads to (23). $\qquad\square$

5.5.9. Remark. We mention briefly two alternative methods of proof for Theorem 5.5.8, which do not presuppose Theorem 5.5.2. One is to carry out a development for the upper half-plane that is analogous to the one for D which culminated in the proof of 5.5.2. Another is to recognize that Example 5.1.5 produced solutions for special cases of Theorem 5.5.8, and that every function a satisfying the hypothesis of 5.5.8 can be approximated by one that has finite range.

Exercises

1. Explain the details of how (8) implies (9) implies (10).

2. Let $C_r(x) = \dfrac{1}{1 - re^{ix}}$. Verify these relationships:

$$P_r(x) = C_r(x) + \overline{C_r(x)} - 1 = \sum_{-\infty}^{\infty} r^{|n|} e^{inx};$$

$$\overline{f(0)} = \frac{1}{2\pi i} \int_{-\pi}^{\pi} \overline{f(e^{i\theta})} C_r(t - \theta) d\theta;$$

$$\overline{f(re^{it})} = \frac{1}{2\pi i} \int_{-\pi}^{\pi} \overline{f(e^{i\theta})} P_r(t - \theta) d\theta.$$

3. Here is a somewhat different way to arrive at the Poisson kernel; provide the details. Let f be holomorphic on the closed unit disk $\overline{D}(0,1))$. For $z := re^{it}$,

$r < 1$, let $z^* := \frac{1}{r}e^{it}$, which is the reflection of z in the unit circle and lies outside it. Let the unit circle be given by $w(\theta) := e^{i\theta}$ for $-\pi \leq \theta \leq \pi$. Then

$$f(z) = \frac{1}{2\pi i} \int_{|w|=1} \frac{f(w)dw}{w - z} \quad \text{and} \quad 0 = \frac{1}{2\pi i} \int_{|w|=1} \frac{f(w)dw}{w - z^*}.$$

Therefore

$$f(z) = \frac{1}{2\pi i} \int_{|w|=1} f(w) \left(\frac{1}{w - z} - \frac{1}{w - z^*} \right) dw$$

$$= \int_{-\pi}^{\pi} f(e^{i\theta}) P_r(t - \theta) d\theta.$$

4. Prove (18).

5. Let h and u be a in 5.5.2. Show that if h is given by an absolutely convergent Fourier series: $h(e^{it}) = \sum_{n=-\infty}^{\infty} a_n e^{int}$, then $u(re^{it}) = \sum_{n=-\infty}^{\infty} a_n r^n e^{int}$.

6. Let $u + iv$ be holomorphic on $\bar{D}$. Let k be continuous on the unit circle. Suppose that $\frac{\partial u}{\partial \vec{n}} = k$ at each point of the unit circle. Show that the Neumann problem for u—that is, the problem of recovering u on D from the values of k on the unit circle—is reducible to a Dirichlet problem for v.

7. Consider the Neumann problem as described in the hypothesis of Theorem 5.5.7. Discuss a procedure for solving it using the ideas of Exercise 6.

8. Explain why u, the function produced by Theorem 5.5.7, is unique up to an additive constant.

9. Define k on the unit circle by

$$k(e^{i\theta}) = \begin{cases} 10 & \text{for } -\frac{\pi}{2} \leq \theta \leq \frac{\pi}{2}, \\ -20 & \text{for } \frac{3\pi}{4} \leq \theta \leq \frac{5\pi}{4}, \\ 0 & \text{otherwise.} \end{cases}$$

Solve the resulting Neumann problem using the method of Theorem 5.5.7. At least set up the integral for u. If you have suitable computing power available, produce a drawing of level curves of u.

10. Interpret Exercise 9 as a fluid flow problem.

11. Interpret Exercise 9 as a temperature distribution problem. What's the significance of the fact that the temperature u is determined only up to an additive constant?

12. Solve Exercise 9 by conformal mapping methods.

13. Explain why condition (19) is really necessary in the hypothesis of Theorem 5.5.7.

14. In the proof of Theorem 5.5.8, carry out the computations to show that (24) leads to (25), and that (25) leads to (23). Discuss and justify the change of variable, which is said to convert an integral on $[-\pi, \pi]$ to an integral on the infinite interval $(-\infty, \infty)$.

15. Show that for functions $a : \mathbb{R} \to \mathbb{R}$ with finite range, the method of Example 5.1.5 gives the same answer as the integral formula (23).

Help on Selected Exercises

4. Suggestion: Show that there is a constant c such that

$$||e^{i\theta_{k+1}} - e^{i\theta_k}| - (\theta_{k+1} - \theta_k)| < c(\theta_{k+1} - \theta_k)^2,$$

and that therefore the change from (17) to (18) does not make a difference in the value of the limit as $n \to \infty$.

6. Let $U(r,\theta) = u(r\cos\theta, r\sin\theta)$ and likewise for V. On the unit circle, $\dfrac{\partial u}{\partial \vec{n}}$ is the same as $\dfrac{\partial U}{\partial r}$. By the Cauchy-Riemann equations (see Exercise 25 in Section 1.6, page 71), $\dfrac{\partial U}{\partial r} = \dfrac{1}{r}\dfrac{\partial V}{\partial \theta}$. Let $h(e^{i\theta}) = \displaystyle\int_0^\theta k(e^{is})ds$. Then find a harmonic function v on D whose boundary values are given by h. Then find u such that $u + iv$ is holomorphic on D.

12. Let T be a Möbius map from D onto $\mathbb{Y}^o$ such that $\text{Arc}(1, -\frac{\pi}{2}, \frac{\pi}{2})$ is mapped to $[1, \infty]$ and $\text{Arc}(1, \frac{3\pi}{4}, \frac{5\pi}{4})$ to $[-\infty, -1]$. Then apply $\sin^{-1}$, which maps $\mathbb{Y}^o$ to a vertical half strip on which the solution is obvious.

5.6   WHEN IS THE SOLUTION UNIQUE?

Theorem 5.5.2 may be restated to include an answer to the question of uniqueness, at least for Dirichlet problems on the unit disk:

5.6.1. Theorem. *Given a piecewise continuous function h on ∂D, there exists a unique bounded harmonic function u on D such that for every point $e^{i\theta_0}$ at which h is continuous,*

$$\lim_{t\to\theta_0, r\to 1^-} u(re^{it}) = h(e^{i\theta_0}). \tag{1}$$

A proof and a discussion of related matters appear in [24, Section 15.4] and [22, Chapter 4].

The statement provides an answer as well for a broad class of Dirichlet problems which are reducible to one on the unit disk.

The proof of uniqueness is easy in the case when h is continuous on ∂D. Then u may be described as harmonic on D, continuous on $\bar{D}$, and equal to h on ∂D. Let v be another such function. Then $u - v$ would attain its maximum and its minimum somewhere on $\bar{D}$. But $u - v \equiv 0$ on ∂D. If $u - v$ were nonconstant, it would take on an extremum somewhere on D, which by 3.1.4 or 1.6.5 cannot be the case. Therefore $u \equiv v$.

We will describe an example to show that Theorem 5.6.1 becomes false if the word "bounded" is omitted from the hypothesis, assuming h has at least one discontinuity $e^{i\theta_1}$. Consider the Möbius map from D onto the open upper half-plane:

$$T : 1, -1, -i \rightarrow \infty, 0, 1; \quad T(z) = i\frac{z+1}{z-1}.$$

The function given by

$$U(z) := \operatorname{Im} T(z) \equiv \operatorname{Im}\left(i\frac{z+1}{z-1}\right) \tag{2}$$

is a real-valued harmonic function on D, continuous and equal to 0 at every point of ∂D except 1. Along every H_α for $0 < \alpha < \pi$, $\operatorname{Im} z \rightarrow \infty$ as $z \rightarrow \infty$. The pre-image $T^{-1}(H_\alpha)$ is a circular arc through -1 and 1. Along each of those arcs, $U(z) \rightarrow \infty$ as $z \rightarrow 1$. If u solves the Dirichlet problem, so does $z \mapsto u(z) + U(e^{-i\theta_1}z)$.

The remarks above are pertinent to Neumann problems. In that case, the boundary values are prescribed for the normal derivative of u and not for u itself. Therefore our conclusion will be not that u is uniquely determined, but that it is of a given form and determined up to one or two constants. In the examples in this book, the Neumann conditions are homogeneous, which means that the normal derivative is specified to equal 0 on all or part of the boundary.

Consider, for example, the fluid flow problem for $\mathbb{Y}^o$, Figure 5.4-1, discussed in Section 5.4. We seek a harmonic function u on $\mathbb{Y}^o$ such that $u_y = 0$, and satisfying other conditions which imply that u_y is bounded. Then the problem of finding u_y on $\mathbb{Y}^o$ is equivalent to that of finding a function harmonic on D, continuous at all points of ∂D with one exception, and bounded; such a function exists and is unique by Theorem 5.6.1. Thus $u_y \equiv 0$ is the only possibility, from which we conclude that $u_{xx} \equiv 0$, hence $u(x, y) = ax + b$ for some constants a and b. Other Neumann problems are reducible to that one.

The uniqueness questions in the Dirichlet-Neumann problems considered in this book can be resolved by reference to the results and methods we have mentioned. For example, in Example 5.1.6, the boundary conditions imply that $u_y \equiv 0$ equals 0 everywhere on the boundary except at the corners. With the added assumption that u_y is bounded, 5.6.1 applies and assures that $u(x, y)$ has the form $x \mapsto ax + b$; and the prescribed values of u on the vertical walls of the strip settle the values of a and b.

Exercise

1. Let U be the function given by (2). Discuss further the behavior of $U(z)$ as z approaches 1 from within the unit disk. For example, consider U restricted to the pre-image under T of the horizontal line $y = 17$.

5.7 THE SCHWARZ REFLECTION PRINCIPLE

At the end of the section we will arrive at a statement of the Reflection Principle which will cover a rather general situation. We will arrive there by easy steps, beginning with a review and application of simple facts. You have three choices. If you study the first Proposition, you will have the basic idea. If you study also the second one, you will have seen a representative version of the Reflection Principle. If you complete the section, you will know its most general form, Theorem 5.7.3. To prepare for the discussion of Schwarz-Christoffel transformations, you should at least understand what the Theorem says.

The mapping $z \mapsto \bar{z}$, known as complex conjugation, is conjugate-holomorphic. So if f is holomorphic, then $z \mapsto f(\bar{z})$ and $z \mapsto \overline{f(z)}$ are conjugate-holomorphic; and the mapping

$$z \mapsto \overline{f(\bar{z})},$$

is holomorphic. Those familiar facts are special cases of 2.3.6.

Complex conjugation is also writeable as $x + iy \mapsto x - iy$, and may be understood as reflection in the real axis. So we may restate what we just asserted as follows. Let f be a holomorphic function given by $f(x, y) = u(x, y) + iv(x, y)$. Then the mapping

$$(x, y) \mapsto u(x, -y) - iv(x, -y)$$

is holomorphic. The procedure that gives this mapping is as follows: Reflect in the real axis; apply f; reflect in the real axis again.

Now we will give another restatement, this time being specific about domains. When O is an open connected subset of the open upper half-plane $\mathbb{Y}^{o}$, let O^{*} denote the reflection of O in the real axis:

$$O^{*} := \{\bar{z} \mid z \in O\} \equiv \{z \mid \bar{z} \in O\}.$$

For example: If O is the disk $D(1 + 2i, 1)$, then O^{*} is the disk $D(1 - 2i, 1)$. If $O = \mathbb{Y}^{o}$, then O^{*} is the open lower half-plane $\mathbb{Y}^{o}_{-}$. Make a few sketches so that you have some simple cases in mind. Let f be holomorphic on O. Then one can extend f to a holomorphic function F on the union $O \cup O^{*}$:

$$F(z) \equiv U(x, y) + iV(x, y)$$
$$:= \begin{cases} f(z) \equiv u(x, y) + iv(x, y) & \text{if } z \equiv x + iy \in O, \\ \overline{f(\bar{z})} \equiv u(x, -y) - iu(x, -y) & \text{if } z \equiv x + iy \in O^{*}. \end{cases} \qquad (1)$$

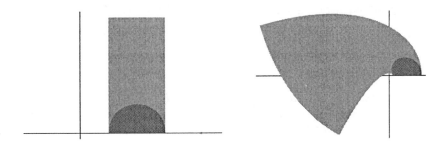

Figure 5.7-1 On the left, a certain rectangle's interior O. On the right, its image under a given function f, as discussed in Proposition 5.7.1. A half-disk within the rectangle and its image are shown with darker shading.

Now suppose that the intersection of the boundaries of O and O^* contains a nonempty open interval I of the real axis, and $O \cup I \cup O^*$ is open. Question: Under what conditions can F be defined also on I so that it is then a holomorphic extension of f to $O \cup I \cup O^*$?

There are cases in which the extension is easily seen to be possible. Let $f(z) :=$ Log z for $z \in O := \mathbb{Y}^o$, and let $I := (0, \infty)$. The extension to $O \cup I \cup O^*$ is provided by $F(z) := $ Log z.

Suppose that there is a holomorphic extension F of f to $O \cup I \cup O^*$. If $\{z_n\}$ is a sequence in O that converges to a point $x \in I$, then $\{\bar{z}_n\}$ also converges to x. So by the assumed continuity of F,

$$\lim_{n \to \infty} F(z_n) = F(x) = \lim_{n \to \infty} F(\bar{z}_n).$$

Since $F(\bar{z}_n) = \overline{F(z_n)}$, the limit on the right must also equal $\overline{F(x)}$. Therefore for all $x \in I$, $F(x) = \overline{F(x)}$, that is, $F(x)$ must be real.

So: A necessary condition for f to extend across I is that f should have a continuous extension to $O \cup I$, with real values on I. Is that condition also sufficient? Suppose that it holds, and let

$$F(z) = \begin{cases} f(z) & \text{if } z \in O \cup I, \\ \overline{f(\bar{z})} & \text{if } z \in O^*. \end{cases}$$

Then F is holomophic on $O \cup O^*$. Because of the way F is defined, it is also continuous on $O \cup I \cup O^*$. But must it be holomorphic on $O \cup I \cup O^*$?

The answer is Yes. In fact, the next Proposition, which is an uncomplicated special case of the Reflection Principle, presents an even better result: The hypothesis need not require f to have a continuous extension to $O \cup I$. It suffices to require only the seemingly weaker condition that v have a continuous extension to $O \cup I$, with $v = 0$ on I. This should not be much of a surprise, since we know that f is determined up to an additive constant by v.

5.7.1. Proposition. *Let $a_0 \in \mathbb{R}$ and let $0 < r < b$. Let I denote the open interval $(a_0 - r, a_0 + r)$. Let $f \equiv u + iv$ be holomorphic on the interior O of the rectangle $I \times (0, b)$:*

$$O := \{x + iy \mid x \in I \text{ and } 0 < y < b\}.$$

Let O^* be the reflection of O across the real axis. Let $F := U + iV$ be the extension of f to $O \cup O^*$ given by (1). Suppose that whenever $\{z_n\}$ is a sequence in O that converges to a point of I, then $v(z_n) \to 0$. Then F can be defined on I so that it is holomorphic on $O \cup I \cup O^*$.

Proof. Define V to be identically 0 on I. We know that V is harmonic and therefore has the mean value property on O and also on O^*. In view of 5.5.6, in order to show that V is harmonic on $O \cup I \cup O^*$, it suffices to show that for each $a \in I$,

$$\frac{1}{2\pi} \int_{-\pi}^{\pi} V(a + se^{it})dt \; = \; V(a) \equiv 0 \tag{2}$$

for all sufficiently small $s > 0$. Since

$$V(x, y) = -V(x, -y) \quad \text{for all } (x, y) \in O \cup O^*,$$

the function $t \mapsto V(a + se^{it})$ is an odd function, from which (2) follows, provided $\bar{D}(a, s) \subset O \cup I \cup O^*$. Therefore V is harmonic on $O \cup I \cup O^*$, and in particular on $D(a_0, r)$.

On $D(a_0, r)$, there is a holomorphic function G, determined up to an additive constant, such that $V \equiv \operatorname{Im} G$.

On $D(a_0, r) \cap \mathbb{Y}^o$, $V \equiv \operatorname{Im} F$ also; therefore $F - G$ is constant. We adjust the choice of G so that $F = G$ on $D(a_0, r) \cap \mathbb{Y}^o$.

Since G is real-valued on I, the coefficients c_k in its Taylor series at a,

$$G(z) = \sum_{k=0}^{\infty} c_k(z - a)^k,$$

are all real. Consequently, $G(\bar{z}) = \overline{G(z)}$. So $G(z) = F(z)$ for all $z \in D(a_0, r) \cap \mathbb{Y}^o_-$. As soon as we define F to agree with G on I, we know that F is a holomorphic extension of f to $O \cup I \cup O^*$. $\qquad \square$

The next step is to deal with a more general set O. In the proof of 5.7.1, there was one disk $D(a_0, r)$ containing I and contained in O (since $r < b$). In the next version of the Reflection Principle, we must allow for the use of more than one disk.

5.7.2. Proposition. *Let $f \equiv u + iv$ be holomorphic on the open connected set $O \subseteq \mathbb{Y}^o$. Let O^* be the reflection of O in the real axis, and let F be the holomorphic extension of f to $O \cup O^*$ given by $F(z) = \overline{F(\bar{z})}$. Let I be a subset of $bO \cap bO^*$ such that $O \cup I \cup O^*$ is open. Suppose that whenever $\{z_n\}$ is a sequence in O that converges to a point of I, then $v(z_n) \to 0$. Then F may be defined also on I so that it is holomorphic on $O \cup I \cup O^*$.*

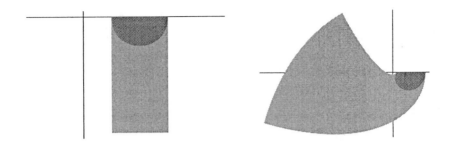

Figure 5.7-2   On the left, the reflection of the set O of Figure 5.7-1 in the real axis. On the right, its image under F.

Proof. For each point $t \in I$, let D_t be an open disk, with center t, that is contained in $O \cup I \cup O^*$. Proceeding as in the proof of 5.7.1, we find that there is a holomorphic function F_t on D_t that agrees with F on $D_t \cap \mathbb{Y}^o$ and also on $D_t \cap \mathbb{Y}^o_-$. If two such disks intersect, $D_t \cap D_s \neq \emptyset$, then F_t and F_s agree on $D_t \cap D_s \cap \mathbb{Y}^o$, since each agrees with F there; and hence they agree on $D_t \cap D_s$. So there is no inconsistency in defining F to equal F_t on D_t for every $t \in I$, and this gives the desired extension of F to $O \cup I \cup O^*$.   □

As presented so far, the Reflection Principle uses reflection in the real axis only, either for the domain or the range. There is a more general version which allows reflection, for both domain and range, in an arbitrary line or circle. Whenever A is a line or a circle, the mapping J_A, called reflection in A, is conjugate-holomorphic (see subsection 2.1I). Thus if each of A and B is a line or a circle, and if g is holomorphic, then $J_B \circ g \circ J_A$ is holomorphic. The general case reduces easily, thanks to Möbius maps, to the case of Proposition 5.7.2, where the only reflection used was $J_{\mathbb{R}} : z \mapsto \bar{z}$.

5.7.3. Theorem. *Let each of A and B be either a line or a circle. Let U be an open connected set which lies on one side of A, and let $U^* := J_A[U]$, its reflection in A. Let g be holomorphic on U, and let $G := J_B \circ g \circ J_A$, the holomorphic extension of g to $U \cup U^*$. Suppose that the intersection of the boundaries of U and of its reflection U^* contains a subset K of A such that $U \cup K \cup U^*$ is open. Suppose that whenever $\{w_n\}$ is a sequence of points in U that converges to a point of K, then $g(w_n)$ approaches B. Then G may be defined also on K so that it is holomorphic on $U \cup K \cup U^*$.*

Proof. Thanks to the availability of Möbius maps, this result reduces to 5.7.2. Figure 5.7-3 diagrams the scheme. Let S be a Möbius map which takes A onto the real axis, such that $S[U]$ lies in the upper half-plane. Let T be a Möbius map which takes B onto the real axis. Let

$$f := T \circ g \circ S^{-1}, \quad O := S[U], \quad \text{and} \quad I := S[K].$$

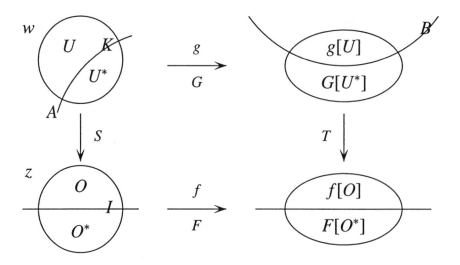

Figure 5.7-3   A schematic diagram referred to in the proof of 5.7.3.

Then O^*, the reflection of O in the real axis, is equal to $S[U^*]$. Let F be the extension of f to $O \cup O^*$ given by $F(\overline{z}) = \overline{f(z)}$, or, in other notation,

$$F = J_{\mathbb{R}} \circ f \circ J_{\mathbb{R}} \quad \text{and thus} \quad f = J_{\mathbb{R}} \circ F \circ J_{\mathbb{R}}.$$

The relation between G and F may be established as follows (read the equations, but follow the diagram):

$$
\begin{aligned}
G &= J_B \circ g \circ J_A && \text{(by the definition of } G) \\
&= J_B \circ T^{-1} \circ f \circ S \circ J_A && \text{(by the definition of } f) \\
&= T^{-1} \circ J_{\mathbb{R}} \circ f \circ J_{\mathbb{R}} \circ S && \text{(by a property of Möbius maps, 5.2.21)} \\
&= T^{-1} \circ F \circ S && \text{(from the definition of } F).
\end{aligned}
$$

If $\{z_n\}$ is a sequence of points in O that tends to a point $x \in I$, then the points $w_n := S^{-1}(z_n)$ tend to the point $S^{-1}(x)$ in K. By hypothesis, then, $g(w_n)$ approaches B. Therefore $f(z_n)$, since it equals $T(g(S^{-1}(z_n)))$, approaches $\mathbb{R}$; that is, $\text{Im}\,(f(z_n)) \to 0$.

We have established that Proposition 5.7.2 applies, so that we can extend F to a function holomorphic on $O \cup I \cup O^*$. With F thus extended, the function $G = T^{-1} \circ F \circ S$ gives the desired extension of g.   □

5.8   SCHWARZ-CHRISTOFFEL FORMULAS

In Sections 5.1–5.4, we worked with a certain repertory of familiar and easily understood mappings. They included powers, Möbius maps, the sine, and the exponential.

Using combinations of those mappings, we were often able to obtain an explicit, usable expression for the needed conformal equivalence between two sets. We would like to be able to do the like in more kinds of situations.

We will now describe a method for expressing, as a definite integral, an equivalence f between $\mathbb{Y}^o$ and the interior of a polygon P. The point, of course, is that if we wish to solve a Dirichlet or Neumann problem on P, then we can use f to reduce it to an equivalent problem on $\mathbb{Y}$—which we can solve. A mapping given as a definite integral may be difficult to describe, but will often allow us to find out what we need to know about the solution, at least with the aid of a computer.

The method, discovered by H. A. Schwarz and E. B. Christoffel independently in the 1860s, applies also with D in the place of $\mathbb{Y}^o$, and to a broader class of domains than just polygons. Engineering applications are many, numerical methods are essential, and the development of efficient techniques is an active research area.

To establish the Schwarz-Christoffel Formula for a conformal equivalence f from the open upper half-plane $\mathbb{Y}^o$ to the interior of a polygon P, one begins with the assumptions that f exists, and that it extends to a one-to-one continuous map from $\mathbb{Y}$ onto P. Those assumptions are justified by the Riemann Mapping Theorem and the Osgood-Taylor-Carathéodory Theorem, which are presented in Chapter 6 for those prepared to study them..

A Triangles

The next theorem establishes the Schwarz-Christoffel Formula for the case when the polygon is a triangle. If you wish to skip the proof, then you should study the statement of the theorem, and then proceed to Example 5.8.3.

On the other hand, if you seek a somewhat deeper understanding of Schwarz-Christoffel Formulas, then you should read the proof of the theorem with care. The proof is beautiful, but it is not short. We deal only with the case of a triangle, because in that way we can make the essential ideas more vivid, and avoid forcing you to keep track of quite so many possibilities. If you master the case of a triangle, then perhaps you can devise a proof for other cases, such as we discuss briefly in subsections B and C.

5.8.1. Theorem. *Let Δ be a triangle with vertices w_k and exterior angles b_k as shown in Figure 5.8-1.*

(i) *There exists a one-to-one continuous mapping f from $\mathbb{Y} \cup \{\infty\}$ onto Δ which is a conformal mapping from $\mathbb{Y}^o$ onto Δ^o.*

(ii) *Select three points m_k on the extended line $\hat{\mathbb{R}}$ such that $-\infty < m_1 < m_2 < m_3 \leq +\infty$. Then there exists a unique mapping f, as described in part (i), such that $f(m_k) = w_k$ for each k. We will call the points m_k the pre-vertices.*

(iii) *For $z \in \mathbb{Y}^o$, f is given by one of the following formulas. If $m_3 < \infty$, then*

$$f(z) = C_1 \int_{z_0}^{z} (\zeta - m_1)^{-b_1/\pi}(\zeta - m_2)^{-b_2/\pi}(\zeta - m_3)^{-b_3/\pi}d\zeta \; + \; C_2; \quad (1)$$

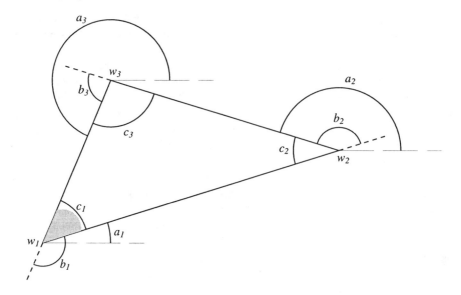

Figure 5.8-1   A triangle labeled to exemplify the notation in our discussion of polygons.

if $m_3 = \infty$, then

$$f(z) = C_1 \int_{z_0}^{z} (\zeta - m_1)^{-b_1/\pi}(\zeta - m_2)^{-b_2/\pi} d\zeta \; + \; C_2, \qquad (2)$$

where C_1 and C_2 are constants. In either case, z_0 may be chosen to be any point of $\mathbb{Y} \setminus \{m_1, m_2, m_3\}$, and the integral may be taken over any path in that set that runs from z_0 to z. In each case, the intended power function $z \mapsto z^\alpha$ is specified by $-\frac{\pi}{2} < \arg z < \frac{3\pi}{2}$.

Proof. We will observe the labeling conventions shown in Figure 5.8-1. We define w_4 to equal w_1. The oriented boundary $[w_1, w_2, w_3, w_4]$ goes counterclockwise around the triangle. The angles are measured counterclockwise. In the case shown in this picture, they are all positive. At vertex w_k, b_k is the exterior angle and c_k is the interior angle. The edge that begins at w_k has parametrization $t \mapsto w_k + te^{ia_k}$ $(0 \leq t \leq |w_{k+1} - w_k|)$. The shaded region near w_1 is intended to represent the image under f of the half-disk $D(m_1, \delta) \cap \mathbb{Y}^o$, for some $\delta > 0$.

In this section, as we mentioned above, we will assume the Riemann Mapping Theorem and the Osgood-Taylor-Carathéodory Theorem. They imply the existence of f as claimed in part (i).

To prove the existence claim in part (ii), we will use Möbius maps. Given f as in (i), there will be distinct points m'_k on the extended line $\hat{\mathbb{R}}$ such that $f(m'_k) = w_k$ for each k. Replace f with $f \circ S$, where S is the Möbius map

$$S : m_1, m_2, m_3 \mapsto m'_1, m'_2, m'_3.$$

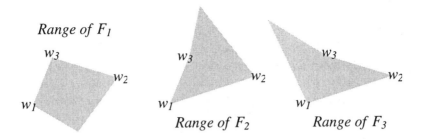

Figure 5.8-2   In the proof of 5.8.1, while f maps $\mathbb{Y}^o$ onto the triangle, each of its extensions F_k maps $\mathbb{Y}^o \cup I_k \cup \mathbb{Y}^o_-$ onto the union of the triangle with its reflection in one of its edges. The ranges of the three extensions are represented here.

Then the new f maps m_k to w_k as claimed.

To prove the uniqueness claim in (ii), suppose that f and φ are two one-to-one continuous maps from $\mathbb{Y} \cup \{\infty\}$ onto Δ, holomorphic from $\mathbb{Y}^o$ onto Δ^o, such that $f(m_k) = \varphi(m_k) = w_k$ for each k. Then $\varphi^{-1} \circ f$ is a one-to-one holomorphic function from $\mathbb{Y}^o$ onto $\mathbb{Y}^o$. By 5.2.20, $\varphi^{-1} \circ f$ agrees on $\mathbb{Y}^o$ with a Möbius map; which, since it has three fixed points m_k, must be the identity, by 5.2.5. Therefore $\varphi = f$.

We now set out to establish the formula (1) for the case when the pre-vertices are all finite: $-\infty < m_1 < m_2 < m_3 < \infty$. Let

$$I_1 := (m_1, m_2), \quad I_2 := (m_2, m_3), \quad \text{and} \quad I_3 := \{x \mid x < m_1 \text{ or } x > m_3\} \cup \{\infty\}.$$

Also, let

$$I := I_1 \cup I_2 \cup I_3.$$

For each k, f maps I_k one-to-one continuously onto (w_k, w_{k+1}), which is one of the triangle's edges (less its endpoints). Let B_k be the line containing that edge. The Reflection Principle 5.7.3 applies, with $\mathbb{Y}^o$, $\mathbb{R}$, I_k, and B_k in the roles of U, A, K, and B, respectively. We obtain a holomorphic function F_k which extends f from $\mathbb{Y}^o \cup I_k$ to $\mathbb{Y}^o \cup I_k \cup \mathbb{Y}^o_-$. On $\mathbb{Y}^o_-$, $F_k = J_{B_k} \circ f \circ J_{\mathbb{R}}$. The extension F_k is one-to-one. Therefore F'_k has no zeros.

Since a reflection is its own inverse, $F_k^{-1} = J_{\mathbb{R}} \circ f^{-1} \circ J_{B_k}$.

The ranges of the three extensions F_1, F_2, F_3 are pictured in Figure 5.8-2. Incidentally, if you will draw the corresponding pictures for a triangle which has an interior angle exceeding $\frac{\pi}{2}$, you will see that the images $F_k[\mathbb{Y}^o_-]$ are not necessarily disjoint.

The three extensions have a close relationship. Let F_j and F_k be any two of the three. Then

$$F_j \circ F_k^{-1} = J_{B_j} \circ f \circ J_{\mathbb{R}} \circ J_{\mathbb{R}} \circ f^{-1} \circ J_{B_k} = J_{B_j} \circ J_{B_k} \quad \text{on } J_{B_k}[\Delta].$$

Because this is the composition of two reflections in lines, it has the form $w \mapsto C_1 w + C_2$, with $C_1 \neq 0$ (see subsection 2.1I). Therefore

$$F_j(z) = C_1 F_k(z) + C_2 \quad \text{for } z \in Y_-^o.$$

Therefore

$$\frac{F_j''}{F_j'} = \frac{F_k''}{F_k'} \quad \text{on } Y_-^o. \tag{3}$$

Accordingly, the function g given by

$$g(z) := \frac{f''(z)}{f'(z)} \quad \text{for } z \in Y^o \tag{4}$$

has a holomorphic extension to $Y^o \cup I \cup Y_-^o$; we simply let $g(z) := F_k''(z)/F_k'(z)$ for $z \in I_k \cup Y_-^o$, and by (3), the definition is consistent.

The function g is important because it has the form of a logarithmic derivative. On any simply connected set on which g is holomorphic, g has an antiderivative, which is $\log f'$. If we can identify g, we have taken one step toward identifying f.

We mention parenthetically that the extension of f itself is a bit more complicated. Let $P_k := \{x + iy \mid x \in I_k \text{ and } y \leq 0\}$, as in Figure 5.8-3. Let $V := Y^o \cup P_1 \cup P_2 \cup P_3$, which is the complex plane with three vertical half-lines removed. Extend f to V by letting $f(z) := F_k(z)$ for $z \in P_k$. The function f, thus extended, may or may not be one-to-one on V, depending on the shape of Δ; and there's no hope of extending f to a punctured neighborhood of the point m_k.

We should point out an alternative way to understand the extension of $g \equiv f''/f'$. With a_k defined as indicated in Figure 5.8-1, we claim that

$$\text{for } x \in I_k, \quad \frac{f'(x)}{e^{ia_k}} \text{ is real and positive and } \quad \frac{f''(x)}{e^{ia_k}} \text{ is real.} \tag{5}$$

Item (a) near the end of this proof justifies the claim. We know that g is holomorphic on Y^o. By (5), it is real-valued on I. By the reflection principle, g extends to a holomorphic function on $Y^o \cup I \cup Y_-^o$, such that

$$g(\bar{z}) = \overline{g(z)}.$$

The isolated singularities of g occur at ∞, m_1, m_2, and m_3. We will show in item (b), near the end of the proof, that the first one is removable, with $g(\infty) = 0$. We will show in item (c) that the others are simple poles, with

$$\text{Res}(g, m_k) = -\frac{b_k}{\pi}.$$

If we subtract from g its principal parts at the three poles, and remove the removable singularities, then the result is a bounded entire function:

$$z \mapsto g(z) + \frac{b_1/\pi}{z - m_1} + \frac{b_2/\pi}{z - m_2} + \frac{b_3/\pi}{z - m_3}.$$

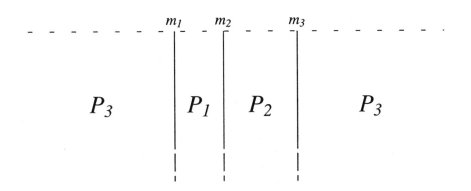

Figure 5.8-3   The set V referred to in the proof of 5.8.1 consists of the plane with three half-lines removed. V contains the upper half-plane, the three regions P_k, and the real axis (shown as a dashed line) except for the points m_k.

By Liouville's Theorem, this function is constant. Since the function tends to 0 at ∞, the constant is 0. Therefore

$$g(z) \;=\; -\,\frac{b_1/\pi}{z - m_1} \;-\; \frac{b_2/\pi}{z - m_2} \;-\; \frac{b_3/\pi}{z - m_3}.$$

On the upper half-plane, $g(z) = \dfrac{d}{dz} \log f'(z)$, hence

$$\log f'(z) = C - \frac{b_1}{\pi} \log(z - m_1) - \frac{b_2}{\pi} \log(z - m_2) - \frac{b_3}{\pi} \log(z - m_3) \quad (6)$$

for some constant C. Each of the three logarithms on the right can be taken to be the logarithm given by

$$\log(re^{it}) = \ln r + it \quad \text{for} \quad -\frac{\pi}{2} < t < \frac{3\pi}{2},$$

and then the expression for $\log f'$ makes good sense for all $z \in V$. It follows that

$$f'(z) \;=\; e^C (z - m_1)^{-b_1/\pi}(z - m_2)^{-b_2/\pi}(z - m_3)^{-b_3/\pi}.$$

If we choose a point $z_0 \in V$, then equation (1) follows for some constants C_1 and C_2. We have not yet made use of the size and position of the triangle; the exterior angles b_k tell us only the shape. That is why the constants remain to be determined.

When $m_3 = \infty$, one may show that f is given by (2). One method is to repeat the procedure above with suitable modifications. Another is to carry out a suitable change of variable in formula (1). We leave the details to Exercise 3.

We postponed certain steps of this proof. Here they are.

(a) *Proof of* (5). Let $x \in I_k$ and consider the difference quotient whose limit is the nonzero complex number $f'(x)$:

$$\frac{f(x+h) - f(x)}{h} \to f'(x) \quad \text{as } h \to 0. \tag{7}$$

This limit is of course the same if we restrict h to real values. The line segment $[w_k, w_{k+1}]$ is at angle a_k from the horizontal, so it may be parametrized by

$$t \mapsto w_k + te^{ia_k} \quad (0 \le t \le |w_{k+1} - w_k|).$$

Whenever h is real, the numerator in (7) is the difference of two points on $[w_k, w_{k+1}]$, and the quotient is therefore always equal to a positive number times e^{ia_k}. Therefore the limit $f'(x)$ is also of that form. It follows that the difference quotient that tends to $f''(x)$,

$$\frac{f'(x+h) - f'(x)}{h} \to f''(x) \quad \text{as } h \to 0,$$

is always a real multiple of e^{ia_k} when h is real. Therefore the limit $f''(x)$ is also a real multiple of e^{ia_k}.

(b) *Classifying the singularity of* (4) *at* ∞. The function F_3 extends f to a holomorphic function on $\mathbb{Y}^o \cup I_3 \cup \mathbb{Y}^o_-$, which is a punctured neighborhood of ∞. Thus extended, f is one-to-one near ∞. Also, $f(z)$ has a finite limit as $z \to \infty$, which equals some number w^* on the segment $[w_3, w_1]$. Therefore f has a removable singularity at ∞. Its series representation for z near ∞ is of this form, with $c_{-1} \ne 0$:

$$f(z) = w^* + \frac{c_{-1}}{z} + \frac{c_{-2}}{z^2} + \cdots. \quad \text{Then}$$

$$f'(z) = -\frac{c_{-1}}{z^2} - \frac{2c_{-2}}{z^3} - \cdots \quad \text{and}$$

$$f''(z) = \frac{2c_{-1}}{z^3} + \frac{6c_{-2}}{z^4} + \cdots.$$

It follows that the first summand in the Laurent series for f''/f' near ∞ is $-2/z$. So g is bounded at ∞, and in fact tends to 0 there. With $g(\infty) := 0$, the singularity is removed.

(c) *Classifying the singularity of* (4) *at* m_k. To reduce clutter in the formulas, let's write $c, m,$ and w instead of $c_k, m_k,$ and w_k, respectively; and also b instead of b_k, noting the relation between interior and exterior angles, $c - \pi = b$.

If we wish to replace Δ with some other triangle similar to Δ, then we need only replace f with $z \mapsto C_1 f(z) + C_2$ for suitable constants C_1, C_2. But then g is unchanged! Therefore we may suppose, which we do for the sake of convenience, that $w = 0$ and that the positive real axis bisects the interior angle c. Choose $\epsilon > 0$, and then $\delta > 0$, such that

$$f[\Upsilon \cap D(m, \delta)] \subseteq \Delta \cap D(0, \epsilon) \subseteq \{re^{i\theta} \mid 0 < r < \epsilon, \ -\tfrac{c}{2} < \theta < \tfrac{c}{2}\}.$$

Thus f maps the upper half of a disk centered at m into a corner of the triangle Δ; a sector of angle π is mapped to one of angle c. The image of the segment $[m - \delta, m + \delta]$ consists of two line segments which meet at 0 with angle c. Make a sketch. Let

$$p(z) := f(z)^{\pi/c} \quad \text{for } z \in \Upsilon \cap D(m, \delta).$$

The $\frac{\pi}{c}$-power function takes $re^{i\theta}$ to $r^{\pi/c}e^{i\pi\theta/c}$ for $-\tfrac{c}{2} < \theta < \tfrac{c}{2}$. So p maps $\Upsilon^{\circ} \cap D(m, \delta)$ conformally onto a set contained in

$$\{re^{i\theta} \mid 0 < r < \epsilon^{\pi/c}, \ -\tfrac{\pi}{2} < \theta < \tfrac{\pi}{2}\},$$

and maps the interval $(m - \delta, m + \delta)$ to an open interval of the imaginary axis contained in $(-ie^{\pi/c}, ie^{\pi/c})$. By the Reflection Principle, p may be extended to a holomorphic function on $D(m, \delta)$. The function p, thus extended, is a one-to-one holomorphic function on $D(m, \delta)$, and

$$f(z) = p(z)^{c/\pi} \quad \text{for } z \in D(m, \delta) \cap \Upsilon^{\circ}.$$

Since p has a zero of order 1 at m, we may write

$$p(z) = (z - m)q(z),$$

where q is holomorphic and has no zeros in $D(m, \delta)$. Therefore the function q has a logarithm on $D(m, \delta)$ (by 3.2.3), and we choose it to be the one such that

$$\mathrm{Log}\, p(z) = \mathrm{Log}\,(z - m) + \log q(z) \quad \text{for } z \in D(m, \delta) \cap \Upsilon^{\circ}.$$

Then

$$
\begin{aligned}
f(z) &= p(z)^{c/\pi} = e^{(c/\pi)\mathrm{Log}\,p(z)} \\
&= e^{(c/\pi)(\mathrm{Log}\,(z-m)+\log q(z))} = (z - m)^{c/\pi}\, \underbrace{e^{(c/\pi)\log q(z)}}_{=:\,d(z)}.
\end{aligned}
$$

The function d is holomorphic and nonzero on $D(m, \delta)$, and

$$f(z) = (z - m)^{c/\pi}d(z) \quad \text{for } z \in \Upsilon^{\circ} \cap D(m, \delta).$$

So

$$
\begin{aligned}
f'(z) &= \tfrac{c}{\pi}(z - m)^{(c/\pi)-1}d(z) + (z - m)^{c/\pi}d'(z) \\
&= (z - m)^{(c/\pi)-1}\underbrace{\left(\tfrac{c}{\pi}d(z) + (z - m)d'(z)\right)}_{=:\,k(z)}.
\end{aligned}
$$

The function k is holomorphic and nonzero on $D(m, \delta_1)$ for some δ_1 with $0 < \delta_1 \le \delta$; and c is not allowed to equal π. A computation shows that

$$\frac{f''(z)}{f'(z)} = \frac{\frac{c}{\pi} - 1}{z - m} + \frac{k'(z)}{k(z)}.$$

Therefore g has a pole of order 1 at m, and $\operatorname{Res}(g, m) = \frac{c}{\pi} - 1$.

$\square$

5.8.2. Remark. The proof works because of the amazingly nice properties of the function g, which provides a scaffolding for the construction of f. The following brief account omits any mention of g, but may make it easier to remember how we arrive at equation (6), and how to write it down. From equation (5), we know that we should have

$$\arg f'(x) = a_k \quad \text{for all } x \in I_k. \tag{8}$$

Observe that for $x \in I$ and for each k,

$$\frac{\arg(x - m_k)}{\pi} = \begin{cases} 0 & \text{if } x > m_k, \\ 1 & \text{if } x < m_k, \end{cases}$$

always using the argument whose value lies between $-\pi/2$ and $3\pi/2$. In view of that information about what happens on the boundary, it is reasonable to expect that for all $z \in V$,

$$\begin{aligned} \arg f'(z) = a_3 &+ (a_2 - a_3) \frac{\arg(z - m_3)}{\pi} \\ &+ (a_1 - a_2) \frac{\arg(z - m_2)}{\pi} \\ &+ (a_3 - a_1) \frac{\arg(z - m_1)}{\pi}, \end{aligned} \tag{9}$$

because if we put a real number $x \in I_k$ into the role of z, then (9) agrees with (8). Thus the harmonic function given by (9) has the correct boundary values on $\mathbb{R}$. And (6), where we have written b_k instead of $a_k - a_{k-1}$, is a holomorphic function on Y^o with $\arg f'$ as its imaginary part.

5.8.3. Example. *Find the Schwarz-Christoffel formula for a conformal mapping from the upper half-plane onto a right triangle with interior angles $\frac{\pi}{2}, \omega,$ and $\frac{\pi}{2} - \omega$.*

Solution. As would be the case in a practical problem, we have a choice of coordinate system for the triangle. A reasonable decision would be to let the vertices be 0, 1, and $i \tan \omega$. (Make a sketch!) The exterior angles would then be $\frac{\pi}{2}, \pi - \omega,$ and $\frac{\pi}{2} + \omega$. We also have a choice of how the vertices are numbered; let's try $w_1 = 1$, $w_2 = i \tan \omega$, and $w_3 = 0$. Finally, we have a choice of the points m_k. For the sake of maximizing symmetry, we let $m_1 = -1, m_2 = 1$, and $m_3 = \infty$. Then the exterior angles are $b_1 = \pi - w, b_2 = \frac{\pi}{2} + w$, and $b_3 = \frac{\pi}{2}$. Equation (2) becomes

$$f(z) = C_1 \int_0^z \frac{d\zeta}{(\zeta+1)^{1-\frac{w}{\pi}}(\zeta-1)^{\frac{1}{2}+\frac{w}{\pi}}} + C_2.$$

If the value of w is specified, then from the conditions $f(-1) = 1$, $f(1) = i\tan w$ it will be possible to solve for C_1 and C_2. In some cases, the definite integrals involved may have to be evaluated by numerical methods.   □

5.8.4. Example. *Find the Schwarz-Christoffel formula for a conformal mapping from the upper half-plane onto a right isosceles triangle.*

Solution. This problem is the case of Example 5.8.3 in which w is specified to be $\pi/4$. Thus

$$f(z) = C_1 \int_0^z \frac{d\zeta}{(\zeta+1)^{3/4}(\zeta-1)^{3/4}} + C_2$$

$$= C_1 \int_0^z \frac{d\zeta}{(\zeta^2-1)^{3/4}} + C_2.$$

The two conditions $f(-1) = 1$ and $f(1) = -1$ give us two linear equations in the constants:

$$C_1 \int_0^{-1} \frac{d\zeta}{(\zeta^2-1)^{3/4}} + C_2 = 1 \quad \text{and}$$

$$C_1 \int_0^1 \frac{d\zeta}{(\zeta^2-1)^{3/4}} + C_2 = i. \tag{10}$$

We have to be careful with the powers. If ζ is real and if $-1 < \zeta < 1$, then $(\zeta+1)^{3/4}$ is real and positive, while $(\zeta-1)^{3/4} = e^{i3\pi/4}|\zeta-1|$. Let k be the real positive value of $\int_0^1 \frac{dx}{(1-x^2)^{3/4}}$, which *Mathematica* computes to be

$$k = \frac{2\sqrt{\pi}\,\Gamma(\frac{5}{4})}{\Gamma(\frac{3}{4})} \approx 2.62206. \tag{11}$$

Then the equations (10) become

$$-C_1 e^{-i3\pi/4}k + C_2 = 1 \quad \text{and}$$

$$C_1 e^{-i3\pi/4}k + C_2 = i,$$

which yield the unique solution $C_1 = \dfrac{1}{i\sqrt{2}k}$, $C_2 = \dfrac{1+i}{2}$. So the desired mapping is given by

$$f(z) = \frac{1}{i\sqrt{2}k} \int_0^1 \frac{d\zeta}{(\zeta^2-1)^{3/4}} + \frac{1+i}{2}.$$

Notice that $f(0)$ is the midpoint of the edge $[1, i]$.   □

5.8.5. Example. *Find the Schwarz-Christoffel formula for a conformal mapping f from the upper half-plane to an isosceles triangle with two interior angles of θ radians.*

Solution. Making our choices to maximize symmetry, we choose vertices $w_1 = -1$, $w_2 = 1$, and $w_3 = i \tan \theta$; and their pre-images under f to be $m_1 = -1$, $m_2 = 1$, and $m_3 = \infty$, respectively. Make a sketch. The exterior angles are $b_1 = b_2 = \pi - \theta$ and $b_3 = 2\theta$. Then, if we let $a := (\pi - \theta)/\pi$, equation (1) becomes

$$f(z) = C_1 \int_0^z \frac{d\zeta}{(\zeta^2 - 1)^a} + C_2. \tag{12}$$

Once again, we must be careful about the power, noting here that $(\zeta^2 - 1)^a = e^{ia\pi}|\zeta^2 - 1|$ for $-1 < \zeta < 1$.   □

5.8.6. Example. *The base of an isosceles triangle with base angles $\pi/4$ is maintained at $100°$, and its other two sides at $0°$. If the height of the triangle is 1, find the steady-state temperature along the centerline at heights .2, .4, .6, and .8.*

Solution. We will make use of the work done in Example 5.8.5. If $\theta = \pi/4$, then $a = 3/4$. Using the conditions $f(-1) = -1$ and $f(1) = 1$, we find that $C_2 = 0$ and $C_1 = \dfrac{e^{i3\pi/4}}{k}$, where k is as in (11); so (12) becomes

$$f(z) = \frac{e^{i3\pi/4}}{k} \int_0^z \frac{d\zeta}{(\zeta^2 - 1)^{3/4}}.$$

This mapping leads us to solve first the equivalent problem on the upper half-plane, in which the interval $[-1, 1]$ is maintained at $100°$ and the rest of the line at $0°$. The solution is given by

$$u(z) \mapsto -100 \arg(z + 1) + 100 \arg(z - 1). \tag{13}$$

From the symmetries in view, it is reasonable to believe that that f maps the points iy with $y > 0$ onto the centerline of the triangle; let's verify it computationally. Here is a pullback:

$$f(iy) = \frac{e^{i3\pi/4}}{k} \int_0^y \frac{i\,dt}{((it)^2 - 1)^{3/4}}. \tag{14}$$

Let's be precise about those powers:

$$(it + 1)^{3/4} = (\sqrt{t^2 + 1})^{3/4} e^{i(\arctan t)\frac{3}{4}} \quad \text{and}$$
$$(it - 1)^{3/4} = (\sqrt{t^2 + 1})^{3/4} e^{i(\pi - \arctan t)\frac{3}{4}}.$$

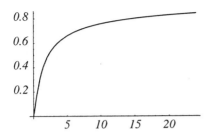 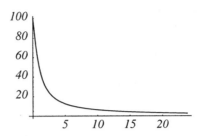

Figure 5.8-4 The function f of Example 5.8.5 maps the upper half-plane $\mathbb{Y}$ to the triangle Δ. It maps the centerline $\{iy \mid y \geq 0\}$ of $\mathbb{Y}$ to the centerline $\{it \mid 0 \leq t \leq 1\}$ of Δ. The graph on the left shows the function $y \mapsto t$. In Example 5.8.6, a temperature-distribution problem is solved for the upper half-plane $\mathbb{Y}$. The graph on the right shows the temperature along the centerline of $\mathbb{Y}$. It decreases from $100°$ toward $0°$.

The denominator in (14) equals $(t^2 + 1)e^{i3\pi/4}$, the product of those two factors. Therefore

$$f(iy) = \frac{i}{k} \int_0^y \frac{dt}{(t^2 + 1)^{3/4}},$$

and indeed the centerline of the upper half-plane is mapped to the centerline of the triangle. Figure 5.8-4, on the left, shows the graph of $y \mapsto \operatorname{Im} f(iy)$.

Now we must return to (13) and study the values of the temperature u when $z = iy$.

$$u(iy) = \frac{100}{\pi}(\arg(iy - 1) - \arg(iy + 1)) = \frac{100}{\pi}(\pi - 2\arctan y).$$

The graph of $y \mapsto u(iy)$ is shown in Figure 5.8-4, on the right.

The temperature $U(w)$ at each point w in the triangle is given by $U(w) = u(f^{-1}(w))$. Applying this rule just to w of the form it where t has one of the four values, we find the answers:

$$U(i0.2) \approx 67°; \quad U(i0.4) \approx 38°; \quad U(i0.6) \approx 17°; \quad U(i0.8) \approx 4°. \qquad (15)$$

$\square$

B Rectangles and Other Polygons

The following Theorem, which is a bit more general than 5.8.1, may be proved in essentially the same way. At the end of the subsection, after giving some examples, we will offer some remarks about what the proof entails.

5.8.7. Theorem. *Let E be a bounded, simply connected open set whose boundary ∂E can be parametrized as a polygonal path with n vertices, with $W_{\partial E}(z)$ equal to 1 for $z \in E$ and 0 for $z \notin \bar{E}$.*

(i) *There exists a continuous mapping f from $\mathbb{Y} \cup \{\infty\}$ onto $\bar{E}$ which is conformal from $\mathbb{Y}^\circ$ onto E.*

(ii) *Let $-\infty < m_1 < m_2 < \cdots < m_n \leq \infty$ be the points on the extended line $\hat{\mathbb{R}}$ that f maps to the vertices. Let $w_k = f(m_k)$ for each k. The points m_k are called the pre-vertices. Any three of the pre-vertices can be prescribed, and then f is uniquely determined.*

(iii) *Let b_k be the angle by which the boundary turns at the vertex w_k; that is, the angle such that*

$$\frac{w_{k+1} - w_k}{|w_{k+1} - w_k|} = e^{ib_k} \frac{w_k - w_{k-1}}{|w_k - w_{k-1}|},$$

and require that

$$\text{either } 0 < b_k < \pi \text{ or } -\pi \leq b_k < 0 \text{ for each } k. \tag{16}$$

Let z_0 be a point of $\mathbb{Y}$ other than a pre-vertex. Then there exist constants C_1 and C_2 such that for $z \in \mathbb{Y}$,

$$f(z) = C_1 \int_{z_0}^{z} (\zeta - m_1)^{-b_1/\pi} (\zeta - m_2)^{-b_2/\pi} \cdots (\zeta - m_{n_0})^{-b_{n_0}/\pi} d\zeta + C_2, \tag{17}$$

where n_0 equals n if $m_n < \infty$ and equals $n - 1$ if $m_n = \infty$. The intended power function $z \mapsto z^x$ is specified by $-\frac{\pi}{2} < \arg z < \frac{\pi}{2}$. The integral may be taken over any path in $\mathbb{Y} \setminus \{m_1, \cdots, m_n\}$ that runs from z_0 to z.

If $n > 3$, then when one arrives at the formula (17), having selected three of the pre-vertices, the remaining $n - 3$ pre-vertices as well as the two constants are determined. We cannot choose them, but we now must find them. Such is the difficult **Schwarz-Christoffel parameter problem**, and one generally needs to proceed by numerical calculations, and by trial and error.

5.8.8. Example. *Find a conformal mapping from $\mathbb{Y}^\circ$ onto a rectangle.*

Solution. We need to cover the case of an arbitrary rectangle, but we are free to decide how to place it in our coordinate system, and free to choose three of the pre-vertices. We do so with a view to making the equations as simple as possible.

Let the vertices of the rectangle E be

$$w_1 = -1 + ic, \quad w_2 = -1, \quad w_3 = 1, \quad w_4 = 1 + ic,$$

where $c > 0$. Make a sketch.

We may choose $m_2 = -1$ and $m_3 = 1$. For the sake of symmetry, we would like to choose $m_1 = -a$ and $m_4 = a$ for some $a > 1$. But we are not free to choose all four pre-vertices. So let's choose $m_1 = -p$ for some $p > 1$. Then f is uniquely determined, and m_4 is determined; we will have $m_4 = q$ for some $q > 1$ (which depends on c and on our choice of p). Happily, when given p and q one may show (see Exercise 4) that there exists a, between p and q, and a Möbius map

$$T : -p, -1, 1 \rightarrow -a, -1, 1, \tag{18}$$

such that $T(q) = a$. Then we replace f by $f \circ T^{-1}$, and the new f enjoys the desired symmetry!

We will find that a is determined by c, and is independent of our choice of p.

The exterior angles b_k are all equal to $\pi/2$. So we can easily write down the formula

$$f(z) = C_1 \int_0^z (\zeta + a)^{-1/2} (\zeta + 1)^{-1/2} (\zeta - 1)^{-1/2} (\zeta - a)^{-1/2} d\zeta \ + C_2.$$

We would like to rewrite it as

$$f(z) = C_1 \int_0^z \frac{d\zeta}{(\zeta^2 - a^2)^{1/2} (\zeta^2 - 1)^{1/2}}, \tag{19}$$

but to do so we need first to ponder the meaning of the square root functions.

The argument assigned to each of the quantities $\zeta \pm a$ for $\zeta \in \mathbb{Y} \setminus \{-a, a\}$ is $\arg_{1 \mapsto 0}$, which varies from 0 to π. So the sum of the two arguments has range equal to the interval $[0, 2\pi]$; the sum equals 0 only on the interval $(a, +\infty)$, and equals 2π only on the interval $(-\infty, -a)$. The computation

$$\begin{aligned}
(\zeta + a)^{1/2} (\zeta - a)^{1/2} &= \exp[\tfrac{1}{2} \ln |\zeta + a| + i\tfrac{1}{2} \arg_{1 \mapsto 0}(\zeta + a) \\
&\quad + \tfrac{1}{2} \ln |\zeta - a| + i\tfrac{1}{2} \arg_{1 \mapsto 0}(\zeta - a)] \\
&= \exp[\tfrac{1}{2} \ln |\zeta^2 - a^2| + i\tfrac{1}{2} \arg_{1 \mapsto 0}(\zeta^2 - a^2)]
\end{aligned}$$

is valid for $\zeta \in \mathbb{Y}^\circ$ and for ζ in the real intervals $(-a, a)$ and (a, ∞), but not for ζ in the real interval $(-\infty, -a)$, where the last line should be

$$\exp[\tfrac{1}{2} \ln |\zeta^2 - a^2| + i\tfrac{1}{2} \arg_{1 \mapsto 2\pi}(\zeta^2 - a^2)].$$

Thus, if we use the radical sign over a positive real number to indicate always its positive square root,

$$(x^2 - a^2)^{1/2} = \begin{cases} \sqrt{x^2 - a^2} & \text{for } x > a, \\ i\sqrt{a^2 - x^2} & \text{for } -a < x < a, \\ -\sqrt{x^2 - a^2} & \text{for } x < -a. \end{cases}$$

A similar discussion holds for the quantities $\zeta \pm 1$. We will keep this in mind when we use the expression (19).

Since we know the value of f at four points, we can find equations that will determine C_1, C_2, and a. From the conditions $1 = f(1)$ and $-1 = f(-1)$, we obtain these equations:

$$1 = C_1 \int_0^1 \frac{dx}{i\sqrt{a^2 - x^2} i\sqrt{1 - x^2}} + C_2 \equiv -C_1 \int_0^1 \frac{dx}{\sqrt{a^2 - x^2} \sqrt{1 - x^2}} + C_2;$$

$$-1 = C_1 \int_0^{-1} \frac{dx}{i\sqrt{a^2 - x^2} i\sqrt{1 - x^2}} + C_2 \equiv C_1 \int_0^1 \frac{dx}{\sqrt{a^2 - x^2} \sqrt{1 - x^2}} + C_2.$$

It follows that $C_2 = 0$ and

$$-1 = C_1 \int_0^1 \frac{dx}{\sqrt{a^2 - x^2}\sqrt{1 - x^2}}. \tag{20}$$

Since $ic = f(a) - f(1)$,

$$ic = C_1 \int_1^a \frac{dx}{i\sqrt{a^2 - x^2}\sqrt{x^2 - 1}}$$

or

$$-c = C_1 \int_1^a \frac{dx}{\sqrt{a^2 - x^2}\sqrt{x^2 - 1}}. \tag{21}$$

When c is specified numerically, so that we know the shape of the rectangle, we can use (20) and (21) to find approximations to C_1 and a. □

5.8.9. Example. *Identify the Schwarz-Christoffel mapping from $\mathbb{Y}^\circ$ to a 2×1 rectangle.*

Solution. Our work on Example 5.8.8 leads us to equations (20) and (21), which now hold with $c = 1$. We need to find a such that the two integrals are equal. Using the numerical integration command of *Mathematica* to evaluate the two integrals for various values of a, we find that equality holds when $a \approx 1.419$; and that the common value is about 1.300. Accordingly, $C_1 \approx -0.769$. Therefore with those values for a and C_1, and with $C_2 = 0$, $f(z)$ is given approximately by (19).

You may wish to undertake as an exercise some drawings by hand to show the behavior of the mapping: Sketch the image under f of a rectangular grid, consisting of horizontal and vertical line segments in $\mathbb{Y}^\circ$. Sketch also the image of a polar grid, consisting of semicircles and rays from the origin. If you wish to check the product of your educated guesses, you can find computer-generated graphics in Figure 1.126 of [28]. Also of interest, Figure 1.127 shows the image of the polar grid under a conformal map from D onto a square. □

You may wish to undertake the substantial exercise of writing up a proof of Theorem 5.8.7. It is a matter of retracing the steps in the proof of 5.8.1, adjusting the reasoning to accommodate a wider range of possibilities at each step. Also involved is the use of a procedure that will be presented in Example 6.4.1. In any event, the following notes may interest you.

- In 5.8.7, the angles b_k are restricted only by (16), but they still add to 2π.

- The set E is convex precisely when $0 < b_k < \pi$ for each k, which is always the case for a triangle.

- If $n > 3$, the list of angles b_k does not determine even the shape of E. But if the angles b_k and the lengths $|w_k - w_{k-1}|$ are specified, then E is determined up to rigid motions (rotations and translations).

- When the integral (17) is taken over a real interval containing the point m_k, the integral is improper only if $0 < b_k < \pi$, but it always converges nicely.

- Figure 5.8-2 was drawn for a triangle in which each exterior angle is less than $\pi/2$. Consider what the corresponding picture might look like in other cases, including when some b_k equals $-\pi$.

- If $b_k = -\pi$, then the intersection of the two segments $[w_{k-1}, w_k]$ and $[w_k, w_{k+1}]$ contains the shorter of the two. Because of this possibility, we cannot assert in the statement of 5.8.7 that f is one-to-one on the boundary; or that E is the interior of $\bar{E}$.

- If $b_k = -\pi$ for one or more values of k, then the Osgood-Taylor-Carathéodory Theorem does not apply, but Example 6.4.1 suggests how to replace it in the proof of 5.8.7.

- A thorough treatment of this material is given in [51, Section IX.5].

C Generalized Polygons

Theorem 5.8.7 generalizes to certain cases when E is unbounded and one or more of the vertices w_k equals ∞. We offer no discussion of the proof here, but indicate briefly some of those cases, as follows. Let B_k be a directed line, half-line, or line segment that runs from w_k to w_{k+1}, such that the boundary of E is the union of the sets B_k. Require that for each point w on B_k, there exists $r > 0$ such that the half of the disk $D(w, r)$ which is on the left of B_k is contained in E. Of course, we can no longer speak of ∂E as a path with winding number 1 for each point of E.

In Figure 5.4-1, the unbounded sets $\mathbb{Y}^o, O_2, O_3$, and O_5 are of the kind just described. We will explain how to use 5.8.7, generalized, to arrive at conformal mappings from $\mathbb{Y}^o$ to each of the four.

For $\mathbb{Y}^o$ the answer is trivial to find; but notice how the Theorem applies, very easily, to any half-plane. Such a set is the left side of some line. There being no vertex, the integrand in the Schwarz-Christoffel Formula is simply the constant 1, so we get $f(z) = C_1 \int_0^z d\zeta + C_1 = C_1 z + C_2$.

For O_2, the vertices may be taken to be $w_1 = 0$, $w_2 = i$, and $w_3 = 0$. The boundary of O_2 consists of the segments $B_1 = [0, i]$ and $B_2 = [i, 0]$, and the line B_3, which consists of the real axis and runs from 0 via ∞ to 0. The turns are $b_1 = \frac{\pi}{2}$, $b_2 = -\pi$, and $b_3 = \frac{\pi}{2}$. Let's take the pre-vertices to be -1, 0, and 1, respectively. Then the Schwarz-Christoffel Formula is

$$f(z) = C_1 \int_0^z \frac{\zeta}{(\zeta + 1)^{1/2}(\zeta - 1)^{1/2}} d\zeta + C_2$$
$$= C_1 \int_0^z \frac{\zeta}{(\zeta^2 - 1)^{1/2}} d\zeta + C_2, \tag{22}$$

which is no surprise, since in Example 5.4.3 we constructed a map from O_2 to $\mathbb{Y}^o$, $z \mapsto \sqrt{z^2 + 1}$, of which the inverse is $z \mapsto \sqrt{z^2 - 1}$.

For O_3, the boundary consists of two lines, which we may take to be given by $x = 0$ downward and $x = \pi$ upward. We identify one vertex $w_1 = \infty$, where there is a turn of π, and take $m_1 = 1$. That gives us

$$f(z) = C_1 \int_1^z \frac{d\zeta}{\zeta} + C_2,$$

and indeed $z \mapsto -i \log z$ does it.

For O_5, we may deal with an isosceles triangle of various base-lengths. Let $c > 0$, and take the vertices to be $w_1 = -c$, $w_2 = i$, and $w_3 = c$. Then $B_1 = [-c, i]$, $B_2 = [i, c]$, and B_3 is the part of the real axis running from c via ∞ to $-c$. The angles of turn are $b_1 = \beta$, where $\beta = \operatorname{arccot} c$; $b_2 = -2\beta$, and $b_2 = \beta$. Let's take $m_1 = -1$, $m_2 = 0$, and $m_3 = 1$. Then

$$f(z) = C_1 \int_0^z \frac{\zeta^{2\beta}}{(\zeta + 1)^\beta (\zeta - 1)^\beta} d\zeta + C_2 = C_1 \int_0^z \frac{\zeta^{2\beta}}{(\zeta^2 - 1)^\beta} d\zeta + C_2. \tag{23}$$

Notice that this formula duplicates (22) when $\beta = \frac{\pi}{2}$.

Exercises

1. Re-do Example 5.8.4, but with f mapping $-1, 1, \infty$ to $0, 1, i$, respectively.

2. In the triangle of Example 5.8.6, sketch the four isotherms corresponding to the four temperatures mentioned in (15). You may do using an educated guess, or with precision, using a computer. Then sketch also a few paths of heat flow.

3. Prove that when $m_3 = \infty$ in Theorem 5.8.1, f is given by (2). One method is by a change of variable: Replace the f of (1) (where m_3 is finite) with $f \circ S$, where S is a Möbius map such that $S(\infty) = m_3$ and $S : \mathbb{Y} \to \mathbb{Y}$. The Möbius map S given by

$$S(\omega) := m_3 - \frac{1}{\omega} \tag{24}$$

is an easy choice to work with.

4. Given $p > 1, q > 1$, show that there exists a Möbius map T such that for some real number a between p and q, equation (18) holds, and also $T(q) = a$.

5. Theorem 5.8.7 can be formulated with D in place of $\mathbb{Y}^o$, and proved in much the same way. Alternatively, the result for D can be derived from 5.8.7 by means of a change of variable in (17). One may use a Möbius map $T : D \to \mathbb{Y}^o$, like for example the one given by

$$T(w) = -i\frac{w + i}{w - i}.$$

Carry out the change of variable, $\zeta = T(w)$, and show that one obtains the following formula for the map $h := f \circ T$ from D to O:

$$h(w) = K_1 \int_0^w \frac{dw}{(w - \tau_1)^{b_1/\pi} \cdots (w - \tau_n)^{b_n/\pi}} + K_2 \quad \text{for } |w| \le 1,$$

where $\tau_k := T^{-1}(m_k)$.

6. In Example 5.8.8, since $f(\infty) - f(a) = ic - (1 + ic) = -1$, we must have

$$-1 = C_1 \int_a^\infty \frac{dx}{(x^2 - a^2)^{1/2}(x^2 - 1)^{1/2}}, \tag{25}$$

as well as

$$-1 = C_1 \int_1^a \frac{dx}{(a^2 - x^2)^{1/2}(x^2 - 1)^{1/2}}. \tag{26}$$

By carrying out a change of variable, show that (25) and (26) are in agreement.

7. Identify the Schwarz-Christoffel mapping from $\mathbb{Y}^o$ to a square.

8. Devise a conformal mapping from $\mathbb{Y}^o$ onto the bounded set whose boundary is $[0, 2, 2 + i, 1 + i, 1 + i2, 2i, 0]$. Suggestion: Try, as the corresponding vertices, $0, 1, a, \infty, -a, -1$, and 0; and hope to determine a value of $a > 1$ that works.

9. Devise a conformal mapping from $\mathbb{Y}^o$ onto the bounded set whose boundary is $[-1, 1, 1 + i2, -1 + i2, -1 + i, i, -1 + i, -1]$.

10. Using numerical methods as necessary, explore the fluid flow problem for the set O_5 shown in Figure 5.4-1, using the Schwarz-Christoffel equation (23).

11. By the method of subsection C, find a conformal mapping f from $\mathbb{Y}^o$ onto the vertical half-strip O_2 shown in Figure 5.1-3. Let $w_1 = -\frac{\pi}{2}, w_2 = \frac{\pi}{2}, w_3 = \infty$; and use as pre-vertices $m_1 = -1, m_2 = 1, m_3 = \infty$.

Help on Selected Exercises

3. If S is given by (24), then when we make the change of variable $\zeta = S(w)$ in (1), we find an equation like (2).

$$f(S^{-1}(z)) =$$

$$C_1 \int_{S^{-1}(z_0)}^{S^{-1}(z)} \frac{1}{(m_3 - \frac{1}{w} - m_1)^{b_1/\pi} (m_3 - \frac{1}{w} - m_2)^{b_2/\pi} (\frac{-1}{w})^{b_3/\pi}} \frac{-dw}{w^2} + C_2$$

$$= C_1' \int_{S^{-1}(z_0)}^{S^{-1}(z)} \frac{dw}{(w - \frac{1}{m_3 + m_1})^{b_1/\pi} (w - \frac{1}{m_3 + m_2})^{b_2/\pi}} + C_2'.$$

In the last step we made use of the fact that $b_1 + b_2 + b_3 = 2\pi$. We find that f has this form:

$$f(z) = C_1' \int_{z_0}^z \frac{d\omega}{(\omega - m_1')^{-b_1/\pi}(\omega - m_2')^{b_2/\pi}} + C_2'.$$

4. Let

$$S : -1, 1, 0 \rightarrow \infty, 0, -1, \quad \text{that is } S(z) = \frac{z-1}{z+1}.$$

For $x > 0$, let $T(z) = S^{-1}(xS(z))$, and solve the equation $T(-p) = -T(q)$ for x in terms of p and q. One obtains

$$x = \frac{\sqrt{qp + p - q - 1}}{\sqrt{qp - p + q - 1}}.$$

For example, if $p = 2$ and $q = 3$, then $x = \sqrt{\frac{2}{3}}$; and the common value of $T(q)$ and $-T(p)$ is

$$a = \frac{\sqrt{6}+1}{\sqrt{6}-1} \approx 2.3798.$$

11. One obtains $f(z) = C_1 \int_0^z \frac{d\zeta}{\sqrt{1-\zeta^2}} + C_2$. As one might remember from 2.2C, $f(z) = \sin^{-1} z$.

6

Lagniappe

When this book is used in the usual one-semester undergraduate course, the subject matter of this chapter need not come into view. A real analysis course beyond the calculus sequence would be an ideal background for reading it. Although we have made the proofs as elementary as we know how, they demand a greater mathematical maturity than do those of the previous chapters. And yet some of even the youngest readers may want to have a look. The sections are independent of each other, and each will lend itself to a partial reading.

Section 6.1 presents Dixon's 1971 proof of Cauchy's Theorem. It is a bit more sophisticated than the proof given in Chapter 3, and has the advantage of an elegant, global point of view.

We will try to persuade you that Runge's Theorem answers neatly some questions which may have occurred to you already, and leads to interesting conclusions. You may be pleased just to read and understand the statement of the Theorem, and to see how Cauchy's Theorem follows from it. The proof presented in Section 6.2 is a proof of many steps. However, it is both elementary and pedestrian. It is based on the manipulations of power series that were introduced in Section 2.6. Step by step, you will probably be able to see where it is going.

From the point of view of mathematical rigor, the Riemann Mapping Theorem and the Osgood-Taylor-Carathéodory Theorem, Sections 6.3 and 6.4, are necessary to make the book complete. Our treatment of the Schwarz-Christoffel Formula depended on them. We began with the knowledge, based on those two results, that there exists a conformal equivalence f with certain behavior at the boundary. Knowing that, we derived an explicit formula for f. Without certainty about the existence of f, we might devise a plausible formula for f, but in most cases we would find it formidably difficult to assure ourselves of its validity.

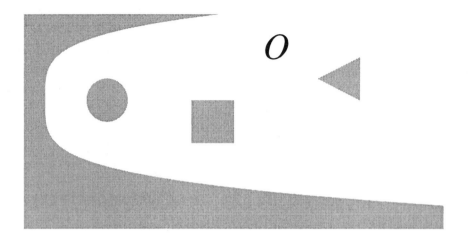

Figure 6.1-1 The open connected set O, indicated in the picture as the unshaded part, is unbounded and has three holes in it.

To be sure, in the conformal mapping problems of Chapter 5, we knew that suitable mappings existed. We could fairly easily devise and describe mappings that clearly worked.

You may be content to study just the statement of the Riemann Mapping Theorem and not the proof. The proof that we will give is deep and abstract. It is a non-constructive existence proof; the mapping is shown to exist but in general is not identified. If you read it, you will be able to tell that a history of clever discoveries lies behind the procedures. To study the proof, to the point of being convinced of the Theorem, is only a first step toward understanding the several important ideas that are in play. The usefulness of those ideas extends to other theorems and theories in higher mathematics.

6.1 DIXON'S 1971 PROOF OF CAUCHY'S THEOREM

Proof. Recall the statement of the Theorem (page 204). Let Γ be a contour in the open set O such that

$$W_\Gamma(z) = 0 \quad \text{for all } z \notin O. \tag{1}$$

Let f be holomorphic on O. In view of 3.1.11, it suffices to prove that

$$f(z) \cdot W_\Gamma(z) = \frac{1}{2\pi i} \int_\Gamma \frac{f(w)}{w-z} dw, \quad \text{provided } z \in O \setminus \Gamma. \tag{2}$$

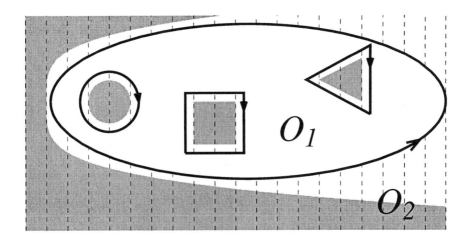

Figure 6.1-2 A contour Γ has been drawn in O. It consists of the large ellipse, taken counterclockwise, and three other closed paths, each taken clockwise. The set O_2, indicated by vertical dashed lines, is the outside of Γ. The sets O and O_2 overlap, and their union is $\mathbb{C}$. The set O_1 is the inside of Γ.

Let

$$
Q(z,w) := \begin{cases} \dfrac{f(w) - f(z)}{w - z} & \text{for } z \neq w; \\[2mm] f'(z) & \text{for } z = w. \end{cases} \tag{3}
$$

Notice the symmetry, $Q(z,w) = Q(w,z)$. If we hold z fixed, then the resulting function of one variable, given by $q(w) := Q(z,w)$, is holomorphic on O (see 3.1.10). But now, both z and w vary at once. We need the result that Q is not only continuous in each variable separately, but also continuous on $O \times O$. We will prove it below as Lemma 6.1.1.

Figures 6.1-1 and 6.1-2 illustrate, for one particular case, the objects that arise in the proof.

The winding number $W_\Gamma(z)$ is defined for every z not on Γ. In view of condition (1), the inside of Γ, by which we mean the set

$$
O_1 := \{z \in \mathbb{C} \mid z \notin \Gamma \text{ and } W_\Gamma(z) \neq 0\},
$$

is contained in O; and the outside of Γ, by which we mean the set

$$
O_2 := \{z \in \mathbb{C} \mid z \notin \Gamma \text{ and } W_\Gamma(z) = 0\},
$$

contains the complement of O. Every point of the plane is inside, outside, or on Γ. Since Γ is a compact set, it is contained in $D(0, r)$ for some $r > 0$. Therefore O_2 contains all z with $|z| \geq r$.

Let

$$h(z) := \int_\Gamma Q(z, w)dw \qquad \text{for } z \in O.$$

We will show that h is continuous. We will use that fact to show that it is in fact holomorphic on O. Then we will show that $h(z) = 0$ for all $z \in O$. That will complete the proof of the Theorem, because in the case when $z \in O$ and $z \notin \Gamma$, it gives us

$$\begin{aligned}
0 &= \int_\Gamma \frac{f(w) - f(z)}{w - z}dw \\
&= \int_\Gamma \frac{f(w)}{w - z}dw - f(z) \cdot \int_\Gamma \frac{dw}{w - z} \qquad (4) \\
&= \int_\Gamma \frac{f(w)}{w - z}dw - f(z) \cdot 2\pi i \cdot W_\Gamma(z);
\end{aligned}$$

(2) follows.

To prove that h is continuous at each point $z \in O$, pick $s > 0$ such that $\bar{D}(z, s) \subset O$. Since Q is continuous on $O \times O$, it is uniformly continuous on the compact subset $\bar{D}(z, s) \times \Gamma$. In particular, if $\{z_n\} \subset \bar{D}(z, s)$ and $z_n \to z$, then $Q(z_n, w) \to Q(z, w)$ uniformly for $w \in \Gamma$, and therefore $h(z_n) \to h(z)$.

Since for each w the function $z \mapsto Q(z, w)$ is holomorphic on O, its integral over the boundary of every triangle T contained in O equals zero. If the same is true for h, then h is holomorphic on O; and indeed,

$$\int_{\partial T} h(z)dz = \int_{\partial T} \int_\Gamma Q(z, w)dwdz = \int_\Gamma \int_{\partial T} Q(z, w)dzdw = \int_\Gamma 0dw = 0.$$

The order of the integrations with respect to z and w can be thus reversed because Q is continuous on $\partial T \times \Gamma$.

So h is holomorphic on O, and it remains to show that h is identically zero. The plan is to define an entire function H such that $H(z) = h(z)$ for all $z \in O$; prove that H is bounded and hence constant; and then show that the constant value must be zero.

In view of (4), and since $W_\Gamma(z) = 0$ for $z \in O_2$,

$$h(z) = \int_\Gamma \frac{f(w)}{w - z}dw \qquad \text{for all } z \in O_2 \cap O. \qquad (5)$$

The integral on the right is well defined and gives a holomorphic function of z, not just for $z \in O_2 \cap O$, but for all z in the complement of Γ (by 2.6.14). Therefore we may use it to extend the definition of h to the whole plane. First, let

$$h_1(z) := \int_\Gamma \frac{f(w)}{w - z}dw \qquad \text{for } z \in O_2.$$

By (5), h and h_1 agree on the intersection of their domains, so that we have a consistently defined entire function H if we let H agree with h on O and with h_1 on O_2. For all z sufficiently large, z is in O_2 and

$$|H(z)| = |h_1(z)| \leq \max_{w \in \Gamma}|f(w)| \max_{w \in \Gamma} \frac{1}{|w - z|} \int_\Gamma |dw|.$$

For each $\epsilon > 0$, the right-hand side is less than ϵ for all z sufficiently large. Therefore H is a bounded entire function and hence constant (by 3.1.19); since it also tends to zero at infinity the constant value must be zero. Therefore h equals zero everywhere on O. □

6.1.1. Lemma. *Let f be holomorphic on the open set O. Then the function Q given by (3) is continuous on $O \times O$.*

Proof. The function Q is continuous at each point (a, b) with $a \neq b$, because near such a point it equals a quotient of two continuous functions in which the denominator is nonzero.

It remains to prove that for each point $a \in O$, Q is continuous at the point (a, a). Let $\epsilon > 0$. Since f' is continuous at a, and since O is open, there exists $\delta > 0$ such that $D(a, \delta) \subseteq O$ and

$$|f'(\zeta) - f'(a)| < \epsilon \quad \text{whenever} \quad \zeta \in D(a, \delta).$$

For every z and w in $D(a, \delta)$,

$$f(w) - f(z) = \int_{[z,w]} f' = \int_0^1 f'((1-t)z + tw)(w - z)dt.$$

It follows that

$$Q(z, w) = \int_0^1 f'((1-t)z + tw)dt \quad \text{whenever} \quad z, w \in D(a, \delta). \tag{6}$$

Yes, the equality holds even when $z = w$, because in that case the integrand in (6) is constant and equal to $f'(z)$, which equals $Q(z, z)$. If $|z - a| < \delta$ and $|w - a| < \delta$ then also

$$|((1-t)z + tw) - a| < \delta \quad \text{for each} \quad t \in [0, 1];$$

therefore each value of the integrand in (6) is close to $f'(a)$:

$$|f'((1-t)z + tw) - f'(a)| < \epsilon \quad \text{for all} \quad t \in [0, 1];$$

therefore $|Q(z, w) - Q(a, a)| \equiv |Q(z, w) - f'(a)| < \epsilon.$ □

6.2 RUNGE'S THEOREM

Carl David Tolmé Runge published this result in 1885. Advanced students will appreciate Robert B. Burckel's thorough discussion of the Theorem, its history, and related results ([9, Notes to Chapter VIII]). After we state the Theorem, and before giving a proof, we will offer a number of remarks, including an explanation of how it implies Cauchy's Theorem.

6.2.1. Theorem. *Let f be holomorphic on the open set $O \subset \mathbb{C}$. Let E be a set that contains one point in each component of $\hat{\mathbb{C}} \setminus O$. Then for every compact set $K_0 \subset O$, and for every $\epsilon > 0$, there is a rational function F whose poles are contained in E such that*

$$|f(z) - F(z)| < \epsilon \quad for \ z \in K_0.$$

There are special cases of the Theorem which offer no surprise at all. For example, consider a holomorphic f on a disk $D(a, r)$. The complement of the disk has only one component. We might take E to be the singleton $\{\infty\}$. The assertion of Runge's Theorem in this case is already established, because we already know that

$$f(z) = \sum_{k=0}^{\infty} \frac{f^{(k)}(a)}{k!}(z - a)^k \quad for \ |z - a| < r,$$

and that the partial sums converge to f subuniformly. We exploit this knowledge as follows. For each integer $n > 0$, the partial sum

$$P_n(z) := \sum_{k=0}^{n} \frac{f^{(k)}(a)}{k!}(z - a)^k$$

is a polynomial—in other words, a rational function whose only pole is at ∞. Every compact set $K_0 \subset D(a, r)$ is contained in $\bar{D}(a, s)$ for some $s < r$. Let $\epsilon > 0$. For sufficiently large n, $|f(z) - P_n(z)| < \epsilon$ for $|z - a| \leq s$, hence for all $z \in K_0$.

Runge's Theorem makes the startling assertion that f can be uniformly approximated on compact subsets by polynomials *whenever* the unbounded component of $\hat{\mathbb{C}} \setminus O$ is the only one—not just when O is a disk. Of course, it is not true in general that the polynomials are generated simply by taking partial sums of a Taylor series.

Here is another special case in which we already know Runge's conclusion. Let f be holomorphic on the annulus $A := \mathrm{Ann}_a(r_1, r_2)$. It of course has a Laurent series representation (3.3.1):

$$f(z) = \sum_{k=1}^{\infty} c_{-k}(z - a)^{-k} + \sum_{k=0}^{\infty} c_k(z - a)^k \qquad for \ all \ z \in A.$$

The complement of A in $\hat{\mathbb{C}}$ has two components. Let $E = \{a, \infty\}$. Every compact subset $K_0 \subset A$ is contained in $\overline{\mathrm{Ann}_a(s_1, s_2)}$ for some s_1, s_2. For every $\epsilon > 0$ there exist m, n such that

$$\left| f(z) - \underbrace{\sum_{k=1}^{m} c_{-k}(z-a)^{-k}}_{=:Q_m(z)} - \underbrace{\sum_{k=0}^{n} c_k(z-a)^k}_{=:P_n(z)} \right| < \epsilon \quad \text{for } z \in \overline{\text{Ann}_a(s_1, s_2)}.$$

Thus f is approximated on K_0 by a rational function $Q_m + P_n$, which is the sum of a rational function whose only pole is at a, and a polynomial whose only pole is at ∞.

We return now to the general situation addressed by Runge's Theorem. The rational function F whose existence is asserted may have any finite number of poles. Call them $b_1, \cdots, b_J$, and let N_j be the order of the pole at b_j. So F has this form:

$$F(z) = c + \sum_{j=1}^{J} P_j(z), \tag{1}$$

where c is a constant and P_j is the principal part of F at the pole b_j. If $b_j = \infty$, then P_j is a polynomial:

$$P_j(z) = \sum_{k=1}^{N_j} c_{j,k} z^k. \tag{2}$$

Otherwise, it has this form:

$$P_j(z) = \sum_{k=1}^{N_j} c_{j,-k}(z - b_j)^{-k}. \tag{3}$$

How Runge's Theorem implies Cauchy's Theorem. Suppose that we know Runge's Theorem and wish to prove Cauchy's Theorem 3.2.1. Let Γ be a contour in the open set $O \subseteq \mathbb{C}$ such that $W_\Gamma(z) = 0$ for all $z \notin O$. Let f be holomorphic on O. What we need to prove is that

$$\int_\Gamma f = 0. \tag{4}$$

Apply Runge's Theorem. Let Γ play the role of K, and let $\epsilon > 0$. Obtain a rational function F whose poles b_j are outside O, such that

$$|f(z) - F(z)| \equiv |f(z) - c + \sum_{j=1}^{J} P_j(z)| < \epsilon \quad \text{for } z \in \Gamma.$$

Then $\left| \int_\Gamma (f - F) \right|$ is less than ϵ times the length of Γ. If we show that $\int_\Gamma F = 0$, it will follow that $\left| \int_\Gamma f \right|$ is less than ϵ times the length of Γ. That being true for arbitrary $\epsilon > 0$, (4) follows.

It remains to show that $\int_\Gamma F = 0$. Equations (1), (2), and (3) tell us that F is a combination of functions which can be a constant, a polynomial, and powers $z \mapsto (z - b_j)^{-k}$ where b_j is outside O. Except for the -1 powers, every such part has an antiderivative on O, and hence its integral over Γ is 0. And $\int_\Gamma \dfrac{dz}{(z - b_j)^{-1}}$ equals $2\pi i W_\Gamma(b_j)$, which by hypothesis is also 0. $\qquad\square$

We approach the proof of Runge's Theorem through a sequence of Lemmas. The first one is called "moving the pole from a to b."

6.2.2. Lemma. *Let K be a compact subset of $\mathbb{C}$. Let $a \in \mathbb{C}$ and $b \in \hat{\mathbb{C}}$ be two points belonging to the same component of $\hat{\mathbb{C}} \setminus K$. Let F_a be a rational function whose sole pole is at a, and let $\epsilon > 0$. Then there exists a rational function F_b whose sole pole is at b such that*

$$\max_{z \in K} |F_a(z) - F_b(z)| < \epsilon.$$

Proof. We will need a definition for the distance from a point $p \in \mathbb{C}$ to the set K:

$$\text{dist}(p, K) := \max\{r \mid D(p, r) \cap K = \emptyset\} \equiv \inf\{|p - z| \mid z \in K\}.$$

Case 1, when $F_a(z) = \dfrac{1}{z - a}$; $a, b \in \mathbb{C}$; and $|a - b| < \text{dist}(b, K)$. For $z \in K$,

$$F_a(z) \;=\; \frac{1}{z - b - (a - b)} \;=\; \frac{1}{(z - b)\left(1 - \frac{a-b}{z-b}\right)} \;=\; \sum_{k=0}^{\infty} \frac{(a - b)^k}{(z - b)^{k+1}}.$$

The series converges uniformly for $z \in K$ because

$$\max_{z \in K} \left| \frac{a - b}{z - b} \right| < 1.$$

Case 2, when F_a is arbitrary; $a, b \in \mathbb{C}$; and $|a - b| < \text{dist}(b, K)$. We start with the general form for F_a and apply the Case 1 procedure:

$$F_a(z) = c + \sum_{k=1}^{n} \frac{a_k}{(z - a)^k}$$

$$= c + \sum_{k=1}^{n} a_k \left(\sum_{j=0}^{\infty} \frac{(a - b)^j}{(z - b)^{j+1}} \right)^k$$

$$= \lim_{J \to \infty} \left[c + \sum_{k=1}^{n} a_k \left(\sum_{j=0}^{J} \frac{(a - b)^j}{(z - b)^{j+1}} \right)^k \right].$$

The limit is uniform for $z \in K$, so we may take F_b to be given by the expression in the brackets for sufficiently large J.

Case 3, when F_a is arbitrary and $a, b \in \mathbb{C}$. There exist points $a_0 := a, a_1, \cdots, a_m := b$ such that $|a_{j+1} - a_j| < \text{dist}(a_{j+1}, K)$ for each j. (Exercise 3 asks you to justify this assertion.) Apply Case 2 to move the pole m times.

Case 4, when F_a is arbitrary and $b = \infty$. For some $r > 0$, the disk $D(0, r/2)$ contains K. By Case 3, there is a rational function F_r with sole pole at r such that $|F_r(z) - f(z)| < \epsilon/2$ for $z \in K$. Since F_r is holomorphic on $D(0, r)$, there exists a polynomial P such that $|F_r(z) - P(z)| < \epsilon/2$ for $z \in D(0, r/2)$. Then, of course, P has its sole pole at ∞, and $|P(z) - f(z)| < \epsilon$ for $z \in K$.

Cases 3 and 4 include all possible cases. $\qquad\qquad\qquad\qquad\qquad\qquad\square$

Some elementary remarks to prepare the way for the next Lemma: A continuous function $f := u + iv$ on $[0, 1]$ is uniformly continuous (2.4.8). Therefore if

$$\omega_f(\delta) := \max\{|f(p) - f(q)| \mid p, q \in [0, 1], |p - q| \le \delta\}, \qquad (5)$$

then $\omega_f(\delta) \to 0$ as $\delta \to 0$. Now consider the partition of $[0, 1]$ into k subintervals of equal length. Applying the Mean Value Theorem for Integrals (subsection 1.5A) to the real-valued function u on each of those subintervals, we know that there exist points $t_j^* \in \left[\frac{j-1}{k}, \frac{j}{k} \right]$ such that

$$\int_0^1 u(t) dt = \sum_{j=1}^k \int_{\frac{j-1}{k}}^{\frac{j}{k}} u(t) dt = \frac{1}{k} \sum_{j=1}^k u(t_j^*).$$

Likewise there exist points $t_j^{**} \in \left[\frac{j-1}{k}, \frac{j}{k} \right]$ such that

$$\int_0^1 v(t) dt = \frac{1}{k} \sum_{j=1}^k v(t_j^{**}).$$

Thus

$$\int_0^1 f(t) dt = \frac{1}{k} \sum_{j=1}^k (u(t_j^*) + iv(t_j^{**})).$$

If for every j we replace t_j^* and t_j^{**} by j/k, how much does the sum change? It is easy to show that

$$\left| \int_0^1 f(t) dt - \frac{1}{k} \sum_{j=1}^k f\left(\frac{j}{k} \right) \right| \le 2\omega_f \left(\frac{1}{k} \right).$$

6.2.3. Lemma. *For a continuous map $h : [0, 1] \times K \to \mathbb{C}$, where K is a compact subset of $\mathbb{C}$, let*

$$H(z) := \int_0^1 h(t, z) dt, \quad \text{and} \quad H_k(z) := \frac{1}{k} \sum_{j=0}^k h\left(\frac{j}{k}, z \right) \quad \text{for } k = 1, 2, \cdots.$$

Then $H_k \to H$ uniformly on K.

Proof. Let $\omega_h^*(\delta)$ be the maximum value of $|h(t, z) - h(t', z')|$, considering all $t, t' \in [0, 1]$ with $|t - t'| \le \delta$; and all $z, z' \in K$ with $|z - z'| \le \delta$. Then

$$\max_{z \in K} |H(z) - H_k(z)| \le 2\omega_h^* \left(\frac{1}{k}\right) \to 0 \quad \text{as} \quad k \to \infty.$$

$\square$

Proof of Runge's Theorem. Let f, O, K_0, ϵ, and E be as given in the hypothesis. There is no loss if we replace the given set K_0 by a larger compact subset of O, and our first step is to do so.

Since K_0 is a compact subset of O, there is a number $\eta > 0$ such that $|z - \zeta| > \eta$ whenever $z \in K_0$ and $\zeta \notin O$; that is, $\text{dist}(z, \mathbb{C} \setminus O) \ge \eta$ for every $z \in K_0$. Also since K_0 is compact, it is contained in $D(0, R)$ for some R. We replace K_0 with the perhaps larger set

$$K := \{z \mid |z| \le R \text{ and } \text{dist}(z, \mathbb{C} \setminus O) \ge \eta\}.$$

The set K has the advantage that it has "no extra holes." More precisely: For every point $z \in \hat{\mathbb{C}} \setminus K$, there is a point $w \in \hat{\mathbb{C}} \setminus O$ such that z and w are in the same component of $\hat{\mathbb{C}} \setminus K$, by the following argument. Either (1) $|z| > R$ and z is in the same component as ∞; or (2) $z \in D(w, \eta)$ for some $w \in \hat{\mathbb{C}} \setminus O$, so that z is in the same component as w.

Apply Lemma 3.2.2, obtaining a contour Γ_0 such that

$$W_{\Gamma_0}(z) = \begin{cases} 0 & \text{for } z \notin O, \\ 1 & \text{for } z \in K; \end{cases}$$

and such that if f is holomorphic on O, then

$$f(z) = \frac{1}{2\pi i} \int_{\Gamma_0} \frac{f(w)}{w - z} dw \quad \text{for } z \in K.$$

Since Γ_0 consists of a finite number P of C^1 curves γ_p, we may write

$$f(z) = \sum_{p=1}^{P} \frac{1}{2\pi i} \int_{\gamma_p} \frac{f(w)}{w - z} dw \quad \text{for } z \in K.$$

It suffices to show that for each p there exists a rational function f_p, with poles all in E, such that

$$\left| \frac{1}{2\pi i} \int_{\gamma_p} \frac{f(w)}{w - z} dw - f_p(z) \right| < \frac{\epsilon}{P} \quad \text{for all } z \in K.$$

Each γ_p can be parametrized by $w : [0,1] \to \mathbb{C}$, so that

$$\frac{1}{2\pi i} \int_{\gamma_p} \frac{f(w)}{w-z} dw = \frac{1}{2\pi i} \int_0^1 \frac{f(w(t))w'(t)}{w(t)-z} dt$$

$$= \lim_{J\to\infty} \frac{1}{2\pi i} \frac{1}{J} \sum_{j=1}^J \frac{f(w(j/J))w'(j/k)}{w(j/J)-z},$$

the limit being uniform for $z \in K$ by Lemma 6.2.3. Thus the integral is approached uniformly for $z \in K$ by sums of the form

$$\sum_{j=1}^J F_{a_j}(z), \quad \text{where } F_{a_j}(z) := \frac{c_j}{a_j - z}$$

for certain complex numbers a_j and c_j. Select J sufficiently large so that

$$\left| \frac{1}{2\pi i} \int_{\gamma_p} \frac{f(w)}{w-z} dw - \sum_{j=1}^J F_{a_j}(z) \right| < \frac{\epsilon}{2P}.$$

Since F_{a_j} is a rational function whose sole pole is at the point $a_j \in \mathbb{C}\backslash K$, by Lemma 6.2.2 there exists a rational function F_{b_j} whose sole pole is at a point $b_j \in E$ such that

$$|F_{a_j}(z) - F_{b_j}(z)| < \frac{\epsilon}{2JP} \quad \text{for } z \in K.$$

Let

$$f_p(z) = \sum_{j=1}^J F_{b_j}(z).$$

$\square$

Exercises

1. Let O be an open connected subset of $\mathbb{C}$. Use Runge's Theorem to show that $\hat{\mathbb{C}} \backslash O$ is connected if and only $W_\Gamma(\alpha) = 0$ for every contour Γ in O and every $\alpha \notin O$.

2. Prove Runge's Theorem with the first sentence replaced by "Let f be holomorphic on the open set $O \subset \hat{\mathbb{C}}$."

3. In the proof of Lemma 6.2.2, Case 3, explain the assertion as to the existence of points $a_0, a_1, \cdots, a_m$.

4. Let $O := D \backslash \bar{D}(\frac{1}{2}, \frac{1}{2})$. Let f be holomorphic on O. Must there exist a sequence of polynomials converging subuniformly on O to f? Must there

exist a sequence of polynomials converging uniformly on O to f? If F is holomorphic on an open set containing the closure of O, must there exist a sequence of polynomials converging uniformly on O to F?

5. Is there a sequence of polynomials P_n such that $P_n(0) = 1$ for all $n = 1, 2, 3, \ldots$, but $\lim_{n \to \infty} P_n(z) = 0$ for every $z \neq 0$?

6.3 THE RIEMANN MAPPING THEOREM

We will write $O_1 \cong O_2$ to mean that the sets O_1 and O_2 are conformally equivalent. The Riemann Mapping Theorem, as we shall present it, asserts that every simply connected, open, proper subset of $\mathbb{C}$ is conformally equivalent to the open unit disk D. It follows immediately that if O_1 and O_2 are simply connected, open, proper subsets of $\mathbb{C}$, then $O_1 \cong O_2$, since if $f_k : O_k \to D$ is a conformal equivalence for $k = 1, 2$, then $f_2^{-1} \circ f_1$ is a conformal equivalence of O_1 and O_2.

There is a gap in the calculus-of-variations proof given in Riemann's 1851 dissertation [58]. The gap was later filled by David Hilbert and others. A variety of proofs are available today; see Burckel's survey [9, Chapter IX]. The proof we shall give is due to L. Fejér and F. Riesz. As you will see, it makes use of several substantial other results.

6.3.1. Theorem. *Let O be a simply connected, open, proper subset of $\mathbb{C}$. Then O is conformally equivalent to the open unit disk D.*

Proof. We may suppose that $0 \in O$, and that O is bounded. Here is why.

- If O is bounded, let a be a point in O. Let
$$O - a := \{z - a \mid z \in O\},$$
which contains 0. Since $O \cong O - a$, it suffices to show $O - a \cong D$.

- If O is unbounded, let b be a complex number that is outside O. Then $O - b$ is simply connected and does not contain 0. By 3.2.3, page 208, there exists a holomorphic function g on $O - b$ such that
$$g(z)^2 = z \quad \text{for all } z \in O - b.$$

Notice that if $z_1, z_2 \in O - b$ and $g(z_1) = -g(z_2)$, then $z_1 = z_2$. Therefore if $w \in g[O - b]$, then $-w \notin g[O - b]$. Since g is an open mapping, the set $g[O - b]$ and hence the set $U := \{-w \mid w \in g[O - b]\}$ are open; and $U \cap g[O - b] = \emptyset$. The function $h : z \mapsto \dfrac{1}{z + b}$ maps U onto a neighborhood of ∞. Let $d \in h[g[O - b]]$. Then
$$z \mapsto h(g(z - b)) - d$$
is a conformal equivalence from O onto a bounded simply connected region that contains 0. Note the use here of Proposition 3.2.5.

So we may suppose that $0 \in O$ and that $O \subset D(0, B)$ for some finite number $B > 0$. Consider the family $\mathcal{F}$ of all holomorphic functions f that map O one-to-one into D, and such that $f(0) = 0$ and $f'(0) > 0$. The family $\mathcal{F}$ is not empty, since $z \mapsto z/B$ belongs to it. Let

$$m = \sup\{f'(0) \mid f \in \mathcal{F}\},$$

which is finite by 3.2.6, page 209. Choose a sequence of functions $f_n \in \mathcal{F}$ such that $f_n'(0) \to m$. By Exercise 5 of Section 3.2, page 211, the family $\mathcal{F}$ is equicontinuous. Therefore, by the Arzelà-Ascoli Theorem[1] (see, for example, [51, Section 4.3] or [61, pp. 245–246]), the sequence $\{f_n\}$ has a subsequence $\{f_{n(k)}\}$ that converges subuniformly on O. By 3.1.22, page 197, its limit f must be holomorphic on O. By 3.2.7, page 210, $f_{n(k)}' \to f'$ subuniformly. Since $f'(0) = m$, the function f is nonconstant. Therefore by 3.4.14, page 248, f is one-to-one. It remains to show that f is onto. The proof we give is called Koebe's trick. One devises, under the assumption that f is not onto, another function $F \in \mathcal{F}$ for which $F'(0) > f'(0)$, which of course cannot exist.

Suppose that f is not onto, so that there exists $a \in D \setminus f[O]$. Let N_a be the Möbius map studied in Example 5.2.18, page 336, and note its derivative:

$$N_a(z) = \frac{a - z}{1 - \bar{a}z}; \qquad N_a'(z) = \frac{|a|^2 - 1}{(1 - \bar{a}z)^2}.$$

Then 0 is not contained in the set $N_a[f[O]]$. Since that set is simply connected, there is a holomorphic function G such that

$$G(z)^2 = z \quad \text{for all } z \in N_a[f[O]].$$

Let $b := G(a)$. It is easy to show that the function

$$F := \frac{b}{|b|} N_b \circ G \circ N_a \circ f$$

belongs to the family $\mathcal{F}$. We may compute its derivative at 0 using the Chain Rule:

$$F'(0) = \frac{b}{|b|} N_b'(b) \, G'(a) \, N_a'(0) \, f'(0)$$

$$= \frac{b}{|b|} \frac{|a| - 1}{(1 - |a|)^2} \frac{1}{2b} (|a|^2 - 1) f'(0) > f'(0) \equiv m.$$

We have reached the promised contradiction and completed the proof.   □

Exercises

1. In proving the Riemann Mapping Theorem, how do we use the hypothesis that makes O a *proper* subset of the plane? Is the Theorem still true if we remove that hypothesis? Suggestion: Think about Liouville's Theorem.

---

[1] In its complex-analysis application, this result is known as Montel's Theorem.

2. In the proof of the Riemann Mapping Theorem, justify in detail the statements regarding the function F.

6.4 THE OSGOOD-TAYLOR-CARATHÉODORY THEOREM

The Riemann Mapping Theorem assures the existence of a one-to-one holomorphic mapping f from any open simply connected set O onto $\mathbb{Y}^o$. It is natural to ask whether or when f extends to a continuous, perhaps one-to-one, mapping from $\bar{O}$ onto $\mathbb{Y}$. Two papers published in 1913, [50] and [11], independently addressed the question and led to a considerable variety of related results, methods of proof, and applications. See [9, Chapter IX] for a thorough survey.

We will use a combination of the ideas in [60, pp. 308–311] and [49] to establish the kinds of claims that we used in Chapter 5 concerning extensions of f to the boundary.

A boundary point β of an open set U is a **simple** boundary point of U if, for every sequence $\{\alpha_n\} \subset U$ such that $\alpha_n \to \beta$, there is a curve γ given by $z : [0, 1] \to U \cup bU$ such that $z(t) \in U$ for $0 \leq t < 1$, $z(1) = \beta$; and a strictly increasing sequence of numbers t_n, with $0 \leq t_n < 1$, such that $z(t_n) = \alpha_n$ for each n. The requirement on the function z is merely that it be continuous, so that $\lim_{t \to 1} z(t) = \beta$.

6.4.1. Example. *Consider the set $O := D \setminus [-1, 0]$ shown on the left in Figure 6.4-1. Identify the simple and non-simple boundary points of O. Devise a conformal equivalence g from O onto $\mathbb{Y}^o$ and discuss its extendability to the boundary.*

Solution. The origin, and the points of the unit circle other than -1, are simple boundary points of O, while the points of the interval $[-1, 0)$ are not. However, each point of $[-1, 0)$ is a simple boundary point of $O \cap \mathbb{Y}^o$, and is also a simple boundary point of $O \cap \mathbb{Y}^o_-$.

We will devise a conformal equivalence g from O onto the open upper half plane $\mathbb{Y}^o$. Because of the presence of non-simple boundary points, it does not extend to a continuous map from $\bar{O}$ onto $\mathbb{Y}$.

Let's construct g in three steps. Make sketches to follow the action. First, the function given by $g_1(z) = \sqrt{z}$ is conformal from O onto the the right-hand half of D, the set $D_1 := \{z \in D \mid \operatorname{Re} z > 0\}$. Second, the Möbius map T given by

$$T(z) = \frac{i - z}{z + i}$$

maps D_1 onto Q_I, the open first quadrant. Finally, the function given by $g_2(z) = z^2$ maps Q_I onto $\mathbb{Y}^o$. Then we may take g to be the composition of those three mappings:

$$g(z) = g_2(T(f_1(z))) = \left(\frac{i - \sqrt{z}}{\sqrt{z} + i} \right)^2.$$

You may verify the following observations.

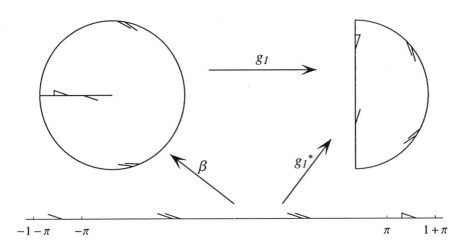

Figure 6.4-1 On the left, an open connected set O, discussed in Example 6.4.1, whose boundary points are not all simple. There is a conformal equivalence g_1 of O onto the half-disk D_1 shown on the right. The boundary of O is a path parametrized by β; the boundary of D_1 is parametrized by g_1^*.

- The map g_2 is a one-to-one continuous map from Q_I onto $\mathbb{Y}$, though of course it is merely continuous, not holomorphic, at the corner. It takes the positive real axis to itself, and the upper imaginary axis to the negative real axis.

- The map T, being a Möbius map, is a one-to-one meromorphic map from $\hat{\mathbb{C}}$ onto $\hat{\mathbb{C}}$, so of course it is a one-to-one continuous map from $\bar{D}_1$ onto Q_I. In particular, it maps the segment $[i, 0]$ to $[0, 1]$; $[0, -i]$ to the infinite real interval $[1, \infty]$; and the semicircle $\mathrm{Arc}(1, -\frac{\pi}{2}, \frac{\pi}{2})$ to the upper imaginary axis.

- The map g_1 is the source of the difficulty. It does not extend to a continuous map from $\bar{O}$ onto $\bar{D}_1$. If we restrict g_1 to $O \cap Y^o$, or to $O \cap \mathbb{Y}^o_-$, then it does extend continuously to the boundary thereof. In the first case, the extension maps the segment $[-1, 0]$ to the segment $[i, 0]$. In the second case, the extension maps $[-1, 0]$ to the segment $[-i, 0]$.

The last observation suggests that although g_1 and hence g fail to extend continuously to $\bar{O}$, they do something almost as nice and just as useful. The boundary of O may be seen as a path consisting of the segment $[0, -1]$, the arc $\mathrm{Arc}(1, -\pi, \pi)$, and the segment $[-1, 0]$. The interval $[-1, 0)$ on the real axis is traversed twice by this path. We may parametrize the path on the interval $[-\pi - 1, \pi + 1]$ as follows. Let

$$\beta(t) = \begin{cases} -t - \pi - 1 & \text{if } -\pi - 1 \le t \le -\pi, \\ e^{it} & \text{if } -\pi \le t \le \pi, \\ t - \pi - 1 & \text{if } \pi \le t \le \pi + 1. \end{cases} \tag{1}$$

Notice that as t increases, $\beta(t)$ moves around the boundary of O with some subset of O on its left at all times. The region on the left can be taken to be $O \cap \Upsilon^o_-$ for the first t-interval, O for the second, and $O \cap \Upsilon^o$ for the third. The t-interval is mapped by β twice onto $[-1, 0)$: $\beta[(-\pi - 1, -\pi]] = \beta[[\pi, \pi + 1)] = [-1, 0)$.

We have the following result: For each t, the function $g_1^*(t)$ may be defined so that if α_n is a sequence that lies in the region on the left and converges to $\beta(t)$, then $g_1^*(t) = \lim_{n\to\infty} g_1(\alpha_n)$. For each point β on bO, if $\beta = \beta(t)$, then putting $g_1(\beta) = g_1^*(t)$ makes g_1 continuous from the left at β.

To put things together and summarize: The extended function $g^* = g_2 \circ T \circ g_1^*$ maps $[-\pi - 1, \pi + 1]$ onto the extended real axis as follows: It maps $[-\pi - 1, -\pi]$ onto $[1, \infty]$, $[-\pi, \pi]$ onto $[-\infty, 0]$, and $[\pi, \pi + 1]$ onto $[0, 1]$.   □

As the Example suggests, many of the specific conformal equivalences that we might actually write down have extensions to the boundary which are easy to understand. We now start work toward a general theorem.

6.4.2. Lemma. *Let $z_0 \in E \subset \mathbb{C}$, where E is open, and let $r > 0$. Suppose that there is an arc of the circle $|z - z_0| = r$ of length greater than πr which does not meet $\bar{E}$. Let g be holomorphic on E and continuous on $\bar{E}$, such that*

$$|g(z)| \le M \quad \text{for all} \quad z \in \bar{E} \quad \text{and} \quad |g(z)| \le \epsilon \quad \text{for all} \quad z \in bE \cap D(z_0, r).$$

Then $|g(z_0)| \le \sqrt{\epsilon M}$.

Proof. We may suppose without loss of generality that $z_0 = 0$. The closure of $E \cap (-E)$ does not meet the circle $|z| = r$. Let

$$E_1 := E \cap (-E) \cap D(0, r).$$

If $\zeta \in bE_1$, then either ζ or $-\zeta$ belongs to $bE \cap D(0, r)$. Let $h(z) := g(z)g(-z)$ for $z \in \bar{E}_1$. Then $|h(z)| \le \epsilon M$ for $z \in bE_1$. Therefore $|h(z)| \le \epsilon M$ for all $z \in E_1$, and the result follows.   □

6.4.3. Lemma. *Consider a bounded open set $U \subset \mathbb{C}$. Let f be a one-to-one holomorphic function from U onto D. Let β be a simple boundary point of U. Then $f(\beta)$ may be defined so that f is continuous on $U \cup \{\beta\}$ and $|f(\beta)| = 1$.*

Proof. We point out first that the condition $|f(\beta)| = 1$ is inescapable once the rest of the statement is proved. If $\{\alpha_n\}$ is a sequence in U, and if $f(\alpha_n) \to w$ where $|w| < 1$, then $\alpha_n \equiv f^{-1}(f(\alpha_n)) \to f^{-1}(w)$, which is not on the boundary of U. Therefore if α_n tends to a boundary point and $f(\alpha_n) \to w$, then $|w| = 1$.

The Lemma is equivalent to the statement that whenever $\{\alpha_n\}$ is a sequence in U converging to β, $f(\alpha_n)$ converges to the same limit. We must then define $f(\beta)$ to be that limit.

Suppose the Lemma were false. Then there would exist a sequence $\{\alpha_n\} \subset U$ converging to β such that $f(\alpha_{2j-1}) \to w_1$, $f(\alpha_{2j}) \to w_2$, and $w_1 \ne w_2$. Let γ be the curve provided by the definition of a simple boundary point. There are two arcs of the unit circle that run between w_1 and w_2. Then $\{f(z(s)) \mid 0 \le s < 1\}$ would accumulate at every point of at least one of those two arcs. Restrict attention to that arc. Let p be the midpoint of the arc, a the midpoint of the subarc from w_1 to p, and c the midpoint of the subarc from p to w_2. Make a sketch.

Let $\delta := |w_1 - w_2|$. We may suppose

$$|f(a_{2j-1}) - w_1| < \frac{\delta}{4} \quad \text{for all } j,$$

$$|f(a_{2j}) - w_2| < \frac{\delta}{4} \quad \text{for all } j, \quad \text{and}$$

$$|f(z(t))| > \frac{1}{2} \quad \text{for all } t.$$

We can find x_j and y_j such that

$$t_{2j-1} < x_j < y_j < t_{2j},$$
$$f(z(x_j)) \text{ is on the radial segment } [0, a],$$
$$f(z(y_j)) \text{ is on the radial segment } [0, b], \quad \text{and}$$
$$f(z(s)) \text{ lies in the sector between those two radii, with}$$
$$\frac{1}{2} < |f(z(s))| < 1 \quad \text{for } x_j < s < y_j.$$

Let ζ_j be the closed path consisting of the segment $[0, f(x_j)]$, followed by the curve

$$s \mapsto f(z(s)) \quad \text{for } x_j \le s \le y_j,$$

followed by the segment $[f(y_j), 0]$. Let E_j be the inside of ζ_j. Then $\frac{p}{2} \in E_j \subset D$. Let $0 < r < \frac{1}{4}|a - b|$, and choose z_0 on the radius $[0, p]$ so near to p that the circle $|z - z_0| = r$ has an arc of length greater than πr that lies outside D. For all sufficiently large j,

$$|f(z(s))| > |z_0| \quad \text{if } s_{2j-1} \le t \le s_{2j},$$

and therefore $z_0 \in E_j$. If $w \in bE_j \cap D(z_0, r)$, then $w \in \{f(z(t)) \mid x_j \le s \le y_j\}$, hence $f^{-1}(w) \in z[[s_{2j-1}, s_{2j}]]$. Recall that $z(s_{2j-1})$ and $z(s_{2j})$ are approaching β. Let

$$\epsilon_j := \sup\{|f^{-1}(w) - \beta| \mid w \in bE_j \cap D(z_0, r)\},$$

which tends to 0 as $j \to \infty$. Let

$$M := \sup\{|f^{-1}(z) - \beta| \mid z \in D\},$$

which is finite because U is bounded. By Lemma 6.4.2, $|f^{-1}(z_0) - \beta| \le \sqrt{\epsilon_j M}$ for all j, so $f^{-1}(z_0) = \beta$, which cannot be the case; we have reached a contradiction. $\quad\square$

6.4.4. Theorem. *Let O be a bounded simply connected open set in the plane whose every boundary point is simple. Let f be a conformal equivalence from O onto D. Then f extends to a one-to-one continuous mapping from $\bar{O}$ onto $\bar{D}$.*

Proof. It follows from Lemma 6.4.3 that f extends to a continuous mapping from $\bar{O}$ into $\bar{D}$, which we will denote also by f. We must show that it is onto, and that it is one-to-one. Since continuous functions preserve compact sets, $f[\bar{O}]$ is a compact, hence closed, subset of $\bar{D}$. But the only closed set between D and $\bar{D}$ is $\bar{D}$. So f is onto.

It remains to show that f is one-to-one. That is, we need to show that if $\beta_1, \beta_2 \in bO$ and $f(\beta_1) = f(\beta_2)$, then $\beta_1 = \beta_2$. We may suppose that $f(\beta_1) = f(\beta_2) = -1$; otherwise, replace f with $-\overline{f(\beta_1)}f$. For $j = 1, 2$, there is a curve γ_j given by $z_j : [0, 1] \to \bar{O}$ such that

$$z_j : [0, 1) \to O, \quad z_j(1) = \beta_j, \quad \text{and} \quad f(z_j(t)) \to -1 \text{ as } t \to 1^-.$$

There exists s_0 such that $0 < s_0 < 1$ and

$$s_0 < s < 1 \Longrightarrow |z_j(s) - \beta_j| < 0.1|\beta_1 - \beta_2| \quad \text{for } j = 1, 2.$$

Thus

$$s_0 < s_1, s_2 < 1 \Longrightarrow |z_1(s_1) - z_2(s_2)| > 0.8|\beta_1 - \beta_2|. \tag{2}$$

There exists $\delta > 0$ such that $D(-1, \delta) \cap f \circ z_j[[0, s_0]] = \emptyset$ for $j = 1, 2$. Let $A(\delta) = D(-1, \delta) \cap D$. Let $0 < r \le \delta$. The boundary of $A(r)$ is formed by an arc of the unit circle and the arc given by

$$\theta \mapsto -1 + re^{i\theta} \quad (-\eta(r) \le \theta \le \eta(r)), \tag{3}$$

where $\eta(r)$ increases to $\frac{\pi}{2}$ as $r \to 0^+$. On $\text{Arc}_{-1}(r, -\eta(r), \eta(r))$, there exist points w_1, w_2 such that $w_j = f(z_j(s_j))$ for some s_1, s_2 as in (2). Therefore

$$0.8|\beta_1 - \beta_2| < |z_1(s_1) - z_2(s_2)| \equiv |f^{-1}(w_1) - f^{-1}(w_2)|.$$

The difference $f^{-1}(w_1) - f^{-1}(w_2)$ equals the complex integral of $(f^{-1})'$ along the part of the circular arc (3) that runs from w_1 to w_2. Therefore

$$c < \int_{-\eta(r)}^{\eta(r)} |(f^{-1})'(-1 + re^{i\theta})| r \, d\theta, \tag{4}$$

where $c := 0.8|\beta_1 - \beta_2|$. Use the Schwarz Inequality (stated in subsection 1.5.A), obtaining

$$\frac{c^2}{\pi r} < \int_{-\eta(r)}^{\eta(r)} |(f^{-1})'(-1 + re^{i\theta})|^2 r \, d\theta.$$

Integrate both sides over the r-interval $[0, \delta]$. The integral of the right-hand side is the area of $A(\delta)$ (see Exercise 7, Section 1.5, 71), which is less than the area of D. The integral of the left-hand side would be infinite if $\beta_1 \ne \beta_2$, so it must be the case that $\beta_1 = \beta_2$. $\qquad \square$

6.4.5. Remark. Another way to understand the inequality (4) is to realize that the mapping $\theta \mapsto f^{-1}(-1 + re^{i\theta})$ gives a curve that contains the points z_1 and z_2. The length of that curve is at least c.

A one-to-one onto continuous function $h : A \to B$ such that h^{-1} is also continuous is a **homeomorphism** of A and B. The sets are **homeomorphic** if such a function exists. Notice that the composition of two homeomorphisms is also a homeomorphism. Thus being homeomorphic is an equivalence relation among sets.

If A and B are compact and h is a one-to-one continuous function from A onto B, it follows that h^{-1} is continuous. *Proof:* It suffices to show that h is open. Let V be open in A. Then $A \setminus V$ is closed, hence compact. Therefore $h[A \setminus V]$ is compact, hence closed. Therefore $B \setminus h[A \setminus V]$, which is the same set as $h[V]$, is open in B.

The function f of Theorem 6.4.4 was extended to a one-to-one continuous function from $\bar{O}$ onto $\bar{D}$. Therefore it is a homeomorphism of those two sets. Restricted to bO, it is a homeomorphism of bO and the unit circle.

A **Jordan curve** is any image of the unit circle under a one-to-one continuous map into the plane. Thus the Riemann Mapping Theorem together with Theorem 6.4.4 imply that if O is a bounded open simply connected set, and if every boundary point of O is simple, then the boundary is a Jordan curve.

The following result is an example of the several variations on Theorem 6.4.4 which are now easy to establish. It dispenses with the boundedness requirement on O, and replaces D by $\mathbb{Y}^o$.

6.4.6. Theorem. *Let O be a simply connected open subset of $\hat{\mathbb{C}}$ whose every boundary point is simple, and such that the complement of O contains a disk $D(a, r)$. Let g be a conformal map from O onto $\mathbb{Y}^o$. Then g extends to a homeomorphism of $\bar{O}$ onto $\mathbb{Y}$.*

Proof. Let S be a Möbius map such that $S[O]$ is bounded, for example $z \mapsto \dfrac{1}{z - a}$. Let T be a Möbius map that takes $\mathbb{Y}^o$ onto D. Let $f := T \circ g \circ S^{-1}$. Theorem 6.4.4 applies to f, which maps $S[O]$ onto D. Extend f as that Theorem provides, and then $T^{-1} \circ f \circ S$ is the desired extension of g. (The nice thing about Möbius maps is that they are homeomorphisms from $\hat{\mathbb{C}}$ to $\hat{\mathbb{C}}$. So they are "already extended.") □

6.4.7. Example. *In Example 6.4.1 we considered a specific set O whose boundary points are not all simple. We made the boundary into a closed path, parametrized as in (1). We considered a specific conformal equivalence $g : O \to D$, and showed that there is a one-to-one function $g^* : [\pi - 1, \pi + 1] \to bD$ such that if $\beta = \beta(t)$, then $g(\beta) = g^*(t)$ makes g continuous at β from the left (in the sense provided by the parametrization). Prove that the same is true for every conformal equivalence from O onto D.*

Solution. Let f be an arbitrary conformal equivalence from O onto D. The proof of Lemma 6.4.3 may be adapted to show that if α_n converges to $\beta(t)$ from the left, then $\lim_{n \to \infty} f(\alpha_n)$ exists; we define $f^*(t)$ to equal the value of the limit.

It remains to show that if $t_1 < t_2$, then $f^*(t_1) \neq f^*(t_2)$. That result follows in the case when $\beta(t_1) \neq \beta(t_2)$ just as in the proof of one-to-oneness for f in Theorem

6.4.4. It remains to deal with the case when $\beta(t_1)$ and $\beta(t_2)$ have the same value β. In that case we modify the proof of Theorem 6.4.4, taking into account Remark 6.4.5 to arrive at (4), as follows. The number β belongs to the interval $[-1, 0)$, and there are two curves γ_j given by $z_j : [0, 1] \to \bar{O}$ such that $z_1[[0, 1)]$ lies above the real axis, and $z_2[[0, 1)]$ lies below the real axis, and $z_j(1) = \beta$ for $j = 1, 2$. We need to show that the two limits

$$f^*(t_1) = \lim_{s \to 1^-} f(z_1(s)) \quad \text{and} \quad f^*(t_2) = \lim_{s \to 1^-} f(z_2(s))$$

are unequal. If they are equal, we may suppose that they both equal -1. There exists s_0 such that $0 < s_0 < 1$ and

$$s_0 < s < 1 \Longrightarrow |z_j(s) - \beta| < 0.1|\beta| \quad \text{for } j = 1, 2.$$

If $s_0 < s < 1$, then, the length of any curve lying in O that runs from $z_1(s)$ to $z_2(s)$ must be at least $c := 1.8|\beta|$. The inequality (4) follows, and the rest of the proof is the same. $\qquad \square$

The Examples of this section demonstrate a method which will work in a variety of cases in which the boundary points are not all simple, including the cases of the generalized polygons considered in Section 5.8. That is, when there is a conformal equivalence from O onto $\mathbb{Y}^o$, the boundary can be parametrized as a closed path, and a one-to-one function f^* defined from the parameter interval onto the real line, and so forth. We leave it to you to formulate and confirm the appropriate result in individual cases.

References

1. Felipe Acker, *The missing link*, The Mathematical Intelligencer **18**, no. 3 (1996) 4–9.

2. M. Ya. Antimirov, A. A. Kolyshkin, and Rémi Vaillancourt, *Complex Variables*, Academic Press, 1998.

3. William E. Boyce and Richard C. DiPrima, *Elementary Differential Equations and Boundary Value Problems*, sixth edition, John Wiley & Sons, Inc., New York, 1997.

4. Bart Braden, *Pólya's geometric picture of complex contour integrals*, Mathematics Magazine **59** (1987), 321–327.

5. Bart Braden, *Picturing functions of a complex variable*, College Mathematics Journal **16** (1985), 63–72.

6. C. Briot and C. Bouquet, *Théorie des fonctions doublement périodiques et, en particulier, des fonctions elliptiques*, Paris, 1859.

7. James Ward Brown and Ruel V. Churchill, *Complex Variables and Applications*, sixth edition, McGraw-Hill, New York, 1996.

8. R. Creighton Buck, *Advanced Calculus*, third edition, McGraw-Hill, New York, 1978.

9. Robert B. Burckel, *An Introduction to Classical Complex Analysis, Volume 1*, Birkhäuser, Basel and Stuttgart, 1979.

10. C. Carathéodory, *Theory of Functions of a Complex Variable*, Volume 1, second edition 1958; Volume 2, second edition 1960 (translated by F. Steinhardt), Chelsea Publishing Co., New York.

11. C. Carathéodory, *Über die gegenseitige Beziehung der Ränder bei der konformen Abbildung des Inneren einer Jordanschen Kurve auf einen Kreis*, Mathematische Annalen **73** (1913), 305–320.

12. Baron Augustin Louis Cauchy, *Mémoire sur les fonctions complémentaires*, Comptes Rendus Acad. Sci. Paris **19** (1844), 1377–1384.

13. I. Černý, *A simple proof of Cauchy theorem*, Časopis pro pěstování matematiky **101** (1976), 366–369.

14. Darrell Desbrow, *On evaluating* $\int_{-\infty}^{+\infty} e^{ax(x-2b)}\,dx$ *by contour integration round a parallelogram*, American Mathematical Monthly **105** (1998), 726–731.

15. John D. Dixon, *A brief proof of Cauchy's integral theorem*, Proceedings of the American Mathematical Society **29** (1971), 625–626.

16. William Dunham, *Euler: The Master of Us All*, Mathematical Association of America, Washington, DC, 1999.

17. J. Chris Fisher and J. Shilleto, *Three aspects of Fubini's Theorem*, Mathematics Magazine **59** (1986), 40–42.

18. B. A. Fuchs and B. V. Shabat, *Functions of a Complex Variable and Some of Their Applications, Volume I*, Pergamon Press, Oxford, 1964.

19. Enrique A. Gonzàlez-Velasco, *Fourier Analysis and Boundary Value Problems*, Academic Press, San Diego, 1995.

20. Robert E. Greene and Steven G. Krantz, *Function Theory of One Complex Variable*, John Wiley & Sons, Inc., New York, 1997.

21. Frederick P. Greenleaf, *Introduction to Complex Variables*, W. B. Saunders Company, Philadelpha, 1972.

22. Maurice Heins, *Selected Topics in the Classical Theory of Functions of a Complex Variable*, Holt, Reinhart, and Winston, New York, 1962.

23. Maurice Heins, *Complex Function Theory*, Academic Press, New York, 1968.

24. Peter Henrici, *Applied and Computational Complex Analysis*, Volume 3, John Wiley & Sons, Inc., New York, 1993

25. Einar Hille, *Analytic Function Theory, Volume I*, first edition, Ginn and Company, Boston, 1959; second edition, AMS/Chelsea, 1982.

26. Einar Hille, *Analytic Function Theory, Volume II*, first edition, Ginn and Company, Boston, 1962; second edition, AMS/Chelsea, 1987.

27. Deborah Hughes-Hallett, Andrew M. Gleason, William G. McCallum *et al*, *Calculus, Single and Multivariable*, second edition, John Wiley & Sons, Inc., New York, 1998.

28. Valentin IA. Ivanov and M. K. Trubetskov, *Handbook of conformal mapping with computer-aided visualization*, CRC Press, Boca Raton, FL, 1995.

29. G. J. O. Jameson, *A First Course on Complex Functions*, Chapman and Hall Ltd., London 1970.

30. B. Frank Jones, Jr., *Rudiments of Riemann Surfaces*, Rice University Lecture Notes in Mathematics, Number 2, Houston, TX, 1971.

31. David W. Kammler, *A First Course in Fourier Analysis*, Prentice-Hall, Englewood Cliffs, NJ, 2000.

32. Yitzhak Katznelson, *An Introduction to Harmonic Analysis*, Second Corrected Edition, Dover, New York, 1976.

33. Konrad Knopp, *Infinite Sequences and Series*, translated by Frederick Bagemihl, Dover, New York, 1956.

34. H. Kober, *Dictionary of Conformal Representations*, Dover, New York, 1957.

35. Heinz König, *The missing past of the not-so-missing link*, in: Letters to the Editor, The Mathematical Intelligencer **19**, no. 3 (1997), 7.

36. Prem K. Kythe, *Computational Conformal Mapping*, Birkhäuser, Boston, 1998.

37. Detlef Laugwitz, *Riemann's dissertation and its effect on the evolution of mathematics*, The American Mathematical Monthly **106** (1999), 463–469.

38. Jerrold E. Marsden and Michael J. Hoffman, *Basic Complex Analysis*, third edition, W. H. Freeman and Company, New York, 1998.

39. John H. Mathews and Russell W. Howell, *Complex Analysis for Mathematics and Engineering*, third edition, Wm. C. Brown Publishers, Dubuque, IA, 1996.

40. Dragoslav S. Mitrinović and Jovan D. Kečkić, *The Cauchy Method of Residues: Theory and Applications*, D. Reidel Publishing Company, Dordrecht, The Netherlands, 1984.

41. Dragoslav S. Mitrinović and Jovan D. Kečkić, *The Cauchy Method of Residues*, Volume 2, Kluwer Academic Publishers, Dordrecht, The Netherlands, 1993.

42. L. J. Mordell, *The definite integral* $\displaystyle\int_{-\infty}^{+\infty} \frac{e^{ax^2+bx}}{e^{cx}+d}dx$ *and the analytic theory of numbers*, Acta Mathematica **61** (1933), 323–360.

43. Giacinto Morera, *Un teorema fondamentale nella teorica delle funzioni di una variabile complessa*, Rend. del R. Istituto Lombardo di Scienze e Lettere (2) **19** (1886), 304–307.

44. R. Kent Nagle and Edward B. Saff, *Fundamentals of Differential Equations and Boundary Value Problems, second edition*, Addison Wesley Publishing Company, Reading, MA, 1996.

45. Tristan Needham, *Visual Complex Analysis*, Clarendon Press, Oxford, 1997.

46. Zeev Nehari, *Introduction to Complex Analysis*, Allyn and Bacon, Inc., Boston, 1961.

47. Zeev Nehari, *Conformal Mapping*, McGraw-Hill, New York, 1952.

48. E. Neuenschwander, *The Casorati-Weierstrass theorem (studies in the history of complex function theory I)*, Historia Mathematica **5** (1978), 139–166.

49. W. P. Novinger, *An elementary approach to the problem of extending conformal maps to the boundary*, American Mathematical Monthly **82** (1975), 279–282.

50. W. F. Osgood and E. H. Taylor, *Conformal transformations on the boundaries of their regions of definition*, Transactions of the American Mathematical Society **14** (1913), 177–198.

51. Bruce P. Palka, *An Introduction to Complex Function Theory*, Springer-Verlag, New York, 1991.

52. Émile Picard, *Sur une propriété des fonctions entières*, Comptes Rendus de l'Academie des Sciences. Paris **88** (1879), 1024–1027.

53. Émile Picard, *Sur les fonctions analytiques uniformes dans le voisinage d'un point singulier essentiel*, Comptes Rendus Acad. Sci. Paris **89** (1879), 745–747.

54. George Pólya and Gordon Latta, *Complex Variables*, John Wiley & Sons, Inc., New York, 1974.

55. George Pólya and Gabor Szegö, *Problems and Theorems in Analysis*, Volume I (translated from German by D. Aeppli), Die Grundlehren der mathematischen Wissenschaften in Einzeldarstellungen, Bd. 193, 1972; and Volume II (translated by C. E. Billingheimer), *ibid.*, Bd. 216, 1976, Springer-Verlag, New York.

56. H. A. Priestley, *Introduction to Complex Analysis*, revised edition, Oxford University Press, Oxford, 1990.

57. Reinhold Remmert, *Theory of Complex Functions*, translated by Robert B. Burckel, Springer-Verlag, New York, 1990.

58. Georg Friedrich Bernhard Riemann, *Grundlagen für eine allgemeine Theorie der Functionen einer veränderlichen complexen Grösse*, Inaugural Dissertation (1851), Göttingen; *Werke*, 5–43.

59. E. B. Saff and A. D. Snyder, *Fundamentals of Complex Analysis for Mathematics, Science, and Engineering*, second edition, Prentice Hall, Englewood Cliffs, NJ, 1993.

60. Walter Rudin, *Real and Complex Analysis*, second edition, McGraw-Hill, New York, 1974.

61. Walter Rudin, *Real and Complex Analysis*, third edition, McGraw-Hill, New York, 1987.

62. Stanislaw Saks and Antoni Zygmund, *Analytic Functions* (translated from Polish by E. J. Scott), third edition, American Elsevier, New York, 1971.

63. Donald Sarason, *Notes on Complex Function Theory*, Henry Helson, Berkeley, CA, 1994.

64. Hermann Amandus Schwarz, *Gesammelte Mathematische Abhandlungen*, Volume II, Chelsea Publishing Company, New York, 1972.

65. R. T. Seeley, *Fubini implies Leibniz implies $F_{yx} = F_{xy}$*, American Mathematical Monthly **68** (1961), 56–57.

66. Frank Smithies, *Cauchy and the Creation of Complex Function Theory*, Cambridge University Press, Cambridge, United Kingdom, 1998.

67. James Stewart, *Calculus, Early Transcendentals*, fourth edition, Brooks/Cole Publishing Company, Pacific Grove, CA, 1999.

68. William A. Veech, *A Second Course in Complex Analysis*, W. A. Benjamin, New York, 1967.

69. J. L. Walsh, *The Cauchy-Goursat Theorem for rectifiable Jordan curves*, Proceedings of the National Academy of Sciences **19** (1933), 540–541.

70. Robert Weinstock, *Elementary Evaluations of $\int_0^\infty e^{-x^2} dx$, $\int_0^\infty \cos x^2 dx$, and $\int_0^\infty \sin x^2 dx$*, American Mathematical Monthly **97** (1990), 39–42.

71. L. Zalcman, *Picard's theorem without tears*, American Mathematical Monthly **85** (1978), 265–268.

Index

A

Abel's partial summation method, 185
Absolute convergence, 167
Accumulation point, 59, 129, 194–195, 238
Alternating Series Test, 167, 183, 284
Analytic function, 6, 121
Antiderivative, 121, 142, 175, 189, 193
    existence of, 208–209
Arc, 139
Argument Principle, 188, 243–244, 247, 250
Argument, 37
    notation, 38
Arzelà-Ascoli Theorem, 405

B

Basis, 39
Bernoulli, lemniscate of, 347
Bessel equation, 184
Bessel functions, 183
Beta function, 301
Bilinear mapping, 327
Binary operation, 20, 34, 83
Bolzano-Weierstrass Theorem, 137
Boundary of a set, 51–54, 59–60, 131, 136–137
    in the extended plane, 237
Boundary value problems, 2–3, 317
Bounded set, 54, 131
Braces, 19
Brackets, in notation for an image, 21
Braden, Bart, 160

Branch
    of the argument, 37
    of the logarithm, 106
    point, 37, 218, 294
Briot and Bouquet, 233
British Admiralty, 317
Bromwich integral, 311
Brown, Kevin Scott, xvi
Burckel, Robert Bruce, xvi, 205, 398, 404, 406

C

Cartesian product, 20
Casorati-Weierstrass Theorem, 218
    history of, 233
Cassinians, 347
Cassini, ovals of, 347
Cauchy, Baron Augustin Louis, 187
Cauchy integral, 178, 188, 216, 355
Cauchy integral representation, 189
    for derivatives, 190
Cauchy product, 198
Cauchy sequence, 133, 166–167
Cauchy's Theorem, xvi, 187, 208, 214
    proof by Černý, 205
    convex sets version, 204, 207
    proof by Dixon, 393–394
    homology version, xvi, 204
    in proof of Residue Theorem, 239
    on a Riemann surface, 303
    implied by Runge's, 205, 399
Cauchy-Hadamard Theorem, 172, 181, 183, 213

Cauchy-Riemann equations, 5, 8, 15, 70, 80, 102, 120, 124, 158
  geometrical significance, 72
Chain Rule, xiii, 77, 123, 127
Chordal distance, 264
Christoffel, Elwin Bruno, 375
Circle, 139
  capital "C", 328
  preserved by Möbius maps, 328
Circles of Apollonius, 100, 346
Circles of Steiner, 346
Circular arc, 139, 162
Circulation, 160
Clairaut's Theorem, 65–66
Closed path
  with 34-component complement, 163
  with winding number zero, 163
Closed set, 54, 130
  that is not compact, 60
Closure of a set, 59
Compact set, xiv, 54, 133
Compactness
  in terms of sequences, 133
  of the extended plane, 237
  preserved by continuous maps, 134
Completeness property of the reals, xiv, 131
Complex conjugation, 98
Complex number system, 83, 115
Component, 147
Composition
  of conformal equivalences, 323
  of functions, 25
  of holomorphic functions, 121
  of homeomorphisms, 411
  of meromorphic functions, 249, 329
  of Möbius maps, 328
  of reflections, 378
Concave up, 27
Conformal equivalence, 323
  preserves simply-connectedness, 209
Conformal mapping, 6, 47, 77, 119, 124
  dictionary, 317
  inversely, 119
  method, 7, 12, 14, 317, 323, 327, 368
Conjugate pair, 29
Conjugate-holomorphic function, 121, 127, 159, 351, 370
Conjugate-meromorphic function, 338
Conjugation, 45, 107
Connected set, 57
  continuous image of, 57
Connectedness, 51
  pathwise, 147
Continuity
  delta-epsilon definition, 51, 55
  open set definition, 56
  in terms of sequences, 130, 137

  uniform, 134
Contour, 138, 161
Convergence
  pointwise, 134
  set of, 181
  subuniform, 135
  uniform, 135, 179
Convex, 149
Cosine, xiii, 20–23, 25, 104, 107, 327
Coulomb, 11, 341
Crescent-shaped set, 334
Critical point, 79
Curl, 68–70, 158
Curve, 138
  closed, 140
  constant, 140
  initial point, 140
  length of, 140
  motion of a particle, 139
  null-homologous, 204
  parameter interval, 140
  parametrization, 140
  piecewise smooth, 147
  smooth, 140
  tangent vector to, 156
  terminal point, 140
Cusp, 140
Cycloid, 140
Cylindrical surface, 343

D

D'Alembert, Jean Le Rond, 199
Dedekind, Richard, 183
Determinant, xiv, 34
Determinism, 194
Diameter, 152
Difference quotient
  shown to be holomorphic, 192
Differentiability, 116–118
  complex sense, 119
  implies continuity, 116
  geometric meaning, 124
  as linear approximability, 116, 124
Differential equations, 176
Differentiation under the integral sign, 65, 154
Dirichlet problem, 2, 6, 8, xiv, 17, 76
  defined, 322
  Poisson integral method, 355, 358
  uniqueness of solution, 368
Dirichlet-Neumann problem, 326
  defined, 322
  uniqueness of solution, 369
Disk, 52
  centered at infinity, 237
  closed, 52
  punctured, 52

Distance, 51
Distortion, 29, 125
Divergence, 68
   of a vector field, 158
Dixon, John D., 205, 393–394
Dot product, 34–35, 45

E

Editors, superb, xvii
Electric field, 10, 341, 344
   in cylinder problem, 348
   in two-cylinder problem, 349
   in two-wire problem, 346
Electric intensity field, 341
Electrostatic field, 341
Electrostatic potential, 12
Ellipse, 50, 107
   in standard position, 107
English sentence, 18
Equicontinuity, 211, 405
Equipotential surfaces, 342
Equivalence relation among curves, 143, 164
Euclidean space, 20
Euler, Leonhard, 87
Eventually, 129
Exponential, xiii, 20, 23, 104, 112, 124, 219, 238, 242, 327
   integrals involving, 273, 282
   Taylor series, 176
   valence of, 242
Extended plane, 237, 240
   visualized as Riemann sphere, 260
Exterior, 59
Extremum Principle, 75

F

Field, 34, 83
Fixed points
   of holomorphic functions, 250
   of Möbius maps, 329, 331, 336, 340
Fluid flow, 158
   circulation-free, 159
   complex potential, 351, 353
   impenetrable wall, 351
   incompressible, 159
   irrotational, 159, 352
   not quite perfect, 352
   stream function, 351
   streamlines, 351
   turbulence, 352
   velocity potential, 351
Flux, 160
Form (1-form), 156
Fourier series, 183, 367
Fourier transforms
   defined, 286

integrals giving, 286
integration techniques based on, 268
Frequently, 129, 135
Fresnel integrals, 273, 289
Fubini's Theorem, 65
   applied in proof of Cauchy's, 208
Function
   continuous, 51
   differentiable, 66
   graph, 24, 27
   inverse, 23
   one-to-one, 22–23, 25
   onto, 22
   piecewise continuous, 63–64
   range, 21
   restriction, 22–23, 32
Fundamental Theorem of Algebra, 196
   proved with Rouché's Theorem, 251
Fundamental Theorem of Calculus, 62, 141

G

Gauss, Carl Friedrich, 73, 187, 196
Geometric series, 355
Geometric view of Poisson integral, 9
Gibbs's phenomenon, 183
Goursat's Lemma, 150–151, 156, 158, 165, 188
Gradient, 69, 102, 118
Greatest lower bound, 131
Green's function, 11
Green's Theorem, 156, 158
Grid
   orthogonal, 4, 11, 33, 41, 73, 99, 102, 124
   polar, 4, 29, 32
   rectangular, 4–5, 8, 10, 29, 31

H

Half-line, 37, 103
Harmonic conjugate, 5, 8, 34, 73, 78, 102, 124
   of a temperature function, 318, 321–322
Harmonic function, 2, 5, 7, 27, 34, 73, 137, 200
   characterized by mean value property, 363
   mean value property, 73, 76, 189
Heat
   conduction, 2
   loss by convection, 315
Hedlund, Gustav, xvi
Heins, Maurice, xvi
Herglotz's Formula, 357
Hessian, 80
Hilbert, David, 404
Hille, Einar, xvi
Holomorphic function, 6, 121–122, 159
   characterization in eight ways, 193
   and conformal map, 121
   at infinity, 226
   mapping properties, 252
   preserves orthogonal grids, 125

Hurwitz's Theorem, 248
  in proof of Riemann Mapping Theorem, 405
Hyperbola, 107
  in standard position, 107
Hyperbolic cosine and sine, xiii, 20, 23, 27, 110, 115
Identity mapping, 327

I

Improper integrals, 63–64, 268, 294
Independence of path, 148, 193, 209
Infimum, 131
Infinite series, 166
Inner product, 35
Inside
  of a closed path, 153
  of a contour, 161
Integral test, 167
Integration
  of complex-valued functions, 138
  review of theory, 61
Interior, 52–53, 59–60
Intersection, 19
  of closed sets, 60
  of open sets, 54, 60
Intervals
  notation for, 19
Inverse trigonometric functions, 127
Isolated point, 59
Isotherm, 7
Ivanov, Valentin, 317

J

Jacobian matrix, 118–119
Jordan curve, 411
Jordan's Lemma, 275
  applied, 287

K

Kammler, David W., xvi
Kober, H., 317
Koebe's trick, 405

L

Laplace transform
  applied to initial value problems, 309
  Bromwich integral for inverse, 311
  complex inversion formula, 311
  inverse, 309
  reason for usefulness, 310
Laplacian, 69–70, 76, 78
  2-dimensional, 342
  3-dimensional, 69, 81
  geometric significance, 72
  in polar coordinates, 74
Laurent series, 212, 216, 219–220, 223, 229, 380

definition, 213
  special case of Runge's Theorem, 398
Law of Cosines, 35–36, 362
Least upper bound, xiv, 58, 131
Legendre polynomials, 184
Length
  of a contour, 161
Limit inferior, 136
Limit superior, 136–137
Linear algebra, xiii
Linear fractional transformation, 327
Linear mapping, 39
  conformal, 46, 48
  inversely conformal, 46, 48
  isogonal, 46, 48
  magnification, 46, 48
  nonsingular, 46
  orientation-preserving, 46
  orientation-reversing, 46
  pure magnification, 45, 47–48
  rotation, 43, 45
Linear mappings, 40, 116
Lines and circles
  complex equations for, 92, 95
  reflections in, 96, 100
Liouville's Theorem, 196–197, 200, 229, 241
  used in proof of Schwarz-Christoffel Formula, 379
Liouville, Joseph, 196
Logarithm, xiii, 20, 23–24, 105, 112
  existence of on simply connected set, 208
  integrals involving a, 293, 275
  notation, 106
  principal, 103
  Riemann surface of, 303
Louisiana State University, xiii
Lower bound, 131
Loxodromic Möbius map, 329

M

Magnification, 120, 124, 327
Mapping properties, 188
Maps-to arrow, 24
Mathematica, 268, 314
  Apart, 222
  ContourPlot, 353
  for generating conformal maps, 317
  graphics commands, xvi
  Integrate, 271
  NIntegrate, 361, 388
  ParametricPlot3D, 361
Mathematics 4036 at LSU, xiii
Matrix, 49
  multiplication not commutative, 40
  notation, 118
Maximum Modulus Principle, 258
Maximum principle, 199

Mean value property, 2, 27
Mean Value Theorem for Integrals, 62, 401
Mellin transforms, 306
Meromorphic function, 236, 238
Metric space, 51
Minimum Modulus Principle, 258
Modulus, 36
Montel's Theorem, 405
Morera's Theorem, 149, 209
Multidimensional calculus, 155
Multiplication, true, 35
Multiplicity, 242, 246
Möbius map, 96, 327–330, 334, 340, 344, 366, 405

N

Needham, Tristan, 361
Negative-power series, 213
Neighborhood, 60
Neumann condition
    homogeneous, 322, 352
Neumann problem, 6, 76
    defined, 322
    Poisson integral method, 364
    reducible to a Dirichlet problem, 367
    uniqueness of solution, 368
Nonsingular, 41, 119
Norm, 51
North Pole, 260
Notation
    for intervals, 19
    for partial derivatives, 66
Novinger, W. P., 406
Numerical analysis, 317

O

Open mapping, 51, 56
    cannot attain extremum, 56
Open sets, 51–52, 54–57
    in the extended plane, 237
Orientation
    -preserving, 41, 45
    -reversing, 41, 43
Orthogonal vectors, 36
Osgood-Taylor-Carathéodory Theorem, 375, 393,
        406, 411
Outside
    of a closed path, 153
    of a contour, 161

P

Parabolic Möbius map, 329
Parallelogram, 35, 41, 44, 97
    degenerate, 35
Partial derivatives, xiii, 117, 120
Partial fractions, 241, 277, 282
Path, 138, 144

initial point, 145
    terminal point, 145
Perimeter, 152
Picard's Great Theorem, 229
Picard's Little Theorem, 229
Piecewise continuous, 178
Pizza, 73
Plane geometry, 100, 112
Poisson integral formula, 8–9, 74, 273, 357
    for a half-plane, 365
    Schwarz's geometric interpretation, 9, 361
Poisson integral
    a geometric view of, 9
Poisson kernel, 357, 366
Polar coordinates, 28, 30, 34, 36–37, 103, 106
Polar grid, 125
Pole, 217, 221, 231, 238
    at infinity, 226
    order of, 217
Potential
    electrostatic, 343
Power mapping, 251
Power series, 172
    method for solving differential equations, 176
    radius of convergence, 173, 182–183
    representation, 190, 193
    set of convergence, 172–173, 181
Powers, 98, 103, 327
Pre-image, 21
Principal argument, 37
Principal part, 219
    at infinity, 226
Principal value of an integral, 61, 65, 285, 289
Product Rule, 77, 123
Ptolemy, 262
Pullback, 141–142, 146, 163, 267
Punctured disk, 105
Punctured plane, 103
Pólya diagrams, 160

Q

Quadrilateral, sum of angles, 101
Quotient of holomorphic functions, 189, 199

R

Ramanujan's Identity, 292
Ratio Test, 182
Rational functions, 124
    integrals of, 275, 277, 287
    valence of, 243
Ray, 104
Real numbers, 34
Reciprocal, 93–94, 96, 98
Rectangular grid, 124
Reflection principle, 338
Reflection, 23, 93, 96–97, 100, 107

Remmert, Reinhold, 187
Removable singularity, 238
Residue, 221, 224
  definition, 219
  at infinity, 226, 231, 241
  finding by integral over arc, 225
  of logarithmic derivative, 244
  finding at a pole of order 1, 223
Residue Calculus, 65, 294, 310
  defined, 267
Residue Theorem, xiv, 188, 282, 285
  in proof of Argument Principle, 246
  statement and proof, 239
Riemann integral, xiv, 61, 65
Riemann Mapping Theorem, 393, 404
  proof by Fejér and F. Riesz, 404
  proof by Riemann, 404
  used in proof of Schwarz-Christoffel Formula,
    375
Riemann sphere, 188, 237, 260
  antipodal points on, 264
  chordal distance between points on, 264
Riemann sums, 62, 362
Riemann surface, 121, 300
  a contour on a, 305
  integration on a, 302
  of the logarithm, 302–303
Riemann zeta function, 211
Roots, 98, 103
  existence of on simply connected set, 208
Rotation, 43, 120, 124
Rouché's Theorem, 247
Runge's Theorem, 205, 393
  implies Cauchy's Theorem, 399
  statement and proof, 398

S

Saddle points, 72
Sawtooth wave, 183
Scalar multiplication, 34–35, 84
Schwarz, Hermann Amandus, 361, 375
Schwarz Inequality, 62
Schwarz Reflection Principle, 370, 372–373
  in proof of Schwarz-Christoffel Formula, 381
  special case, 371
Schwarz's Lemma, 258
Schwarz-Christoffel Formula, 375, 393
  for generalized polygon, 389
  for isosceles triangle, 384
  for polygon, 385
  for rectangle, 386
  for right isosceles triangle, 383
  for right triangle, 382
  for triangle, 376
Schwarz-Christoffel parameter problem, 386
Sequence, 128
  divergence of a, 130

of functions, 134
  limit of a, 130
Series, 166
  convegence, 166
  derived, 172–173
  geometric, 168, 171
  partial sums of a, 166
  summands of a, 166
  Taylor, 174
Set notation, 18
Simply connected, 208
  equivalent conditions, 208
Sine, xiii, 20–23, 25, 104, 106, 113, 238, 242, 327
  Taylor series for, 177, 220
  valence of, 242
Singleton, 19, 21
Singularity
  classifying a, 219
  classifying, 230
  essential, 217, 220
  essential at infinity, 226
  at infinity, 225
    for rational function, 227
  isolated, 216
  removable at infinity, 226
  removable, 217, 220
Smithies, Frank, 187
Smooth curve, 138, 259
Square root function
  used in proof of Riemann Mapping Theorem,
    404
Starlike set, 165
  hub of, 165
Steady flow of a perfect fluid, 350
Stereographic projection, 262
Stewart, Loc Thi, xvii
Stirling's Formula, 186
Stoltzfus, Neal, xvii
Subsequence, 128, 135
Subuniform boundedness, 209
Subuniform convergence, 197, 209
Supremum, 131
Symmetry principle for Möbius maps, 338

T

Tangent line, 117
Tango dancers, 6
Taylor series, xiv, 174, 191, 217, 219–220, 222,
  230
Temperature distribution problem, 2, 5–7, 318,
  347, 367
  in a half-plane, 321
  with insulation, 322, 324
Translation, 327
Triangle Inequality, 51, 167
Triangle, 146

sum of angles, 101
Trigonometric functions
    integrals of, 268
Trigonometric identities, 25, 45, 107, 110

U

Undamped vibrating system, 310
Uniform convergence, xiv, 168
Union, 19
    of closed sets, 60
    of open sets, 54
Uniqueness
    of power series representation, 176, 194, 199
    of the solution in a boundary value problem, 318
Upper bound, 131

V

Valence, 242, 246
Vector addition, 84
Vector field
    incompressible, 70
    irrotational, 70
Vector notation, 35
Velocity vector field, 158
Visualization, 3, xvi, 25, 30, 102
Volts, 342

W

Web site, xv
Weierstrass's Theorem, 195
Weierstrass, Karl Theodor Wilhelm, 187, 197
Wild and crazy, 218, 220
Winding number, 153, 161, 163, 207, 214, 244
    constant on each component, 155

Y

Yale University, xvi

Z

Zero of a function, 217
    at infinity, 226